AF729325

Protein Engineering

Protein Engineering

Tools and Applications

Edited by
Huimin Zhao

WILEY-VCH

Volume Editor

Professor Huimin Zhao
University of Illinois at Urbana
Chemical & Biomolecular Engineering
600 South Mathews Avenue
215 Roger Adams Laboratory
61801 Urbana IL
USA

Series Editors

Prof. Dr. Sang Yup Lee
KAIST
373-1; Guseong-Dong
291 Daehak-ro,Yuseong-gu
305-701 Daejon
South Korea

Prof. Dr. Jens Nielsen
Chalmers University
Department of Biology and Biological Engineering
Kemivägen 10
412 96 Göteborg
Sweden

Prof. Dr. Gregory Stephanopoulos
Massachusetts Institute of Technology
Department of Chemical Engineering
Massachusetts Ave 77
Cambridge, MA 02139
USA

Cover Culture Flasks in microbiological laboratory / science photo, fotolia

All books published by **WILEY-VCH** are carefully produced. Nevertheless, authors, editors, and publisher do not warrant the information contained in these books, including this book, to be free of errors. Readers are advised to keep in mind that statements, data, illustrations, procedural details or other items may inadvertently be inaccurate.

Library of Congress Card No.:
applied for

British Library Cataloguing-in-Publication Data
A catalogue record for this book is available from the British Library.

Bibliographic information published by the Deutsche Nationalbibliothek
The Deutsche Nationalbibliothek lists this publication in the Deutsche Nationalbibliografie; detailed bibliographic data are available on the Internet at <http://dnb.d-nb.de>.

© 2021 WILEY-VCH GmbH, Boschstr. 12, 69469 Weinheim, Germany

All rights reserved (including those of translation into other languages). No part of this book may be reproduced in any form – by photoprinting, microfilm, or any other means – nor transmitted or translated into a machine language without written permission from the publishers. Registered names, trademarks, etc. used in this book, even when not specifically marked as such, are not to be considered unprotected by law.

Print ISBN: 978-3-527-34470-3
ePDF ISBN: 978-3-527-81509-8
ePub ISBN: 978-3-527-81511-1
oBook ISBN: 978-3-527-81512-8

Cover Design Adam-Design, Weinheim, Germany
Typesetting Straive, Chennai, India
Printing and Binding CPI Group (UK) Ltd, Croydon, CR0 4YY

C9783527344703_270924

Contents

Part I

Directed Evolution

1

Continuous Evolution of Proteins *In Vivo*

Alon Wellner[1], Arjun Ravikumar[1], and Chang C. Liu[1,2,3]

[1] *University of California, Department of Biomedical Engineering, 3201 Natural Sciences II, Irvine, CA, 92697, USA*
[2] *University of California, Department of Chemistry, 1102 Natural Sciences 2, Irvine, CA, 92697, USA*
[3] *University of California, Department of Molecular Biology and Biochemistry, 3205 McGaugh Hall, Irvine, CA, 92697, USA*

1.1 Introduction

Directed evolution is a powerful approach for engineering new biomolecular and cellular functions [1–3]. In contrast to rational design approaches, directed evolution exploits diversity and evolution to shape the behavior of biological matter by applying the Darwinian cycle of mutation, selection, and amplification of genes and genomes. By doing so, the field of directed evolution has generated important insights into the evolutionary process [4–6] as well as useful RNAs, proteins, and systems with wide-ranging applications across biotechnology and medicine [7–11].

To mimic the evolutionary process, classical directed evolution approaches carry out cycles of *ex vivo* diversification on genes of interest (GOIs), transformation of the resulting gene libraries into cells, and selection of the desired function (Figure 1.1). Each iteration of this cycle is defined as a round of evolution, and as selection stringency increases over rounds, either automatically through competition or manually through changing conditions (or both), this process can lead GOIs closer and closer to the desired function. This overall process makes practical sense for a number of reasons, especially for the goal of protein engineering (i.e. GOI encodes a protein). First, *ex vivo* diversification is appropriate, because test tube molecular biology techniques such as DNA shuffling, site-directed saturation mutagenesis, and error-prone (ep) polymerase chain reaction (PCR) [2] are capable of generating exceptionally high and precise levels of sequence diversity for any GOI. Second, transforming diversified libraries of the GOI into cells is appropriate, because each GOI variant needs to be translated into a protein in order to express its function, and cells, especially model microbes, are naturally robust hosts for protein expression. Third, carrying out selection inside cells is appropriate, because (i) cells automatically maintain the genotype–phenotype connection between the GOI and expressed protein that is necessary for amplification of desired variants,

Protein Engineering: Tools and Applications, First Edition. Edited by Huimin Zhao.
© 2021 WILEY-VCH GmbH. Published 2021 by WILEY-VCH GmbH.

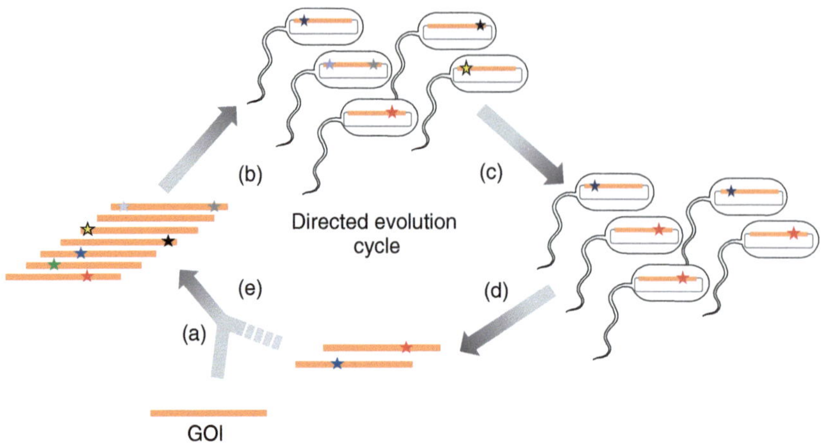

Figure 1.1 A schematic illustration of a typical directed evolution setup. (a) A GOI is diversified *ex vivo*, typically by applying an error-prone PCR to generate a GOI library. (b) The library is then cloned into an expression vector and transformed/transfected into cells that are subjected to (c) outgrowth and selection for enhanced protein activity. (d) Plasmid DNA that is enriched for library members with increased properties is extracted and (e) subjected again to diversification and selection. The directed evolution cycle is iterated until the desired outcome is achieved or until diminishing returns (a plateau is reached).

(ii) we often care about a GOIs function within the context of a cell, especially as metabolic engineering and cell-based therapy applications mature, and (iii) the use of cell survival as the output for a desired protein function allows millions or billions of GOI variants to be simultaneously tested – it is easy to culture billions of cells under selection conditions – in contrast to *ex vivo* screens that are much lower throughput. Survival-based selections are not always immediately available, but one can often find a way to reliably link the desired function of a protein to cellular fitness.

While sensible, the practical requirement that diversification should occur *in vitro* but expression and selection should occur *in vivo* in this classical directed evolution pipeline creates significant suboptimalities. First, the number of steps that can be taken along an adaptive path becomes few, since each round of *in vitro* mutation, transformation, and *in vivo* selection takes several days or weeks to carry out. Second, limited DNA transformation efficiencies result in strong bottlenecking of diversity that can mitigate the probability of finding the most optimal solutions in sequence space. Third, the number of evolution experiments that can be run simultaneously is minimal, because *in vitro* mutagenesis, cloning, and transformation are experimentally onerous, demanding extensive researcher intervention [12]. These shortcomings keep two highly promising categories of experiments largely outside the grasp of classical methods: first is the directed evolution of genes towards highly novel functions that likely require long mutational paths to reach (e.g. the optimization of multi-gene metabolic pathways or the *de novo* evolution of enzyme activity); and second is the large-scale replication of directed evolution experiments, needed in cases when many different functional variants of a gene are desired (e.g. the evolution of multiple synthetic receptors for a collection of ligands) or when statistical power is required in order to understand outcomes in experimental evolution (e.g. probing the scope of adaptive trajectories leading to resistance in a drug target).

An emerging field of *in vivo* continuous directed evolution seeks to overcome these shortcomings by performing both continuous diversification of the GOI and selection entirely within living cells [13]. In this way, GOIs can be rapidly and continuously evolved through basic serial passaging of cells under selective conditions. This removes the labor-intensive cycling between *in vitro* and *in vivo* steps and the DNA transformation bottlenecks associated with the classical pipeline, creating a new paradigm for directed evolution that is limited only by the generation time of the host cell and the number of cells that can be cultured. These limitations are usually negligible – in most host organisms for directed evolution such as *Escherichia coli* and *Saccharomyces cerevisiae*, generation time is fast (20–100 minutes) and the number of cells that can be cultured is massive (10^8–10^9 ml^{-1}) – so the potential power of continuous systems is enormous. Moreover, *in vivo* continuous directed evolution is amenable to high-throughput experiments, because serial passaging is straightforward and can be automated at scale or converted to continuous culture using bioreactors [14–16]. In this chapter, we discuss various systems that partially or fully achieve *in vivo* continuous evolution (ICE).

1.2 Challenges in Achieving *In Vivo* Continuous Evolution

Before discussing how ICE can be realized, we shall first clarify why this is a challenging problem. The difficulty of achieving ICE of GOIs lies in the fundamental relationship between how fast one can mutate an information polymer and its length. Several theories predict that organisms face an "error threshold" at mutation rates on order $1/L$ (where L is the length of the genome), near which selection cannot maintain fitness, leading to gradual decline towards low fitness, or above which one is nearly guaranteed a lethal mutation every cycle of replication, leading to rapid extinction [17–20]. Because cellular genomes are large (e.g. $\sim 5 \times 10^6$ in *E. coli*, $\sim 1.2 \times 10^7$ in *S. cerevisiae*, and $\sim 3 \times 10^9$ in humans), this implies that evolution strongly favors low genomic mutation rates (e.g. $\sim 5 \times 10^{-8}$ substitutions per base [s.p.b.] in *E. coli*, $\sim 10^{-10}$ s.p.b. in *S. cerevisiae* and $\sim 3 \times 10^{-9}$ in human somatic cells) [21–23]. Experiment confirms this prediction. Drake observed empirically that mutation rates scale as $1/L$ across many organisms [17]; evolution experiments have shown that when mutator phenotypes do arise, they are accompanied by fitness costs and only transiently persist [19, 24–26]; and more direct tests in yeast find that there is indeed a mutation-induced extinction threshold at $\sim 1/L$, above which yeast cannot propagate [18]. Yet individual GOIs are small in comparison with genomes, so they are capable of tolerating much higher error rates. In fact, they *require* much higher per base error rates than genomes to generate the same amount of total mutational diversity, because they have fewer bases. Following the $1/L$ scaling, a typical 1 kb GOI should be able to tolerate mutation rates on order $\sim 10^{-3}$ s.p.b.

Therefore, the primary challenge in achieving rapid ICE is how to develop molecular machinery or other strategies that target rapid mutagenesis to only GOIs, allowing the host genome to replicate at mutation rates below its low error thresholds but driving the GOI at the high mutation necessary for fast generation of sequence diversity. When considering the level of targeting in the ideal case, the formidability of this challenge becomes quite apparent. Ideally, one should continuously mutate GOIs at rates close to their error threshold ($\sim 10^{-3}$) to maximize diversification but leave the genomic error rate completely unchanged, as the genome's error rate is evolutionarily optimized for host fitness. In *E. coli*, *S. cerevisiae*, and human cells, this means that on-target versus off-target mutagenesis must differ by 10^6-fold, 10^7-fold, and 10^7-fold, respectively, which is much more than the 10- to 1000-fold targeting required in most synthetic biology problems involving molecular recognition. *How can we achieve such extreme precision in mutational targeting in the cell?*

There is yet another hard challenge in realizing ICE, which has to do with the durability of mutagenesis. Ideally, one wants a high rate of mutagenesis on the GOI to persist indefinitely (or at least for as long as the experimenter cares), so that a protein can traverse long mutational pathways towards desired functions. Because one needs to achieve mutational targeting to the GOI, there is almost always a risk to durability: any mechanism for targeting the GOI over the rest of the genome will necessarily rely on some cis-elements in or surrounding the GOI to mediate the targeting. If these cis-elements become mutated, which is quite likely since they are usually in or near the GOI undergoing rapid mutation, then mutagenesis will slow or stop. Ideally, a continuous evolution system will limit the chance that a cis-element for mutational targeting gets degraded. In the case that it does, an ideal system will remove the GOI containing the mutated cis-element from the population so that it can't fix in the population (through gradual mutational accumulation or a selective sweep if mutagenesis comes with a fitness cost) and end the continuous evolution process prematurely. *How do we achieve architectures for durability?*

Other challenges for ICE include generality across host organisms, the ability to mutate many genes simultaneously, and fine control over mutation rate and spectra; but the most defining ones are targeting and durability. In the remainder of this chapter, we review several *in vivo* continuous directed evolution platforms within the framework of these challenges. We highlight in Section 1.4.4 and note here that our recently developed orthogonal DNA replication (OrthoRep), among systems for ICE, seems uniquely capable of complete precision in mutational targeting (as far as we can tell), and is a highly durable architecture for enforcing prolonged mutagenesis in GOIs. We also highlight, in Section 1.3, phage-assisted continuous evolution (PACE), which has been remarkably successful for continuous biomolecular evolution. Although PACE is not an entirely *in vivo* system, it achieves complete precision in mutational targeting and durability – in fact by not being entirely *in vivo,* as we will explain. We do not discuss several powerful technologies for non-continuous *in vivo* diversification or streamlined diversification methods, such as MAGE [27], CREATE [28], DiVERGE [29], and CPR [30], but note that these are also promising approaches to protein evolution as they address some of the constraints of classical directed evolution methods. A summary of various characteristics of the systems we discuss is provided in Table 1.1.

Table 1.1 Comparison among approaches for *in vivo* continuous evolution.

Approach	Systems	Mutation rate	Targeting of mutagenesis	Durability of mutagenesis	Number (simple estimates) and location of genes that can be evolved simltaneously	Generality across host organisms	Mutational spectrum	References
Continuous rounds of evolution with a conditionally replicating bacteriophage	PACE	Mutates GOIs at $\sim 10^{-3}$ s.p.b.	Complete targeting to the bacteriophage genome, since *E. coli* are constantly replaced	Indefinitely continuous since mutagenesis is enforced. In practice, this method is typically implemented for 1–3 weeks	1–10 genes encoded on bacteriophage genome.	Currently in *E. coli*. Could be implemented with mammalian cells using non-integrating viruses (e.g. adenovirus).	Fairly unbiased mutational spectrum.	[33]
Targeted mutagenesis in *E. coli* with error-prone DNA polymerase I	ep Pol I/ColE1-based systems, CRISPR-guided DNA polymerases (EvolvR)	GOIs encoded near the ColE1 origin are mutated by ep Pol I at $\sim 10^{-3}$ s.p.b. CRISPR-guided Pol I can induce rates as high as 10^{-2}, but this quickly drops off after the guide region	Targeting with unfused ep Pol I is maximally only ~400-fold. Fusion to nCas9 generally improves targeting to ~1000-fold	Durability remains to be tested. Ep Pol I/ColE1 incurs significant off-target mutagenesis, which could quickly abrogate mutagenesis. EvolvR risks breaking down because it rapidly mutates the gRNA target region	1–5 genes encoded on a plasmid with ep Pol I/ColE1. 1–20 genes on plasmids or at their endogenous genomic loci with EvolvR, depending on how many targeting sgRNAs one can stably encode.	Both systems are currently in *E. coli*. EvolvR should be fairly general across hosts, especially with the use of Phi29 DNAP.	ep Pol I mutates ColE1 plasmids with a bias towards transition mutations. EvolvR generates substitutions of all four nucleotide types, in a relatively unbiased manner. If needed, this can be improved through DNAP engineering.	[47–50, 54]

(*Continued*)

Table 1.1 (Continued)

Approach	Systems	Mutation rate	Targeting of mutagenesis	Durability of mutagenesis	Number (simple estimates) and location of genes that can be evolved simltaneously	Generality across host organisms	Mutational spectrum	References
Yeast systems that do not use engineered DNA polymerases for mutagenesis	TaGTEAM, ICE	For TaGTEAM, ~10^{-4} s.b.p. at 10 kb regions on both sides of the tetO array. For ICE, 1.5×10^{-4}, if excluding the rate of retro-transposition needed to induce mutagenesis	TaGTEAM offers targeting of genomic GOIs, however with low accuracy. ICE's targeting is theoretically good since off-target regions are not reverse transcribed	Durability remains to be tested. Off-target mutation and the requirement that retro-transposition occurs back into the original locus for continued evolution with ICE will likely affect durability	1–10 genes on plasmids or at engineered genomic loci	Both systems are currently in yeast. ICE has been demonstrated in several diverged yeast species. TaGTEAM should function in *E. coli* and mammalian cells. ICE could be implemented in new hosts using retrotrans-posable elements similar to Ty1.	TaGTEAM generates a broad spectrum of both transitions and transversions. In addition, 25% of mutations are single base deletions. In ICE there is a 1 : 1 ratio between transitions and transversions.	[55, 58]

Somatic hypermutation as a means for targeted mutagenesis of GOIs	Hypermutator B cell line, Ramos cell line, dCas9-AID fusions (such as CRISPRx), T7 RNAP-AID fusion	CRISPRx mutates GOIs at ∼5×10^{-4} s.p.b.	Efficient targeting. No increase in mutagenesis rate was detected in an off-target locus. The hyperactive AID variant can create dense, highly variable point mutations within a region of 100 bp surrounding an sgRNA target site	Durability remains to be tested.	1–10 genes on plasmids or at engineered genomic loci with the hypermutator B cell line, Ramos cell line, or T7 RNAP-AID fusion. Dozens of genes at endogenous genomic loci with dCas9-AID fusions	Systems depending on natural SHM are limited to mammalian cells. AID-fusions are currently available in mammalian systems or *E. coli*, depending on the system. AID fusions should be extensible to all host-types.	AID generates point mutations rather than insertions and deletions, and it favors transitions over transversions. However, repair pathways operate at AID-mutated loci to extend the scope of mutagenesis.	[67–73]
Orthogonal DNA replication	OrthoRep	Mutates GOIs at ∼10^{-5} s.p.b.	Complete orthogonality (at least 100 000-fold targeting)	Indefinitely continuous since mutagenesis is enforced. This method has been implemented for up to 300 generations without any sign of erosion	1–10 genes encoded on a special orthogonal plasmid	Currently in yeast. Should be extensible to bacteria and mammalian systems using related protein-priming DNAPs.	TP-DNAP1-4-2 strongly favors transition mutations. This can be readily improved through DNAP engineering.	[74, 75, 78]

Source: Esvelt et al. [33]; Fabret et al. [47]; Alexander et al. [49].

1.3 Phage-Assisted Continuous Evolution (PACE)

The most successful method for continuous protein evolution thus far is the PACE system developed in the lab of David Liu (Figure 1.2) [2, 12, 14, 15, 31–37]. PACE reimagines traditional "rounds" of directed evolution as generations of the M13 bacteriophage life-cycle, thereby transforming a step-wise and labor-intensive procedure into a continuous biological process. In PACE, GOIs are encoded in the M13 genome, and the resulting phage continuously replicate in a vessel (termed "lagoon") that experiences a constant influx of *E. coli* cells. To create a selection pressure for GOIs to evolve, the activity of interest is coupled to phage survival. This is achieved by deleting the essential gene III (gIII), encoding coat protein III (pIII), from the M13 genome. The host *E. coli* strain is engineered to encode gIII in a genetic circuit that makes pIII expression dose-dependent on the desired activity of the GOI (see the following text for examples); so only phage that successfully evolves the GOI can trigger pIII expression and continue propagating. Due to the rapid generation time of M13 (~10 minutes without selection), evolution in this manner can iterate hundreds of times in just a few days.

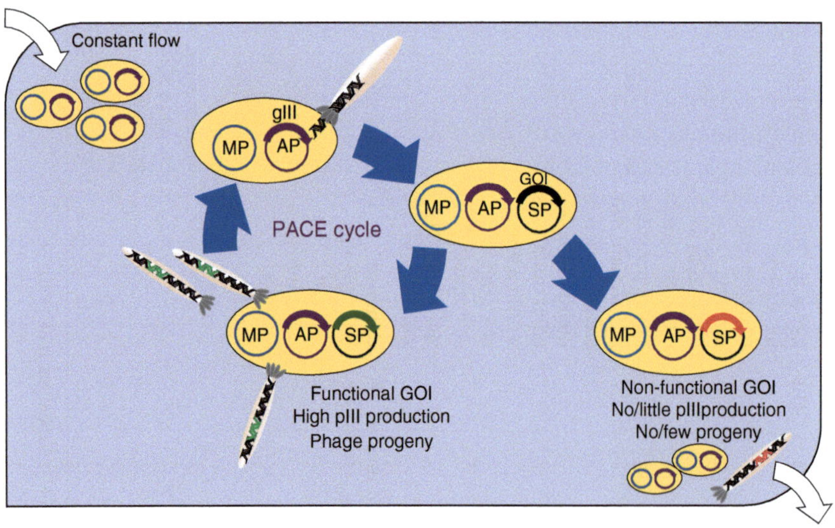

Figure 1.2 Pace. Phage carrying the selection plasmid (SP) encoding the GOI propagates on *E. coli* cells which are constantly flowing into the "lagoon" at a rate that does not permit their propagation but is longer than the phage life cycle, thus permitting phage replication. Upon infection, the SP (as well as the bacterial genome) experiences a high degree of mutagenesis due to the presence of a mutator plasmid (MP). In a PACE experiment, high GOI activity (green) is linked to drive strong gIII expression, resulting in progeny that can then infect incoming *E. coli*. No GOI activity (or a weak one, red) results in poor progeny production, becoming washed away from the lagoon at a larger rate (alongside bacterial cells). The system is designed to run for hundreds of generations without human intervention and result in the evolution of the GOI towards the desired activity. Source: Packer and Liu [2]; Badran and Liu [12]; Carlson et al. [14]; Dickinson et al. [15].

A key parameter in PACE is the *E. coli* flowrate, which should exceed their doubling time but be slower than the phage life cycle, allowing only phage to replicate in the lagoon (on average). Consequently, only phage accumulates mutations, whereas *E. coli* are physically prevented from doing so. High rates of mutation on the phage (and *E. coli*) genome are driven by a mutator plasmid (MP) that is carried by the *E. coli* cells and induced in the lagoon for error-prone M13 replication. The latest version of the MP is able to drive potent mutagenesis at $>10^{-3}$ s.p.b. by combining the effects of six different mutagenesis drivers [38].

Esvelt et al. first demonstrated proof of concept by evolving T7 RNA polymerase (RNAP) to initiate transcription from new promoter sequences [33]. pIII expression was bottlenecked at the level of transcription by encoding promoter sequences unrecognized by wild-type (wt) T7 RNAP (or any *E. coli* RNAPs), thus driving the selection to favor T7 RNAP variants that are able to efficiently recognize the new promoters. After eight days and 200 "rounds" of PACE, new T7 RNAPs emerged that could transcribe from the distant T3 RNAP promoter as efficiently as wt T7 RNAP does from its cognate promoter [33]. Similarly, T7 RNAP variants that efficiently initiate transcription with ATP or CTP, instead of GTP, were evolved. Since that landmark study, the ability to couple T7 RNAP activity to PACE has been exploited in a number of ways, ranging from basic adaptation studies to selections for split T7 RNAP [14, 15, 35–38].

In principle, PACE is applicable for the evolution of any biomolecular function that can be linked to pIII expression; and in just a few years since its inception, this has been realized in a wide range of applications beyond RNAP evolution. A notable example is the evolution of new DNA binding domains. Hubbard et al. employed the classic one-hybrid selection with PACE to evolve transcription activator-like effector nucleases (TALENs) with broadly improved DNA cleavage specificity [34]. Although TALENs are highly promising for gene editing, their major limitation is that they require the 5′ nucleotide of the target sequence to be T [39]. New TALEs (TALENs without the fused nuclease) were evolved with PACE by fusing the DNA binding domain of the canonical CBX8-targeting TALE to the ω subunit of *E. coli* RNAP. The PACE system was designed to include the TALE target sequence upstream of gIII. TALEs that successfully bind the target DNA recruit holoenzyme RNAP around the ω subunit, resulting in subsequent pIII expression. With this TALE selection, the identity of the target sequence can be custom-tailored, in this case, to encode noncanonical 5′ nucleotides. After using an additional negative selection (see below) that inhibited variants with promiscuous substrate specificity, Hubbard et al. were able to evolve TALE variants that displayed two- to fourfold increases in specificity for 5′ A, 5′ C, or 5′ G versus 5′ T, relative to wt TALE.

The one-hybrid PACE format was also used for overcoming one of Cas9's main limitations, restricted protospacer adjacent motif (PAM) compatibility. This time, Hu et al. fused a catalytically dead variant of *Streptococcus pyogenes* Cas9 (dCas9) to the ω subunit of *E. coli* RNAP [40]. Then, the authors cleverly fed the lagoon with a mixture of host *E. coli* cells bearing a library of target sequences that covers all 64 possible PAM sequences, to select for broadened PAM compatibility. After PACE, several variants were isolated that could efficiently recognize NG, GAA, and

GAT as PAMs. Upon restoration of nuclease catalytic activity to these evolved dCas9 variants, the authors remarkably found that one of them, xCas9, exhibited greater DNA specificity than wt Cas9, even with its newly-gained broad PAM compatibility. This result challenges the widely-held assumption that there must be a trade-off between editing specificity and PAM compatibility and suggests that Cas9 can be improved through laboratory evolution to meet the most demanding challenges of CRISPR-Cas9 applications.

Another important form of PACE is its use with two-hybrid selection for the evolution of high-affinity protein-binders [31]. In the bacterial two-hybrid system, the ω subunit of *E. coli* RNAP is fused to a protein of interest, which is recruited to DNA through its interaction with a target protein. This target protein is fused to a DNA binding domain that localizes the complex at its cognate sequence encoded upstream of a reporter gene. If the protein of interest binds the target protein, then the RNAP holoenzyme can reconstitute around the ω subunit and drive expression of the downstream reporter. Badran et al. adapted this system for PACE using gIII as the reporter. After extensive optimization, Badran et al. were able to use this PACE format to evolve the insecticidal protein, *Bacillus thuringiensis* δ-endotoxin (Bt toxin) Cry1Ac, to bind and inhibit a new receptor in the gut of the insect pest *Trichoplusia ni* (TnCAD) [31]. Although wt Cry1Ac did not detectably bind TnCAD, the evolved variants were able to bind with nM affinity. Significantly, this strategy could overcome widespread Bt toxin resistance, which primarily occurs through mutational changes that inhibit binding to the native receptor of wt Cry1Ac. Badran et al. demonstrated this by showing that evolved Cry1Ac is highly potent at killing *T. ni* that are resistant to wt Cry1Ac. An exciting possibility for the future would be to evolve TnCAD to resist the new Cry1Ac variant, and then iterate this cycle in a study of molecular co-evolution.

Additional positive selections developed for PACE have enabled evolution of proteases that are drug resistant [32] or have altered substrate specificities [41], aminoacyl-tRNA synthetases (aaRSs) that can accept noncanonical amino acids [42], and protein variants with improved soluble expression [43]. Negative selections are also compatible with PACE, and are useful in cases where it is desirable to evolve high specificity towards the target substrate and restrict promiscuity towards others (especially the native substrate). This can be achieved by introducing a dominant negative allele of pIII, pIII-neg, that inhibits phage propagation [14]. The expression of pIII-neg can then be linked to the unwanted activity (e.g. recognition of the T7 promoter by T7 RNAP) for negative selection. (This strategy was successfully employed during TALEN and aaRS evolution.) Selection stringency and mutation rate are also important determinants of PACE outcomes and can be titrated [14, 35]. Lastly, we note that the Isalan lab developed a system related to PACE that accommodates the evolution of multiple genes, starting from combinatorial libraries. With this system, they were able to evolve a panel of orthogonal dual promoter-transcription factor pairs that were used to make multi-input logic gates [44, 45].

Clearly, PACE is a powerful method for continuous protein evolution, but as noted early in this chapter, it is not an entirely *in vivo* system. Rather, M13 serves as a biological carrier of the GOI from one *E. coli* host cell to the next, with a given cell serving as a host of error-prone replication just once (on average). This ingenious design circumvents the challenges of *in vivo* mutational targeting. Since mutagenesis is induced in the lagoon, where *E. coli* briefly reside without doubling, mutation rates can be elevated entirely through untargeted mechanisms (and temporarily induced to be as high as desired), without consideration for replication of the *E. coli* genome. Even if *E. coli* cells stochastically replicate in the lagoon and become a source of cheater mutations (e.g. constitutive gIII expression), the flow rate ensures that any progeny are quickly diluted out. What's left in the lagoon is a population of M13 that selectively undergoes error-prone replication. In effect, targeting of mutations to the phage genome containing the GOI is complete, as the host *E. coli* is constantly replaced.

PACE also achieves durable mutagenesis by enforcing continuity. Replication of GOIs is intrinsically coupled to mutagenesis, through error-prone replication of the M13 genome. Any phage that escapes mutagenesis through a mutation in the phage genome's origin of replication, for example, must do so at the expense of being replicated. Only variants that continue to accumulate mutations can survive and propagate. And since *E. coli* cells do not persist long enough in the lagoon to evolve, the mutation rate experienced by phage remains unchanged. The durability of PACE is best evidenced by the long mutational trajectories traversed during evolution experiments, which have yielded protein variants with up to 16 mutations [46].

However, because PACE is not entirely *in vivo*, it suffers two major limitations. First, it requires continuous propagation of phage in a population of freshly diluted *E. coli* cells, which has been achieved thus far with a chemostat or turbidostat setup. This greatly limits the throughput and accessibility of PACE experiments, typically to fewer than ten replicates or experiments especially when different selection environments are desired across replicates. Second, PACE is restricted to selections that are linked to phage propagation. This precludes selections for *in vivo* phenotypes like tolerance or metabolism, as well as cell-based selections like fluorescence-activated cell sorting (FACS) or droplet sorting. These limitations motivate the need for continuous directed evolution systems that operate entirely *in vivo*.

1.4 Systems That Allow *In Vivo* Continuous Directed Evolution

1.4.1 Targeted Mutagenesis in *E. coli* with Error-Prone DNA Polymerase I

The first system that was able to perform continuous targeted mutagenesis *in vivo* was published in 2000 by Fabret et al. [47]. It was designed based on the developments in understanding the mechanism of ColE1 plasmid replication in *E. coli*. For plasmids that contain a ColE1 origin of replication, DNA polymerase (DNAP) I (Pol I) is responsible for elongating from the RNA primer that initiates replication at the origin. Pol I will extend for about 400–2000 bp, after which DNAP III

(Pol III), responsible for bulk DNA replication in *E. coli*, replaces Pol I [48]. When using a genome-encoded proofreading-deficient Pol I, genes that were cloned near the ColE1 origin experienced a 6- to 20-fold higher degree of mutagenesis over genes at more remote areas in the plasmid, showing targeting. The system's components were further combined with mismatch repair mutants to raise the mutation rate on GOIs yet another 20- to 40-fold, although significant increases in genomic mutation rates of at least several hundred-fold were observed. As a proof of concept, the authors evolved dominant negative variants of LacI that would outcompete a genomically-encoded wt LacI in binding its cognate operator, LacO. After 30 generations, LacI mutants that caused complete abolishment of wt LacI's binding to LacO were isolated. These variants were altered in their DNA binding domain but still formed tetramers with wt LacI, thereby abolishing LacI's repression at LacO.

Further improvement of the Pol I/ColE1 system was demonstrated in 2003 (Figure 1.3a) [46, 49]. Camps et al. modified the system to express the ep Pol I from a plasmid with a Pol I-independent origin of replication. Then, they used a host *E. coli* strain (J2000) whose genomically-encoded wt Pol I was temperature sensitive (ts) [49]. At restrictive temperatures, the ts Pol I becomes inactive such that only the ep Pol I acts, preventing the high-fidelity ts Pol I from competing for replication at the ColE1 origin. Based on prior studies of Pol I from the same lab [50], Camps et al. engineered a Pol I variant that was exceptionally error-prone, leading to mutation rates as high as 8.1×10^{-4} s.p.b at the GOI when the ts Pol I was inactivated. Mutagenesis expanded to about 3 kb from the ColEI origin and

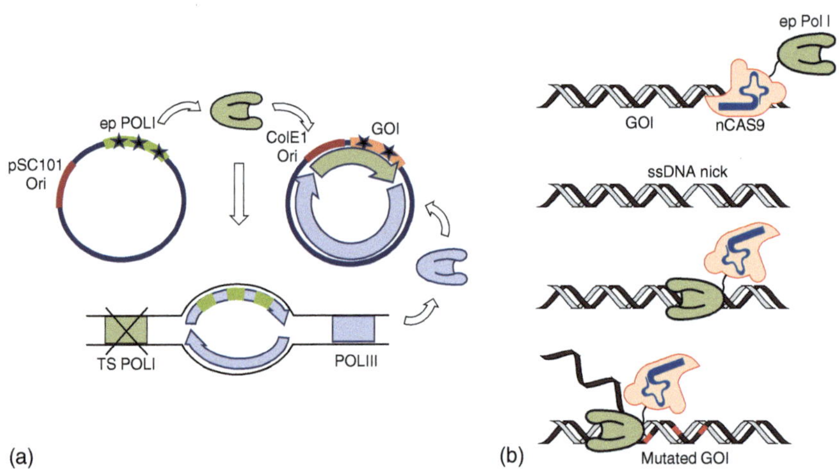

Figure 1.3 Targeted mutagenesis in *E. coli* with error-prone DNA polymerase I. (a) An ep version of Pol I is expressed from a plasmid whose replication is driven by a non-ColE1 origin of replication (ori). The GOI is placed on the target plasmid near the ColE1 ori and thus targeted for mutagenesis. After 1–3 kb of ep replication, Pol III replaces Pol I to replicate the remainder of the plasmid with high fidelity. The genomic allele of POL I is temperature sensitive, such that enhanced mutagenesis can be induced by growth at the restrictive temperature. Source: Alexander et al. [49]; Camps et al. [46]. (b) The EvolvR system is composed of a CRISPR-guided nickase that nicks the target GOI, fused to ep Pol I that performs nick translation.

was evenly distributed within this region, albeit with certain biases in mutational preference. As a proof of concept experiment, Camps et al. demonstrated that their system could be used to evolve enzymes with diverged function by generating TEM-1 β-lactamase mutants that were able to hydrolyze a third-generation lactam antibiotic, aztreonam.

The ep Pol I/ColE1 system has subsequently been applied in a handful of additional directed evolution experiments. For example, Koch et al. used the system to prepare a library of terminal alkane hydroxylases with the aim of evolving variants that can oxidize butane [51]. Although they only used the system for the preparation of mutant libraries (i.e. as a mutator strain) and not for continuous evolution involving serial passaging under prolonged selection conditions, they demonstrated that one can create large libraries of GOI variants directly *in vivo*. In another application, an M13 phagemid with a ColE1 origin was made to encode LuxR and infect *E. coli* harboring the ep Pol I [52]. LuxR is a transcriptional activator and drove the transcription of an antibiotic resistance gene (β-lactamase) controlled by the *lux* promoter in the *E. coli*. Through several cycles of infecting fresh *E. coli*, antibiotic selection, lysis of *E. coli*, and phage isolation, LuxR evolved a 17-fold higher binding affinity to the *lux* promoter sequence.

While the ep Pol I/ColE1 system approaches ICE, it is limited by off-target mutagenesis and low durability. Because Pol I is responsible for Okazaki fragment mending throughout the genome and also participates in DNA repair [53], expressing an ep Pol I causes substantial mutagenesis genome-wide. Targeting of mutations to the GOI does occur – owing to the ColEI origin, the limited role of Pol I in lagging strand replication, and special growth conditions optimized to time ep Pol I action with growth phases where genome replication activity is low – but is maximally only ~400-fold. Therefore, when highly ep Pol Is are used, it is possible that off-target mutagenesis will lower the fitness of the cell, causing fixation of suppressor mutations that abrogate the activity of ep Pol I. Still, the Pol I/ColE1 system represents a landmark development that encouraged the field to pursue new strategies for realizing ICE.

Perhaps the closest conceptual descendant of the ep Pol I/ColE1 system is a new *E. coli* continuous evolution system called EvolvR, which uses CRISPR-guided ep DNAPs to continuously target mutations to GOIs (Figure 1.3b). Rather than rely on the natural targeting of Pol I to ColE1, Halperin et al. [54] fused ep Pol I variants (and other DNAPs) to a nickase Cas9 (nCas9) that would serve two purposes. First, nCas9 would bring the ep Pol I to any GOI encoded on a plasmid or the genome using a guide RNA (gRNA). Second, the nCas9 would nick the target strand, creating a free 3′-OH substrate from which the ep Pol I could extend. Once nCas9 releases the nicked product, it is believed that ep Pol I then latches on and carries out error-prone extension from the nick. This highly clever idea was demonstrated in *E. coli* with a number of ep Pol I variants spanning different mutation rates and activities, as well as with a moderately ep Phi29 DNAP with high processivity. Using the most mutagenic ep Pol I, Halperin et al. measured a mutation rate approaching 10^{-2} s.p.b. (a 7.7 million-fold elevation compared to wt cells) at the first nucleotide 3′ of the nCas9-induced nick. While this extreme mutation rate quickly dropped when moving away from the nick, other Pol I and Phi29 DNAP variants with moderate error rates could achieve mutagenesis windows up to 350 bp. With these characteristics and with the potential to use multiple gRNAs to simultaneously target multiple parts of a gene, EvolvR could readily and efficiently generate sequence diversity on a GOI *in vivo* to support continuous evolution. Indeed, in a proof of principle experiment, Halperin et al. used EvolvR to rapidly evolve spectinomycin resistance by targeting mutagenesis to the *rpsE* gene and found new resistance mutations that were previously unknown.

Future studies and improvements on EvolvR will clarify how well it drives ICE for prolonged periods of time, needed to traverse long mutational pathways. Durability may be difficult in the current architecture, because the mutation rate is maximal at nucleotides within the target region of the gRNA, which if mutated, will reduce the ability of the system to continue inducing mutagenesis. Since the GOI can still be replicated (by high-fidelity host systems) in the absence of EvolvR function, this may result in the fixation of partially adapted GOI mutants that stop mutating, leading to premature cessation of evolution. In addition, EvolvR still has off-target elevations in mutation rate, presumably because ep Pol I or Phi29 can participate in genomic replication and/or because Cas9 has off-target binding. Strategies that use more processive ep DNAPs with no activity in normal genome replication and alternative CRISPR systems that nick outside the critical regions for gRNA targeting may overcome potential issues of targeting and durability. We also anticipate that this system should readily transfer to cell-types other than *E. coli*. Therefore, EvolvR is a highly promising new system for ICE with enormous potential, especially for the multiplexed evolution of genes at their endogenous genomic loci rather than on a plasmid.

1.4.2 Yeast Systems That Do Not Use Engineered DNA Polymerases for Mutagenesis

The first demonstration of continuous targeted mutagenesis *in vivo* in yeast was published in 2013 under the name TaGTEAM (Figure 1.4a), which stands for targeting glycosylases to embedded arrays for mutagenesis [55]. In TaGTEAM, mutagenesis at the GOI is initiated by recruiting a DNA glycosylase, which normally functions as the first step in the base excision repair (BER) pathway responsible for removing chemically altered DNA bases [56]. The authors adopted the yeast 3-methyladenine glycosylase, Mag1p, and fused it to the tet repressor (tetR) that binds a 19-bp operator sequence, tetO. By introducing a non-recombinogenic tetO array (with each tetO site separated by 10–30 bp of random sequence), the tetR-Mag1p fusion could be targeted to GOIs in the chromosome or plasmid. It is presumed that tetR-Mag1p targeting generates a build-up of unprocessed abasic sites at target loci, leading to replication fork stalling and recruitment of ep translesion polymerases [57]. This faulty repair can lead to both point mutations and frameshifts. To test their system for its ability to generate mutagenesis at a GOI, Finney-Manchester et al. introduced a 240X tetO array upstream of a *URA3* auxotrophic marker in a region of chromosome 1 that does not contain nearby essential genes. The distance between the tetO array and the marker was titrated to assess the size of the area subjected to mutagenesis. The

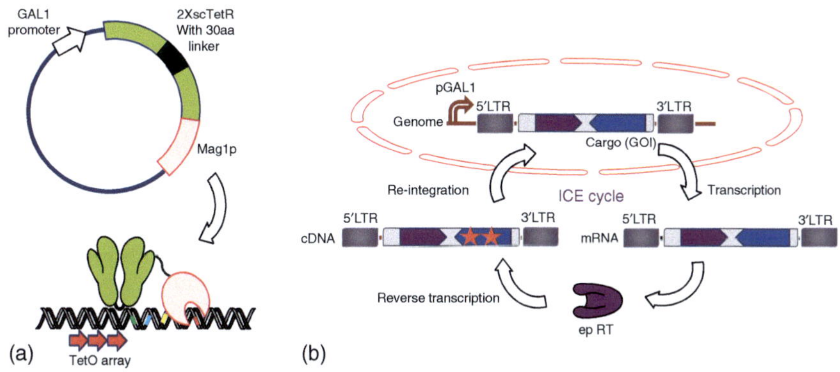

Figure 1.4 Yeast systems for targeted mutagenesis of GOIs. (a) TaGTEAM is achieved by fusing the yeast 3-methyladenine DNA glycosylase, Mag1p, to a tetR DNA-binding domain. Upon expression of the fusion from an inducible galactose promoter, the 20 kb region that is proximal to the tetO array experiences a high degree of mutagenesis. (b) In ICE, the GOI is cloned into an inducible Ty1 retrotransposon in the genome. The ICE cycle begins with inducible transcription of the retroelement followed by ep reverse transcription driven by Ty1's encoded rt. The cycle ends upon re-integration of the mutated cDNA into the genome. Source: Based on Crook et al. [58].

presence of tetR-Mag1p resulted in a >800-fold increase in mutation rate spanning a 10 kb region. However, the off-target mutation rate was also increased 40-fold in the absence of the array, indicating genome-wide mutagenesis by tetR-Mag1p. No direct applications of the system have been published to date, but this mutagenic strategy was important for opening new avenues of thought in the field.

ICE is another notable example of continuous evolution in yeast (Figure 1.4b), introduced in 2016 [58]. ICE adopts a strategy for DNA diversification that is based on the mutagenic properties of the Ty1 retrotransposon element. A GOI is cloned into the Ty1 cassette, which then gets transcribed into an RNA. Next, the RNA is reverse transcribed to form cDNA and reintegrated into the chromosome [59]. The mutagenic properties of the system stem from Ty1's self-encoded reverse transcriptase (rt), which introduces mutations at a rate of $\sim 2.5 \times 10^{-5}$ to $\sim 1.5 \times 10^{-4}$ per base per retrotransposition event [58, 60], thus allowing rapid mutagenesis of Ty1 and its embedded GOI. However, since mutagenesis depends on retrotransposition and the retrotransposition rate of Ty1 with a GOI inserted is low, the high mutation rate of Ty1's rt is only occasionally experienced on the GOI. Therefore, the authors carried out a series of experiments to increase the retrotransposition rate. By fine-tuning various parameters including the cargo's promoter strength, host genotype (i.e. deletions of certain host genes), cell density, temperature, initiator methionine tRNA expression (which acts to prime Ty1 replication), and inclusion of terminators, the authors were able to significantly increase retrotransposition rate. Altogether, the optimization process reached a mutation rate capable of generating up to 1.6×10^7 distinct mutants of a GOI per round per liter cultured [58]. Crook et al. then used ICE in three independent experiments to test the system's ability to evolve genetic material. In the first demonstration, URA3 was evolved for increased resistance to 5-fluoroorotic acid (5-FOA); in the second example, the Spt15p global transcription regulator was evolved to confer a complex cellular phenotype of butanol resistance; and in the third example, a multigene pathway spanning 4.6 kb and containing two enzymes and a regulatory region was evolved for increased xylose catabolism. Additional experiments will clarify the extent to which ICE continuously mutates GOIs, as the ability for Ty1 elements to semi-randomly spread throughout the yeast genome [61, 62] could potentially complicate analysis, reduce mutational accumulation for the GOI, and diffuse the target of evolution. These issues could potentially be solved by somehow limiting Ty1 integration to a single location in the genome, turning the retrotransposon into a "retrocisposon," and then increasing the "retrocisposition" rate to access high levels of diversification. In fact, the ability to achieve "retrocisposition" would also be important for reaching continuous evolution in other systems based on retroelement-mediated mutagenesis, such as a recently reported bacterial approach for *in vivo* genome editing and evolution [63]. Nevertheless, ICE is an important example of continuous evolution in yeast.

1.4.3 Somatic Hypermutation as a Means for Targeted Mutagenesis of GOIs

Some groups have harnessed one of nature's built-in mechanisms for generating targeted DNA diversity, somatic hypermutation (SHM). In SHM, B cells create point mutations in their immunoglobulins (Igs) to drive antibody affinity maturation [64]. The enzyme responsible for SHM is Activation Induced cytidine Deaminase (AID), which deaminates cytidine (C) to generate uridine (U). This triggers various mismatch repair mechanisms resulting in a mutation rate of $\sim 10^{-3}$ s.p.b. at Ig loci [65]. Several researchers have successfully hijacked this natural mechanism for diversifying and evolving non-antibody proteins. In 2001, Bachl et al. set the stage for SHM-based protein directed evolution [66]. They demonstrated a high rate of reversion of a premature stop codon in a green fluorescent protein (GFP) cloned into a hypermutator B cell line (18–81) that expresses endogenous AID. They concluded that elevated reversion rates depended on AID and were rate limited by transcriptional levels of the target gene, in agreement with previous findings on SHM mechanisms [67, 68]. In 2004, Wang et al. applied SHM to the directed evolution of an entire open reading frame [69] by integrating a single copy of red fluorescent protein (RFP) into Ramos cells, which express endogenous AID, using a lentivirus. Through iterative SHM and FACS, RFP mutants with enhanced photostability and far-red emissions were evolved. The study was conducted in the pre-CRISPR era, and thus the RFP GOI was not targeted to an Ig locus but was rather integrated at various genomic locations within their cell population. However, the authors noted that the most evolved RFP variant, which they called mPlum, was located in the Ig heavy chain locus of chromosome 14, indicating that there is indeed a target locus where mutagenesis rates are highest, and that SHM is responsible for high levels of mutagenesis at the GOI. Yet it is expected that this targeting is incomplete, as mutation rates readily occur outside the Ig domain in cell lines that express endogenous AID [70].

Recently, a major development that avoids the use of hypermutator cell lines that express endogenous AID to mutate GOIs was independently published by

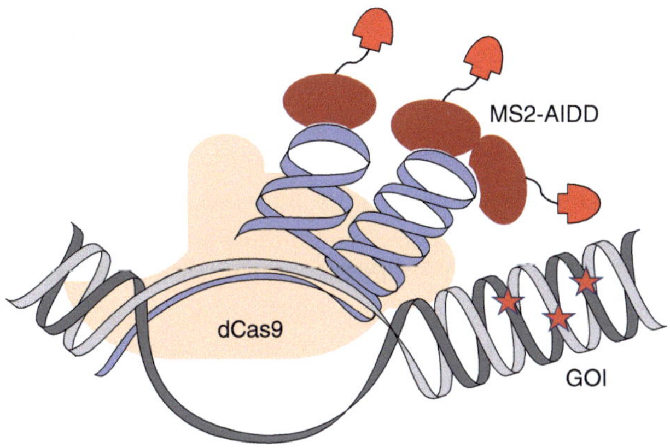

Figure 1.5 Targeted mutagenesis by somatic hypermutation (CRISPRx). A region of ~100 bp is targeted for mutagenesis by a dCas9 complexed with sgRNA-MS2 to a hyperactive cytidine deaminase (AID). Source: Hess et al. [71]; Ma et al. [72].

two groups (Figure 1.5) [71, 72]. Hess et al. linked AID to a catalytically inactive dCas9 using MS2-modified sgRNAs, which achieved precise targeting of SHM to defined loci in HEK293 cells [71]. The system, which they called CRISPRx, allowed targeted mutagenesis of multiple genomic locations simultaneously. Their reported mutation rate was $\sim 5 \times 10^{-4}$ s.b.p., which is similar to that observed for SHM [65]. In their first application, Hess et al. evolved GFP (excitation, 395 nm; emission, 509 nm) into enhanced green fluorescent protein (EGFP) (490/509 nm) by selecting for spectrum-shifted variants. Later, they mutated the target of the cancer therapeutic bortezomib, PSMB5, and identified known and novel mutations that confer bortezomib resistance. At the same time, Ma et al. developed a dCas9-AID fusion and targeted BCR-ABL for mutagenesis to efficiently identify known and new mutations conferring imatinib resistance mutations in chronic myeloid leukemia cells [72].

In both of these CRISPR-guided AID strategies, induction of mutagenesis at the GOI was followed directly by a single round of enrichment for the selected phenotype. Therefore, these studies do not directly demonstrate continuous evolution. However, multi-generation continuous directed evolution could be carried out using cell lines stably transcribing sgRNAs that tile the GOI. Although Hess et al. observed some limited off-target mutagenesis, owing both to off-target activity of AID and off-target binding of sgRNAs [71], the durability of this system, while untested, may be reasonably high, as the positions that are most prone to mutagenesis are outside of the spacer and PAM needed for sgRNA binding and multiple sgRNAs targeted to the same locus can be used. In addition, these methods are capable of introducing diversity at endogenous genomic loci, since CRISPR targeting is programmable.

Another strategy for targeting AID to GOIs is based on fusing AID to T7 RNAP [73]. The main advantage is that T7 RNAP induction could be precisely controlled in *E. coli*, and although not demonstrated by More et al., could be largely transferred between various organisms. Mutations accumulate during induction of transcription, as the T7 RNAP carries AID over large stretches of DNA. Indeed, due to its high processivity, T7 RNAP can direct mutagenesis over several kb.

1.4.4 Orthogonal DNA Replication (OrthoRep)

Our lab recently developed a system for ICE, termed OrthoRep, based on orthogonal DNA replication [74, 75] (Figure 1.6). Fundamentally, OrthoRep can be described as a cell harboring a synthetic DNA replication system that propagates without affecting endogenous replication of the host genome. We implemented this additional replication system in the form of an orthogonal DNAP/plasmid pair, where orthogonality means that the DNAP is dedicated to the cognate plasmid and does not participate in genomic replication (unlike Pol I in the Pol I/ColE1 systems). This property allows us, broadly, to engineer DNA replication *in vivo* for user-defined purposes without harming the host. For the purpose of ICE, we can make the orthogonal DNAP as error-prone as desired, since the genome is completely spared from mutation. Then, GOIs can simply be encoded on the orthogonal plasmid and rapidly and continuously mutated by the orthogonal ep DNAP during evolution.

To create OrthoRep, we developed an orthogonal DNAP/plasmid pair in *S. cerevisiae* by leveraging the unique pGKL1 and pGKL2 (or p1 and p2) selfish elements [76, 77]. p1 and p2 are linear, high copy, DNA plasmids that can replicate autonomously in the cytoplasm of yeast. P1 and p2 each encode dedicated DNAPs,

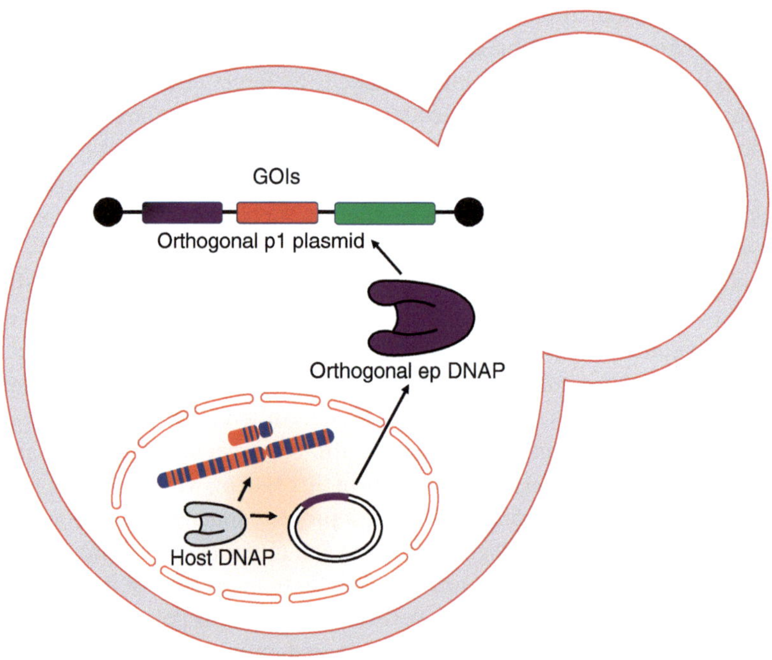

Figure 1.6 OrthoRep. In OrthoRep, GOIs encoded on the orthogonal p1 plasmid are replicated by the orthogonal ep DNAP. The genome is fully spared from mutation by the orthogonal ep DNAP. Source: Ravikumar et al. [74]; Ravikumar et al. [75].

TP-DNAP1, and TP-DNAP2, respectively, and rely on additional p2-encoded replication and transcription factors to propagate. For p1 and p2 replication, TP-DNAP1 and TP-DNAP2 recognize terminal proteins, TP1 and TP2, which are covalently linked to the 5′ termini of p1 and p2. These terminal proteins act as origins of replication and serve as primers for initiation (in contrast to canonical RNA primers). This unique mechanism of protein-priming combined with the compartmental isolation of cytoplasmic p1 and p2 replication from nuclear DNA make these elements orthogonal to genomic replication. In fact, we demonstrated that TP-DNAP1/p1 and TP-DNAP2/p2 are mutually orthogonal DNAP/plasmid pairs, showing that TP-DNAPs are highly specific for their cognate TP-bound plasmid [78]. For OrthoRep, we repurposed the TP-DNAP1-p1 pair, while leaving the TP-DNAP2-p2 pair intact.

Since the native p1 mutation rate of $\sim 1 \times 10^{-9}$ s.p.b. was far too low for continuous evolution experiments, a large protein engineering effort was undertaken to reduce the fidelity of TP-DNAP1. First, the *TP-DNAP1* gene was deleted from p1 and TP-DNAP1 was expressed in trans from a nuclear expression vector. This enabled facile characterization of TP-DNAP1 variants and prevents ep TP-DNAP1s from mutating their own gene. Then, after mixed successes in predicting TP-DNAP1 mutators from related DNAPs, we cloned a scanning mutagenesis library of *TP-DNAP1* covering all single amino acid variants, and from this, screened ~14 000 clones (~onefold theoretical coverage) for elevated mutation rates. This effort yielded a set of moderate mutators, which were then used to clone and screen combinatorial libraries, eventually leading to the discovery of a variant (TP-DNAP1-4-2) that mutates p1 at $\sim 1 \times 10^{-5}$ s.p.b. At this rate, OrthoRep mutates GOIs 100 000-fold faster than the *S. cerevisiae* genome.

The high p1 mutation rate driven by TP-DNAP1-4-2 showed no sign of erosion over extensive serial culturing (at least 90 generations and in unpublished experiments, at least 300 generations) and we found that the genomic rate ($\sim 10^{-10}$ s.p.b.) remained unchanged in the presence of TP-DNAP1-4-2, demonstrating continuous mutagenesis with complete orthogonality. Notably, p1 replication with TP-DNAP1-4-2 was experimentally determined to exceed the error-induced extinction threshold of the host genome (at most, 4.7×10^{-6} s.p.b.). Even moderate genomic rates of 1.6×10^{-7}–5.2×10^{-7} s.p.b. were observed to be unstable over short durations, confirming that targeted mutagenesis with OrthoRep bypasses genomic error thresholds in a sustainable manner.

The utility of OrthoRep as a scalable directed evolution platform was demonstrated in an experiment that repeated the evolution of *Plasmodium falciparum* DHFR (*Pf*DHFR) resistance to the antimalarial drug pyrimethamine, in 90 independent lines. By encoding *PfDHFR* on p1 in a *S. cerevisiae* strain lacking the endogenous dihydrofolate reductase (DHFR) gene (*DFR1*), we were able to evolve *Pf*DHFR for strong resistance to pyrimethamine inhibition. This was done simply by serially passaging small volume (0.5 ml) cultures a few times under drug selection, allowing for the experiment to be done at the scale of 90 replicates. Adapted populations primarily converged on a previously undiscovered, but highly fit, region of the *Pf*DHFR resistance landscape. Moreover, by repeating evolution many times, we were able to capture rare stochastic events that steered populations away from the common path towards alternative high- and moderate-fitness outcomes. In fact, some of the events occurred in just 1/90 replicates, providing insight into the (ir)reproducibility of molecular adaptation that would not have been captured with smaller experiments. We also showed that evolution is highly dependent on the starting genotype by repeating evolution with a variant of *Pf*DHFR containing a synonymous codon change, and finding a drastically different set of evolutionary outcomes.

Moving forward, we believe that the properties of OrthoRep are uniquely positioned to address the long-term technological challenges associated with ICE (Table 1.1). Although many approaches are capable of elevating mutagenesis of GOIs over genomic genes, the level of targeting in OrthoRep (at least ~100 000-fold) is currently unmatched. Furthermore, continuous mutagenesis in OrthoRep should be extraordinarily durable. Because replication of GOIs occurs exclusively through the action of the ep DNAP, inheritance of GOIs is intrinsically coupled to mutagenesis. Put in more specific terms, cells may acquire a disabling mutation in p1's origin of replication that ceases its mutagenesis, but these mutant p1s will no longer get replicated, and are immediately removed from the population. In addition, p1's origins of replication are mostly proteinaceous (the TPs) and cannot be mutated by the orthogonal DNAP, so the chance of a disabling mutation at the origin is low to begin with. In short, continuity of mutagenesis is enforced in the orthogonal DNA replication architecture. This durability will become increasingly important as the field makes headway towards more and more difficult protein functions that require many mutations to access. Such problems include the *de novo* evolution of enzymes, evolution of protein–protein interactions, and evolution of high-affinity therapeutic antibodies against difficult targets, all of which are at the forefront of the protein engineering field.

1.5 Conclusion

While directed evolution has been an extraordinarily powerful approach to protein engineering, the predominant approach of subjecting a GOI to rounds of *in vitro* PCR-based mutagenesis, transformation into host cells, and selection limits both the scale and depth of protein functions that can be evolved. By achieving *in vivo* continuous mutation at high rates targeted to only GOIs, the emerging field of ICE promises a fundamental transformation in the power and accessibility of directed evolution. We look forward to the continued application and improvement of the systems described in this chapter to ambitions protein engineering problems for years to come.

References

1 Turner, N.J. (2009). Directed evolution drives the next generation of biocatalysts. *Nat. Chem. Biol.* 5 (8): 567–573.

2 Packer, M.S. and Liu, D.R. (2015). Methods for the directed evolution of proteins. *Nat. Rev. Genet.* 16 (7): nrg3927.

3 Cobb, R.E., Chao, R., and Zhao, H. (2013). Directed evolution: past, present, and future. *AlChE J.* 59 (5): 1432–1440.

4 Bloom, J.D. and Arnold, F.H. (2009). In the light of directed evolution: pathways of adaptive protein evolution. *Proc. Nat. Acad. Sci.* 106 (Suppl. 1): 9995–10000.

5 Wellner, A., Gurevich, M.R., and Tawfik, D.S. (2013). Mechanisms of protein sequence divergence and incompatibility. *PLos Genet.* 9 (7): e1003665.

6 Soskine, M. and Tawfik, D.S. (2010). Mutational effects and the evolution of new protein functions. *Nat. Rev. Genet.* 11 (8): 572–582.

7 Jespers, L.S., Roberts, A., Mahler, S.M. et al. (1994). Guiding the selection of human antibodies from phage display repertoires to a single epitope of an antigen. *Nat. Biotechnol.* 12 (9): 899–903.

8 Coelho, P.S., Brustad, E.M., Kannan, A., and Arnold, F.H. (2013). Olefin cyclopropanation via carbene transfer catalyzed by engineered cytochrome P450 enzymes. *Science* 339 (6117): 307–310.

9 McIsaac, R.S., Engqvist, M.K.M., Wannier, T. et al. (2014). Directed evolution of a far-red fluorescent rhodopsin. *Proc. Nat. Acad. Sci.* 111 (36): 13034–13039.

10 Goldsmith, M., Eckstein, S., Ashani, Y. et al. (2016). Catalytic efficiencies of directly evolved phosphotriesterase variants with structurally different organophosphorus compounds in vitro. *Arch. Toxicol.* 90 (11): 2711–2724.

11 Joyce, G.F. (2004). Directed evolution of nucleic acid enzymes. *Annu. Rev. Biochem.* 73 (1): 791–836.

12 Badran, A.H. and Liu, D.R. (2015). In vivo continuous directed evolution. *Curr. Opin. Chem. Biol.* 24: 1–10.

13 d'Oelsnitz, S. and Ellington, A. (2018). Continuous directed evolution for strain and protein engineering. *Curr. Opin. Biotechnol.* 53: 158–163.

14 Carlson, J.C., Badran, A.H., Guggiana-Nilo, D.A., and Liu, D.R. (2014). Negative selection and stringency modulation in phage-assisted continuous evolution. *Nat. Chem. Biol.* 10 (3): 216–222.

15 Dickinson, B.C., Leconte, A.M., Allen, B. et al. (2013). Experimental interrogation of the path dependence and stochasticity of protein evolution using phage-assisted continuous evolution. *Proc. Nat. Acad. Sci.* 110 (22): 9007–9012.

16 Wong, B.G., Mancuso, C.P., Kiriakov, S. et al. (2018). Precise, automated control of conditions for high-throughput growth of yeast and bacteria with eVOLVER. *Nat. Biotechnol.* 36 (7): 614.

17 Drake, J.W. (1991). A constant rate of spontaneous mutation in DNA-based microbes. *Proc. Nat. Acad. Sci.* 88 (16): 7160–7164.

18 Herr, A.J., Ogawa, M., Lawrence, N.A. et al. (2011). Mutator suppression and escape from replication error–induced extinction in yeast. *PLos Genet.* 7 (10): e1002282.

19 Wilke, C.O., Wang, J.L., Ofria, C. et al. (2001). Evolution of digital organisms at high mutation rates leads to survival of the flattest. *Nature* 412 (6844): 331.

20 Nowak, M. and Schuster, P. (1989). Error thresholds of replication in finite populations mutation frequencies and the onset of muller's ratchet. *J. Theor. Biol.* 137 (4): 375–395.

21 Lang, G.I. and Murray, A.W. (2008). Estimating the per-base-pair mutation rate in the yeast *Saccharomyces cerevisiae. Genetics* 178 (1): 67–82.

22 Lee, H., Popodi, E., Tang, H., and Foster, P.L. (2012). Rate and molecular spectrum of spontaneous mutations in the bacterium *Escherichia coli* as determined by whole-genome sequencing. *Proc. Nat. Acad. Sci.* 104 (41): E2774–E2783.

23 Milholland, B., Dong, X., Zhang, L. et al. (2017). Differences between germline and somatic mutation rates in humans and mice. *Nat. Commun.* 8: 15183–15183.

24 Doming, E. (2006). *Quasispecies: Concept and Implications for Virology*, 299.

25 Giraud, A., Matic, I., Tenaillon, O. et al. (2001). Costs and benefits of high mutation rates: adaptive evolution of bacteria in the mouse gut. *Science* 291 (5513): 2606–2608.

26 Notley-McRobb, L., Seeto, S., and Ferenci, T. (2002). Enrichment and elimination of mutY mutators in *Escherichia coli* populations. *Genetics* 162 (3): 1055–1062.

27 Wang, H.H., Isaacs, F.J., Carr, P.A. et al. (2009). Programming cells by multiplex genome engineering and accelerated evolution. *Nature* 460 (7257): 894.

28 Liang, L., Liu, R., Garst, A.D. et al. (2017). CRISPR EnAbled trackable genome engineering for isopropanol production in *Escherichia coli. Metab. Eng.* 41: 1–10.

29 Nyerges, Á., Csörgő, B., Draskovits, G. et al. (2018). Directed evolution of multiple genomic loci allows the prediction of antibiotic resistance. *Proc. Nat. Acad. Sci.* 115 (25): 201801646.

30 Abil, Z., Ellefson, J.W., Gollihar, J.D. et al. (2017). Compartmentalized partnered replication for the directed evolution of genetic parts and circuits. *Nat. Protoc.* 12 (12): 2493–2512, nprot.2017.2119.

31 Badran, A.H., Guzov, V.M., Huai, Q. et al. (2016). Continuous evolution of *Bacillus thuringiensis* toxins overcomes insect resistance. *Nature* 533 (7601): 58.

32 Dickinson, B.C., Packer, M.S., Badran, A.H., and Liu, D.R. (2014). A system for the continuous directed evolution of proteases rapidly reveals drug-resistance mutations. *Nat. Commun.* 5 (1): 5352.

33 Esvelt, K.M., Carlson, J.C., and Liu, D.R. (2011). A system for the continuous directed evolution of biomolecules. *Nature* 472 (7344): 499.

34 Hubbard, B.P., Badran, A.H., Zuris, J.A. et al. (2015). Continuous directed evolution of DNA-binding proteins to improve TALEN specificity. *Nat. Methods* 12 (10): 939–942.

35 Leconte, A.M., Dickinson, B.C., Yang, D.D. et al. (2013). A population-based experimental model for protein evolution: effects of mutation rate and selection stringency on evolutionary outcomes. *Biochemistry* 52 (8): 1490–1499.

36 Pu, J., Zinkus-Boltz, J., and Dickinson, B.C. (2017). Evolution of a split RNA polymerase as a versatile biosensor platform. *Nat. Chem. Biol.* 13 (4): 432–438.

37 Pu, J., Kentala, K., and Dickinson, B.C. (2017). Multidimensional control of Cas9 by evolved RNA polymerase-based biosensors. *ACS Chem. Biol.* 13 (2): 431–437.

38 Badran, A.H. and Liu, D.R. (2015). Development of potent *in vivo* mutagenesis plasmids with broad mutational spectra. *Nat. Commun.* 6 (1): 8425.

39 Deng, D., Yan, C., Pan, X. et al. (2012). Structural basis for sequence-specific recognition of DNA by TAL effectors. *Science* 335 (6069): 720–723.

40 Hu, J.H., Miller, S.M., Geurts, M.H. et al. (2018). Evolved Cas9 variants with broad PAM compatibility and high DNA specificity. *Nature* 556 (7699): 57.

41 Packer, M.S., Rees, H.A., and Liu, D.R. (2017). Phage-assisted continuous evolution of proteases with altered substrate specificity. *Nat. Commun.* 8 (1): 956.

42 Bryson, D.I., Fan, C., Guo, L.-T. et al. (2017). Continuous directed evolution of aminoacyl-tRNA synthetases. *Nat. Chem. Biol.* 13 (12): 1253.

43 Wang, T., Badran, A.H., Huang, T.P., and Liu, D.R. (2018). Continuous directed evolution of proteins with improved soluble expression. *Nat. Chem. Biol.* 13: 1253–1260.

44 Brödel, A.K., Isalan, M., and Jaramillo, A. (2018). Engineering of biomolecules by bacteriophage directed evolution. *Curr. Opin. Biotechnol.* 51: 32–38.

45 Brödel, A.K., Jaramillo, A., and Isalan, M. (2016). Engineering orthogonal dual transcription factors for multi-input synthetic promoters. *Nat. Commun.* 7: 13858.

46 Camps, M., Naukkarinen, J., Johnson, B.P., and Loeb, L.A. (2003). Targeted gene evolution in *Escherichia coli* using a highly error-prone DNA polymerase I. *Proc. Nat. Acad. Sci.* 100 (17): 9727–9732.

47 Fabret, C., Poncet, S., Danielsen, S. et al. (2000). Efficient gene targeted random mutagenesis in genetically stable *Escherichia coli* strains. *Nucleic Acids Res.* 28 (21): e95.

48 Troll, C., Yoder, J., Alexander, D. et al. (2014). The mutagenic footprint of low-fidelity Pol I ColE1 plasmid replication in *E. coli* reveals an extensive interplay between Pol I and Pol III. *Curr. Genet.* 60 (3): 123–134.

49 Alexander, D.L., Lilly, J., Hernandez, J. et al. (2014). Random mutagenesis by error-prone Pol I plasmid replication in *Escherichia coli*. *Methods Mol. Biol.* 1179: 31–44.

50 Shinkai, A. and Loeb, L.A. (2001). *In vivo* mutagenesis by *Escherichia coli* DNA polymerase I. Ile(709) in motif A functions in base selection. *J. Biol. Chem.* 276 (50): 46759–46764.

51 Koch, D.J., Chen, M.M., Beilen, J.B.v., and Arnold, F.H. (2009). *In vivo* evolution of butane oxidation by terminal alkane hydroxylases AlkB and CYP153A6. *Appl. Environ. Microbiol.* 75 (2): 337–344.

52 Na, D., Lee, S., Yi, G.-S., and Lee, D. (2011). Synthetic inter-species cooperation of host and virus for targeted genetic evolution. *J. Biotechnol.* 153 (1–2): 35–41.

53 Allen, J.M., Simcha, D.M., Ericson, N.G. et al. (2011). Roles of DNA polymerase I in leading and lagging-strand replication defined by a high-resolution mutation footprint of ColE1 plasmid replication. *Nucleic Acids Res.* 39 (16): 7020–7033.

54 Halperin, S.O., Tou, C.J., Wong, E.B. et al. (2018). CRISPR-guided DNA polymerases enable diversification of all nucleotides in a tunable window. *Nature* 560 (7717): 248–252.

55 Finney-Manchester, S.P. and Maheshri, N. (2013). Harnessing mutagenic homologous recombination for targeted mutagenesis in vivo by TaGTEAM. *Nucleic Acids Res.* 41 (9): e99.

56 Nilsen, H. and Krokan, H.E. (2001). Base excision repair in a network of defence and tolerance. *Carcinogenesis* 22 (7): 987–998.

57 Boiteux, S. and Guillet, M. (2004). Abasic sites in DNA: repair and biological consequences in *Saccharomyces cerevisiae. DNA Repair* 3 (1): 1–12.

58 Crook, N., Abatemarco, J., Sun, J. et al. (2016). *In vivo* continuous evolution of genes and pathways in yeast. *Nat. Commun.* 7: 13051.

59 Wilhelm, F.X., Wilhelm, M., and Gabriel, A. (2005). Reverse transcriptase and integrase of the *Saccharomyces cerevisiae* Ty1 element. *Cytogenet. Genome Res.* 110 (1–4): 269–287.

60 Boutabout, M., Wilhelm, M., and Wilhelm, F.-X. (2001). DNA synthesis fidelity by the reverse transcriptase of the yeast retrotransposon Ty1. *Nucleic Acids Res.* 29 (11): 2217–2222.

61 Carr, M., Bensasson, D., and Bergman, C.M. (2012). Evolutionary genomics of transposable elements in *Saccharomyces cerevisiae. PLoS One* 7 (11): e50978.

62 Devine, S.E. and Boeke, J.D. (1996). Integration of the yeast retrotransposon Ty1 is targeted to regions upstream of genes transcribed by RNA polymerase III. *Genes Dev.* 10 (5): 620–633.

63 Simon, A.J., Morrow, B.R., and Ellington, A.D. (2018). Retroelement-based genome editing and evolution. *ACS Synth. Biol.* 11.

64 Wu, X., Feng, J., Komori, A. et al. (2003). Immunoglobulin somatic hypermutation: double-strand DNA breaks, AID and error-prone DNA repair. *J. Clin. Immunol.* 23 (4): 235–246.

65 Rajewsky, K., Forster, I., and Cumano, A. (1987). Evolutionary and somatic selection of the antibody repertoire in the mouse. *Science* 238 (4830): 1088–1094.

66 Bachl, J., Carlson, C., Gray-Schopfer, V. et al. (2001). Increased transcription levels induce higher mutation rates in a hypermutating cell line. *J. Immunol.* 166 (8): 5051–5057.

67 Chaudhuri, J., Tian, M., Khuong, C. et al. (2003). Transcription-targeted DNA deamination by the AID antibody diversification enzyme. *Nature* 422 (6933): 726–730.

68 Yu, K., Huang, F.-T., and Lieber, M.R. (2004). DNA substrate length and surrounding sequence affect the activation-induced deaminase activity at cytidine. *J. Biol. Chem.* 279 (8): 6496–6500.

69 Wang, L., Jackson, W.C., Steinbach, P.A., and Tsien, R.Y. (2004). Evolution of new nonantibody proteins via iterative somatic hypermutation. *Proc. Natl. Acad. Sci. U.S.A.* 101 (48): 16745–16749.

70 Wang, C.L., Harper, R.A., and Wabl, M. (2004). Genome-wide somatic hypermutation. *Proc. Natl. Acad. Sci. U.S.A.* 101 (19): 7352–7356.

71 Hess, G.T., Frésard, L., Han, K. et al. (2016). Directed evolution using dCas9-targeted somatic hypermutation in mammalian cells. *Nat. Methods* 13 (12): 1036–1042.

72 Ma, Y., Zhang, J., Yin, W. et al. (2016). Targeted AID-mediated mutagenesis (TAM) enables efficient genomic diversification in mammalian cells. *Nat. Methods* 13 (12): 1029–1035.

73 Moore, C.L., Papa, L.J. 3rd, and Shoulders, M.D. (2018). A processive protein chimera introduces mutations across defined DNA regions *in vivo*. *J. Am. Chem. Soc.* 140: 11560–11564.

74 Ravikumar, A., Arrieta, A., and Liu, C.C. (2014). An orthogonal DNA replication system in yeast. *Nat. Chem. Biol.* 10 (3): 175–177.

75 Ravikumar, A., Arzumanyan, G.A., Obadi, M.K.A. et al. (2018). Scalable, continuous evolution of genes at mutation rates above genomic error thresholds. *Cell* 175 (7): 1946–1957.

76 Klassen, R. and Meinhardt, F. (2007). Linear protein-primed replicating plasmids in eukaryotic microbes. *Microb. Linear Plasmids* 7: 187–226.

77 Gunge, N. and Sakaguchi, K. (1981). Intergeneric transfer of deoxyribonucleic acid killer plasmids, pGKl1 and pGKl2, from *Kluyveromyces lactis* into *Saccharomyces cerevisiae* by cell fusion. *J. Bacteriol.* 147 (1): 155–160.

78 Arzumanyan, G.A., Gabriel, K.N., Ravikumar, A. et al. (2018). Mutually orthogonal DNA replication systems *in vivo*. *ACS Synth. Biol.* 7 (7): 1722–1729.

2

In Vivo Biosensors for Directed Protein Evolution

Song Buck Tay and Ee Lui Ang

Singapore Institute of Food and Biotechnology Innovation, Agency for Science, Technology and Research 31 Biopolis Way, Singapore 138669, Singapore

2.1 Introduction

Biosensors are biological elements that detect cellular analytes and, thereupon, deliver measurable signals that correlate to the stimuli. Since the inaugural demonstration of glucose monitoring using oxidases contained in an electrode form [1], a range of nucleic acids, antibodies, cell receptors, and even organelles have been progressively integrated into bio-sensing roles related to healthcare, food control, and environmental monitoring [2, 3]. The early interests of biosensor development are mainly focused on medical diagnostic and quality assurance applications [4]. More recently, utilization of biosensors for the detection of metabolites has gained momentum due to advances in technologies for biosensor design and discovery, as well as the potential for protein and metabolic engineering to produce a variety of value-added chemicals [5, 6]. In this context, the accessibility to optimal enzyme candidates is fundamental to productive biosynthesis.

Directed evolution has become an important tool for protein functional development [7]. The capacity to deliberately introduce mutations at a preferred frequency and region has shortened the time-scale for protein optimization, especially when the evolved characteristics are obvious. Basically, a strong phenotype–genotype association is at the heart of directed evolution and the methodical screening or selection of a diversified population is crucial in maintaining this link to allow for recovery of important genetic information [8]. In the event where the desired phenotypes are conveniently perceptible, useful genotypes can be isolated by physical partitioning of the mutant populations. Some of these characteristics include protein thermostability, working pH conditions, and tolerance against certain compounds [9–11]. A direct assay of protein or enzymatic properties can also be applied if the substrates and products of catalysis are available or readily detectable [12, 13]. However, when phenotypic output is not evident and there is no affiliation between the desired product and growth, establishing a correlation between phenotype and genotype becomes a challenge.

Protein Engineering: Tools and Applications, First Edition. Edited by Huimin Zhao.
© 2021 WILEY-VCH GmbH. Published 2021 by WILEY-VCH GmbH.

Biosensors can be used to bridge this association gap by coupling an obscure trait to a fitness advantage through expression of a gene required for host proliferation [14]. Importantly, a selective pressure is imposed against deleterious interactions between protein of interest and host in an intracellular sensing-reporter system, thereby demanding positive candidates to function within the context of an entire host proteome [15]. Similarly, biosensors can be linked to reporter systems that transduce activity levels or chemical concentrations into measurable colorimetric, luminescence, and fluorescence responses [16]. Over the years, a variety of methods have been developed to push the limits of mutant libraries analysis. In particular, the application of biosensors within cellular chassis has been a significant enabling tool for various directed protein evolution platforms, by facilitating the implementation of high throughput screening and selection approaches for improving biocatalysts [17–19]. In addition, an *in vivo* characterization strategy can bypass difficulties related to protein purification and *in vitro* validation [20]. As enzyme optimization within a biosynthetic pathway becomes more relevant, *in vivo* biosensors also offer an attractive, multiplexed phenotypic screening solution for evolution of related protein targets concurrently [21].

In general, a biosensor consists of two functional components: a bioreceptor domain for sensing and a biotransducer module for signal amplification [22]. For initiation of biosensing, the bioreceptor recognizes specific analytes and the downstream biotransducer component is activated upon complementary interactions. Riboswitches, aptamers, and ligand-binding domains are some common examples of a bioreceptor. A certain degree of selectivity for the analyte is necessary for a bioreceptor and the level of stringency can determine the scope of biosensor application [2]. The biotransducer serves as a regulatory element, which mediates mechanisms such as mRNA stability and transcription initiation, and this component is usually linked to a signal amplification effector that can be reporter or selection marker-based (Figure 2.1). Typically, biosensors detect products of biocatalysis or outcome of catalytic events in directed protein evolution modalities; although lately, biosensors have been positioned to report on improvement in protein folding as well as transmembrane transport [23–25].

While the history of biosensors dates back almost 60 years, the majority of *in vivo* biosensors have been devised based on a small number of naturally occurring regulatory elements such as riboswitches and transcription factors [26]. More recently, novel biosensors are rapidly being developed via modification of binding specificities and orthogonality using rational design or random mutagenesis methods [27, 28]. Nevertheless, the variety of biocatalysts and their corresponding products far exceed the sensing and recognition elements available [29], thus retaining the need for sensor discovery and engineering. Currently, biosensors used for directed protein evolution *in vivo* can be broadly categorized into two groups: nucleic acid and protein-based biosensors (Table 2.1). This chapter highlights the different types of *in vivo* biosensors used in directed protein evolution, the considerations for biosensor design, and the evolving trends in biosensor application. An emphasis will be placed on recent research efforts toward the improvement of biocatalysts.

Table 2.1 List of *in vivo* biosensors associated with directed protein evolution.

Biosensor	Analyte/activity detected	Type of biosensor	References
glmS ribozyme	Glucosamine-6-phosphate	RNA	[30]
N-acetyl neuraminic acid-specific aptazyme	*N*-acetyl neuraminic acid	RNA	[31]
Theophylline-specific aptazyme	Theophylline	RNA	[15]
Lysine-specific riboswitch	Lysine	RNA	[32]
Thiamine-specific riboswitch	Thiamine	RNA	[24]
pA promoter	Lysine	DNA	[33]
P_{ERG1} promoter	Ergosterol	DNA	[34]
C4-LysR transcriptional activator	3-hydroxypropionic acid	Transcription factor	[35]
DmpR transcriptional activator	Phenols	Transcription factor	[36]
LysG transcriptional activator	Arginine	Transcription factor	[37]
Bacteriophage 434-derived synthetic repressor	Tobacco Etch virus protease activity	Transcription factor	[20]
PRO_1 hybrid transcriptional activator	Progesterone	Transcription factor	[38]
TtgR transcriptional repressor	Resveratrol	Transcription factor	[39]
Split LexA-B42	Glycosynthase activity	Transcription factor	[40]
Dexamethasone-cephem-methotrexate hybrid sensor	β-Lactamase activity	Transcription factor	[41]
TnaC peptide	Tryptophan	Transcription factor	[42]
3,4-dihydroxy-phenylalanine (DOPA) dioxygenase	DOPA	Enzyme	[43]
T7 lysozyme-T7 RNA polymerase hybrid sensor	Hepatitis C virus protease activity	Enzyme	[44]
Split T7 RNA polymerase	Cytidine deaminase expression	Enzyme	[25]

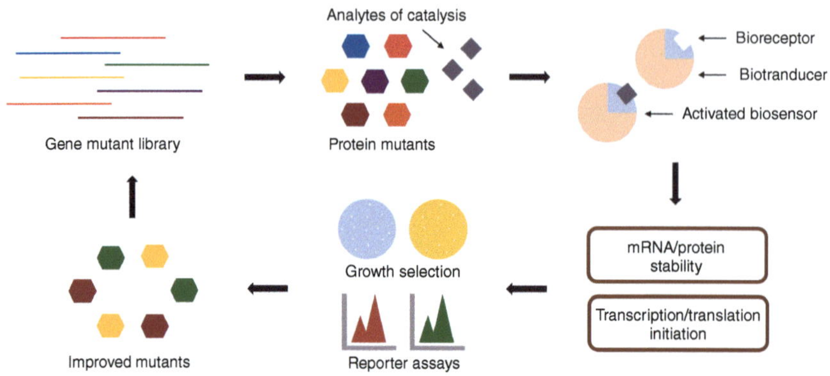

Figure 2.1 Overview of *in vivo* biosensor-mediated directed protein evolution. Desired protein variants from a mutant library produce analytes that bind to the bioreceptor component of a biosensor. The activated biosensor complex proceeds to regulate transcription and translation rates, which can be modulated by mRNA and protein stability. Biosensors are typically linked to reporter genes or selection markers to facilitate a quantifiable evaluation of protein candidates. Improved mutants can be used as templates for subsequent rounds of libraries development and protein evolution.

2.2 Nucleic Acid-Based *In Vivo* Biosensors for Directed Protein Evolution

Natural biological systems have evolved to utilize nucleic acids for detection of intracellular metabolites to effect regulatory functions [45]. The binding of a ligand at the sensing domain modulates the activity of the transducer component, leading to a change in gene expression level. Nucleic acids are usually involved in biorecognition and regulatory response, although there are instances where these nucleotide molecules operate as an actuator, particularly in the form of a ribozyme [46]. Compared to protein biosensors, the range of detectable metabolites by nucleic acid biosensors is still limited. However, nucleic acid biosensors have the advantage of faster response time and the possibility of being used in tandem to mediate tighter control of gene expression [47, 48]. Both RNA and DNA biosensors have been developed for the evolution of proteins and biocatalysts.

2.2.1 RNA-Type Biosensors

In nature, RNA genetic control elements exist in the form of riboswitches, which are often associated with mRNAs to regulate their expression levels. Upon ligand binding at the aptamer region of a riboswitch, the expression platform typically undergoes a conformational change that can disrupt the stability of the transcript or affect the rate of mRNA translation (Figure 2.2a) [49, 50]. In some cases, metabolite interactions can activate self-cleaving mechanisms in RNA transcripts and this has been particularly well-studied in *glmS* ribozyme (Figure 2.2b); whereby the metabolite,

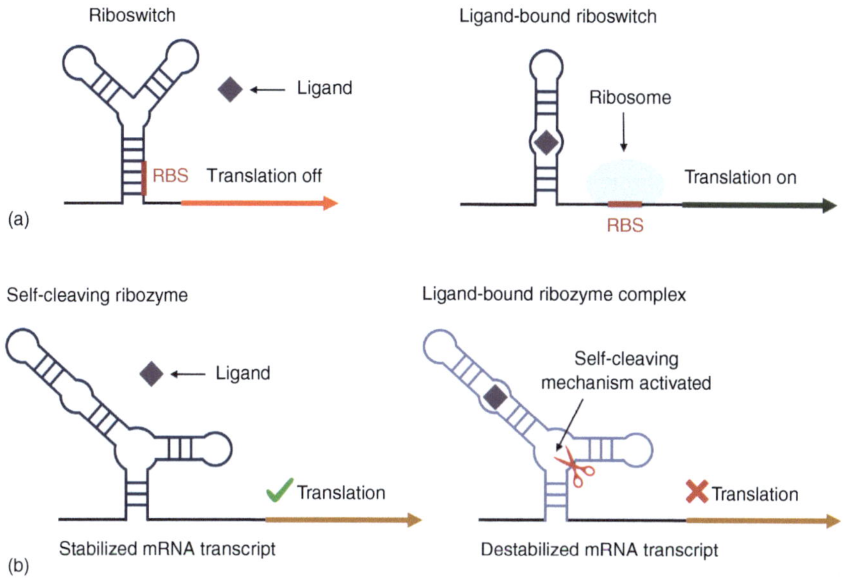

Figure 2.2 RNA-based *in vivo* biosensors. Ligand-binding at a riboswitch induces a conformational change in the mRNA transcript that enables ribosome-binding for translation initiation (a); ligand-binding at a self-cleaving ribozyme destabilizes the mRNA transcript, which leads to mRNA degradation and inhibition of translation (b).

glucosamine-6-phosphate (GlcN6P), acts as a coenzyme to facilitate RNA degradation [51]. Given its dependency on ligand-binding for activity in a feedback repression model, the *glmS* ribozyme has served as a useful biosensor template for directed evolution of proteins of interest. Besides full-length riboswitches, natural aptamer domains have also been coupled to reporter systems or essential genes to construct synthetic RNA biosensors for high-throughput screening or selection of biocatalysts. The identification and characterization of riboswitches that respond to various core metabolites and cofactors have extended the scope of protein optimization [52]. Meanwhile, the development of synthetic aptamers has also expanded the repertoire of biosensing domains that can be used for *in vivo* detection of secondary metabolites and ligands [53].

Bacteria and fungi secondary metabolism offers a rich source of natural products that exhibit a range of anti-microbial, anti-cancer, and immunosuppressant properties [54, 55]. Biosensor-mediated directed evolution approaches can offer a convenient route to optimize catalytic elements of key intermediates in some of these relevant alternative pathways. For instance, *N*-acetyl glucosamine (GlcNAc) is an essential precursor molecule for signal initiation in Gram-positive *Streptomyces* antibiotics biosynthesis [56]; additionally, GlcNAc has also been highlighted as a sole carbon source for bioethanol production in a strain of fermenting yeast [57]. To enhance unnatural GlcNAc biosynthesis in an eukaryotic system, a biosensor was constructed by fusing the *glmS* ribozyme to the 3′-untranslated region (UTR) of a cytosine deaminase gene *FCY1* to select for better producers of GlcNAc [30]. In the presence of fluorocytosine, cytosine deaminase catalyzes the formation of fluorouracil, which is cytotoxic and leads to growth inhibition [58]. By coupling this enzyme to *glmS* ribozyme, binding of GlcN6P (a precursor of GlcNAc) results in the cleavage of *FCY1* transcripts, thereby rescuing cellular toxicity and increasing growth rate. Initially, this suicide synthetic riboswitch was used to evolve glutamine–fructose-6-phosphate transaminase (Gfa1p), an upstream enzyme that regulates a rate-limiting step involved in GlcN6P production. Through fluorocytosine selection of an error-prone mutant library, colonies that exhibited better growth rates were isolated and two point mutations that improved GlcN6P production were identified. With an understanding that Gfa1p is allosterically regulated by uridine 5′-diphospho-GlcNAc, ameliorating mutations were expected to be within proximity of the enzyme's allosteric binding pocket [59]. Instead, this synthetic riboswitch-mediated selection revealed mutations targeted to the glutamine amidotransferase domain, suggesting that allosteric binding in the enzyme could be influenced by inter-domain interactions. Interestingly, this biosensor was also deployed for selection of an enzyme downstream of GlcN6P, which is not endogenous to a yeast system. Using a negative growth assay, a library of haloacid dehalogenase-like phosphatases (HAD phosphatase) that catalyze the dephosphorylation of *N*-acetyl glucosamine-6-phosphate (GlcNAc6P) to GlcNAc was introduced into the best performing GlcN6P-producing strain and selected. Based on further end-product quantitation, two HAD phosphatase candidates that showed reduced growth rates were able to achieve elevated heterologous production of GlcNAc in yeast from improved turnover of GlcNAc6P. Although a negative growth method is usually not used to select for positive mutants in directed evolution studies, the high specificity of *glmS* ribozyme accommodated for such a design as the GlcN6P biorecognition domain does not interact strongly with analogs such as GlcNAc6P and GlcNAc [60, 61].

With established RNA biosensor design principles, riboswitches can be readily engineered to respond to a range of desired metabolites [62]. In fact, the expression platform of a riboswitch has been shown to be modular and can accept a diversity of natural and synthetic aptamers to create new chimeric RNA-based biosensors [63]. For the purpose of detecting sialic acids present on the surface of metastatic cancer cells, an *N*-acetyl neuraminic acid (Neu5Ac)-specific aptamer was conjugated to a hammerhead ribozyme to create a novel RNA biosensor for diagnostic use [64]. Subsequently, this synthetic aptazyme was adapted to evolve *N*-acetyl neuraminate synthase *in vivo* via a directed evolution approach, to obtain over-producers of Neu5Ac in *Escherichia coli*. Neu5Ac is the predominant sialic acid and its biosynthesis has gathered significant interest due to potential uses in the pharmaceutical industry [65]. To select for *N*-acetyl neuraminate synthase variants with enhanced catalytic efficiencies, the Neu5Ac aptazyme was coupled to a gene encoding for tetracycline/H^+ antiporter (*tetA*) and expressed in host cells harboring the synthase mutant library. The use of *tetA* gene as a selection marker allows for both positive and negative selection, such that it confers cells tetracycline resistance while rendering them more sensitive to toxic metal salts such as $NiCl_2$ [66]. However, since Neu5Ac binding would lead to cleavage of *tetA* transcripts, an optimal concentration of $NiCl_2$ was applied for enrichment of improved *N*-acetyl neuraminate synthase mutants. Through sequencing of randomly selected colonies and evaluating the proportion of mutant genotypes, the best Neu5Ac-producing strain ($2.61\,g\,l^{-1}$) was identified and this corresponded to the most common genotype in the mutant population [31]. The selection of a high Neu5Ac producer was possible due to the broad detection range (0–50 mM) of the Neu5Ac aptazyme biosensor.

The modularity of biosensors is not limited to the core sensing and signal transduction elements, but a range of reporter systems can also be incorporated depending on the sensitivity demands [67]. In addition to growth-coupled selection markers, fluorescence genes are routinely used for high-throughput reporting of ligand–receptor interactions for enzyme functional development [68]. Likewise, a riboswitch or ribozyme linked to a fluorescent reporter can offer a rapid and sensitive readout of coupling reactions. By using a theophylline-specific aptazyme biosensor conjugated to a green fluorescent protein (GFP), a caffeine demethylase variant with 33-fold enhancement in catalytic activity was isolated via a high-throughput screening methodology [15]. Upregulation in caffeine demethylase activity increased the production of theophylline and led to higher signal detection. As this aptamer is highly specific and binds theophylline 10 000-fold more tightly than caffeine, no false positive was identified despite surveying a large library of mutants [69]. Notably, the biosensor also has a sensitivity range that spans roughly 2 orders of magnitude (0.01–5 mM); hence when the screen was performed using an upper limit of substrate concentration, a small improvement in activity level could elicit a distinguishable change in fluorescence readout to allow for efficient identification of positive variants. Overall, modular RNA biosensors can provide a flexible and generalizable screening or selection platform for facilitating protein optimization *in vivo*, and these devices can be used to evolve other relevant properties of important biocatalysts. In more recent studies, RNA riboswitches have been employed to assist in the directed evolution of a chimeric aspartate kinase that is resistant to end product inhibition [32]; as well as a switch in substrate specificity, from nicotinamide riboside to thiamine, of a membrane transporter involved in uptake of structurally diverse B-vitamins [24].

2.2.2 DNA-Type Biosensors

Compared with RNA biosensors, stand-alone DNA biosensors are considerably less common for the purpose of directed protein evolution. Single-stranded DNA aptamers have been used for metabolites detection *in vitro*, but the stability of genetically-encoded DNA aptamers in host cells remains a significant challenge [70, 71]. Currently, *in vivo* DNA biosensing elements are limited to metabolite-responsive promoters linked to reporter genes and these devices generally form part of transcription factor-enabled biosensor setups (Figure 2.3a) [72]. Through the development of transcriptomic analysis tools and algorithm-based chromatin immunoprecipitation assays, numerous metabolite-responsive promoters have since been characterized on a genome-wide scale [73, 74]. Some of these

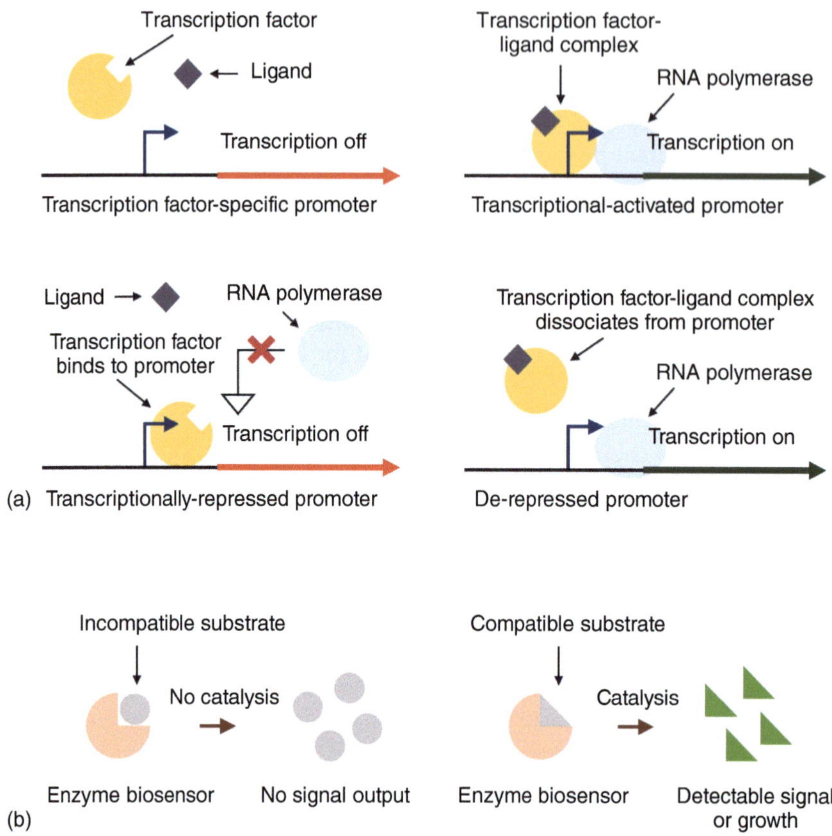

Figure 2.3 Protein-based *in vivo* biosensors. Ligand-binding at a transcription factor constitutes an activated complex that either localizes to or dissociates from an associated promoter region to mediate gene expression (a). An enzyme-coupled biosensor catalyzes the formation of an intermediate or product that confers growth advantage or detectable signal (b).

promoter elements have been incorporated into biosensor designs for detection of metabolites and compounds such as lysine, ergosterol, and 1-butanol, with the aim of catalytic enhancement in their respective pathways [33, 34, 75]. A crucial consideration, however, is the choice of host system that the promoter is implemented as this can affect its level of responsiveness. For example, pA, pN, and LysE-related promoters were found to be sensitive to lysine in *Corynebacterium glutamicum* up to several grams per liter [72]; however, in an *E. coli* system, only the pA promoter was able to elicit a dynamic range that was applicable for high-throughput screening of lysine over-producers [34]. To achieve an efficient transduction of signal input to reporter output, the selection of an appropriate promoter element can be crucial in a biosensor design with regards to the chassis used for screening or selection. Moving forward, the capacity to identify novel metabolite-responsive promoters and design context-responsive regulatory elements should expand on the use of DNA-based biosensors in protein optimization efforts [76, 77].

2.3 Protein-Based *In Vivo* Biosensors for Directed Protein Evolution

Over the past decades, protein-based biosensors are commonly being used to detect a wide range of intracellular compounds and metabolites [23]. Like nucleic acid-based biosensors, the bioreceptor component of protein-based biosensors usually consists of a ligand-binding subunit, while the biotransducer module can constitute a binding or activity domain that is activated through structural rearrangement [22]. In general, protein-based biosensors are likely to exhibit a broader range of specificities due to the existence of various functional groups within binding pockets, which can foster interactions with structurally similar substrates [78]. Nonetheless, despite the diversity of detectable molecules, *in vivo* protein-based biosensors used toward directed protein evolution are predominantly grouped under two different types: transcription factor and enzyme-based biosensors.

2.3.1 Transcription Factor-Type Biosensors

Across biological systems, transcription factors play a major role in the control of gene expression, primarily by regulating RNA polymerase association with DNA for transcription initiation (Figure 2.3a) [79]. Since allosteric control of transcription factors by small molecules is a sensing mechanism ubiquitous in nature, these regulatory proteins can be readily exploited to monitor the concentrations of compounds and metabolites within cellular environment [80]. Unlike RNA riboswitches and aptazymes, a key advantage of transcription factor-based biosensors is the propensity for signal amplification [71]; whereby a small change in ligand concentration can effect downstream expression of multiple genetic transcripts, leading to an augmented readout for enhanced sensitivity. Such a characteristic is highly relevant for low level detection of biocatalytic products and evolution of new activities in proteins or enzymes. Nonetheless, transcription factors can have multiple binding sites within a genomic milieu [81]; hence a comprehensive survey of the target localities and an understanding of regulatory networks are often necessary prior to the implementation of transcription factors in biosensing applications [82, 83].

Given the meaningful amount of work done on the discovery of novel transcription factors and characterization of related promoter elements, a growing repertoire of ligand-binding tools has become available for utility in protein optimization [6, 84]. For instance, recently identified 3-hydroxypropionic acid (3-HP) and phenol-specific transcription factors have been used in biosensor designs for high-throughput screening and selection of enzymatic variants in their respective biosynthetic pathways [85, 86]. 3-HP is an important platform chemical that serves as a feedstock for polymer production and an economical biosynthetic route from glycerol has been established in recombinant *E. coli* [87, 88]. However, a critical limitation lies in the low catalytic activity of an important α-ketoglutaric semi-aldehyde dehydrogenase (KGSADH) enzyme, which results in the accumulation of 3-hydroxy-propionaldehyde and eventual cellular toxicity [89]. By conjugating the 3-HP biosensor (C4-LysR) to a *tetA* selection marker, an improved aldehyde dehydrogenase mutant with 2.79-fold higher catalytic efficiency was identified from a pool of KGSADH variants following tetracycline selection [35]. Importantly, this improved variant was rapidly obtained over two serial selection cultures, indicating that the 3-HP biosensor can be a powerful tool for the engineering of essential biocatalyst in related 3-HP biosynthetic pathways. In a separate study, a switch in substrate specificity, from tryptophan indole to tyrosine phenol, was enabled in a lyase enzyme via the generation of a substrate binding site-specific mutant library from a related template and a phenol-specific transcription factor (DmpR) biosensor [36]. The outcome of ligand-binding was linked to fluorescence readout and a subsequent flow cytometry-based cell sorting process managed to isolate a lyase variant that displayed 13-fold lower catalytic efficiency ($k_{cat}/K_m = 0.58 \pm 0.05$) for tyrosine phenol as compared with a wild-type enzyme. This finding demonstrated the utility of a hypersensitive biosensor for directed evolution of new catalytic activity in enzyme templates and the use of a transcription factor-based biosensor likely facilitated an amplification of the reporter signal despite a low concentration of phenol substrates.

As enzymes involved in the synthesis of cellular building blocks are often inhibited by the end-products of their respective pathways, negation of feedback inhibition has been a common target for protein optimization [90]. So far, protein engineering methods involving random mutagenesis have been able to produce feedback resistant enzyme variants, and these efforts were further enhanced by the use of end-product biosensors that can translate inconspicuous phenotypes into detectable optical outputs [91, 92]. In the context of amino acids biosynthesis, transcription factors that detect these building blocks are common in nature due to their role in maintaining homeostasis [93]. Accordingly, transcription factor-based biosensors are implicated in the development of feedback resistant enzymes, particularly in the directed evolution of an *N*-acetyl glutamate kinase (ArgB) to achieve arginine overproduction in *C. glutamicum* [37]. To effect this change, an arginine-specific biosensor (LysG) was co-expressed with a library of ArgB mutants *in vivo* and elevated levels of arginine production were correlated to higher fluorescence signals. As a result of the screen, ArgB mutants that are at least 20-fold less sensitive to arginine inhibition have been identified. Interestingly, two ArgB mutants that displayed similar product accumulation profiles had remarkably different levels of catalytic efficiencies. In retrospect, such an observation emphasizes the strength of an *in vivo* selection for protein optimization, since an *in vitro* kinetic assessment is fundamentally unable to integrate the total cellular characteristics needed for overproduction.

Besides the usual approach of sensing changes in enzymatic performance through product concentrations, functional domains of a transcription factor can also be designed to report on the event of biocatalysis directly. Notably, in the directed evolution of a Tobacco Etch Virus (TEV) protease, the peptide sequence of a nominally poor substrate was inserted between two DNA binding domains of a transcription factor to form a single chain repressor, which would serve to regulate the expression of an auxotrophic genetic marker (*HIS3*) required for host survival [20]. Without proteolytic cleavage of the peptide sequence by an active TEV protease mutant, binding of the single chain repressor would hinder expression of the essential gene and lead to cell death in a histidine-deficient *E. coli* auxotroph. By further introducing a secondary kanamycin selection marker in tandem to *HIS3*, proteolytic efficiency can be amplified in the form of growth rates, thereby facilitating the discovery of a TEV protease variant with 39-fold improvement over wild-type enzyme. Seemingly, the incorporation of tandem selection markers in this biosensor design allowed for a tunable control for improved stringency, as well as a higher dynamic range for lower limits of detection. More importantly, this genetic reporter system can be used to engineer changes in substrate recognition for enzymes with high specificity and expand on the molecular targets of endoproteases, often used as therapeutic agents for the degradation of peptides involved in pathogenic onset [94].

As a strategy to expand the scope of transcription factor-based biosensors, ligand-binding units have also been combined with independent activation domains to create hybrid biosensors for substrates and small molecule detection [95]. In addition, a control element can be incorporated by engineering conditional stability into these ligand-binding domains [96]. Without cognate ligand-binding, these recognition domains will be destabilized and degraded by the endogenous ubiquitin proteasome system [97]. Conversely, activation domains fused to ligand-binding units can be stabilized in the presence of cognate ligands, thereby constituting sensor responses. As a proof-of-concept for protein optimization using a conditionally stabilized hybrid biosensor, a ligand-binding domain specific for progesterone (PRO_1) was fused to a transcription activation domain and linked to the expression of an auxotrophic selection marker [38]. When used to select an error-prone library of 3β-hydroxysteroid dehydrogenase (3β-HSD), which converts pregnenolone to progesterone as part of steroid biosynthesis, 3β-HSD variants with up to twofold increase in progesterone production were rapidly obtained. As this progesterone-binding domain was designed from a scaffold protein using a method that requires minimal modification while maintaining high selectivity, these design principles could possibly enable biosensor generation for ligands with unknown receptors [98]. Alternatively, known transcription factors can also be evolved toward the intended ligand specificities, by applying directed evolution methods upon the biosensors. In a recent study, a TtgR transcriptional repressor was engineered to recognize non-native resveratrol as an activating ligand through a single amino acid substitution using a random mutagenesis method [39]. The identification of a resveratrol-responsive TtgR biosensor was facilitated by a fluorescence reporter, where improved resveratrol binding led to the release of TtgR from the promoter region to enable transcription [99]. Following the switch in ligand-specificity, the TtgR biosensor was subsequently used to evolve *p*-coumarate CoA ligase (4CL), an important enzyme in the resveratrol biosynthesis [100]. From a random mutagenesis library of 10^5 transformants, the biosensor was able to identify a 4CL mutant that exhibited 1.7-fold higher catalytic activity against *p*-coumaric acid, which led to a 4.7-fold increase in overall resveratrol production when expressed in a biosynthetic pathway. As 4CL functions as a key enzyme in the production of diverse phenylpropanoid compounds [101], the resveratrol biosensor can be indirectly used to enhance the biosynthesis of useful end products in other relevant pathways.

For all the diversity in protein biocatalysts, there is still an entire range of enzymatic reactions that does not lead to the production of a metabolite with a known transcription factor or end in the degradation of a substrate [102], making the observation of such phenomena rather difficult. An enzyme that falls under this category could catalyze a bond formation or generate a short-lived intermediate, and for that matter, conventional modes of detection generally do not apply; hence necessitating the development of more intricate biosensors as a means to optimize this group of biocatalysts. To this end, a chemical complementation assay that requires the reconstitution of two separate transcription factor domains was established to facilitate the survey of less discernible reactions [103]. This assay exploits the yeast three-hybrid system, which links biocatalysis to the expression of a reporter gene *in vivo*, via the use of small molecules or substrates that bridge or induce dimerization between two halves of a receptor [104, 105]. In a canonical setup, the substrate of interest is conjugated to fusion proteins methotrexate (Mtx) and dexamethasone (Dex), each bound to its respective dihydrofolate reductase (DHFR) and glucocorticoid receptor (GR). Correspondingly, DHFR is fused to a transcriptional DNA binding domain, while GR is linked to an analogous activation domain to mediate gene expression. Notably, this chemical complementation assay was adapted for directed evolution of a glycosynthase in a bond formation reaction, such that two carbohydrate moieties were independently attached to Mtx and Dex, and coupling of the transcription factor domains drive the expression of an *LEU2* gene to promote survival in leucine-deficient auxotrophs [40]. Under the appropriate selection conditions, improved glycosynthase mutants should increase the growth rates of the auxotrophic hosts and as a result, a glycosynthase variant with fivefold increased activity was isolated in the process. Likewise, chemical complementation has also been used to monitor bond cleavage reactions, particularly in the hydrolysis of cephalosporin for directed evolution of the β-lactamase [41]. Since only the investigated ligand chemistry was varied for each assay, chemical complementation can likely serve as a generalizable strategy for examining enzymatic reactions with inconspicuous phenotypes or those with unknown product-receptor association. Nevertheless, these ligand-protein conjugates have to be synthesized *in vitro* and introduced into biological systems across membranes and compartments [22]. Hence, this integration process could be a potential limitation that needs to be overcome before broad utilization can become feasible.

Presently, the majority of curated transcription factors are known to be specific for ligands or small molecules associated with the central metabolic pathways [106]. On the other hand, regulatory elements that recognize the products of secondary metabolism are considerably less extensive and this has remained a constraint for protein evolution despite ongoing engineering efforts [107, 108]. A strategy was proposed to use a biosensor that is specific for an intermediate, which is both a product of the core pathway and a precursor of a secondary pathway, to balance upstream and downstream metabolic fluxes for efficient biosynthesis and enzymatic development [42]. This approach, termed intermediate sensor-assisted push–pull strategy (InterSPPS), was subsequently applied in a proof-of-concept study to optimize deoxyviolacein production, using a tryptophan-specific transcription factor-based biosensor [109]. Biosynthesis of tryptophan was improved through the optimization of upstream ribosomal binding sites, while upregulation of deoxyviolacein production was accomplished via directed evolution of a rate-limiting downstream enzyme (VioB) and coupling the readout to fluorescence detection [110]. From the Inter-SPPS approach, it was proven that a biosensor specific for an upstream intermediate can be used to evolve a downstream enzyme; hence suggesting that biosensors associated with the central metabolic pathways could mitigate the gaps in secondary metabolic pathway regulation through inventive biosensor placement. In any case, the existing number of known transcription factors is still exceeded by the variety of metabolites, and their application in protein engineering might be further complicated by incompatibility issues in non-native hosts [19, 111]. Thus, in our continual pursuit to explore this vast enzyme chemistry space, the modularity and promiscuity encoded within transcription factors can perhaps be the building blocks for forthcoming development of novel and designer biosensors [112, 113].

2.3.2 Enzyme-Type Biosensors

Compared with transcription factor-based biosensors, enzymatic biosensing tools are relatively less common in nature [114]. Enzyme-type biosensors typically detect analytes that are precursors or intermediates of metabolic pathways. Moreover, the reactions generally need to produce a detectable phenotype or confer a survival advantage in the absence of an associated reporter and this could restrict their compatibility for *in vivo* directed protein evolution (Figure 2.3b) [23]. An enzyme that fulfils these criteria is a 3,4-dihydroxy-phenylalanine (DOPA) dioxygenase (DOD), which can convert DOPA into a yellow and highly fluorescent pigment, known as betaxanthin [115]. For the purpose of evolving a suitable tyrosine hydroxylase in yeast for overproduction of (*S*)-reticuline, DOD was used as a biosensor to report the catalytic activities of a corresponding mutagenic library [43]. Tyrosine hydroxylase, which catalyzes tyrosine to DOPA, is a rate-limiting step in the production of (*S*)-reticuline [116]; itself a key precursor for major branches of benzylisoquinoline alkaloids biosynthesis [117]. From a colony-based fluorometric survey of the tyrosine hydroxylase library, an improved variant with up to 80% lower side-product formation was obtained and this has also translated into a 2.8-fold increase in DOPA production. However, with an understanding that this biosensor exhibited a 110-fold dynamic range for DOPA, further tyrosine hydroxylase variants with higher catalytic efficiency can possibly be identified using more sensitive methods of detection. Given that betaxanthin is a fluorescent molecule with good solubility, this enzyme-based biosensor can also potentially be repositioned to optimize other biocatalysts along the tyrosine biosynthetic pathway.

Although enzymes are recognized for their unique specificity toward substrates, polymerases are one of the few exceptions that can work with generic templates. In particular, RNA polymerases like transcription factors regulate gene expression and can be suitable vectors for signal transduction [118]. A certain constraint, however, involves RNA polymerase-mediated transcriptions being initiated by specific sets of regulatory factors instead of assorted metabolic ligands [119]. Hence, in order to integrate RNA polymerases for biosensing purposes, modest understanding of the mechanisms is presumably required for engineering design to bridge this divide. A study focused on revealing the vulnerabilities of protease inhibitors in resistance evolution was notable for using T7 RNA polymerase as a biosensor for identification of key mutations implicated during drug resistance [44]. N-terminal fusions to T7 RNA polymerases are well tolerated and the enzyme is known to be naturally inhibited when bound to a T7 lysozyme [120, 121]. By introducing selected cleavage polypeptides as linker sequences, this T7 RNA polymerase–lysozyme complex was engineered to be proteolytically-activated, with respect to the different proteases of interest (Figure 2.4a). The biosensor was further linked to the expression of an essential phage propagation gene (gIII) under the control of a T7 promoter, and phage-assisted continuous evolution (PACE) was used to probe resistance development. In this setup, cleavage of the linker sequences would liberate T7 RNA polymerases for gIII expression and promote the replication of phages carrying the evolved proteases of interest within a fixed-volume lagoon [122]. Since the lagoon was continuously diluted with a steady influx of bacteria host cells, non-infectious phages harboring inactive proteases were rapidly diluted out, leaving an enriched population of propagative phages in the system. This selection method is feasible due to the fact that dilution happens faster than bacteria cell division but slower than phage replication, thus ensuring that mutations are accumulated only in the phage genome [44]. Subsequent addition of protease-inhibiting drug candidates provided the selection pressures required for enzyme evolution and drug-specific variants were obtained within 1–3 days of selection. In essence, the T7 RNA polymerase biosensor established a robust linkage between activity of interest and gIII expression, thereby enabling such rapid characterization of key resistance mutations. Seemingly, other enzyme activities or properties with no direct connection to gene expression can be associated via this approach and exploit the high-throughput capacity of PACE for expedited optimization.

In a recent study, RNA polymerase-mediated PACE was further redesigned to enhance the solubility of a rat apolipoprotein B mRNA editing catalytic subunit 1 (rAPOBEC1) in *E. coli* by coupling protein expression to polymerase activity [25]. The eukaryotic rAPOBEC1 protein is a potent cytidine deaminase for both RNA and DNA templates, although poor solubility in *E. coli* has restricted its utilization in Cas9-related base editing [123, 124]. To evolve better expression using a dual selection strategy, rAPOBEC1 variants were first fused to the N-terminal segment of a split T7 RNA polymerase (T7n), which upon recombination would facilitate downstream gene expression. T7 RNA polymerase can be separated into two distinct inactive segments between amino acids 179 and 180, yet retaining the affinity to spontaneously associate to restore function (Figure 2.4b) [125]. In the subsequent step, an affinity tag (GCN4 leucine zipper peptide epitope) was linked to the N-terminal of rAPOBEC1-T7n, followed by the attachment of an *E. coli* RNA polymerase omega subunit (RpoZ) to the C-terminal of the biosensor construct. For the purpose of a dual selection, the gIII genetic sequence was also split into two and inserted into separate plasmid systems to create an AND gate for phage propagation.

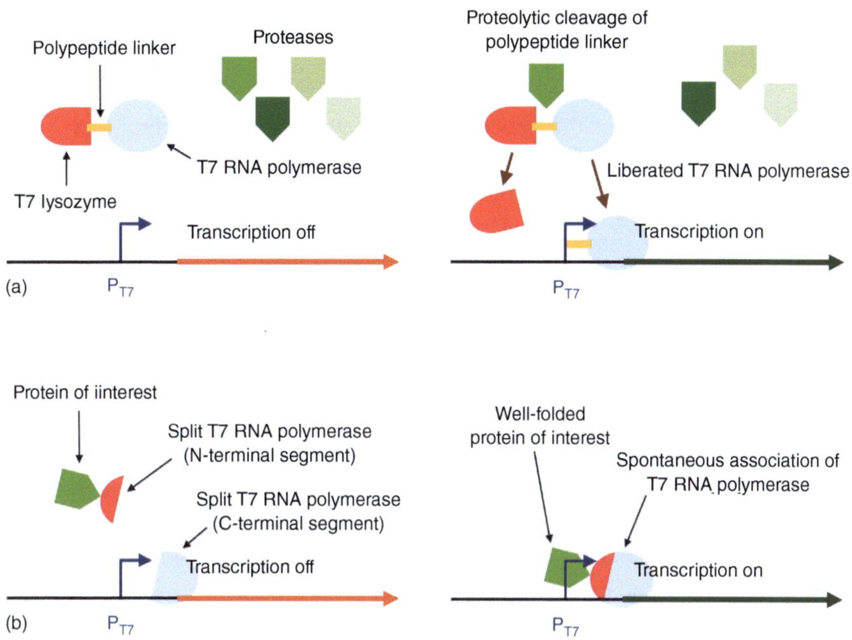

Figure 2.4 RNA-polymerase-based *in vivo* biosensors. T7 RNA polymerase linked to a T7 lysozyme is naturally inhibited. Proteolytic cleavage of the polypeptide linker by a suitable protease liberates the T7 RNA polymerase for transcription initiation (a). A split T7 RNA polymerase can spontaneously re-associate to form a functional enzyme to mediate gene expression. Formation of the N-terminal segment of T7 RNA polymerase can be dependent on the conjugated protein of interest (b).

As such, one part of the biosensor was designed for the selection of well-folded, soluble rAPOBEC1 proteins that can reconstitute T7 RNA polymerase activity and promote partial expression of the gIII selection marker in one of the plasmid systems. The second selection component was introduced to ensure full-length translation of the quadruple fusion construct as truncated T7n-containing false positives have earlier been found to mediate phage propagation [25]. For this part of the biosensor, binding of the affinity tag to its promoter-localized protein followed by the association of RpoZ with *E. coli* RNA polymerase would ensure that the entire sequence has been translated before partial gIII expression can occur in the other plasmid system. Collectively, successful binding and expression of the hybrid rAPOBEC1 protein construct would result in full-length selection marker expression and favor selected phage replication. Using this dual selection-based PACE strategy, rAPOBEC1 variants with up to fourfold increase in expression were obtained when co-expressed with a mutagenesis plasmid that dramatically elevates phage mutation rates. Due to the stringency integrated into this biosensor design, rAPOBEC1 protein expression was enhanced without a loss or regression of native catalytic activity. Likewise, this dual selection-based RNA polymerase biosensor system can be adapted to improve the solubility of proteins of interest while preserving desired pre-existing functions.

2.4 Characteristics of Biosensors for *In Vivo* Directed Protein Evolution

As analytical coverage of the generated sequence space is a limiting factor, the development of efficient methods that distinguish favorable variants from mutant libraries remain one of the most crucial part of directed protein evolution [35, 126]. In the context of protein engineering, *in vivo* biosensors have a vital role in high-throughput screening and selection of positive mutants from large mutagenic libraries [127]. Biosensors establish a link between genotypes and phenotypes by converting discreet enzymatic outcomes into detectable responses for facile segregation of useful genetic information [21]. Importantly, an understanding of the protein molecular properties such as catalytic or structural mechanism can facilitate the selection of a suitable biosensor [128, 129]. In principle, a biosensor has to detect molecules of interest at a level of practical relevance and correlate the signal to a proportionate phenotypic readout [4]. To this end, the bioreceptor of choice should possess exquisite specificity, detection range, and sensitivity, while the associated response or reporter element should exhibit an appropriate dynamic range to realize optimal biosensor function [80, 130].

For ligand-binding subunits and enzymes, specificity is defined by the range of small molecules or substrates that can bind and elicit a signaling response [131]. In the event that a bioreceptor recognizes a variety of analytes, the level of selectivity for an analyte of interest would affect its detection capability in the presence of other competing species [132]. High specificity or selectivity for an analyte can minimize the frequency of false positives, which can be critical in high-throughput evaluation of large mutant libraries [15]. Moreover, a highly specific biosensor can be advantageous for implementing a negative screen or selection, as superfluous interactions that contribute to a perplexed diminishing signal can potentially be avoided [30, 42, 128]. While specificity determines an assay's precision, the detection range and sensitivity of a bioreceptor can have an impact on the quality of protein variants isolated. A wide detection range for analytes typically increases the practicality of a biosensor and facilitates the identification of high-performing evolved proteins of interest [31]. On the other hand, the ability to differentiate variations in protein performance is determined by the sensitivity of a biosensor, which is described as the change in response output with respect to the change in the amount of analyte [133]. Ideally the change in responses is proportional to the concentration of analytes and biosensors with a broad sensitive linear range are generally preferred in most applications [134]. In addition, a biosensor with high sensitivity can be useful for evolution of new specificity or catalytic activity in proteins [36]. As novel characteristics being evolved usually exhibit a lower starting threshold, small increments in activity can be detected more readily using a biosensor with higher sensitivity.

The capacity for analyte detection is also influenced by a reporter dynamic range, given that low analyte concentrations can go unnoticed without appropriate signal amplification [135]. Notably, transcription factor-based biosensors can possess a high dynamic range as small changes in analyte concentrations are amplified through gene expression into noticeable phenotypes [38]. To further expand on the dynamic range, transcription factor-associated promoter regions are compelling targets of engineering strategies to advance signal transduction [136]. Besides enhancing response intensity, promoter choices can also affect the level of background noise originating from reporter components. For example, a switch from an inducible promoter to a constitutive version has been found to reduce noise by approximately one third, thereby improving biosensor performance [15]. Furthermore, as biosensors are often expressed in heterologous hosts, plasmid copies and genetic regulatory elements are vital considerations for minimizing potential metabolic burdens, which can have a negative impact on protein expression [35, 79]. Fluorescent proteins are the most common type of reporter systems used in the screening methodologies for directed evolution due to the fact that protein fusions normally do not interfere with a biosensor's function [18, 137]. More importantly, fluorescent proteins are usually non-toxic to living cells, and therefore serve as convenient tools for intracellular feedback [138]. For selection-based methods, antibiotic resistance markers are still widely utilized although the enforced selection pressure needs to be optimized for each distinct biosensor application [18, 139]. Apparently, the use of selection markers in tandem can increase a sensor dynamic range in the form of growth rates and offer an additional means of control for improved stringency of selection [20]. Moving forward, the process for biosensor optimization is being revolutionized by novel *in silico* design approaches such that previously intractable analytes could be duly ascertained using a new generation of synthetically-constructed hybrid biosensor systems [140].

2.5 Conclusions and Future Perspectives

With the development of advanced biotechnological tools, biocatalysts are no longer limited by their intrinsic properties, as function-specific variants can be engineered to suit the intended applications [141]. Directed evolution has been a vital enabling tool for this paradigm shift and the integration of *in vivo* biosensors into protein engineering has further expanded the scope of biocatalysts that can be readily optimized [71, 142]. Presently, *in vivo* biosensors are often used to interrogate changes in catalytic activities, which can be mediated by modification of active site geometry or allosteric interaction [30]; as well as the evolution of new specific activities through progressive alterations of substrate-binding affinity [24]. By exploiting the inherent attributes of biosensors, other important qualities such as structural folding and solubility can also be distinguished [25]. As for enzymes that catalyze products with no associated receptors, the reactions can potentially be linked to established biosensors via inventive redesign of regulatory components to achieve indirect signaling correlations [40]. So far, *in vivo* biosensors have been particularly effective for rapid evaluation of large protein mutant libraries, although their utility is still limited by the number of candidates accessible from designated assays. Conventional methods of screening and selection are typically batch-based, which involve microplates or agar media platforms for differentiating potential protein variants. However, with the recent development of high capacity fluorescence-activated cell sorting and PACE methods, *in vivo* biosensors are being adapted for even higher throughput evaluation methodologies [43, 44]. As with the case for all biosensors used in protein engineering, the more samples surveyed the faster desirable variants can be identified.

While directed evolution strategies are primarily involved in single protein optimization, the recognition of a need for pathway context has since initiated a rethink of individual-centric approaches [143, 144]; particularly with a growing interest to increase the compatibility of biocatalysts for industrial-relevant applications [7]. In certain instances, efforts focused on evolving a single key enzyme of interest have encountered difficulties related to overall productivity, due to the accumulation of deleterious intermediates and inhibition by end-products [145, 146]. As a strategy to evolve coveted enzymatic properties without compromising eventual objectives, directed evolution of multiple enzymes within a biosynthesis pathway has since been explored and shown to be a promising method [147, 148]. Clearly, *in vivo* biosensors play an important role in the development of independent proteins, but were also found to be highly relevant in a multivariate optimization process [30, 42]. Hence, as directed protein evolution moves toward a more holistic outlook, *in vivo* biosensors can be part of even more intricate biological monitoring networks and their scope of utility is perhaps only limited by originality, given that versatility for adaptation time and again.

Acknowledgments

This work was supported by the National Research Foundation Singapore Competitive Research Program grant [NRF-CRP19-2017-05-095] and the RIE2020 Advanced Manufacturing and Engineering IAF-PP grant [A19B3a0009].

References

1 Clark, L.C. and Lyons, C. (1962). Electrode systems for continuous monitoring in cardiovascular surgery. *Ann. N.Y. Acad. Sci.* 102 (1): 29–45.
2 Rajpoot, K. (2017). Recent advances and applications of biosensors in novel technology. *Biosensors J.* 6 (2): 145.
3 Ali, J., Najeeb, J., Ali, M. et al. (2017). Biosensors: their fundamentals, designs, types and most recent impactful applications: a review. *J. Biosens. Bioelectron.* 8 (1): 235.
4 Bhalla, N., Jolly, P., Formisano, N., and Estrela, P. (2016). Introduction to biosensors. *Essays Biochem.* 60 (1): 1–8.
5 Zhang, F. and Keasling, J. (2011). Biosensors and their applications in microbial metabolic engineering. *Trends Microbiol.* 19 (7): 323–329.
6 Liu, D., Evans, T., and Zhang, F. (2015). Applications and advances of metabolite biosensors for metabolic engineering. *Metab. Eng.* 31: 35–43.

7 Cobb, R.E., Chao, R., and Zhao, H. (2012). Directed evolution: past, present, and future. *AlChE J.* 59 (5): 1432–1440.

8 Tizei, P.A.G., Csibra, E., Torres, L., and Pinheiro, V.B. (2016). Selection platforms for directed evolution in synthetic biology. *Biochem. Soc. Trans.* 44 (4): 1165–1175.

9 Tian, K., Tai, K., Chua, B.J.W., and Li, Z. (2017). Directed evolution of Thermomyces lanuginosus lipase to enhance methanol tolerance for efficient production of biodiesel from waste grease. *Bioresour. Technol.* 245: 1491–1497.

10 Li, Y.-x., Yi, P., Yan, Q.-j. et al. (2017). Directed evolution of a β-mannanase from *Rhizomucor miehei* to improve catalytic activity in acidic and thermophilic conditions. *Biotechnol. Biofuels* 10 (1): 143.

11 Soh, L.M.J., Mak, W.S., Lin, P.P. et al. (2017). Engineering a thermostable keto acid decarboxylase using directed evolution and computationally directed protein design. *ACS Synth. Biol.* 6 (4): 610–618.

12 Li, W., Xu, S., Zhang, B. et al. (2017). Directed evolution to improve the catalytic efficiency of urate oxidase from *Bacillus subtilis*. *PLoS One* 12 (5): e0177877.

13 Luo, Y., Chen, Y., Ma, H. et al. (2017). Enhancing the biocatalytic manufacture of the key intermediate of atorvastatin by focused directed evolution of halohydrin dehalogenase. *Sci. Rep.* 7: 42064.

14 Mahr, R., Gätgens, C., Gätgens, J. et al. (2015). Biosensor-driven adaptive laboratory evolution of L-valine production in *Corynebacterium glutamicum*. *Metab. Eng.* 32: 184–194.

15 Michener, J.K. and Smolke, C.D. (2012). High-throughput enzyme evolution in *Saccharomyces cerevisiae* using a synthetic RNA switch. *Metab. Eng.* 14 (4): 306–316.

16 Lin, J.-L., Wagner, J.M., and Alper, H.S. (2017). Enabling tools for high-throughput detection of metabolites: metabolic engineering and directed evolution applications. *Biotechnol. Adv.* 35 (8): 950–970.

17 Xiao, H., Bao, Z., and Zhao, H. (2015). High throughput screening and selection methods for directed enzyme evolution. *Ind. Eng. Chem. Res.* 54 (16): 4011–4020.

18 Zeymer, C. and Hilvert, D. (2018). Directed evolution of protein catalysts. *Annu. Rev. Biochem.* 87 (1): 131–157.

19 Zhang, J., Jensen, M.K., and Keasling, J.D. (2015). Development of biosensors and their application in metabolic engineering. *Curr. Opin. Chem. Biol.* 28: 1–8.

20 Verhoeven, K.D., Altstadt, O.C., and Savinov, S.N. (2012). Intracellular detection and evolution of site-specific proteases using a genetic selection system. *Appl. Biochem. Biotechnol.* 166 (5): 1340–1354.

21 Rogers, J.K., Taylor, N.D., and Church, G.M. (2016). Biosensor-based engineering of biosynthetic pathways. *Curr. Opin. Biotechnol.* 42: 84–91.

22 Michener, J.K., Thodey, K., Liang, J.C., and Smolke, C.D. (2012). Applications of genetically-encoded biosensors for the construction and control of biosynthetic pathways. *Metab. Eng.* 14 (3): 212–222.

23 Shi, S., Ang, E.L., and Zhao, H. (2018). In vivo biosensors: mechanisms, development, and applications. *J. Ind. Microbiol. Biotechnol.* 45 (7): 491–516.

24 Bali, A.P., Genee, H.J., and Sommer, M.O.A. (2018). Directed evolution of membrane transport using synthetic selections. *ACS Synth. Biol.* 7 (3): 789–793.

25 Wang, T., Badran, A.H., Huang, T.P., and Liu, D.R. (2018). Continuous directed evolution of proteins with improved soluble expression. *Nat. Chem. Biol.* 14 (10): 972–980.

26 Vigneshvar, S., Sudhakumari, C.C., Senthilkumaran, B., and Prakash, H. (2016). Recent advances in biosensor technology for potential applications – an overview. *Front. Bioeng. Biotechnol.* 4: 11.

27 Robinson, C.J., Vincent, H.A., Wu, M.-C. et al. (2014). Modular riboswitch toolsets for synthetic genetic control in diverse bacterial species. *J. Am. Chem. Soc.* 136 (30): 10615–10624.

28 Taylor, N.D., Garruss, A.S., Moretti, R. et al. (2016). Engineering an allosteric transcription factor to respond to new ligands. *Nat. Methods* 13 (2): 177–183.

29 Younger, A.K.D., Dalvie, N.C., Rottinghaus, A.G., and Leonard, J.N. (2017). Engineering modular biosensors to confer metabolite-responsive regulation of transcription. *ACS Synth. Biol.* 6 (2): 311–325.

30 Lee, S.-W. and Oh, M.-K. (2015). A synthetic suicide riboswitch for the high-throughput screening of metabolite production in *Saccharomyces cerevisiae*. *Metab. Eng.* 28: 143–150.

31 Yang, P., Wang, J., Pang, Q. et al. (2017). Pathway optimization and key enzyme evolution of *N*-acetylneuraminate biosynthesis using an in vivo aptazyme-based biosensor. *Metab. Eng.* 43: 21–28.

32 Wang, J., Gao, D., Yu, X. et al. (2015). Evolution of a chimeric aspartate kinase for L-lysine production using a synthetic RNA device. *Appl. Microbiol. Biotechnol.* 99 (20): 8527–8536.

33 Shi, S., Choi, Y.W., Zhao, H. et al. (2017). Discovery and engineering of a 1-butanol biosensor in *Saccharomyces cerevisiae*. *Bioresour. Technol.* 245: 1343–1351.

34 Wang, Y., Li, Q., Zheng, P. et al. (2016). Evolving the L-lysine high-producing strain of *Escherichia coli* using a newly developed high-throughput screening method. *J. Ind. Microbiol. Biotechnol.* 43: 1227–1235.

35 Seok, J.Y., Yang, J., Choi, S.J. et al. (2018). Directed evolution of the 3-hydroxypropionic acid production pathway by engineering aldehyde dehydrogenase using a synthetic selection device. *Metab. Eng.* 47: 113–120.

36 Kwon, K.K., Lee, D.-H., Kim, S.J. et al. (2018). Evolution of enzymes with new specificity by high-throughput screening using DmpR-based genetic circuits and multiple flow cytometry rounds. *Sci. Rep.* 8: 2659.

37 Schendzielorz, G., Dippong, M., Grünberger, A. et al. (2014). Taking control over control: use of product sensing in single cells to remove flux control at key enzymes in biosynthesis pathways. *ACS Synth. Biol.* 3 (1): 21–29.

38 Feng, J., Jester, B.W., Tinberg, C.E. et al. (2015). A general strategy to construct small molecule biosensors in eukaryotes. *eLife* 4: e10606.

39 Xiong, D., Lu, S., Wu, J. et al. (2017). Improving key enzyme activity in phenylpropanoid pathway with a designed biosensor. *Metab. Eng.* 40: 115–123.

40 Lin, H., Tao, H., and Cornish, V.W. (2004). Directed evolution of a glycosynthase via chemical complementation. *J. Am. Chem. Soc.* 126 (46): 15051–15059.

41 Sengupta, D., Lin, H., Goldberg, S.D. et al. (2004). Correlation between catalytic efficiency and the transcription read-out in chemical complementation: a general assay for enzyme catalysis. *Biochemistry* 43 (12): 3570–3581.

42 Fang, M., Wang, T., Zhang, C. et al. (2016). Intermediate-sensor assisted push–pull strategy and its application in heterologous deoxyviolacein production in *Escherichia coli*. *Metab. Eng.* 33: 41–51.

43 DeLoache, W.C., Russ, Z.N., Narcross, L. et al. (2015). An enzyme-coupled biosensor enables (*S*)-reticuline production in yeast from glucose. *Nat. Chem. Biol.* 11 (7): 465–471.

44 Dickinson, B.C., Packer, M.S., Badran, A.H., and Liu, D.R. (2014). A system for the continuous directed evolution of proteases rapidly reveals drug-resistance mutations. *Nat. Commun.* 5: 5352.

45 Du, Y. and Dong, S. (2017). Nucleic acid biosensors: recent advances and perspectives. *Anal. Chem.* 89 (1): 189–215.

46 Fedor, M.J. and Williamson, J.R. (2005). The catalytic diversity of RNAs. *Nat. Rev. Mol. Cell Biol.* 6 (5): 399–412.

47 Liu, J., Cao, Z., and Lu, Y. (2009). Functional nucleic acid sensors. *Chem. Rev.* 109 (5): 1948–1998.

48 Shin, I., Ray, J., Gupta, V. et al. (2014). Live-cell imaging of Pol II promoter activity to monitor gene expression with RNA IMAGEtag reporters. *Nucleic Acids Res.* 42 (11): e90.

49 Bastet, L., Chauvier, A., Singh, N. et al. (2017). Translational control and Rho-dependent transcription termination are intimately linked in riboswitch regulation. *Nucleic Acids Res.* 45 (12): 7474–7486.

50 Rinaldi, A.J., Lund, P.E., Blanco, M.R., and Walter, N.G. (2016). The Shine–Dalgarno sequence of riboswitch-regulated single mRNAs shows ligand-dependent accessibility bursts. *Nat. Commun.* 7: 8976.

51 Ferré-D'Amaré, A.R. (2010). The *glmS* ribozyme: use of a small molecule coenzyme by a gene-regulatory RNA. *Q. Rev. Biophys.* 43 (4): 423–447.

52 Montange, R.K. and Batey, R.T. (2008). Riboswitches: emerging themes in RNA structure and function. *Annu. Rev. Biophys.* 37 (1): 117–133.

53 Mok, W. and Li, Y. (2008). Recent progress in nucleic acid aptamer-based biosensors and bioassays. *Sensors (Basel, Switzerland)* 8 (11): 7050–7084.

54 Tyc, O., Song, C., Dickschat, J.S. et al. (2017). The ecological role of volatile and soluble secondary metabolites produced by soil bacteria. *Trends Microbiol.* 25 (4): 280–292.

55 Alberti, F., Foster, G.D., and Bailey, A.M. (2017). Natural products from filamentous fungi and production by heterologous expression. *Appl. Microbiol. Biotechnol.* 101 (2): 493–500.

56 Świątek, M.A., Tenconi, E., Rigali, S., and van Wezel, G.P. (2012). Functional analysis of the *N*-acetylglucosamine metabolic genes of *Streptomyces coelicolor* and role in control of development and antibiotic production. *J. Bacteriol.* 194 (5): 1136–1144.

57 Inokuma, K., Hasunuma, T., and Kondo, A. (2016). Ethanol production from *N*-acetyl-D-glucosamine by *Scheffersomyces stipitis* strains. *AMB Express* 6 (1): 83.

58 Longley, D.B., Harkin, D.P., and Johnston, P.G. (2003). 5-Fluorouracil: mechanisms of action and clinical strategies. *Nat. Rev. Cancer* 3 (5): 330–338.

59 Smith, R.J., Milewski, S., Brown, A.J., and Gooday, G.W. (1996). Isolation and characterization of the GFA1 gene encoding the glutamine:fructose-6-phosphate amidotransferase of *Candida albicans*. *J. Bacteriol.* 178 (8): 2320–2327.

60 Klein, D.J., Ferré, D., and Amaré, A.R. (2006). Structural basis of *glmS* ribozyme activation by glucosamine-6-phosphate. *Science* 313 (5794): 1752–1756.

61 Lim, J., Grove, B.C., Roth, A., and Breaker, R.R. (2006). Characteristics of ligand recognition by a *glmS* self-cleaving ribozyme. *Angew. Chem. Int. Ed.* 45 (40): 6689–6693.

62 Win, M.N. and Smolke, C.D. (2007). A modular and extensible RNA-based gene-regulatory platform for engineering cellular function. *Proc. Natl. Acad. Sci. U.S.A.* 104 (36): 14283–14288.

63 Ceres, P., Garst, A.D., Marcano-Velázquez, J.G., and Batey, R.T. (2013). Modularity of select riboswitch expression platforms enables facile engineering of novel genetic regulatory devices. *ACS Synth. Biol.* 2 (8): 463–472.

64 Cho, S., Lee, B.-R., Cho, B.-K. et al. (2012). In vitro selection of sialic acid specific RNA aptamer and its application to the rapid sensing of sialic acid modified sugars. *Biotechnol. Bioeng.* 110 (3): 905–913.

65 Tao, F., Zhang, Y., Ma, C., and Xu, P. (2010). Biotechnological production and applications of *N*-acetyl-D-neuraminic acid: current state and perspectives. *Appl. Microbiol. Biotechnol.* 87 (4): 1281–1289.

66 Muranaka, N., Sharma, V., Nomura, Y., and Yokobayashi, Y. (2009). An efficient platform for genetic selection and screening of gene switches in *Escherichia coli*. *Nucleic Acids Res.* 37 (5): e39.

67 González-Vera, J.A. and Morris, M.C. (2015). Fluorescent reporters and biosensors for probing the dynamic behavior of protein kinases. *Proteomes* 3 (4): 369–410.

68 Rogers, J.K., Guzman, C.D., Taylor, N.D. et al. (2015). Synthetic biosensors for precise gene control and real-time monitoring of metabolites. *Nucleic Acids Res.* 43 (15): 7648–7660.

69 Jenison, R.D., Gill, S.C., Pardi, A., and Polisky, B. (1994). High-resolution molecular discrimination by RNA. *Science* 263 (5152): 1425–1429.

70 McKeague, M., Velu, R., Hill, K. et al. (2014). Selection and characterization of a novel DNA aptamer for label-free fluorescence biosensing of ochratoxin A. *Toxins* 6 (8): 2435–2452.

71 Carpenter, A.C., Paulsen, I.T., and Williams, T.C. (2018). Blueprints for biosensors: design, limitations, and applications. *Genes* 9 (8): 375.

72 Binder, S., Schendzielorz, G., Stäbler, N. et al. (2012). A high-throughput approach to identify genomic variants of bacterial metabolite producers at the single-cell level. *Genome Biol.* 13 (5): R40.

73 Datta, V., Siddharthan, R., and Krishna, S. (2018). Detection of cooperatively bound transcription factor pairs using ChIP-seq peak intensities and expectation maximization. *PLoS One* 13 (7): e0199771.

74 Yin, X., Shin, H.-D., Li, J. et al. (2017). P*gas*, a low-pH-induced promoter, as a tool for dynamic control of gene expression for metabolic engineering of *Aspergillus niger*. *Appl. Environ. Microbiol.* 83 (6): e03222–e03216.

75 Yuan, J. and Ching, C.-B. (2015). Dynamic control of *ERG9* expression for improved amorpha-4,11-diene production in *Saccharomyces cerevisiae*. *Microb. Cell Fact.* 14 (1): 38.

76 Shang, X., Chai, X., Lu, X. et al. (2018). Native promoters of *Corynebacterium glutamicum* and its application in L-lysine production. *Biotechnol. Lett* 40: 383–391.

77 Brown, A.J., Gibson, S.J., Hatton, D., and James, D.C. (2017). In silico design of context-responsive mammalian promoters with user-defined functionality. *Nucleic Acids Res.* 45 (18): 10906–10919.

78 Copley, S.D. (2015). An evolutionary biochemist's perspective on promiscuity. *Trends Biochem. Sci* 40 (2): 72–78.

79 Mahr, R. and Frunzke, J. (2016). Transcription factor-based biosensors in biotechnology: current state and future prospects. *Appl. Microbiol. Biotechnol.* 100: 79–90.

80 Cheng, F., Tang, X.-L., and Kardashliev, T. (2018). Transcription factor-based biosensors in high-throughput screening: advances and applications. *Biotechnol. J.* 13 (7): 1700648.

81 Lefrançois, P., Euskirchen, G.M., Auerbach, R.K. et al. (2009). Efficient yeast ChIP-Seq using multiplex short-read DNA sequencing. *BMC Genomics* 10: 37.

82 Guo, W.-L. and Huang, D.-S. (2017). An efficient method to transcription factor binding sites imputation via simultaneous completion of multiple matrices with positional consistency. *Mol. Biosyst.* 13 (9): 1827–1837.

83 Grove, A. (2017). Regulation of metabolic pathways by MarR family transcription factors. *Comput. Struct. Biotechnol. J.* 15: 366–371.

84 Schallmey, M., Frunzke, J., Eggeling, L., and Marienhagen, J. (2014). Looking for the pick of the bunch: high-throughput screening of producing microorganisms with biosensors. *Curr. Opin. Biotechnol.* 26: 148–154.

85 Lee, D.-H., Choi, S.-L., Rha, E. et al. (2015). A novel psychrophilic alkaline phosphatase from the metagenome of tidal flat sediments. *BMC Biotech.* 15 (1): 1.

86 Zhou, S., Ainala, S.K., Seol, E. et al. (2015). Inducible gene expression system by 3-hydroxypropionic acid. *Biotechnol. Biofuels* 8 (1): 169.

87 Kumar, V., Ashok, S., and Park, S. (2013). Recent advances in biological production of 3-hydroxypropionic acid. *Biotechnol. Adv.* 31 (6): 945–961.

88 Chu, H.S., Kim, Y.S., Lee, C.M. et al. (2014). Metabolic engineering of 3-hydroxypropionic acid biosynthesis in *Escherichia coli*. *Biotechnol. Bioeng.* 112 (2): 356–364.

89 Celińska, E. (2010). Debottlenecking the 1,3-propanediol pathway by metabolic engineering. *Biotechnol. Adv.* 28 (4): 519–530.

90 Goyal, S., Yuan, J., Chen, T. et al. (2010). Achieving optimal growth through product feedback inhibition in metabolism. *PLoS Comput. Biol.* 6 (6): e1000802.

91 Lee, K.H., Park, J.H., Kim, T.Y. et al. (2007). Systems metabolic engineering of *Escherichia coli* for L-threonine production. *Mol. Syst. Biol.* 3 (1): 149.

92 Mustafi, N., Grünberger, A., Kohlheyer, D. et al. (2012). The development and application of a single-cell biosensor for the detection of l-methionine and branched-chain amino acids. *Metab. Eng.* 14 (4): 449–457.

93 Brasse-Lagnel, C., Lavoinne, A., and Husson, A. (2009). Control of mammalian gene expression by amino acids, especially glutamine. *FEBS J.* 276 (7): 1826–1844.

94 Craik, C.S., Page, M.J., and Madison, E.L. (2011). Proteases as therapeutics. *Biochem. J.* 435 (1): 1–16.

95 Guntas, G. and Ostermeier, M. (2004). Creation of an allosteric enzyme by domain insertion. *J. Mol. Biol.* 336 (1): 263–273.

96 Banaszynski, L.A., Sellmyer, M.A., Contag, C.H. et al. (2008). Chemical control of protein stability and function in living mice. *Nat. Med.* 14 (10): 1123–1127.

97 Egeler, E.L., Urner, L.M., Rakhit, R. et al. (2011). Ligand-switchable substrates for a ubiquitin-proteasome system. *J. Biol. Chem.* 286 (36): 31328–31336.

98 Tinberg, C.E., Khare, S.D., Dou, J. et al. (2013). Computational design of ligand-binding proteins with high affinity and selectivity. *Nature* 501 (7466): 212–216.

99 Alguel, Y., Meng, C., Terán, W. et al. (2007). Crystal structures of multidrug binding protein TtgR in complex with antibiotics and plant antimicrobials. *J. Mol. Biol.* 369 (3): 829–840.

100 Lim, C.G., Fowler, Z.L., Hueller, T. et al. (2011). High-yield resveratrol production in engineered *Escherichia coli*. *Appl. Environ. Microbiol.* 77 (10): 3451–3460.

101 Ehlting, J., Büttner, D., Wang, Q. et al. (1999). Three 4-coumarate:coenzyme A ligases in *Arabidopsis thaliana* represent two evolutionarily divergent classes in angiosperms. *Plant J.* 19 (1): 9–20.

102 Martínez Cuesta, S., Rahman, S.A., Furnham, N., and Thornton, J.M. (2015). The classification and evolution of enzyme function. *Biophys. J.* 109 (6): 1082–1086.

103 Baker, K., Bleczinski, C., Lin, H. et al. (2002). Chemical complementation: a reaction-independent genetic assay for enzyme catalysis. *Proc. Natl. Acad. Sci. U.S.A.* 99 (26): 16537–16542.

104 Lin, H., Abida, W.M., Sauer, R.T., and Cornish, V.W. (2000). Dexamethasone–methotrexate: an efficient chemical inducer of protein dimerization in vivo. *J. Am. Chem. Soc.* 122 (17): 4247–4248.

105 Schwimmer, L.J., Rohatgi, P., Azizi, B. et al. (2004). Creation and discovery of ligand-receptor pairs for transcriptional control with small molecules. *Proc. Natl. Acad. Sci. U.S.A.* 101 (41): 14707–14712.

106 Siddiqui, M.S., Thodey, K., Trenchard, I., and Smolke, C.D. (2011). Advancing secondary metabolite biosynthesis in yeast with synthetic biology tools. *FEMS Yeast Res.* 12 (2): 144–170.

107 Chen, W., Zhang, S., Jiang, P. et al. (2015). Design of an ectoine-responsive AraC mutant and its application in metabolic engineering of ectoine biosynthesis. *Metab. Eng.* 30: 149–155.

108 Sun, Y.-Q., Busche, T., Rückert, C. et al. (2017). Development of a biosensor concept to detect the production of cluster-specific secondary metabolites. *ACS Synth. Biol.* 6 (6): 1026–1033.

109 Gong, F., Ito, K., Nakamura, Y., and Yanofsky, C. (2001). The mechanism of tryptophan induction of tryptophanase operon expression: tryptophan inhibits release factor-mediated cleavage of TnaC-peptidyl-tRNA(Pro). *Proc. Natl. Acad. Sci. U.S.A.* 98 (16): 8997–9001.

110 Jiang, P.-x., Wang, H.-s., Xiao, S. et al. (2012). Pathway redesign for deoxyviolacein biosynthesis in *Citrobacter freundii* and characterization of this pigment. *Appl. Microbiol. Biotechnol.* 94 (6): 1521–1532.

111 Tang, S.-Y. and Cirino, P.C. (2010). Design and application of a mevalonate-responsive regulatory protein. *Angew. Chem. Int. Ed.* 50 (5): 1084–1086.

112 Galvão, T.C., Mencía, M., and De Lorenzo, V. (2007). Emergence of novel functions in transcriptional regulators by regression to stem protein types. *Mol. Microbiol.* 65 (4): 907–919.

113 Kasey, C.M., Zerrad, M., Li, Y. et al. (2018). Development of transcription factor-based designer macrolide biosensors for metabolic engineering and synthetic biology. *ACS Synth. Biol.* 7 (1): 227–239.

114 Wilson, G.S. and Hu, Y. (2000). Enzyme-based biosensors for in vivo measurements. *Chem. Rev.* 100 (7): 2693–2704.

115 Gandía-Herrero, F., García-Carmona, F., and Escribano, J. (2005). A novel method using high-performance liquid chromatography with fluorescence detection for the determination of betaxanthins. *J. Chromatogr. A* 1078 (1): 83–89.

116 Nakagawa, A., Minami, H., Kim, J.-S. et al. (2011). A bacterial platform for fermentative production of plant alkaloids. *Nat. Commun.* 2: 326.

117 Fossati, E., Ekins, A., Narcross, L. et al. (2014). Reconstitution of a 10-gene pathway for synthesis of the plant alkaloid dihydrosanguinarine in *Saccharomyces cerevisiae*. *Nat. Commun.* 5: 3283.

118 Kireeva, M.L., Kashlev, M., and Burton, Z.F. (2013). RNA polymerase structure, function, regulation, dynamics, fidelity, and roles in GENE EXPRESSION. *Chem. Rev.* 113 (11): 8325–8330.

119 Nikolov, D.B. and Burley, S.K. (1997). RNA polymerase II transcription initiation: a structural view. *Proc. Natl. Acad. Sci. U.S.A.* 94 (1): 15–22.

120 Jeruzalmi, D. and Steitz, T.A. (1998). Structure of T7 RNA polymerase complexed to the transcriptional inhibitor T7 lysozyme. *EMBO J.* 17 (14): 4101–4113.

121 Kumar, A. and Patel, S.S. (1997). Inhibition of T7 RNA polymerase: transcription initiation and transition from initiation to elongation are inhibited by T7 lysozyme via a ternary complex with RNA polymerase and promoter DNA. *Biochemistry* 36 (45): 13954–13962.

122 Esvelt, K.M., Carlson, J.C., and Liu, D.R. (2011). A system for the continuous directed evolution of biomolecules. *Nature* 472 (7344): 499–503.

123 Galloway, C.A., Ashton, J., Sparks, J.D. et al. (2010). Metabolic regulation of APOBEC-1 complementation factor trafficking in mouse models of obesity and its positive correlation with the expression of ApoB protein in hepatocytes. *Biochim. Biophys. Acta* 1802 (11): 976–985.

124 Komor, A.C., Kim, Y.B., Packer, M.S. et al. (2016). Programmable editing of a target base in genomic DNA without double-stranded DNA cleavage. *Nature* 533 (7603): 420–424.

125 Shis, D.L. and Bennett, M.R. (2013). Library of synthetic transcriptional AND gates built with split T7 RNA polymerase mutants. *Proc. Natl. Acad. Sci. U.S.A.* 110 (13): 5028–5033.

126 Lane, M.D. and Seelig, B. (2014). Advances in the directed evolution of proteins. *Curr. Opin. Chem. Biol.* 22: 129–136.

127 Dietrich, J.A., McKee, A.E., and Keasling, J.D. (2010). High-throughput metabolic engineering: advances in small-molecule screening and selection. *Annu. Rev. Biochem.* 79 (1): 563–590.

128 Rossum, T., Kengen, S.W.M., and Oost, J. (2013). Reporter-based screening and selection of enzymes. *FEBS J.* 280 (13): 2979–2996.

129 Packer, M.S. and Liu, D.R. (2015). Methods for the directed evolution of proteins. *Nat. Rev. Genet.* 16 (7): 379–394.

130 Richman, S.A., Kranz, D.M., and Stone, J.D. (2009). Biosensor detection systems: engineering stable, high-affinity bioreceptors by yeast surface display. *Methods Mol. Biol.* 504: 323–350.

131 Mehrvar, M. and Abdi, M. (2004). Recent developments, characteristics, and potential applications of electrochemical biosensors. *Anal. Sci.* 20 (8): 1113–1126.

132 Evtugyn, G.A., Budnikov, H.C., and Nikolskaya, E.B. (1998). Sensitivity and selectivity of electrochemical enzyme sensors for inhibitor determination. *Talanta* 46 (4): 465–484.

133 Armbruster, D.A. and Pry, T. (2008). Limit of blank, limit of detection and limit of quantitation. *Clin. Biochem. Rev.* 29 (Suppl 1): S49–S52.

134 Asal, M., Özen, Ö., Şahinler, M., and Polatoğlu, İ. (2018). Recent developments in enzyme, DNA and immuno-based biosensors. *Sensors* 18 (6): 1924.

135 Zhang, F., Carothers, J.M., and Keasling, J.D. (2012). Design of a dynamic sensor-regulator system for production of chemicals and fuels derived from fatty acids. *Nat. Biotechnol.* 30 (4): 354–359.

136 Blazeck, J. and Alper, H.S. (2012). Promoter engineering: recent advances in controlling transcription at the most fundamental level. *Biotechnol. J.* 8 (1): 46–58.

137 DeBlasio, S.L., Sylvester, A.W., and Jackson, D. (2010). Illuminating plant biology: using fluorescent proteins for high-throughput analysis of protein localization and function in plants. *Brief. Funct. Genom.* 9 (2): 129–138.

138 Chalfie, M., Tu, Y., Euskirchen, G. et al. (1994). Green fluorescent protein as a marker for gene expression. *Science* 263 (5148): 802–805.

139 Rennig, M., Martinez, V., Mirzadeh, K. et al. (2018). TARSyn: tunable antibiotic resistance devices enabling bacterial synthetic evolution and protein production. *ACS Synth. Biol.* 7 (2): 432–442.

140 Jester, B.W., Tinberg, C.E., Rich, M.S. et al. (2018). Engineered biosensors from dimeric ligand-binding domains. *ACS Synth. Biol.* 7 (10): 2457–2467.

141 Bornscheuer, U.T., Huisman, G.W., Kazlauskas, R.J. et al. (2012). Engineering the third wave of biocatalysis. *Nature* 485 (7397): 185–194.

142 Martínez, R. and Schwaneberg, U. (2013). A roadmap to directed enzyme evolution and screening systems for biotechnological applications. *Biol. Res.* 46: 395–405.

143 Arnold, F.H. and Volkov, A.A. (1999). Directed evolution of biocatalysts. *Curr. Opin. Chem. Biol.* 3 (1): 54–59.

144 Rohlin, L., Oh, M.-K., and Liao, J.C. (2001). Microbial pathway engineering for industrial processes: evolution, combinatorial biosynthesis and rational design. *Curr. Opin. Microbiol.* 4 (3): 330–335.

145 Chen, Z., Rappert, S., Sun, J., and Zeng, A.-P. (2011). Integrating molecular dynamics and co-evolutionary analysis for reliable target prediction and deregulation of the allosteric inhibition of aspartokinase for amino acid production. *J. Biotechnol.* 154 (4): 248–254.

146 Paddon, C.J., Westfall, P.J., Pitera, D.J. et al. (2013). High-level semi-synthetic production of the potent antimalarial artemisinin. *Nature* 496 (7446): 528–532.

147 Biggs, B.W., De Paepe, B., Santos, C.N.S. et al. (2014). Multivariate modular metabolic engineering for pathway and strain optimization. *Curr. Opin. Biotechnol.* 29: 156–162.

148 Schmidt-Dannert, C., Umeno, D., and Arnold, F.H. (2000). Molecular breeding of carotenoid biosynthetic pathways. *Nat. Biotechnol.* 18 (7): 750–753.

3

High-Throughput Mass Spectrometry Complements Protein Engineering

Tong Si[1,], Pu Xue[1,2,*], Kisurb Choe[1,3,4], Huimin Zhao[1–3,5,#], and Jonathan V. Sweedler[1,4,5,#]*

[1] *University of Illinois at Urbana–Champaign, Carl R. Woese Institute for Genomic Biology, 1206 W Gregory Drive, Urbana, IL 61801, USA*
[2] *University of Illinois at Urbana–Champaign, Department of Chemical and Biomolecular Engineering, 600 S. Mathews Avenue, Urbana, IL 61801, USA*
[3] *University of Illinois at Urbana–Champaign, Department of Biochemistry, 600 S. Mathews Avenue, Urbana, IL 61801, USA*
[4] *University of Illinois at Urbana–Champaign, Beckman Institute for Advanced Science and Technology, 405 N. Mathews Avenue, Urbana, IL 61801, USA*
[5] *University of Illinois at Urbana–Champaign, Department of Chemistry, 505 S Mathews Avenue, Urbana, IL 61801, USA*

3.1 Introduction

For proteins, mass spectrometry (MS) is mainly applied to characterize three properties including catalytic activity, ligand binding, and structure. MS is well suited to study substrate-to-product conversion that results in mass differences [1–8]. Ligand binding can be studied by bind-and-elute strategies by measuring residual and/or eluted molecules [9], or by direct monitoring of mass shift due to formation of a protein-ligand complex [10]. Structural analysis (reviewed elsewhere [11]) often requires advanced MS methodologies such as use of ion mobility MS to differentiate overall shape and packing of a protein [12]. Because of these analytical capabilities, MS measurements can be used to assist protein engineering campaigns that target ligand-binding and catalytic properties.

When one has samples containing complex mixtures of proteins, associated ligands, and enzyme substrates/products, MS provides unparalleled information by simultaneous monitoring of various reaction species, thanks to its mass-resolving capability. Also, MS measurements are generally label-free, and hence eliminate the need of expensive surrogate substrates or indirect assays that couple reaction progress with spectrophotometric or radioactive signals. The use of naturally occurring substrates and ligands is desirable, considering the general observation that

Zhao and Sweedler are corresponding authors
* These authors contributed equally

Protein Engineering: Tools and Applications, First Edition. Edited by Huimin Zhao.
© 2021 WILEY-VCH GmbH. Published 2021 by WILEY-VCH GmbH.

"you get what you screen for" in high-throughput screens (HTSs). Moreover, high selectivity and sensitivity of MS enable the use of multiplexed and miniaturized reactions. Compared with other analytical modalities, these features of MS-based assays reduce analytical time and cost while increase information content of assay readout, and therefore facilitate the increasing use of MS as a general approach to study and engineer proteins [1–8].

In this chapter, we first introduce the main steps in MS-based protein assays, followed by discussions on how technology advances enhance analytical throughput. It is important to select the appropriate MS-based approaches because no individual platform provides all chemical information of interest on the proteins and associated molecules in a sample. We therefore briefly introduce different MS approaches and provide references to direct the interested reader for more detail. A glossary of the measurement approaches described throughout the chapter is also provided (Box 3.1). We then discuss notable applications that study and engineer proteins using MS approaches with a focus primarily on enzymes. Finally, we conclude the chapter by providing some future perspectives.

Box 3.1 An Alphabetical List of MS Terms

CE	capillary electrophoresis
DESI	desorption electrospray ionization
EI	electron impact
ESI	electrospray ionization
GC	gas chromatography
ICR	ion cyclotron resonance
LC	liquid chromatography
LDI	laser desorption/ionization
MALDI	matrix-assisted laser desorption/ionization
NIMS	nanostructure-initiator mass spectrometry
QQQ	triple quadrupole
SAMDI	self-assembled monolayers for MALDI
SIMS	secondary ion mass spectrometry
SIM	selected ion monitoring
SPE	solid-phase extraction
SRM	selected reaction monitoring
TOF	time-of-flight
UHPLC	ultra-high performance liquid chromatography

3.2 Procedures and Instrumentation for MS-Based Protein Assays

There are four main steps to perform a typical MS-based protein assay: assay creation, sample preparation, MS measurement, and data analysis. Each step can be performed in various formats, which are dependent on the nature of target proteins, information to be obtained, throughput, and available instrumentation.

Assays can be set up either in solution or on surfaces. For solution-based assays, microtiter plates are widely used (Figure 3.1a). Segmented flow is also used to carry out reactions at a large scale, in the form of aqueous plugs or droplets separated by air or an immiscible fluid, respectively (Figure 3.1b). On the other hand, assays can be performed on the surfaces of microarrays (Figure 3.1c), columns, and particles [15]. Immobilization of enzymes or substrates on solid support can be achieved via covalent binding, affinity binding, or encapsulation within porous materials [15, 16]. Whereas most assays are performed *in vitro* using purified proteins, cell lysates, or cell-free translation systems, cells are increasingly utilized to study proteins properties *in vivo*. For example, microbial colonies on agar surfaces were used as vessels to study single-step biotransformation [17] and multistep biosynthesis [18].

Because abundant ions are more easily detected by MS, sample preparation is critical to limit the number of molecules competing for ionization at any given time. To separate target analytes from complex mixtures, a variety of strategies can be employed, including solvent extraction, solid-phase extraction (SPE), gas

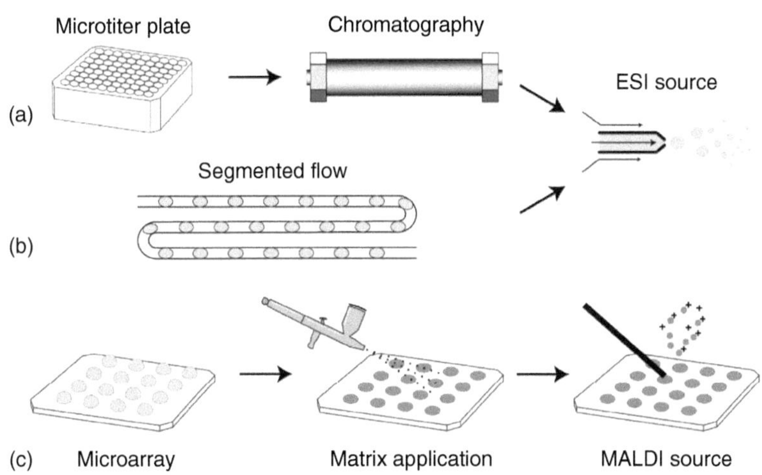

Figure 3.1 Typical procedures of MS-based protein assays.
(a) Solution assays are set up in microtiter plates followed by chromatographic separation before introduction to an ESI source. (b) Segmented flow–ESI–MS setting. Direct infusion of microfluidic droplets to MS has been demonstrated [13, 14]. (c) Assays are set up on microarray surfaces followed by matrix application for MALDI–MS analysis. Sources: (b) Sun et al. [13, 14].

chromatography (GC), liquid chromatography (LC), capillary electrophoresis (CE), and so on. These separations are usually time-consuming and can potentially become the rate-limiting step in MS-based assays and as such, have been the focus for throughput improvement. For example, fast chromatography can be achieved via ultra-high performance liquid chromatography (UHPLC) [19], multiplex injections in a single run [20, 21], and the use of multiple columns in parallel [22]. To further improve throughput, chromatography can be eliminated via direct sample infusion, partially thanks to the ever-increasing mass resolving power of MS instrumentation. However, without a separation, MS measurements tend to have reduced dynamic range. In addition to separation, sample preparation steps also help to bridge various assay formats and MS approaches. For example, approaches have been developed to deposit liquid reaction mixtures on surfaces for desorption-based MS analysis [23, 24].

In a mass spectrometer, gas-phase ions are generated from neutral molecules in the ion source, separated based on mass-to-charge (m/z) ratios in the mass analyzer, and quantified by the ion detector (Figure 3.2a). Relative ion abundance is then plotted versus m/z values to generate a mass spectrum, which contains quantitative and qualitative information of measured molecules (Figure 3.2a). Mass spectrometers are often categorized based ion sources (Figure 3.2b–d) and mass analyzers, and most common types for protein studies are summarized in Box 3.2. To perform an MS measurement, samples need to be firstly introduced into the ion source of a mass spectrometer. Electron impact (EI) ion sources are often coupled with GC. Electrospray ionization (ESI) sources can be coupled with LC, CE, direct infusion, or desorption/ionization approaches from surfaces (i.e. desorption electrospray ionization (DESI) [25], NanoDESI [26], and laser ablation ESI(LAESI) [27]), although

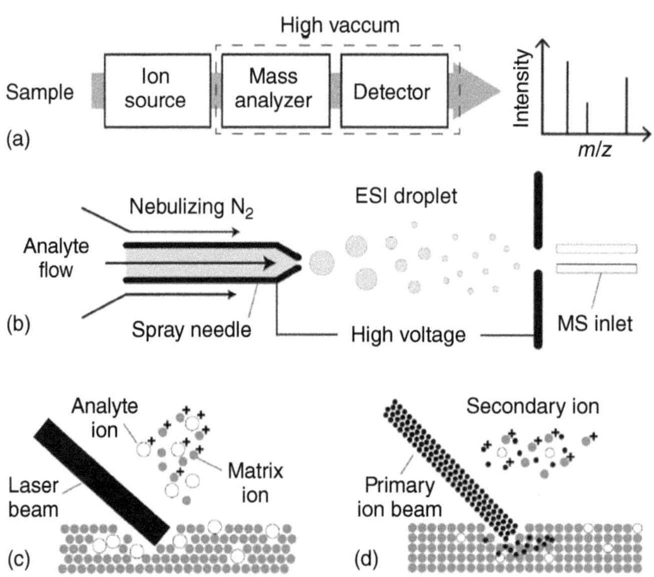

Figure 3.2 (a) Scheme of a mass spectrometer. (b–d) ESI, MALDI and SIMS ion sources.

desorption/ionization processes are more often coupled with matrix-assisted laser desorption/ionization (MALDI, Figure 3.2c) or secondary ion mass spectrometry (SIMS, Figure 3.2d).

Box 3.2 Ion Sources and Mass Analyzers in Common MS-Based Protein Assays

EI can ionize a wide range of volatile organic molecules as enzyme substrates/ products and protein binding ligands. EI generates extensive fragmentation and is considered "hard." MALDI and ESI are two "soft" approaches for the ionization of intact biological molecules (Figure 3.1b,c). SIMS utilizes a focused, accelerated primary ion beam to sputter sample surfaces (Figure 3.1d). When cluster ion sources are used, intact metabolites, lipids, and small peptides can be detected by SIMS.

The mass analyzer determines the detection limit, mass resolution, and quantitation capability of an MS-based platform. The time-of-flight (TOF) mass analyzer has been widely used in protein assays because of its relatively low cost, large *m/z* detection window, and fast scan rates. Detection limits on the zeptomole to attomole range are achieved with TOF-MS while maintaining a mass resolution above 20 000 [28, 29]. Ion cyclotron resonance (ICR) [30] and Orbitrap mass analyzers [31] offer superior mass accuracy, where mass resolution in excess of 100 000 is routine with an acquisition frequency of about 1 Hz. In hybrid instruments, mass analyzers are coupled to collision cells, enabling selection of precursor ions and mass measurements on the fragments (tandem MS). Triple quadrupole (QQQ) MS consisting of three quadrupoles in series with the central one acting as a collision cell is often used for targeted quantitation via selected reaction monitoring (SRM). Multistage fragmentation of ions (MS^n) and analysis of fragments are essential for structural characterization of the analytes.

While the choice of ion sources depends on the nature of target analytes, the mass analyzer used depends on information one wishes to obtain as outlined in Box 3.2. For example, targeted quantitation for monitoring enzyme reactions can be achieved by (i) selected ion monitoring (SIM) of either the substrate or product relative to a chemically similar internal standard that is exogenously provided (Figure 3.3a, P and P*) or a congener (Figure 3.3b, P_1 and P_2); (ii) SRM of both the substrate and product without the need of internal standard (Figure 3.3a, S and P); and (iii) SRM of either the substrate or product with the help of external standard calibration. Whereas (i) and (ii) can be performed in almost any type of mass analyzer with sufficient resolution separating the substrate, product, and standard, (iii) is often carried out in a QQQ analyzer with tandem MS capabilities. Furthermore, when "omics" coverage or characterization of unknowns is needed, such as ligand screening using complex mixtures, high-accuracy, high-resolution ICR or Orbitrap analyzers with tandem MS capabilities are often employed.

Data analysis of raw mass spectra generally starts with signal preprocessing steps such as baseline correction, noise reduction, spectral alignment, and

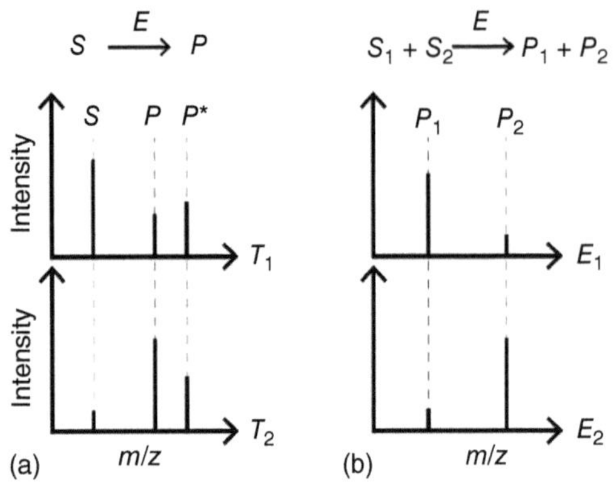

Figure 3.3 Correlation of MS signals and enzyme properties. (a) Enzyme conversion resulting in a mass shift between substrate (S) and product (P) is well suited for MS analysis. Research progress over time (T_1 and T_2) can be quantitated by the relative ion intensities between the product and substrate (I_P/I_S), or between the product and a chemically similar standard (I_P/I_{P*}). (b) Relative product ion intensities (I_{P1} and I_{P2}) can be utilized to measure enzyme selectivity.

normalization [32]. The main purpose of these steps is to reduce variations resulted from sample preparation, instrumental drift, and random errors. Then, because high-throughput MS experiments generate large-size, high-dimensional data sets, it is essential to perform data reduction to extract relevant information. A typical data-reduction pipeline includes peak picking, feature selection, and multivariate statistical analysis such as principal component analysis (PCA) [32]. These steps can be achieved by commercial software often provided by MS instrument vendors or open-source platforms such as OpenMS [33].

3.3 Technology Advances Focusing on Throughput Improvement

Miniaturization, automation, and multiplexity are the main approaches to reduce cost and time associated with MS-based protein assays. Here, we discuss prominent HTS settings highlighting these principles for sample/protein throughput improvement.

To perform miniaturized protein assays on a massive scale, femto- to nanoliter-scale plugs or droplets have been generated in capillary or microfluidic channels (Figure 3.1b). Operations such as reagent addition, dilution, splitting, and sorting can be performed via microfluidic manipulation. For readouts, in addition to optical detection [34], these plug and droplet reactors can also be coupled to MS platforms including ESI [35, 36] and MALDI [37]. Furthermore, miniaturized assays on surfaces can be applied, mainly in the format of substrate or enzyme microarrays (Figure 3.1c). Various surface chemistries have been developed to facilitate substrate/enzyme immobilization to achieve parallel manipulations such as washing/desalting. Such microfluidics-MS or microarray-MS settings have been applied in ligand/inhibitor screening and protein engineering, and notable applications are discussed in Sections 3.4.1 and 3.4.2.

Automation can be applied at different steps in MS-based protein assay for throughput improvement. Robotic liquid handling approaches based on pipette tips [38, 39], pins [40], inkjet, and acoustic deposition [24, 41] have been applied to accelerate assay setup and sample preparation. Notably, robotic automation sometimes requires adaption or modification of manual protocols for sample preparation. For example, Bligh-and-Dyer method [42] is considered the "gold-standard" of lipid extraction for MS-based lipidomic studies. This procedure extracts lipids into the lower organic phase of an aqueous/organic mixture, which is problematic because lower phase recovery can be challenging if performed by liquid-handling robots. To solve this challenge, a new solvent system consisting of methyl-tert-butyl ether/methanol/water was developed, so that lipid-containing organic phase forms the upper layer for facile robotic collection [43].

Automatic introduction of samples into a mass spectrometer substantially improves analytic throughput. Single-probe autosampler is now a standard component for most GC–MS and LC–MS instrumentation for automatic flow injection, and multiprobe autosamplers have also been developed [44]. The Agilent RapidFire system [45, 46] automates sample aspiration, SPE, and ESI-MS injection steps to achieve a cycling time of ~10 seconds. The TriVersa NanoMate system by Adivon [47, 48] automates sample delivery from a 96-well plate to a chip-based nanoESI source. Notably, mechanics that deliver samples from microtiter plates to an ESI source can even be completely omitted by using electrostatic spray ionization [49, 50] or acoustic loading [51]. On the other hand, desorption/ionization-based MS approaches can be performed in an imaging mode for rapid profiling of a spatially defined array of samples [1].

In addition, multiplex assays can be performed to reduce cost and increase information content of individual reactions [52–58], thanks to the superior mass resolving power of MS. For enzymes, one early example utilized five peptide substrate/product pairs to investigate various classes of protease activities in fractionated snake venom using ESI–MS [58]. Distinct types of enzymes can also be combined in the same inhibitor screening campaign, such as a kinase and an esterase [57]. Such multiplexity can be further scaled to metabolomic levels when using complex substrate mixtures such as cellular metabolites [59, 60]. For example, a mixture of cellular metabolites was incubated *in vitro* with candidate proteins. Untargeted, comparative analysis by CE–MS pinpointed substrate(s) and/or product(s) whose levels exhibited substantial changes before and after enzyme treatment [59]. Moreover, mixture screening can be performed using continuous flow through immobilized proteins in columns coupled with MS detection. For example, frontal affinity chromatography (FAC)–MS ranks the relative binding strengths of ligands to an immobilized protein target with a throughput up to 10 000 compounds per day [61, 62]. Immobilized enzyme reactor (IMER)–MS, on the other hand, enables rapid identification of enzyme inhibitors from compound mixtures when coupled with ESI–MS/MS [63, 64].

3.4 Applications of MS-Based Protein Assays: Summary

Recent applications of MS-based protein assays targeting select protein classes are summarized in Table 3.1. We noted two major protein classes: enzymes catalyzing

Table 3.1 Recent, select applications of MS-based protein assays.

Protein type	MS platform
Protease	Segmented flow–ESI [65], LC–ESI [66–69], other ESI [70, 71], MALDI [52, 72–76], SAMDI [77]
Kinase	SIMS [78], LC–ESI [79–85], other ESI [86], MALDI [53, 87–91], CE–MALDI [92]
Other posttranslational modification enzymes	Tyrosine phosphatase (MALDI [93]), (De)methyl/(de)acetyltransferase (LC–ESI [94], segmented flow–ESI [95], MALDI [96–99], SAMDI [100, 101]), deubiquitylases (MALDI [102])
Carbohydrate-active enzymes	Glycotransferases (MALDI [103, 104], SAMDI [100, 105, 106]), glycosidases (other ESI [39], MALDI [107, 108]), glycosyl hydrolases (NIMS [24, 109, 110], MALDI [111], LDI [40], SIMS [112]), sugar nucleotidyltransferases (LC–MS [54]), glycan-binding proteins (MALDI [113])
Acetylcholinesterase (AChE)	LC–ESI [56, 114], DESI [115], segmented flow-ESI [116], other ESI [117–120], MALDI [121–123]
Other proteins	Cytochrome P450s (LC–ESI [124, 125], DESI [17], segmented flow–ESI [126], other ESI [50], SAMDI [127]), other oxidoreductases (CE–MS [128, 129], LC–MS [130]), transferases (LC–ESI [131], other ESI [132], MALDI [133], CE–MS [134]), lyase (flow injection–ESI [135], NIMS [136], LDI [137]), lipase (LC–ESI [138], flow injection–ESI [139], GC–MS [140]), ligase (CE–MS [141]), transporter (CE–MS [142])

post-translational modifications (PTMs) and carbohydrate-active enzymes. This is likely due to the prevalent use of ESI and MALDI MS approaches to analyze peptides and glycans. In comparison, small-molecule substrates, products, and ligands are more challenging targets for MS analyses. As one example, MS assays are developed to profile a model reaction by acetylcholinesterase (AChE) that converts the acetylcholine neurotransmitter into choline and acetate. On the other hand, high-throughput MS-based assays are also applied to protein engineering, although only a few studies have been reported in literature. In this section, we will first cover notable examples that profile wild-type proteases, carbohydrate-active enzymes, and AChE to study their properties. Then, we will discuss how MS is applied to screen enzyme mutant libraries for engineering purposes.

3.4.1 Applications of MS-Based Assays: Protein Analysis

Proteases play essential roles in many important biological processes such as protein turnover, cellular signaling, and blood clotting [143]. Both microarray-MS and microfluidics-MS settings are developed for high-throughput protease assays. For example, peptide microarrays were utilized to analyze multiplex proteases activities when coupling with MALDI–TOF MS [52, 72]. For different proteases, peptide substrates are designed to generate products with different molecular weights upon cleavage. By this way, a unique m/z value is associated with each type of proteolytic activity, which can be quantified using ion intensity ratios between product and standard [72] or between substrate and product [52]. In addition to microarrays, segmented flow ESI–MS was applied by the Kennedy group for label-free inhibitor screening targeting cathepsin B protease [14]. Inhibitor-containing droplets were segmented by perfluorodecalin in a tube and pumped through microfluidic channels. Enzyme, peptide substrate, and quenchant were added sequentially via microfluidic manipulation. Reaction mixtures were then infused into a metal-coated, fused silica ESI emitter for MS detection at a rate of 1.2 seconds per sample [14]. The same group further interfaced compound libraries with segmented flow-MS screening [13], where test compounds stored in 384-well plates were reformatted as nanoliter droplets at 4.5 samples s^{-1} and then analyzed by ESI–MS at 0.5 Hz. This new approach enables the screening of more than 145 000 samples in a single day [13].

For glycan enzymes, MS imaging-based HTS approaches are widely used and reviewed recently [1]. Chemically derived glycan substrates are often performed for analyte immobilization on MS target surfaces, which allows washing steps prior to MS analysis for enhanced sensitivity. For example, oligosaccharide substrates can be covalently immobilized to a self-assembled monolayer (SAM) [105]. SAM is further attached to MALDI targets via gold–thiolate bonds that withstand washing but break upon laser radiation [144]. Using this method, a total of 14 280 combinations of an enzyme, an acceptor/donor pair in four buffers was rapidly screened to discover new glycosyltransferases [105]. A related approach, nanostructure-initiator mass spectrometry (NIMS), utilizes non-covalent, fluoro-phase interactions to capture perfluorinated analytes in a liquid "initiator" phase on porous silicon surfaces [145]. Subsequent washing permits enzyme activity screening within complex context such as cell lysates [145] and microbial communities [146]. One impressive application profiled a diverse set of glycosylhydrolases under 10 000 conditions in triplicates for ionic liquid tolerance and thermostability [110]. In addition to analyte immobilization, chemical derivatization also enhances glycan ionization efficiency. For example, conjugation to an arginine-containing nonapeptide resulted in ~1000-fold signal enhancement for MALDI–TOF MS detection of glycans such as maltoheptaose and oligogalacturonic acids [147].

AChE inhibitors are considered promising drug candidates to treat neuropsychiatric symptoms such as Alzheimer's disease and ataxia. AChE substrate and products (acetylcholine, choline, and acetate) are small molecules (<500 Da), whose measurement by MALDI–TOF MS is considered difficult due to strong matrix background signals in such a mass window [148]. To overcome matrix interference, high-resolution MS analyzers such as Fourier transform mass spectrometry (FTMS) were coupled with MALDI, which enables rapid screening of AChE inhibitors from large compound libraries and complex plant extracts [122, 123]. Another approach to identify AChE inhibitors from complex mixtures is IMER–MS [118–120] as discussed in Section 3.3. For example, plant extracts were loaded on an AChE-immobilized IMER in monolithic capillaries. Unbound compounds were washed off, and bound ligands were eluted, separated, and identified using LC–MS.

Control capillaries with no AChE immobilization were applied as negative control to rule out non-specific binding to monolith support. By this way, several AChE inhibitors were isolated from complex plant extracts [119].

3.4.2 Applications of MS-Based Assays: Protein Engineering

So far, only a few studies apply high-throughput MS assays for protein engineering. To screen a large library of protein mutants, time-consuming chromatographic separation is often replaced with microfluidics [134], surface desorption [17, 18, 149], and direct infusion [139] approaches that interface with MS.

Microfluidics-MS was applied to engineer the catalytic activity of two transaminases [134]. Transaminase ATA-036 catalyzes the formation of β-leucine from its corresponding ketoester (Figure 3.4a). For library creation, *Escherichia coli* was transformed with mutation-encoding plasmids, and then cultivated, induced, and lysed in microtiter plates. Subsequently, the transamination reaction was performed in the plates, followed by acetonitrile precipitation of proteins, filtration, and dilution before MS analysis. The inlet of a Teflon tube was moved from well to well to draw samples from the assay plates to generate droplets (Figure 3.4b), which were then infused into ESI-MS at a rate of 3 Hz to monitor product ion intensity (Figure 3.4c). Quantification via microfluidics-MS exhibited high correlation with that produced by LC–MS, which was performed 700 times fold slower at a rate of four minutes per sample [134]. Such setting was also successfully applied to the other transaminase ATA-117. To further improve throughput, cell-free protein synthesis was performed to eliminate the transformation, cultivation and lysis steps for cell-based approaches. The use of cell-free system required further optimization such as the addition of surfactant to the carrier oil to stabilize droplets, and the replacement of stainless steel ESI needle with a Teflon needle within a CE source to reduce sample carryover [134]. In this study, microfluidics served as a sample introduction interface to MS, and enzyme assays were set up in microtiter plates. To further accelerate the overall

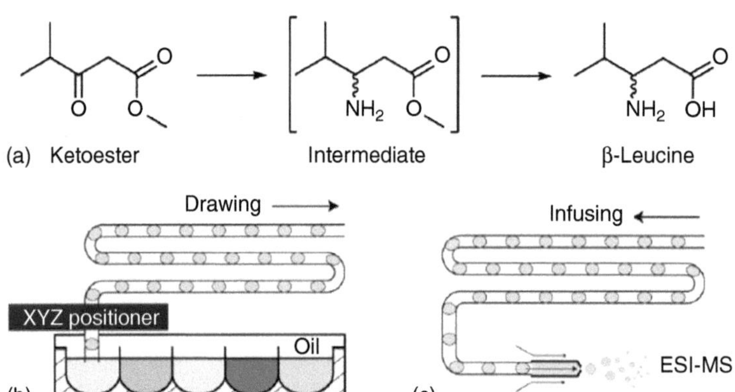

Figure 3.4 Engineer transaminase ATA-036 activity using microfluidics-MS screening. Source: Based on Diefenbach et al. [134].

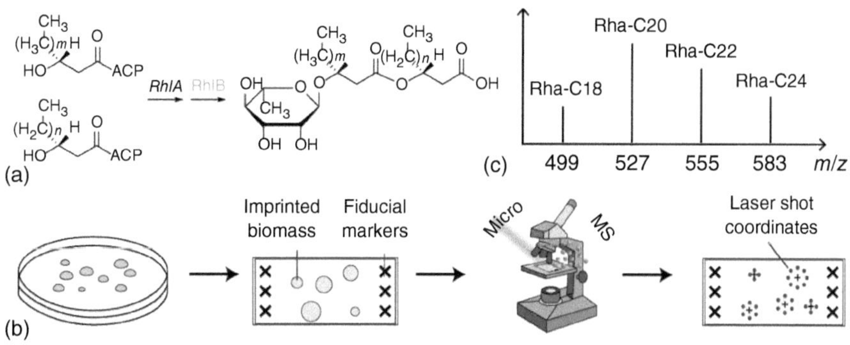

Figure 3.5 Optically guided MALDI MS to profile microbial colonies to engineer enzyme specificity. Sources: Comi et al. [150], Si et al. [18].

workflow, one may envision to directly create mutant libraries in droplets as reaction vessels, which has been previously demonstrated [34], followed by MS screening as described earlier.

For surface desorption-based approaches, our group reported the use of MALDI MS in engineering the substrate specificity of a rhamnolipid synthetase RhlA [18]. RhlA acts as a "molecular ruler" that determines the acyl chain lengths of the lipid moieties (Figure 3.5a). In MALDI mass spectra, the heights of target rhamnolipid peaks with distinct m/z values were used to estimate relative congener abundance (Figure 3.5b), which exhibited a good correlation with LC–MS/MS quantification [18]. To simplify mutant creation and screening, we sought to directly profile randomly distributed bacterial colonies transformed with error-prone polymerase chain reaction (PCR) libraries. Using machine vision algorithms, MALDI laser coordinates were calculated from optical images registering relative locations of fiducial markers and colonies [150] (Figure 3.5c). This innovation eliminated the need of reformatting samples into regular patterns, which is time-consuming but required for a typical MS imaging experiment. This new optically-guided MALDI MS method enabled rapid screening of microbial colonies at a rate of ~1 mutant second^{-1} [18]. This optically guided method may be adapted to other surface MS analyses, such as DESI [17], which complements the analytical capability of MALDI MS.

Furthermore, MS has been applied to engineer more challenging enzyme properties such as enantioselectivity. Unlike rhamnolipid congeners with different molecular weights, enantiomers have identical mass and cannot be differentiated by MS. To overcome this challenge, the Reetz group isotopically labeled one enantiomer in an optically pure form, which was then mixed with the other unlabeled enantiomer in a 1 : 1 manner [139]. The use of *pseudo*-enantiomers or *pseudo*-prochiral compounds as substrates allows MS resolution of chiral substrates and resulting products, whose ion intensities can be used to monitor asymmetric reactions. For example, lipase-catalyzed stereoselective esterification of 2-phenylpropionic acid was investigated using deuterium-labeled substrates in microtiter plates [139] (Figure 3.6). Coupling an automated liquid sampler for microtiter plates with an ESI-MS system, 1000 catalytic or stoichiometric

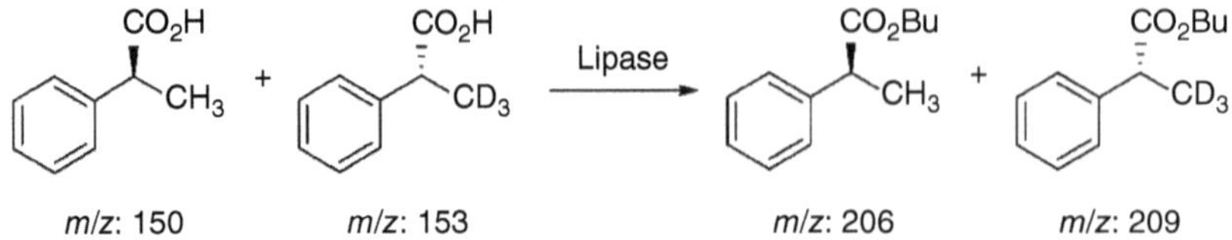

Figure 3.6 Isotopic labeling for resolving chiral substrates/products by MS to engineer enzyme enantioselectivity. Source: Reetz et al. [139]. © 1999, John Wiley & Sons.

asymmetric reactions were screened per day. The conversion, enantioselectivity, and selectivity factor can be calculated using corresponding ion intensities in resulting mass spectra. Moreover, isotopic labeling was also adapted to engineer the regioselectivity of a rebeccamycin halogenase via MALDI MS screening [149].

3.5 Conclusions and Perspectives

MS permits label-free analysis of enzyme reactions and protein–ligand interactions with high specificity and sensitivity. These features make MS a generally applicable approach to study and engineer proteins without the use of optically active surrogate substrates or indirect assays. Because it is unbiased, it allows unexpected products to be characterized. Miniaturized and multiplexed assays, parallel and automated sample preparation, together with instrumental advances, will continue to enhance the throughput, accuracy, quantitation, information richness, and application scope of MS-based protein assays. Currently, most HTS assays screen compound libraries against a select protein target, primarily aiming at ligand and inhibitor discovery for biomedical applications. Screening of protein libraries created via genome mining and mutagenesis approaches, on the other hand, is an emerging area for MS-based assays, considering the huge potential of protein biocatalysts and therapeutics in industrial biotechnology, agriculture, bioremediation, and medicine. However, the wide use of MS in protein research and engineering faces non-technical hurdles. For example, special instrumentation, highly trained personnel, and bioinformatic pipelines are required to set up MS-based protein assays and interpret complex data sets. Possible solutions may include cloud MS laboratory and servers that provide user-friendly access to protein scientists and engineers.

Acknowledgments

The authors acknowledge the funding of the DOE Center for Advanced Bioenergy and Bioproducts Innovation (U.S. Department of Energy, Office of Science, Office of Biological and Environmental Research under Award Number DE-SC0018420). Any opinions, findings, and conclusions or recommendations expressed in this publication are those of the author(s) and do not necessarily reflect the views of the U.S. Department of Energy.

References

1 de Rond, T., Danielewicz, M., and Northen, T.R. (2015). High throughput screening of enzyme activity with mass spectrometry imaging. *Curr. Opin. Biotechnol.* 31: 1–9.

2 Lv, H.T. (2013). Mass spectrometry-based metabolomics towards understanding of gene functions with a diversity of biological contexts. *Mass Spectrom. Rev.* 32 (2): 118–128.

3 Scheerle, R.K. and Grassmann, J. (2011). LC-MS for the determination of the enzymatic activity of proteins. In: *Protein and Peptide Analysis by LC-MS: Experimental Strategies* (ed. T. Letzel), 133–141. Cambridge: Royal Society of Chemistry.

4 Lee, D.Y., Bowen, B.P., and Northen, T.R. (2010). Mass spectrometry-based metabolomics, analysis of metabolite-protein interactions, and imaging. *Biotechniques* 49 (2): 557.

5 Baran, R., Reindl, W., and Northen, T.R. (2009). Mass spectrometry based metabolomics and enzymatic assays for functional genomics. *Curr. Opin. Microbiol.* 12 (5): 547–552.

6 Greis, K.D. (2007). Mass spectrometry for enzyme assays and inhibitor screening: an emerging application in pharmaceutical research. *Mass Spectrom. Rev.* 26 (3): 324–339.

7 Liesener, A. and Karst, U. (2005). Monitoring enzymatic conversions by mass spectrometry: a critical review. *Anal. Bioanal.Chem.* 382 (7): 1451–1464.

8 Bothner, B., Chavez, R., Wei, J. et al. (2000). Monitoring enzyme catalysis with mass spectrometry. *J. Biol. Chem.* 275 (18): 13455–13459.

9 Jonker, N., Kool, J., Irth, H. et al. (2011). Recent developments in protein-ligand affinity mass spectrometry. *Anal. Bioanal.Chem.* 399 (8): 2669–2681.

10 Benesch, J.L., Ruotolo, B.T., Simmons, D.A. et al. (2007). Protein complexes in the gas phase: technology for structural genomics and proteomics. *Chem. Rev.* 107 (8): 3544–3567.

11 Ben-Nissan, G. and Sharon, M. (2011). Capturing protein structural kinetics by mass spectrometry. *Chem. Soc. Rev.* 40 (7): 3627–3637.

12 Uetrecht, C., Rose, R.J., van Duijn, E. et al. (2010). Ion mobility mass spectrometry of proteins and protein assemblies. *Chem. Soc. Rev.* 39 (5): 1633–1655.

13 Sun, S.W. and Kennedy, R.T. (2014). Droplet electrospray ionization mass spectrometry for high throughput screening for enzyme inhibitors. *Anal. Chem.* 86 (18): 9309–9314.

14 Sun, S., Slaney, T.R., and Kennedy, R.T. (2012). Label free screening of enzyme inhibitors at femtomole scale using segmented flow electrospray ionization mass spectrometry. *Anal. Chem.* 84 (13): 5794–5800.

15 Gray, C.J., Weissenborn, M.J., Eyers, C.E. et al. (2013). Enzymatic reactions on immobilised substrates. *Chem. Soc. Rev.* 42 (15): 6378–6405.

16 Forsberg, E.M., Sicard, C., Brennan, J.D. et al. (2014). Solid-phase biological assays for drug discovery. *Annu. Rev. Anal. Chem.* 7: 337–359.

17 Yan, C., Parmeggiani, F., Jones, E.A. et al. (2017). Real-time screening of biocatalysts in live bacterial colonies. *J. Am. Chem. Soc.* 139 (4): 1408–1411.

18 Si, T., Li, B., Comi, T.J. et al. (2017). Profiling of microbial colonies for high-throughput engineering of multistep enzymatic reactions via optically guided matrix-assisted laser desorption/ionization mass spectrometry. *J. Am. Chem. Soc.* 139 (36): 12466–12473.

19 De Vos, J., Broeckhoven, K., and Eeltink, S. (2016). Advances in ultrahigh-pressure liquid chromatography technology and system design. *Anal. Chem.* 88 (1): 262–278.

20 Kuehnbaum, N.L., Kormendi, A., and Britz-McKibbin, P. (2013). Multisegment injection-capillary electrophoresis-mass spectrometry: a high-throughput platform for metabolomics with high data fidelity. *Anal. Chem.* 85 (22): 10664–10669.

21 Schafer, W., Wang, H., and Welch, C.J. (2016). Multiple-injection high-throughput gas chromatography analysis. *J. Sep. Sci.* 39 (15): 2978–2985.

22 de Biasi, V., Haskins, N., Organ, A. et al. (1999). High throughput liquid chromatography/mass spectrometric analyses using a novel multiplexed electrospray interface. *Rapid Commun. Mass Spectrom.* 13 (12): 1165–1168.

23 Kuster, S.K., Fagerer, S.R., Verboket, P.E. et al. (2013). Interfacing droplet microfluidics with matrix-assisted laser desorption/ionization mass spectrometry: label-free content analysis of single droplets. *Anal. Chem.* 85 (3): 1285–1289.

24 Greving, M., Cheng, X., Reindl, W. et al. (2012). Acoustic deposition with NIMS as a high-throughput enzyme activity assay. *Anal. Bioanal.Chem.* 403 (3): 707–711.

25 Takats, Z., Wiseman, J.M., Gologan, B. et al. (2004). Mass spectrometry sampling under ambient conditions with desorption electrospray ionization. *Science* 306 (5695): 471–473.

26 Roach, P.J., Laskin, J., and Laskin, A. (2010). Nanospray desorption electrospray ionization: an ambient method for liquid-extraction surface sampling in mass spectrometry. *Analyst* 135 (9): 2233–2236.

27 Nemes, P. and Vertes, A. (2007). Laser ablation electrospray ionization for atmospheric pressure, in vivo, and imaging mass spectrometry. *Anal. Chem.* 79 (21): 8098–8106.

28 Rubakhin, S.S., Garden, R.W., Fuller, R.R. et al. (2000). Measuring the peptides in individual organelles with mass spectrometry. *Nat. Biotechnol.* 18 (2): 172–175.

29 Nemes, P., Rubakhin, S.S., Aerts, J.T. et al. (2013). Qualitative and quantitative metabolomic investigation of single neurons by capillary electrophoresis electrospray ionization mass spectrometry. *Nat. Protoc.* 8 (4): 783–799.

30 Nikolaev, E.N., Kostyukevich, Y.I., Vladimirov, G.N. et al. (2016). Fourier transform ion cyclotron resonance (FT ICR) mass spectrometry: theory and simulations. *Mass Spectrom. Rev.* 35 (2): 219–258.

31 Eliuk, S. and Makarov, A. (2015). Evolution of orbitrap mass spectrometry instrumentation. *Annu. Rev. Anal. Chem.* 8: 61–80.

32 Rafols, P., Vilalta, D., Brezmes, J. et al. (2018). Signal preprocessing, multivariate analysis and software tools for MALDI-TOF mass spectrometry imaging for biological applications. *Mass Spectrom. Rev.* 37 (3): 281–306.
33 Rost, H.L., Sachsenberg, T., Aiche, S. et al. (2016). OpenMS: a flexible open-source software platform for mass spectrometry data analysis. *Nat. Methods* 13 (9): 741–748.
34 Agresti, J.J., Antipov, E., Abate, A.R. et al. (2010). Ultrahigh-throughput screening in drop-based microfluidics for directed evolution. *Proc. Natl. Acad. Sci. U.S.A.* 107 (9): 4004–4009.
35 Fidalgo, L.M., Whyte, G., Ruotolo, B.T. et al. (2009). Coupling microdroplet microreactors with mass spectrometry: reading the contents of single droplets online. *Angew. Chem. Int. Ed.* 48 (20): 3665–3668.
36 Pei, J., Li, Q., Lee, M.S. et al. (2009). Analysis of samples stored as individual plugs in a capillary by electrospray ionization mass spectrometry. *Anal. Chem.* 81 (15): 6558–6561.
37 Hatakeyama, T., Chen, D.L., and Ismagilov, R.F. (2006). Microgram-scale testing of reaction conditions in solution using nanoliter plugs in microfluidics with detection by MALDI-MS. *J. Am. Chem. Soc.* 128 (8): 2518–2519.
38 Gurard-Levin, Z.A., Scholle, M.D., Eisenberg, A.H. et al. (2011). High-throughput screening of small molecule libraries using SAMDI mass spectrometry. *ACS Comb. Sci.* 13 (4): 347–350.
39 Peng, T., Nagy, G., Trinidad, J.C. et al. (2017). A high-throughput mass-spectrometry-based assay for identifying the biochemical functions of putative glycosidases. *ChemBioChem* 18 (23): 2306–2311.
40 Chang, S.-H., Han, J.-L., Tseng, S.Y. et al. (2010). Glycan array on aluminum oxide-coated glass slides through phosphonate chemistry. *J. Am. Chem. Soc.* 132 (38): 13371–13380.
41 Chin, J., Wood, E., Peters, G.S. et al. (2016). Acoustic sample deposition MALDI-MS (ASD-MALDI-MS): a novel process flow for quality control screening of compound libraries. *J. Lab. Autom.* 21 (1): 204–207.
42 Folch, J., Lees, M., Sloane Stanley, G.H. et al. (1957). A simple method for the isolation and purification of total lipides from animal tissues. *J. Biol. Chem.* 226 (1): 497–509.
43 Matyash, V., Liebisch, G., Kurzchalia, T.V. et al. (2008). Lipid extraction by methyl-tert-butyl ether for high-throughput lipidomics. *J. Lipid Res.* 49 (5): 1137–1146.
44 Morand, K.L., Burt, T.M., Regg, B.T. et al. (2001). Techniques for increasing the throughput of flow injection mass spectrometry. *Anal. Chem.* 73 (2): 247–252.
45 Lowe, D.M., Gee, M., Haslam, C. et al. (2014). Lead discovery for human kynurenine 3-monooxygenase by high-throughput rapidfire mass spectrometry. *J. Biomol. Screening* 19 (4): 508–515.
46 Rohman, M. and Wingfield, J. (2016). High-throughput screening using mass spectrometry within drug discovery. *Methods Mol. Biol.* 1439: 47–63.
47 Benkestock, K., van Pelt, C.K., Akerud, T. et al. (2003). Automated nano-electrospray mass spectrometry for protein-ligand screening by

noncovalent interaction applied to human H-FABP and A-FABP. *J. Biomol. Screening* 8 (3): 247–256.

48 Schwudke, D., Hannich, J.T., Surendranath, V. et al. (2007). Top-down lipidomic screens by multivariate analysis of high-resolution survey mass spectra. *Anal. Chem.* 79 (11): 4083–4093.

49 Qiao, L., Sartor, R., Gasilova, N. et al. (2012). Electrostatic-spray ionization mass spectrometry. *Anal. Chem.* 84 (17): 7422–7430.

50 Qiao, L., Zhong, X., Belghith, E. et al. (2017). Electrostatic spray ionization from 384-well microtiter plates for mass spectrometry analysis-based enzyme assay and drug metabolism screening. *Anal. Chem.* 89 (11): 5983–5990.

51 Sinclair, I., Stearns, R., Pringle, S. et al. (2016). Novel acoustic loading of a mass spectrometer: toward next-generation high-throughput MS screening. *J. Lab. Autom.* 21 (1): 19–26.

52 Hu, J.J., Liu, F., Ju, H.X. et al. (2016). MALDI-MS patterning of caspase activities and its application in the assessment of drug resistance. *Angew. Chem. Int. Ed.* 55 (23): 6667–6670.

53 Smith, A.M.E. and Brennan, J.D. (2014). Simultaneous inhibition assay for human and microbial kinases via MALDI-MS/MS. *ChemBioChem* 15 (4): 587–594.

54 Ko, K.S., Mizanur, R.M., Jackson, J.M. et al. (2013). A mass-differentiated library strategy for identification of sugar nucleotidyltransferase activities from cell lysates. *Anal. Biochem.* 441 (1): 8–12.

55 Scheerle, R.K., Grassmann, J., Letzel, T. et al. (2012). Real-time ESI-MS of enzymatic conversion: impact of organic solvents and multiplexing. *Anal. Sci.* 28 (6): 607–612.

56 Cai, T.T., Zhang, L., Wang, H.Y. et al. (2012). Dual enzyme activities assay by quantitative electrospray ionization quadrupole-time-of-flight mass spectrometry. *Eur. J. Mass Spectrom.* 18 (6): 521–530.

57 Rathore, R., Pribil, P., Corr, J.J. et al. (2010). Multiplex enzyme assays and inhibitor screening by mass spectrometry. *J. Biomol. Screening* 15 (8): 1001–1007.

58 Liesener, A., Perchuc, A.M., Schoni, R. et al. (2005). Screening for proteolytic activities in snake venom by means of a multiplexing electrospray ionization mass spectrometry assay scheme. *Rapid Commun. Mass Spectrom.* 19 (20): 2923–2928.

59 Saito, N., Robert, M., Kitamura, S. et al. (2006). Metabolomics approach for enzyme discovery. *J. Proteome Res.* 5 (8): 1979–1987.

60 Tang, Z., Martin, M.V., and Guengerich, F.P. (2009). Elucidation of functions of human cytochrome P450 enzymes: identification of endogenous substrates in tissue extracts using metabolomic and isotopic labeling approaches. *Anal. Chem.* 81 (8): 3071–3078.

61 Calleri, E., Temporini, C., and Massolini, G. (2011). Frontal affinity chromatography in characterizing immobilized receptors. *J. Pharm. Biomed. Anal.* 54 (5): 911–925.

62 Ng, E.S., Chan, N.W., Lewis, D.F. et al. (2007). Frontal affinity chromatography-mass spectrometry. *Nat. Protoc.* 2 (8): 1907–1917.
63 Forsberg, E.M. and Brennan, J.D. (2014). Bio-solid-phase extraction/tandem mass spectrometry for identification of bioactive compounds in mixtures. *Anal. Chem.* 86 (16): 8457–8465.
64 de Boer, A.R., Letzel, T., van Elswijk, D.A. et al. (2004). On-line coupling of high-performance liquid chromatography to a continuous-flow enzyme assay based on electrospray ionization mass spectrometry. *Anal. Chem.* 76 (11): 3155–3161.
65 Iyer, J.K., Otvos, R.A., Kool, J. et al. (2016). Microfluidic chip-based online screening coupled to mass spectrometry: identification of inhibitors of thrombin and factor Xa. *J. Biomol. Screening* 21 (2): 212–220.
66 Musharraf, S.G., Bhatti, M.S., Choudhary, M.I. et al. (2017). Screening of inhibitors of angiotensin-converting enzyme (ACE) employing high performance liquid chromatography-electrospray ionization triple quadrupole mass spectrometry (HPLC-ESI-QqQ-MS). *Eur. J. Pharm. Sci.* 101: 182–188.
67 Pandya, V., Ekka, M.K., Dutta, R.K. et al. (2011). Mass spectrometry assay for studying kinetic properties of dipeptidases: characterization of human and yeast dipeptidases. *Anal. Biochem.* 418 (1): 134–142.
68 Siemerink, M., Schebb, N.H., Liesener, A. et al. (2010). Development of a fast liquid chromatography/mass spectrometry screening method for angiotensin-converting enzyme (ACE) inhibitors in complex natural mixtures like snake venom. *Rapid Commun. Mass Spectrom.* 24 (5): 687–697.
69 Schebb, N.H., Vielhaber, T., Jousset, A. et al. (2009). Development of a liquid chromatography-based screening methodology for proteolytic enzyme activity. *J. Chromatogr. A* 1216 (20): 4407–4415.
70 Yu, Z., Chen, L.C., Mandal, M.K. et al. (2012). Online electrospray ionization mass spectrometric monitoring of protease-catalyzed reactions in real time. *J. Am. Soc. Mass. Spectrom.* 23 (4): 728–735.
71 Huang, F. and Murray, K.K. (2010). Continuous flow infrared matrix-assisted laser desorption electrospray ionization mass spectrometry. *Rapid Commun. Mass Spectrom.* 24 (19): 2799–2804.
72 Hu, J.J., Liu, F., and Ju, H.X. (2015). Peptide code-on-a-microplate for protease activity analysis via MALDI-TOF mass spectrometric quantitation. *Anal. Chem.* 87 (8): 4409–4414.
73 Doiphode, P.G., Malovichko, M.V., Mouapi, K.N. et al. (2014). Evaluating factor XIII specificity for glutamine-containing substrates using a matrix-assisted laser desorption/ionization time-of-flight mass spectrometry assay. *Anal. Biochem.* 457: 74–84.
74 Gregorius, B., Jakoby, T., Schaumloffel, D. et al. (2013). Monitoring of protease catalyzed reactions by quantitative MALDI MS using metal labeling. *Anal. Chem.* 85 (10): 5184–5190.
75 Grobe, N., Elased, K.M., Cool, D.R. et al. (2012). Mass spectrometry for the molecular imaging of angiotensin metabolism in kidney. *Am. J. Physiol. Endocrinol. Metab.* 302 (8): E1016–E1024.

76 Na, Y.R., Kim, S.Y., Gaublomme, J.T. et al. (2013). Probing enzymatic activity inside living cells using a nanowire-cell "sandwich" assay. *Nano Lett.* 13 (1): 153–158.

77 Patrie, S.M., Roth, M.J., Plymire, D.A. et al. (2013). Measurement of blood protease kinetic parameters with self-assembled mono layer ligand binding assays and label-free MALDI-TOF MS. *Anal. Chem.* 85 (21): 10597–10604.

78 Cho, Y.L., Kim, Y.P., Son, J.G. et al. (2017). On-chip peptide mass spectrometry imaging for protein kinase inhibitor screening. *Anal. Chem.* 89 (1): 799–806.

79 Simithy, J., Gill, G., Wang, Y. et al. (2015). Development of an ESI-LC-MS-based assay for kinetic evaluation of Mycobacterium tuberculosis shikimate kinase activity and inhibition. *Anal. Chem.* 87 (4): 2129–2136.

80 Simithy, J., Reeve, N., Hobrath, J.V. et al. (2014). Identification of shikimate kinase inhibitors among anti-mycobacterium tuberculosis compounds by LC-MS. *Tuberculosis* 94 (2): 152–158.

81 Chen, N., Yang, H.M., Niu, J. et al. (2013). Determination of kinetic parameters and structureactivity relationships of ginsenosides as inhibitors of cyclin-dependent kinase 5/p25 using ultra-pressure liquid chromatography with triple quadrupole tandem mass spectrometry. *Rapid Commun. Mass Spectrom.* 27 (9): 985–992.

82 Chen, N., Sun, Y., Yue, H. et al. (2013). Screening the inhibitors of Tau protein kinase 2 by rapid resolution liquid chromatogrophy-mass spectrometry. *Chem. J. Chin. Univ. Chin.* 34 (5): 1111–1114.

83 Chen, H., Adams, E., and van Schepdael, A. (2013). Study of Abl1 tyrosine kinase inhibitors by liquid chromatography-electrospray ionization-mass spectrometry. *Talanta* 107: 88–94.

84 Chen, H., Adams, E., and van Schepdael, A. (2012). LC-ESI-MS method for the monitoring of Abl 1 tyrosine kinase. *J. Chromatogr. B* 897: 17–21.

85 Mulabagal, V. and Calderon, A.I. (2010). Development of an ultrafiltration-liquid chromatography/mass spectrometry (UF-LC/MS) based ligand-binding assay and an LC/MS based functional assay for mycobacterium tuberculosis shikimate kinase. *Anal. Chem.* 82 (9): 3616–3621.

86 Krappmann, M., de Boer, A.R., Kool, D.R.W. et al. (2016). Mass spectrometric real-time monitoring of an enzymatic phosphorylation assay using internal standards and data-handling freeware. *Rapid Commun. Mass Spectrom.* 30 (8): 1019–1030.

87 Beeman, K., Baumgartner, J., Laubenheimer, M. et al. (2017). Integration of an in situ MALDI-based high-throughput screening process: a case study with receptor tyrosine kinase c-MET. *SLAS Discov.* 22 (10): 1203–1210.

88 Smith, A.M.E., Awuah, E., Capretta, A. et al. (2013). A matrix-assisted laser desorption/ionization tandem mass spectrometry method for direct screening of small molecule mixtures against an aminoglycoside kinase. *Anal. Chim. Acta* 786: 103–110.

89 Zhou, G.C., Khan, F., Dai, Q. et al. (2012). Photocleavable peptide-oligonucleotide conjugates for protein kinase assays by MALDI-TOF MS. *Mol. BioSyst.* 8 (9): 2395–2404.

90 Ji, L.Y., Wu, J.H., Luo, Q. et al. (2012). Quantitative mass spectrometry combined with separation and enrichment of phosphopeptides by titania coated magnetic mesoporous silica microspheres for screening of protein kinase inhibitors. *Anal. Chem.* 84 (5): 2284–2291.

91 Zhou, G.C., Sylvester, J.E., Wu, D. et al. (2011). A magnetic bead-based protein kinase assay with dual detection techniques. *Anal. Biochem.* 408 (1): 5–11.

92 Lu, S., Luo, Q., Li, X.C. et al. (2010). Inhibitor screening of protein kinases using MALDI-TOF MS combined with separation and enrichment of phosphopeptides by TiO_2 nanoparticle deposited capillary column. *Analyst* 135 (11): 2858–2863.

93 Winter, M., Bretschneider, T., Kleiner, C. et al. (2018). Establishing MALDI-TOF as versatile drug discovery readout to dissect the PTP1B enzymatic reaction. *SLAS Discov.* 23 (6): 561–573.

94 Zwick, V., Simoes-Pires, C., and Cuendet, M. (2016). Cell-based multi-substrate assay coupled to UHPLC-ESI-MS/MS for a quick identification of class-specific HDAC inhibitors. *J. Enzyme Inhib. Med. Chem.* 31: 209–214.

95 Sun, S.W., Buer, B.C., Marsh, E.N.G. et al. (2016). A label-free sirtuin 1 assay based on droplet electrospray ionization mass spectrometry. *Anal. Methods* 8 (17): 3458–3465.

96 Guitot, K., Drujon, T., Burlina, F. et al. (2017). A direct label-free MALDI-TOF mass spectrometry based assay for the characterization of inhibitors of protein lysine methyltransferases. *Anal. Bioanal.Chem.* 409 (15): 3767–3777.

97 Richardson, S.L., Hanjra, P., Zhang, G. et al. (2015). A direct, ratiometric, and quantitative MALDI-MS assay for protein methyltransferases and acetyltransferases. *Anal. Biochem.* 478: 59–64.

98 Guitot, K., Scarabelli, S., Drujon, T. et al. (2014). Label-free measurement of histone lysine methyltransferases activity by matrix-assisted laser desorption/ionization time-of-flight mass spectrometry. *Anal. Biochem.* 456: 25–31.

99 Haslam, C., Hellicar, J., Dunn, A. et al. (2016). The evolution of MALDI-TOF mass spectrometry toward ultra-high-throughput screening: 1536-well format and beyond. *J. Biomol. Screening* 21 (2): 176–186.

100 Gurard-Levin, Z.A., Kilian, K.A., Kim, J. et al. (2010). Peptide arrays identify isoform-selective substrates for profiling endogenous lysine deacetylase activity. *ACS Chem. Biol.* 5 (9): 863–873.

101 Kuo, H.-Y., DeLuca, T.A., Miller, W.M. et al. (2013). Profiling deacetylase activities in cell lysates with peptide arrays and SAMDI mass spectrometry. *Anal. Chem.* 85 (22): 10635–10642.

102 Ritorto, M.S., Ewan, R., Perez-Oliva, A.B. et al. (2014). Screening of DUB activity and specificity by MALDI-TOF mass spectrometry. *Nat. Commun.* 5: 11.

103 Yang, H. and Cheng, Q. (2017). Chemoselective ligation reaction of *N*-acetylglucosamine (NAG) with hydrazide functional probes to determine galactosyltransferase activity by MALDI mass spectrometry. *Analyst* 142 (14): 2654–2662.

104 Beloqui, A., Calvo, J., Serna, S. et al. (2013). Analysis of microarrays by MALDI-TOF MS. *Angew. Chem. Int. Ed.* 52 (29): 7477–7481.

105 Ban, L., Pettit, N., Li, L. et al. (2012). Discovery of glycosyltransferases using carbohydrate arrays and mass spectrometry. *Nat. Chem. Biol.* 8 (9): 769–773.

106 Sanchez-Ruiz, A., Serna, S., Ruiz, N. et al. (2011). MALDI-TOF mass spectrometric analysis of enzyme activity and lectin trapping on an array of N-glycans. *Angew. Chem. Int. Ed.* 50 (8): 1801–1804.

107 Yang, H., Chan, A.L., LaVallo, V. et al. (2016). Quantitation of alpha-glucosidase activity using fluorinated carbohydrate array and MALDI-TOF-MS. *ACS Appl. Mater. Interfaces* 8 (4): 2872–2878.

108 Beloqui, A., Sanchez-Ruiz, A., Martin-Lomas, M. et al. (2012). A surface-based mass spectrometry method for screening glycosidase specificity in environmental samples. *Chem. Commun.* 48 (11): 1701–1703.

109 Deng, K., Takasuka, T.E., Heins, R. et al. (2014). Rapid kinetic characterization of glycosyl hydrolases based on oxime derivatization and nanostructure-initiator mass spectrometry (NIMS). *ACS Chem. Biol.* 9 (7): 1470–1479.

110 Heins, R.A., Cheng, X., Nath, S. et al. (2014). Phylogenomically guided identification of industrially relevant GH1 beta-glucosidases through DNA synthesis and nanostructure-initiator mass spectrometry. *ACS Chem. Biol.* 9 (9): 2082–2091.

111 Price, N.P.J. and Naumann, T.A. (2011). A high-throughput matrix-assisted laser desorption/ionization time-of-flight mass spectrometry-based assay of chitinase activity. *Anal. Biochem.* 411 (1): 94–99.

112 Goacher, R.E., Tsai, A.Y., and Master, E.R. (2013). Towards practical time-of-flight secondary ion mass spectrometry lignocellulolytic enzyme assays. *Biotechnol. Biofuels* 6 (1): 132.

113 Gray, C.J., Sanchez-Ruiz, A., Sardzikova, I. et al. (2017). Label-free discovery array platform for the characterization of glycan binding proteins and glycoproteins. *Anal. Chem.* 89 (8): 4444–4451.

114 Fu, Q., Tang, J., Cui, M. et al. (2015). Development of ESI-MS-based continuous enzymatic assay for real-time monitoring of enzymatic reactions of acetylcholinesterase. *J. Chromatogr. B* 990: 169–173.

115 Cheng, S., Wu, Q.H., Xiao, H. et al. (2017). Online monitoring of enzymatic reactions using time-resolved desorption electrospray ionization mass spectrometry. *Anal. Chem.* 89 (4): 2338–2344.

116 Pei, J.A., Li, Q.A., Kennedy, R.T. et al. (2010). Rapid and label-free screening of enzyme inhibitors using segmented flow electrospray ionization mass spectrometry. *J. Am. Soc. Mass. Spectrom.* 21 (7): 1107–1113.

117 Ziemianin, A., Ronco, C., Dole, R. et al. (2012). Screening of new huprines-inhibitors of acetylcholinesterases by electrospray ionization ion trap mass spectrometry. *J. Pharm. Biomed. Anal.* 70: 1–5.

118 Forsberg, E.M., Green, J.R.A., and Brennan, J.D. (2011). Continuous flow immobilized enzyme reactor-tandem mass spectrometry for screening of AChE inhibitors in complex mixtures. *Anal. Chem.* 83 (13): 5230–5236.

119 Wang, L., Zhao, Y., Zhang, Y. et al. (2018). Online screening of acetylcholinesterase inhibitors in natural products using monolith-based immobilized capillary enzyme reactors combined with liquid chromatography-mass spectrometry. *J. Chromatogr. A* 1563: 135–143.

120 Vanzolini, K.L., Vieira, L.C., Correa, A.G. et al. (2013). Acetylcholinesterase immobilized capillary reactors-tandem mass spectrometry: an on-flow tool for ligand screening. *J. Med. Chem.* 56 (5): 2038–2044.

121 Su, X.M., Zhang, J.X., Guo, Y.L. et al. (2012). Study of effects of cigarette smoke condensates on acetylcholinesterase activity in human lung epithelial cells by matrix-assisted laser desorption/ionization-fourier transform mass spectrometry. *Anal. Lett.* 45 (18): 2687–2696.

122 Cai, T.T., Zhang, L., Wang, H.Y. et al. (2011). Assisted inhibition effect of acetylcholinesterase with n-octylphosphonic acid and application in high sensitive detection of organophosphorous pesticides by matrix-assisted laser desorption/ionization fourier transform mass spectrometry. *Anal. Chim. Acta* 706 (2): 291–296.

123 Xu, Z., Yao, S.J., Wei, Y.L. et al. (2008). Monitoring enzyme reaction and screening of inhibitors of acetylcholinesterase by quantitative matrix-assisted laser desorption/ionization fourier transform mass spectrometry. *J. Am. Soc. Mass. Spectrom.* 19 (12): 1849–1855.

124 Wang, X.L., Zhu, Y., and Fang, Q. (2014). Coupling liquid chromatography/mass spectrometry detection with microfluidic droplet array for label-free enzyme inhibition assay. *Analyst* 139 (1): 191–197.

125 de Vlieger, J.S.B., Kolkman, A.J., Ampt, K.A.M. et al. (2010). Determination and identification of estrogenic compounds generated with biosynthetic enzymes using hyphenated screening assays, high resolution mass spectrometry and off-line NMR. *J. Chromatogr. B* 878 (7-8): 667–674.

126 Su, Y., Zhu, Y., and Fang, Q. (2013). A multifunctional microfluidic droplet-array chip for analysis by electrospray ionization mass spectrometry. *Lab Chip* 13 (10): 1876–1882.

127 Helal, K.Y., Alamgir, A., Berns, E.J. et al. (2018). Traceless immobilization of analytes for high-throughput experiments with SAMDI mass spectrometry. *J. Am. Chem. Soc.* 140 (26): 8060–8063.

128 Zhao, S.L., Ji, X.W., Lin, P.T. et al. (2011). A gold nanoparticle-mediated enzyme bioreactor for inhibitor screening by capillary electrophoresis. *Anal. Biochem.* 411 (1): 88–93.

129 Saito, N., Robert, M., Kochi, H. et al. (2009). Metabolite profiling reveals YihU as a novel hydroxybutyrate dehydrogenase for alternative succinic semialdehyde metabolism in *Escherichia coli*. *J. Biol. Chem.* 284 (24): 16442–16451.

130 Preeti, R., Pandey, A., and Binod, P. (2017). Microbial production of ketoreductases: development of a novel high-throughput screening method. *Bioresour. Technol.* 242: 319–323.

131 Paul, P., Suwan, J., Liu, J. et al. (2012). Recent advances in sulfotransferase enzyme activity assays. *Anal. Bioanal.Chem.* 403 (6): 1491–1500.

132 Furdui, C.M. and Anderson, K.S. (2010). *Detection of Novel Enzyme Intermediates*, 663–688. Amsterdam: Elsevier Science Bv.

133 Shimazaki, Y., Sugawara, Y., Manabe, T. et al. (2004). Nondenaturing two-dimensional electrophoresis enzyme profile involving activity and sequence structure of cytosol proteins from mouse liver. *Proteomics* 4 (5): 1406–1411.

134 Diefenbach, X.W., Farasat, I., Guetschow, E.D. et al. (2018). Enabling biocatalysis by high-throughput protein engineering using droplet microfluidics coupled to mass spectrometry. *ACS Omega* 3 (2): 1498–1508.

135 Li, Y.J., Scott, C.R., Chamoles, N.A. et al. (2004). Direct multiplex assay of lysosomal enzymes in dried blood spots for newborn screening. *Clin. Chem.* 50 (10): 1785–1796.

136 de Rond, T., Peralta-Yahya, P., Cheng, X. et al. (2013). Versatile synthesis of probes for high-throughput enzyme activity screening. *Anal. Bioanal.Chem.* 405 (14): 4969–4973.

137 Karas, M. and Hillenkamp, F. (1988). Laser desorption ionization of proteins with molecular masses exceeding 10000 daltons. *Anal. Chem.* 60 (20): 2299–2301.

138 Cardoso, C.L., de Moraes, M.C., and Cass, Q.B. (2009). Immobilization of the enzymes on chromatographic supports: a tool to research of inhibitor compounds. *Quim. Nova* 32 (1): 175–187.

139 Reetz, M.T., Becker, M.H., Klein, H.W. et al. (1999). A method for high-throughput screening of enantioselective catalysts. *Angew. Chem. Int. Ed. Engl.* 38 (12): 1758–1761.

140 Reetz, M.T., Kuhling, K.M., Wilensek, S. et al. (2001). A GC-based method for high-throughput screening of enantio selective catalysts. *Catal. Today* 67 (4): 389–396.

141 Benito-Cambra, M., Gareil, P., Badet, B. et al. (2018). First investigations for the characterization of glucosamine-6-phosphate synthase by capillary electrophoresis. *J. Chromatogr. B* 1072: 130–135.

142 Purcell, A.L. and Hoard-Fruchey, H.M. (2007). A capillary electrophoresis method to assay catalytic activity of botulinum neurotoxin serotypes: implications for substrate specificity. *Anal. Biochem.* 366 (2): 207–217.

143 Lopez-Otin, C. and Bond, J.S. (2008). Proteases: multifunctional enzymes in life and disease. *J. Biol. Chem.* 283 (45): 30433–30437.

144 Su, J. and Mrksich, M. (2002). Using mass spectrometry to characterize self-assembled monolayers presenting peptides, proteins, and carbohydrates. *Angew. Chem. Int. Ed. Engl.* 41 (24): 4715–4718.

145 Northen, T.R., Lee, J.C., Hoang, L. et al. (2008). A nanostructure-initiator mass spectrometry-based enzyme activity assay. *Proc. Natl. Acad. Sci. U.S.A.* 105 (10): 3678–3683.

146 Reindl, W., Deng, K., Gladden, J.M. et al. (2011). Colloid-based multiplexed screening for plant biomass-degrading glycoside hydrolase activities in microbial communities. *Energy Environ. Sci.* 4 (8): 2884–2893.

147 Chang, Y.L., Liao, S.K.S., Chen, Y.C. et al. (2011). Tagging saccharides for signal enhancement in mass spectrometric analysis. *J. Mass Spectrom.* 46 (3): 247–255.

148 Rzagalinski, I. and Volmer, D.A. (2017). Quantification of low molecular weight compounds by MALDI imaging mass spectrometry – a tutorial review. *BBA* 1865 (7): 726–739.

149 Andorfer, M.C., Park, H.J., Vergara-Coll, J. et al. (2016). Directed evolution of RebH for catalyst-controlled halogenation of indole C–H bonds. *Chem. Sci.* 7 (6): 3720–3729.

150 Comi, T.J., Neumann, E.K., Do, T.D. et al. (2017). microMS: a python platform for image-guided mass spectrometry profiling. *J. Am. Soc. Mass. Spectrom.* 28 (9): 1919–1928.

4

Recent Advances in Cell Surface Display Technologies for Directed Protein Evolution

Maryam Raeeszadeh-Sarmazdeh[1] *and Wilfred Chen*[2]

[1] *University of Nevada, Chemical and Materials Engineering Department, Reno, 1664 N. Virginia St, Reno, NV 89557, USA*
[2] *University of Delaware, Department of Chemical and Biomolecular Engineering, Newark, DE 19711, USA*

4.1 Cell Display Methods

Since the introduction of cell-based techniques in the directed evolution of biomolecules, various cells have been used as display platforms. Peptides and proteins were engineered using phage, yeast, or mammalian cell display; each of these techniques has its advantages and limitations [1, 2]. Selecting the suitable cell-display platform is important for a successful directed protein evolution (Figure 4.1 and Table 4.1).

4.1.1 Phage Display

Phage display was first described by Smith in 1985 with peptides fused to the minor coat protein of filamentous phage (pIII) [16]. The phage-display technique has been used for displaying and engineering a variety of different peptides and proteins. In a phage display, protein or peptide of interest (POI) is usually displayed on the surface of filamentous M13 bacteriophage as a fusion to either the major (pVIII) or minor (pIII) phage coat protein [3]. Phage display using fusion to the major coat protein (pVIII), with nearly 3000 copies on the phage surface, is used to enhance signal detection. In contrast, minor coat proteins (pIII) located at the phage tip provide limited copy numbers (3–5) of the protein displayed on the phage surface [4] (Figure 4.1a). Phage display using fusion to the pVIII protein is limited to short peptides (six to seven amino acids) not containing cysteine, whereas fusion to pIII protein is commonly and widely used phage-display approach for directed evolution of proteins and peptides [4].

A broad spectrum of proteins has been engineered using phage display including antibodies [17], protein scaffolds [18], DNA-binding proteins [19], and enzymes [20]. Phage display was used to isolate and evolve antibodies. A library of over 40 billion human antibody clones in single-chain fragment variable (scFv) format

Protein Engineering: Tools and Applications, First Edition. Edited by Huimin Zhao.
© 2021 WILEY-VCH GmbH. Published 2021 by WILEY-VCH GmbH.

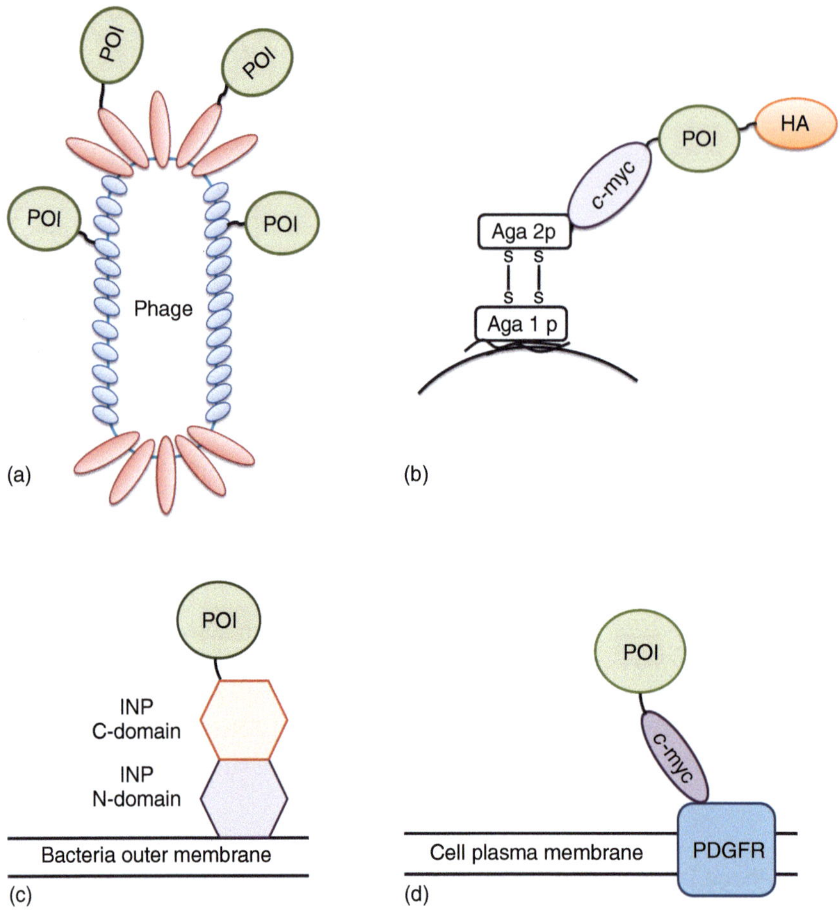

Figure 4.1 Examples of cell surface display system. (a) *Phage surface display*. In phage display, POI (green circle) is usually displayed on the surface of filamentous M13 bacteriophage as a fusion to either the major (pVIII) (small blue ovals on the side) or minor (pIII) (large pink ovals on the tip) phage coat protein. (b) *Bacterial cell display*. POI is anchored to the bacteria outer membrane via fusion to the N-terminus of ice nucleation protein (INP). (c) *Yeast surface display (YSD)*. In the common YSD method, the yeast displayed protein is genetically fused at the C-terminus of Aga2p, yeast surface protein, between two epitope tags (HA and *c*-myc). After secretion, the recombinant Aga2p genetically fused to the protein of interest binds to the anchored Aga1p on the yeast surface via two disulfide bonds. (d) *Mammalian cell display*. POI is fused to the N-terminus of PDGFR, HEK293 cell wall protein. The c-myc epitope tag at the C-terminus of protein is used for detecting protein expression.

with high functionality was designed, synthesized, and characterized using phage display. This phage-display synthetic antibody library is a useful source for developing therapeutics or isolating desired antibodies against a variety of target proteins [21]. Synthetic antibodies that target specific fusion-relevant intermediates of Ebola virus glycoproteins [22] and humanized antibodies that retain high affinity for their original antigens were also developed using phage display [23–25].

Table 4.1 Specifications of cell display systems for directed evolution of proteins.

Display system	Display copy number	Library size (upper limit)	Cell anchor protein	Amenability of protein fusion	References
Phage	(pVIII) ~3000 (pIII) 3–5	10^{12}–10^{13}	pIII, pVIII	C-terminus	[3, 4]
Bacterial	10^5	10^8–10^{10}	INP, autotransporter proteins, OmpA, OmpC, OmpX	C-terminus	[5–9]
Yeast	10^4–10^5	10^8–10^9	Aga1, Aga2, Flo1p, α-agglutinin, Cwp1p, Cwp2p, Tip1p, Sed1p1, YCR89W, Tir1	N- and C-terminus	[10–12]
Mammalian	NS[a)]	10^6	PDGFR	C-terminus	[13–15]

a) Not specified.
Source: Frei and Lai [3]; Bazan et al. [4]; Francisco et al. 5.

Several other protein scaffolds such as Zinc finger domains [26], human growth hormone [27], a lipocalin for fluorescein binding [28], and T lymphocyte-associated antigen 4 [29] variants were engineered for improving in binding affinity and selectivity to their specific targets using directed evolution and phage display. Phage display was also used for engineering the substrate specificity of enzymes. Using directed evolution and phage display, *Staphylococcus aureus* sortase A (SrtA) with a randomized β6 and β7 loop, a loop with a significant role in substrate binding was engineered for substrate specificity [30, 31]. SrtA variants recognizing APxTG or FPxTG peptide tags as substrate were discovered using phage display, thereby broadening the substrate recognition list from the natural specific SrtA substrate, an LPxTG motif [31].

A large size of protein and peptide libraries of variants can be displayed on the phage surface. Therefore, several libraries of variants ranging from short peptides to proteins with molecular weights up to 100 kDa were engineered using phage display. However, a major limitation of phage display is the lack of eukaryotic post-translational machinery, which is often required for developing human therapeutics or other yeast or mammalian proteins that are either glycosylated or contain other post-translational modifications required for their binding or activity [32].

4.1.2 Bacterial Display Systems

To bypass some limitations associated with phage display, several bacterial display systems have been developed. The major advantage of bacterial display is its ability to display larger proteins, and the ease of cell screening using fluorescence activated cell sorting (FACS) [5, 6]. *Escherichia coli* has emerged as the most widely used host due to the large genetic toolbox and the ability to achieve a high-density surface display of full-length target proteins. While Lpp-OmpA is among the most widely used carrier for bacterial surface display [5], recent efforts have focused on two carrier proteins: autotransporter proteins [7] and ice nucleation proteins (INPs) [8], which are more robust and have less impact on growth. Both anchor system have been reported to provide a high level of display (>1 00 000 copies per cell). The key benefit of using *E. coli* is the high transformation efficiency, allowing the construction of libraries >10^8 diversity [9]. Even display platforms based on *Bacillus* spores [10] or bacterial outer membrane vesicle (OMV) [33] are now feasible. However, bacterial display systems are not devoid of their own issues as they are not capable of performing complex post-translational modifications required for many mammalian proteins.

4.1.3 Yeast Surface Display

In yeast surface display (YSD) technique, the protein of interest is fused to the N- or C-terminus of one of the cell wall proteins of *Saccharomyces cerevisiae* including Aga1p, Aga2p, α-agglutinin, Cwp1p, Cwp2p, Tip1p, Flo1p, Sed1p1, YCR89W, and Tir1 [34]. The most common and well-established YSD method was developed by Boder and Wittrup [13], which uses an Aga1p-Aga2p system. In this method, the displayed protein is genetically fused at the C-terminus of Aga2p between two epitope tags (hemagglutinin [HA] and c-myc). Upon secretion, the recombinant Aga2p fused to the protein of interest binds to the β-1,6-glucan anchored Aga1p domain of **a**-agglutinin via two disulfide bonds [13]. Alternative YSD platforms developed for C-terminal fusion of protein of interest to the N-terminus of Aga2p [35]. This is critical for proteins with having their N-terminal domain as the functional domain and the intact N-terminus plays a key role in protein–protein interaction or enzyme inhibition [11]. Multiple copies of protein of interest, around ten thousand to one hundred thousand copies, are displayed on the yeast surface which facilitates quantification using flow cytometry [36]. The other yeast mating factor, α-agglutinin, was also used to display peptides/proteins on the yeast surface via attachment to glycosylphosphatidylinositol (GPI) cell wall anchor. In α-agglutinin display system, the target protein is fused at the N-terminus of α-agglutinin yeast cell wall protein which is bound to the GPI cell wall anchor at the C-terminus [37, 38].

After first use of YSD for affinity maturation of scFv antibody against fluorescein [13], the yeast display technique was widely used for engineering antibodies [13, 36, 39, 40]. Several scFv (single chain fragment) antibodies, as well as Fab fragments and whole IgG antibodies, were engineered using YSD [36, 40]. YSD benefits from intracellular proteolysis quality control system by the endoplasmic reticulum (ER) where the unfolded and unstable proteins get degraded upon secretion and display on the yeast surface [41, 42]. Therefore, YSD was used to improve thermostability of protein binders using directed evolution [12, 43, 44]. Single chain T cell receptors (scTCRs) mutants with improved thermal stability correlated with their expression and display level were isolated using YSD [41].

Recently, several non-antibody binders have also been identified and engineered using YSD [45, 46]. Highly stable scaffolds with unique structural features are great candidates for protein engineering for developing therapeutics. Among these scaffolds, a cysteine knot peptide (knottin) was engineered to target $\alpha_v\beta_3$, $\alpha_v\beta_5$, and $\alpha_5\beta_1$ integrins with high affinities, without altering cysteines responsible for several disulfide bonds to maintain stability of knottin variants [14]. Hyperthermophilic Sso7d scaffolds were also engineered as highly stable binders against wide ranges of pH, temperature, and chemical reagents, using YSD against various targets from a library of mutants randomizing 10 residues of Sso7 [15]. The evolved Sso7 scaffold had melting temperatures over 89 °C, and showed resistance to guanidine hydrochloride [15].

Yeast display has rapidly become a viable alternative to bacterial and phage display as it offers the ability to target complex heterologous proteins that require post-translational modifications [13]. Directed evolution of proteins using YSD benefits from eukaryotic post-translational machinery for functional expression of yeast displayed glycoproteins and DNA homologous recombination for spontaneous insertion of amplified genes of mutant library into the digested yeast display vector after transformation into the yeast cells [36]. The large size of yeast cells also makes it more amenable for cell sorting using FACS more amenable. Among the limitations of YSD are slower growth and transformation efficiency compared with phage and bacteria which limits the size of library. A wide range of anchors are available to provide both N- and C-terminal display, allowing more flexibility in controlling the protein orientation [6]. More examples of protein and enzyme engineering using modified YSD techniques will be discussed in Section 4.3.

4.1.4 Mammalian Display

In mammalian screening system, nonviral and viral expression vectors systems are used to generate and engineer antibody display libraries, particularly human antibodies [47] since this system offers total post-translational modifications such as glycosylation, and some antibodies screened using this system have shown higher affinity compared with other systems [48]. scFv and whole IgG antibodies were displayed on human embryonic kidney 293T (HEK 293T) cells using transient transfection. The scFv CD22 antibody was displayed on HEK293T cells by genetic fusion to the N-terminus of the transmembrane domain of human platelet-derived growth factor receptor (PDGFR) [49] (Figure 4.1d). This system was used to improve the binding affinity of CD22 antibody to target overexpressed the CD22 adhesion molecules in B cell leukemias and lymphomas [49, 50]. For complex antibody libraries, inconsistent expression of heavy chain and light chains, or incorrect antibody assembly, may result in less stable display systems. One other limitation of mammalian cell display is the difficulty in acquisition of antibody sequences [48]. In contrast, a basic single-chain variable fragment (scFv) display appears to be stable, simple, and robust, which may rapidly yield antibody sequences. Mammalian display system offers complex secretory and post-translational machinery required for protein stability and activity; however, one of the drawbacks of the mammalian cell display is the limited size of the library of protein variants compared to phage, bacterial and yeast display.

4.2 Selection Methods and Strategies

In protein engineering using directed evolution, selecting a fast and reliable selection technique and strategy to screen library of protein variants is critical. High-throughput screening methodologies use the link between protein function and quantified measurable values for sorting the cells, which display protein variants. Screening techniques are usually one of the rate-limiting steps in directed evolution and selecting the right method or combination of two has a great significance in successful selection of protein variants with desired function. A summary of some of the high-throughput screening techniques and selection strategies used in directed evolution using cell-based platforms is discussed in this section.

4.2.1 High-Throughput Cell Screening

4.2.1.1 Panning

Biopanning is a commonly used method for screening cell displayed libraries, particularly for phage display, to selectively enrich populations of protein mutants with improved binding affinity, selectivity, or enzymatic activity [51]. In solid phase panning, the displayed library is added to the immobilized target onto a surface (capture), such as the surface of a 96-well plate or magnetic beads, followed by a wash step to remove nonspecific interactions. The selected clones, usually binders with improved affinity, are amplified for sequential rounds of selection (capture and wash) [16].

4.2.1.2 FACS

FACS provides high-throughput screening technology for sorting large library of mutants with single cell resolution [52, 53]. FACS takes advantage of capabilities of flow cytometry by acquiring multiple fluorescent quantities correlated with different properties of proteins displayed on the cell surface such as expression, binding, or activity for a single cell and utilizing this information to evaluate and screen cells displaying libraries of protein variants [54] (Figure 4.2b). FACS is the most widely used high-throughput screening technique for sorting bacterial, yeast, and mammalian cells, where the desired phenotype of protein mutant such as binding or activity is coupled to a fluorescent signal.

FACS instruments can interrogate multiple fluorescent signals from proteins displayed on single cells with rates up to 10^7 cells per hour [52, 56]. Quantification of fluorescent intensity and size of single droplets usually containing an individual cell obtained from a focused laser beam is recorded and used for cell sorting based on applying a charge on a deflection plate that leads to selection of single cells of interest into the collection tube [52]. Using FACS as a screening method provides ultra-high-throughput cell screening for protein mutants with larger library sizes of 10^9 variants or more, and is able to overcome challenges of traditional methods such as agar or 96- or 384-well plate-based assays [57] for protein engineering using directed evolution [52].

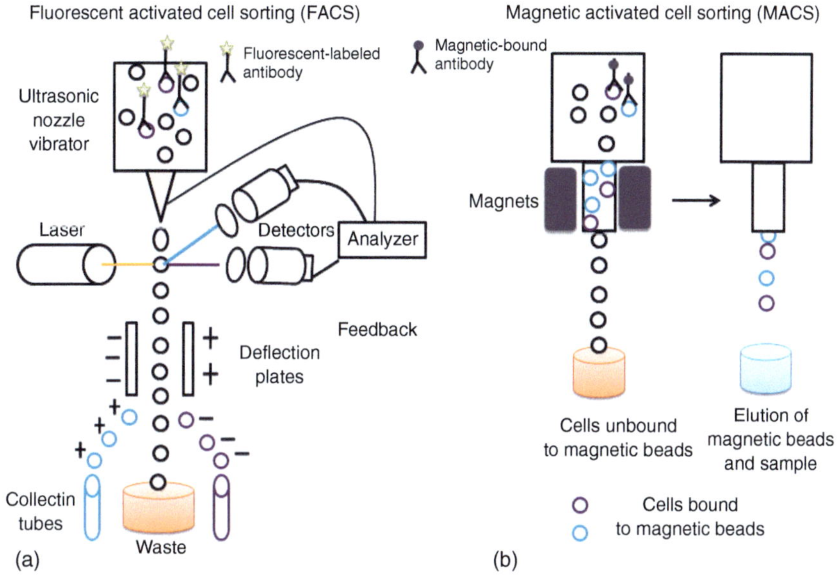

Figure 4.2 High-throughput screening methods. (a) Fluorescent-activated cell sorting (FACS). Fluorescent-labeled proteins (here antibody) on the surface the cell pass through ultrasonic nozzle vibrator which provide a stream of a single drop, which usually contains a single cell. The emission beam from each cell resulted from an excitation laser is detected and analyzed. The feedback from analyzer will guide the cells through collection tubes after getting corresponding charges from deflection plates. (b) Magnetic-activated cell sorting (MACS). Magnetic bound proteins (e.g. antibody) displayed on the cell surface are trapped through a magnetic field while the unbound cells pass through the channel. After removing the magnet, the screened cells displaying proteins of interest, or antibodies bound to the magnetic beads are eluted to the collection tubes.

Protein expression and binding affinity or enzyme activity are usually correlated and measured using two different fluorescent colors. Protein variants with desired phenotypes are then screened using a diagonal FACS sorting gate that includes the protein variants with the highest ratio of binding or activity to protein expression (Figure 4.3b). Multiple rounds of FACS sorting are usually required for enrichment of libraries of protein variants before isolating clones with improved binding affinity or selectivity or enzymatic activity. In an effort to optimize the success of sequential FACS sorting, a support vector machine (SVM) algorithm that used machine learning for creating an optimal gate based on positive and negative control populations was developed [58].

4.2.1.3 MACS

Magnetic-assisted cell sorting (MACS) provides a fast technique for sorting cells displaying protein libraries of mutants, desirable for screening very large cell populations. In most cases, MACS cannot totally replace FACS cell sorting for directed evolution of proteins, however, using streptavidin-coated magnetic beads for capturing biotinylated ligands can reduce the background of undesired library clones prior to FACS [59, 60] (Figure 4.2). Larger populations of proteins can be screened

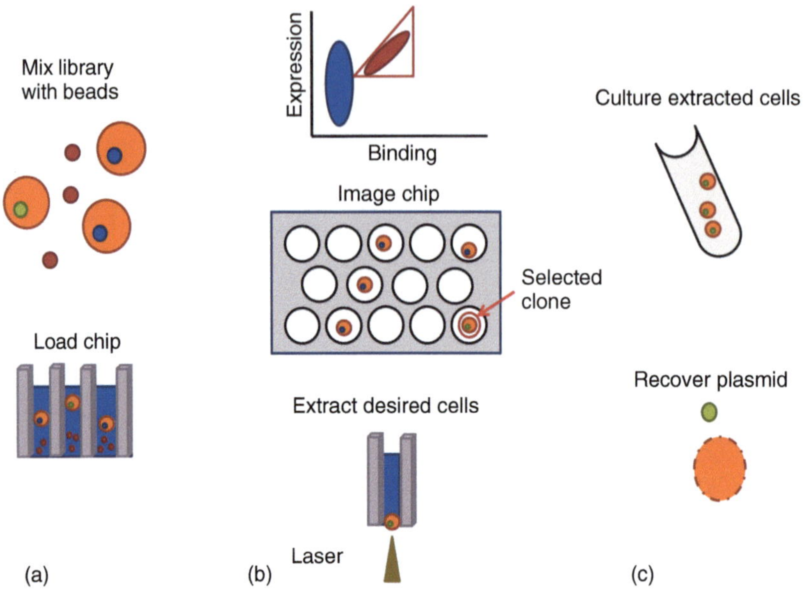

Figure 4.3 Microcapillary single-cell analysis and laser extraction (μSCALE). (a) A mixture of library of protein variants with opaque microbeads is pipetted into the array at a concentration that results in single-cell occupancy in each microcapillary well. (b) (Top panel) the flow cytometry data of cells displaying a protein of interest were dual labeled for binding and expression using two different fluorescent-conjugated antibodies. The FACS diagonal sorting gate is used to screen the clones with the variants with highest binding to expression ratio. (Middle and bottom panel) Each well is quantified based on the fluorescence signal and clones with the desired property will be extracted using a laser-based extraction method and cultured. (c) The plasmid DNA from the isolated clones with improved function is extracted and further analyzed using DNA sequencing. The amplification and screening continues until the library of protein variants are enriched for the desired phenotype. Source: Adapted from Chen et al. [55].

using MACS, and binder variants with lower binding affinity, due to lower expression, will not be eliminated in the first rounds of cell screening. In fact, MACS has been used to isolate weak binders that likely be eliminated in the first rounds of FACS screening. All of these features make MACS more suitable for isolating novel binders from Naïve library of ligands [59], or as initial rounds of screens to reduce the size of library variants before FACS for a more successful screening output.

4.2.2 Selection Strategies

4.2.2.1 Competitive Selection (Counter Selection)

Directed evolution using a counter-selection screening strategy is often used to improve binding selectivity of a ligand for a specific target, or substrate specificity of an enzyme. In a counter-selection screening strategy, an alternative competitor, either a competitive ligand or an enzyme substrate [61] is usually added in excess in incremental amounts in sequential rounds of screening. *S. aureus* sortase A (SrtA) substrate specificity was engineered by screening a library of SrtA mutants against non-natural biotinylated substrates (LPESG and LAETG) in the presence of an excessive amount of a non-biotinylated natural substrate (LPETG). The evolved SrtA variants showed up to 120-fold improvement for substrate specificity [62].

4.2.2.2 Negative/Positive Selection

The negative selection strategy in directed evolution is usually performed to eliminate the off-target binders. In the negative selection approach, protein variants that bind to one target with no or low binding affinity to an undesired target are selected. A combination of positive and negative screen is usually performed to screen for clones that retain binding to a desired target and lack binding to an undesired target, or for altering substrate specificity of an enzyme. This strategy was used to identify horseradish peroxidase (HRP) variants with high enantioselectivity for either D- or L-tyrosinol through altering rounds of negative and positive FACS screen [63]. The engineered HRP variants were either reversed for enantioselectivity compared with wild-type HRP that has a slight preference for L-tyrosinol to a substantial preference for D-tyrosinol by fourfolds, or improved for L-tyrosinol compared with wild-type HRP by eightfold [63].

4.3 Modifications of Cell Surface Display Systems

Cell surface engineering is a well-known technology for displaying functional proteins and peptides onto the surface of bacterial spores, prokaryotic and eukaryotic cells [5, 13, 64]. In all cases, a protein anchor is used to target proteins to surface using the native secretory pathway, and the anchor choice very often impact the display efficiency. Recent advances in modifications of cell display systems with applications in directed evolution will be discussed in this section.

4.3.1 Modification of YSD for Enzyme Engineering

In engineering enzymes, display of both the enzyme library and the substrate on the cell surface is often required. The challenge in designing such a system that links enzyme activity and/or selectivity to a detectable phenotype limits the use of cell surface display for enzyme evolution. Similar to other cell display methods, despite the extensive use of YSD for improving affinity and selectivity of protein binders [36], the use of YSD for engineering enzymes remained limited. However, modified YSD strategies that translate the product of enzymatic reaction to a detectable flow cytometry signal provided robust quantitative methods to evolve enzyme activity and/or selectivity.

In one example, the activity of HRP was linked to addition of a fluorescent-labeled tyrosine on the yeast surface. Active HRP variants were able to bind Alexa Fluor 488-conjugated HRP substrates (L- or D-tyrosinol) to the yeast surface. Using this strategy, HRP variants with higher selectivity for either L- or D-tyrosinol were screened using FACS [63]. Glucose oxidase (Gox) activity was also improved using the HRP-linked addition of tyrosine to the yeast surface in combination with an emulsion technique using directed evolution. Yeast cells displaying Gox variants along with glucosidase, HRP, and tyramide-fluorescent were isolated in water-in-oil single emulsions. β-Octylglucosidase was then cleaved using glucosidases in the water–oil interface, which resulted in the formation of glucose in water [65]. The active Gox variants perform glucose oxidation, and therefore, hydrogen peroxide (H_2O_2) formation, which then leads to the HRP-catalyzed conjugation of tyramide-fluorescein to the tyrosines on the yeast surface [65]. The Gox activity is quantified using fluorescent labeled tyrosines. The two separate phases (water and oil) provided in this modified technique can be used as a general strategy for designing cell display systems, which require to separate manipulation of substrate and product. However, designing such a system finding suitable components that stay in either phase or at the interface is challenging and limits the use of this modified YSD technique.

In another modified YSD strategy, the yeast endoplasmic reticulum sequestration screening (YESS) technique was used for engineering the tobacco etch virus protease (TEV-P) and similar proteases [66]. In this approach, a protease substrate was genetically fused to the C-terminus of Aga2p, followed by epitope tags, for detection of peptide expression and display, and a C-terminal ER retention sequence. The engineered protease mutants with high activity and selectivity for a specific substrate were able to cleave the C-terminal ER-retention tag and allow the secretion and yeast display of the Aga2p-substrate fusion protein, and the corresponding epitope tags for a quantitative detection of the TEV-P activity by flow cytometry [66].

Engineering transpeptidases with more than one substrate is also challenging as a three-component (an enzyme and two substrates) cell display system needs to be developed. Therefore, at least two of these elements (a library of the enzyme variants and one of the substrates) need to be displayed on the yeast surface. Modifying the YSD system to enable a dual protein display on the yeast surface broadens the protein engineering landscape using YSD for more complex systems required to engineer transpeptidases. A library of *S. aureus* Sortase A (SrtA) mutants was genetically fused to the C-terminus of Aga2p and one of the SrtA substrates (LPETG) was conjugated to Aga1p using a reactive peptide handle (S6) via Sfp phosphopantetheinyl transferase reaction. Active SrtA mutants were able to catalyze the transpeptidation reaction between the second biotinyalted SrtA substrate (Biot-GGG) and the LEPTG peptide anchored to Aga1p. The extent of SrtA reaction was then measured using streptavidin-conjugated fluorophores via flow cytometry. SrtA variants were screened using FACS with up to 140-fold in improvement in enzyme activity using this strategy [67]. Fusion of two proteins on the yeast surface has another advantage of eliminating the need for expression of an epitope tag and immunolabeling for quantifying protein expression on the yeast surface. By dual fusion of a green fluorescent protein and a protein of interest (a ligand, receptor, or antibody fragment) to the two termini of Aga2p, simultaneous quantification of protein expression using flow cytometry was performed [68]. Moreover, the recent YSD modified platform providing genetically fusion of heterologous proteins to both N- and C-termini of Aga2p [68] and eliminates the extra steps of chemical modification for conjugating proteins to the yeast cell wall proteins, Aga1p or Aga2p [67].

4.3.2 Yeast Co-display System

Initial YSD systems were based on the protein–protein interactions or enzyme reactions on the cell surface. However, YSD systems based on intracellular association of a multimeric protein and the binding peptide were developed [69]. In the yeast co-display system that was developed in Boder lab, human leukocyte antigen-D Related 1 (HLA-DR1), a model MHC-II, and its bound peptide FLU were co-displayed on the yeast cell surface [69]. In this approach, the protein complex between HLA-DR subunits (HLA-DR1α and HLA-DR1β) and FLU peptide was formed in the yeast ER prior to secretion and display on the surface. Functional HLA-DR1 heterodimers formed in the yeast ER were then recognized by the FLU peptide. The yeast co-display technique provides a closer natural mimic of peptide loading of MHCII in antigen presenting cells (APCs) while overcome the challenge of expressing the multimeric protein (MHCII) and peptide binding intracellularly. The binding affinity of the FLU peptide variants for HLA-DR1 heterodimer was measured by immunolabeling of the V5 and HA epitope tags fused to the two termini of the FLU peptide on the yeast surface. Formation of the HLA-DR1-FLU complex displayed on the cell surface was measured quantitatively by anti-DR antibody [69]. Using the yeast co-display technique, amino acids at P1 position of FLU peptide interacting with wild-type HLA-DR1 heterodimer were identified. Directed evolution of HLA-DR1 library of mutants targeting FLU peptide with Val, Ala, or Glu at P1 position, not recognized by wild-type HLA-DR1, resulted in discovery of HLA-DR1 variants that switch binding selectivity from Tyr in wild-type FLU peptide to these residues (Val, Ala, or Glu) at P1 position [69]. The yeast co-display system is a powerful tool for quantitative measurement of MHC-II antigen-binding motifs, and can be extended for engineering the binding affinity and specificity of other multimeric proteins and their binding peptides.

4.3.3 Surface Display of Multiple Proteins

Although early efforts on cell surface engineering focused on displaying a single protein or peptide, primarily due to the ease of secretion, recent studies have clearly demonstrated the feasibility to display multiple proteins onto the same host [70–72]. Simultaneous protein display enables cells to carry multiple functionalities and significantly expands the number of applications that can be targeted. One of the first examples was to co-display organophosphorus hydrolase (OPH) and a cellulose-binding domain onto the surface of *E. coli* in order to enable simultaneous immobilization and organophosphorus pesticides (OPs) degradation on any cellulose matrix [70]. Inspired by this report, other researchers have adapted similar strategies to display multiple enzymes for degradation of more complex OP pesticide mixtures [73, 74]. Other than using a single anchor to display multiple proteins, different anchors are used to reduce competition for the same secretory machinery.

Attempts to display multiple proteins have also been reported for yeast cells [71, 75]. The initial efforts centered on displaying several cellulases for efficient cellulose hydrolysis and ethanol production. Display of a small protein called expansin with cellulases became a breakthrough for the direct degradation of crystalline cellulose by disrupting the hydrogen bonding between cellulose microfibrils and increases the accessibility of cellulases towards the fiber surfaces [76, 77]. The resulting strain produced up to 3.4 g/l of ethanol directly from phosphoric acid-swollen cellulose (PASC) [77].

Unfortunately, display of multiple proteins on the cell surface typically reduces cell growth and viability, leading to the search for an alternative approach for displaying multiple proteins. In nature, anaerobic bacteria evolved the ability to display a multi-enzyme complex, named cellulosome, in order to efficiently degrade biomass for growth [78]. The main feature of the cellulosome is a structural scaffolding consisting of at least one cellulose-binding module (CBM) and repeating cohesin domains, which are docked individually with different cellulases tagged with the corresponding dockerin domain [79]. This highly ordered structure allows the assembly of multiple cellulases in close proximity, resulting in a high level of enzyme-substrate-microbe synergy towards cellulose and hemicellulose hydrolysis [80]. Because of the highly specific interaction between the dockerin and cohesin domain from the same species, chimeric scaffoldings composed of cohesins from three or four different species have been created, resulting in the formation of synthetic cellulosomes bearing the matching dockerin-tagged enzymes [81, 82].

Efforts have been made to display synthetic cellulosome structures onto the yeast surface [83–86]. Chen group is the first to functionally display a mini-cellulosome composed of three dockerin-tagged enzymes on the yeast surface by *in vitro* assembly [83]. A synthetic yeast consortium was further designed to enable the functional assembly of a mini-cellulosome by intracellular complementation, enabling simultaneous cellulose hydrolysis and ethanol production [84, 87]. More complex cellulosome structures were displayed by using adaptive assembly into highly branched structures for enhanced efficiencies [85, 88]. Although the native function for dockerin/cohesin pair for cellulose assembly, the striking selectivity can be easily exploited for the site-specific assembly of other enzymatic cascade reactions [89, 90]. One pioneer example is the ordering of hydrogenases responsive for the complete oxidation of methanol to CO_2 using a synthetic scaffolding displayed on the yeast surface. Enzyme clustering enhances methanol conversion by fivefold and can be used to significantly improve the overall efficiency of the methanol-based enzyme fuel cell system. Similar enhancements have been reported for other related cascade reactions relying on nicotinamide adenine dinucleotide + hydrogen (NADH) recycling [91]. The flexibility and the ease of assembling larger, more complex enzyme structures is the most exciting feature of this scaffold-based approach. The scaffolding-based surface display system has also been reported on bacterial cells such as *Bacillus subtilis* and *E. coli* [33, 92–94]. It is easy to envision that other related systems such as spore display can also be used for this strategy [10].

4.4 Recent Advances to Expand Cell-Display Directed Evolution Techniques

Cell surface display techniques for directed evolution usually require several rounds of screening of cell populations displaying libraries of protein mutant on the cell surface, followed by screening colonies and sequencing analysis [56]. High-throughput directed evolution screening methods such as FACS and automated systems such as robotic plate readers facilitated the time-consuming and complicated screening process of protein library of random mutants [1]. However, to overcome the battle of time and labor, more recent directed evolution technologies were developed to optimize directed evolution techniques for engineering proteins.

4.4.1 μSCALE (Microcapillary Single-Cell Analysis and Laser Extraction)

μSCALE (microcapillary single-cell analysis and laser extraction) is a multipurpose platform for screening millions of individual bacteria or yeast cells within minutes as well as quantification of the single cells carrying individual protein mutant that was first introduced at Cochran laboratory. μSCALE technique combines a microcapillary array platform with a laser-based technology to provide a powerful methodology that opens a new horizon to interrogation of the sequence-structure-activity relationship of proteins with rapid automated biochemical and biophysical measurements (Figure 4.3). Using the laser-based technology, microcapillary platform offers robust spatial segregation of single cells and real-time measurement and recovering of the microcapillary content, with 5 orders of magnitude greater high-throughput screening capacity compared to a microwell plate or a standard petri dish. μSCALE allows isolation and growth of each individual cell after analysis, eliminating the need for growth and screening single colonies after each high-throughput screening round in a conventional cell-display based directed evolution [95]. In the μSCALE method, a mixture of library of protein variants with opaque microbeads is pipetted into the array at a concentration that results in single-cell occupancy in each microcapillary well (Figure 4.3a). Each well is quantified based on the fluorescence signal detection, and clones with the desired property weae extracted using a laser-based extraction method and cultured (Figure 4.3b). The plasmid DNA is extracted from the selected variants with improved protein function such as binding affinity or enzymatic activity and analyzed via DNA sequencing (Figure 4.3c).

Various classes of proteins have been engineered using μSCALE including binding affinity of an antibody, intensity of fluorescent proteins, and an enzyme activity [95]. Use of μSCALE technology for directed evolution of proteins did not grow out of its original lab to date, which might point the need for complex machinery and expenses that are associated with this technique. However, since microcapillary high-throughput screening provides a fast robust system for protein engineering and cell-based directed evolution and is flexible for being adopted to various assay requirements and different host organisms, it has a great potential to replace conventional screening methods [55].

4.4.2 Combining Cell Surface Display and Next-Generation Sequencing

The deep mutational scanning or next-generation sequencing (NGS) techniques were coupled with the cell display techniques to analyze the amino acid substitution in a library of protein mutants before and after high-throughput screening. The deep mutational scanning strategy provides a more stringent survey through amino acid mutations and allows detection of individual mutations responsible for subtle changes in desired property of protein mutants, which broadens the fitness landscape and decreases the chances of mutant clones being missed in several rounds of high-throughput screening due to the low expression of protein variants or other technical limits that overlooks small changes in protein properties such as binding or activity [96]. Using deep mutational scanning in combination with cell-display techniques can bypass the need for several rounds of screen that might eliminates binders with higher affinity due to their low expression or display level.

Deep mutational scanning was used in combination with YSD to improve affinity of meditope peptide for developing the therapeutic antibody, cetuximab [97], and TCR-peptide major histocompatibility complex (MHC) [98]. Computational analysis of amino acid substitution was usually coupled with deep sequencing, and bioinformatics tools play an important role in the outcome of this approach. NGS provides DNA sequencing capacity of up to 10 million sequences (about 10 000-fold higher than Sanger sequencing). However, it usually provides shorter reads (100 bps) and may read up to 300–400 bps, which makes this technology limited to peptides or smaller size proteins. Another limitation of this approach is the need for strong bioinformatics software and algorithms [96]. However, considering the high-speed of the advances in DNA sequencing technology and bioinformatics tools, these limitations are expected to be overcome in the future.

4.4.3 PACE (Phage-Assisted Continuous Evolution)

Phage-assisted continuous evolution (PACE) was first developed as its comprehensive form in Liu laboratory. PACE exploits the use of continuous culture and selection of the M13 filamentous bacteriophage and provides a platform for continuous directed evolution of proteins at a rate of about 100-fold faster than conventional methods [99, 100]. *E. coli* host cells continuously flow through a fixed-volume vessel containing a replicating population of phage DNA vector encoding the gene of interest [101] (Figure 4.4). Continuous selection is based on linking the desired activity to the production of pIII gene, a significant factor in phage infectious [101]. *E. coli* host cell carries two plasmids: an accessory plasmid (AP), which links specific activity of protein to phage infectivity through expression of gene III and pIII production, and a mutagenesis plasmid (MP), which provides arabinose-induced elevated levels of mutagenesis. The third main component of PACE system is selection phage (SP),

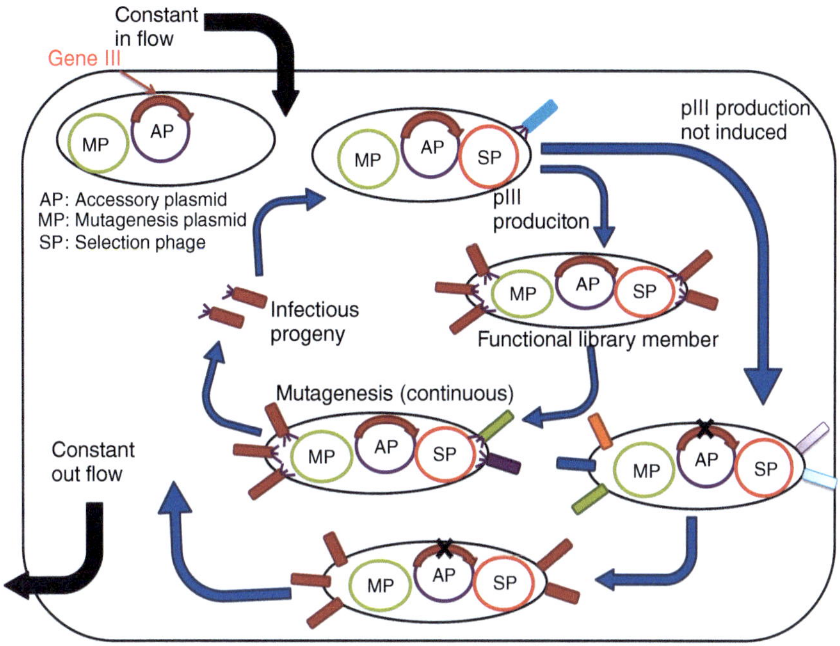

Figure 4.4 Phage-assisted continuous evolution (PACE). Continuous selection of the M13 filamentous bacteriophage is based on linking the desired activity to the production of pIII gene, a significant factor in phage infectious. *E. coli* host cell carries two plasmids: an accessory plasmid (AP), which links specific activity of protein to phage infectivity through expression of gene III and pIII production, and a mutagenesis plasmid (MP), which provides arabinose-induced elevated levels of mutagenesis. The third main component of PACE system is SP (selection phage), which encodes an evolving protein capable of linking protein property to expression of gene III from AP. SP usually carries an antibiotic resistance marker, allowing for selection of host cells carrying the bacteriophage. PACE is usually performed in a fixed-volume vessel. The functional variants lead to the pIII production which results in an infectious M13 phage and continuous process of mutagenesis and gene amplification. The *E. coli* cells carrying AP with mutations that do not lead to the pIII production (undesired phenotype) are eliminated through the constant flow out of the vessel. Source: Source: Adapted from Esvelt et al. [101].

which encodes an evolving protein capable of linking protein property to expression of gene III from AP (Figure 4.4). SP usually carries an antibiotic resistance marker, allowing for selection of host cells carrying the bacteriophage. Activity or selectivity of different classes of proteins such as polymerases, proteases, and genome-editing proteins has been evolved using PACE [102]. Dozens of rounds of selection can be performed in a fixed-volume vessel using PACE with minimal personal supervision [1].

Site-specific incorporation of noncanonical amino acids (ncAA) to proteins requires specific orthogonal aminoacyl-tRNA synthetases (AARSs) for each designed ncAA. These engineered AARSs are required to have high selectivity for orthogonal tRNA (o-tRNA) versus endogenous RNA for *in vivo* incorporation of ncAA. In directed evolution of AARSs, several sequential rounds of positive and negative screening are performed to ensure this high selectivity. To overcome the routine challenge of reducing AARS activity after rounds of screening, rapid selection of library of orthogonal AARSs using PACE provided highly selective and active AARSs for site-specific installation of ncAA [100].

Bacillus thuringiensis toxin (Bt toxin) was also evolved using PACE through 500 generations of mutagenic replication to overcome the issue of insect resistance, which is common for these endotoxins [102], proving PACE potential as a robust protein engineering tool for directed evolution of protein binding interactions. This extensive level of mutagenesis and screen is hard to achieve through any other directed evolution method. Bt toxin Cry1Ac protein variants, screened through PACE selection, provided up to 335-fold increase in insecticidal activity against sensitive and Bt-resistant insect larvae by binding to a cadherin-like receptor from the insect pest *Trichoplusia ni* (TnCAD) that is not natively bound by wild-type Cry1Ac.

Using a bacterial two-hybrid system, Badran et al. established a system to link protein binding to gene III expression and infection of progeny phage [102]. In order to overcome the challenge of having a sensitive system for detecting protein–protein interactions using PACE, several factors of bacterial two-hybrid system such as transcriptional activation and DNA-binding domains, protein expression level, and interaction binding affinity were optimized, yielded to improving the transcriptional activity up to 17-fold compared with a previously reported system (HA4 monobody and SH2 domain of ABL1 kinase) to more than 200-fold using the same pair of proteins. This improved bacterial two-hybrid system was recruited to evolve a HA4 monobody mutant (Y87A) in 48h using PACE, where phage clones survived in lagoons subjected to mutagenesis carried mutations of Ala87 to either Tyr or Trp. Bt toxin Cry1Ac mutants targeting toxin binding regions (TBRs) of cadherin-like receptor from the insect (TnCAD) were subjected to PACE evolution by varying mutagenesis level and stringency [102]. Bt toxin Cry1Ac variants isolated after 528 hours possessed up to 22 consensus mutations, most of them located at domain II, the predicted cadherin-binding domain of Cry1Ac. While most mutations were important in improving binding to cadherin, a subset of these mutations was contributing to decreasing stability of variants [102].

Despite several unique features of PACE such as rapid evolution of proteins, this method has not been widely used due to some limitations. PACE is not easily adaptable for engineering different proteins due to the complex system containing multiple components for linking protein phenotype to phage propagation. Moreover, the need for the assembly and maintaining the bacterial cultures system in large fixed volume flasks cannot easily adapted to every lab space.

4.5 Conclusion and Outlook

Tremendous advance in cell surface engineering and directed evolution has now enabled the rapid optimization of protein and enzyme properties in a high-throughput manner. However, the ability to rapidly co-evolve multiple enzyme phenotypes remains elusive. Moving forward, new strategies must be devised to provide the ability to evaluate several enzyme properties simultaneously. This will likely rely on improved design of synthetic libraries combining both computation modeling and rational prediction in order to survey a boarder set of residues of interest.

References

1 Packer, M.S. and Liu, D.R. (2015). Methods for the directed evolution of proteins. *Nat. Rev. Genet.* 16 (7): 379–394.

2 Tizei, P.A., Csibra, E., Torres, L., and Pinheiro, V.B. (2016). Selection platforms for directed evolution in synthetic biology. *Biochem. Soc. Trans.* 44 (4): 1165–1175.

3 Frei, J.C. and Lai, J.R. (2016). Protein and antibody engineering by phage display. *Methods Enzymol.* 580: 45–87.

4 Bazan, J., Calkosinski, I., and Gamian, A. (2012). Phage display -a powerful technique for immunotherapy: 1. Introduction and potential of therapeutic applications. *Hum. Vaccin. Immunother.* 8 (12): 1817–1828.

5 Francisco, J.A., Campbell, R., Iverson, B.L., and Georgiou, G. (1993). Production and fluorescence-activated cell sorting of *Escherichia coli* expressing a functional antibody fragment on the external surface. *Proc. Natl. Acad. Sci.* 90 (22): 10444–10448.

6 Smith, M.R., Khera, E., and Wen, F. (2015). Engineering novel and improved biocatalysts by cell surface display. *Ind. Eng. Chem. Res.* 54 (16): 4021–4032.

7 Daleke-Schermerhorn, M.H., Felix, T., Soprova, Z. et al. (2014). Decoration of outer membrane vesicles with multiple antigens by using an autotransporter approach. *Appl. Environ. Microbiol.* 80 (18): 5854–5865.

8 Shimazu, M., Mulchandani, A., and Chen, W. (2001). Cell surface display of organophosphorus hydrolase using ice nucleation protein. *Biotechnol. Progr.* 17 (1): 76–80.

9 Chen, W. and Georgiou, G. (2002). Cell-surface display of heterologous proteins: from high-throughput screening to environmental applications. *Biotechnol. Bioeng.* 79 (5): 496–503.

10 Wu, I.L., Narayan, K., Castaing, J.-P. et al. (2015). A versatile nano display platform from bacterial spore coat proteins. *Nat. Commun.* 6: 6777.

11 Raeeszadeh-Sarmazdeh, M., Greene, K.A., Sankaran, B. et al. (2019). Directed evolution of the metalloproteinase inhibitor TIMP-1 reveals that its N- and C-terminal domains cooperate in matrix metalloproteinase recognition. *J. Biol. Chem.*

12 Pavoor, T.V., Wheasler, J.A., Kamat, V., and Shusta, E.V. (2012). An enhanced approach for engineering thermally stable proteins using yeast display. *Protein Eng. Des. Sel.* 25 (10): 625–630.

13 Boder, E.T. and Wittrup, K.D. (1997). Yeast surface display for screening combinatorial polypeptide libraries. *Nat. Biotechnol.* 15 (6): 553–557.

14 Kimura, R.H., Cheng, Z., Gambhir, S.S., and Cochran, J.R. (2009). Engineered knottin peptides: a new class of agents for imaging integrin expression in living subjects. *Cancer Res.* 69 (6): 2435–2442.

15 Gera, N., Hussain, M., Wright, R.C., and Rao, B.M. (2011). Highly stable binding proteins derived from the hyperthermophilic Sso7d scaffold. *J. Mol. Biol.* 409 (4): 601–616.

16 Smith, G.P. (1985). Filamentous fusion phage: novel expression vectors that display cloned antigens on the virion surface. *Science* 228 (4705): 1315–1317.

17 Clackson, T., Hoogenboom, H.R., Griffiths, A.D., and Winter, G. (1991). Making antibody fragments using phage display libraries. *Nature* 352 (6336): 624–628.

18 Chang, H.J. and Yang, A.S. (2014). Design of phage-displayed cystine-stabilized mini-protein libraries for proteinaceous binder engineering. *Methods Mol. Biol.* 1088: 1–17.

19 Isalan, M. and Choo, Y. (2001). Engineering nucleic acid-binding proteins by phage display. *Methods Mol. Biol.* 148: 417–429.

20 O'Neil, K.T. and Hoess, R.H. (1995). Phage display: protein engineering by directed evolution. *Curr. Opin. Struct. Biol.* 5 (4): 443–449.

21 Weber, M., Bujak, E., Putelli, A. et al. (2014). A highly functional synthetic phage display library containing over 40 billion human antibody clones. *PLoS One* 9 (6): e100000.

22 Koellhoffer, J.F., Chen, G., Sandesara, R.G. et al. (2012). Two synthetic antibodies that recognize and neutralize distinct proteolytic forms of the ebola virus envelope glycoprotein. *ChemBioChem* 13 (17): 2549–2557.

23 Chen, G. and Sidhu, S.S. (2014). Design and generation of synthetic antibody libraries for phage display. *Methods Mol. Biol.* 1131: 113–131.

24 Frei, J.C., Kielian, M., and Lai, J.R. (2015). Comprehensive mapping of functional epitopes on dengue virus glycoprotein E DIII for binding to broadly neutralizing antibodies 4E11 and 4E5A by phage display. *Virology* 485: 371–382.

25 Huovinen, T., Syrjanpaa, M., Sanmark, H. et al. (2014). The selection performance of an antibody library displayed on filamentous phage coat proteins p9, p3 and truncated p3. *BMC Res. Notes* 7: 661.

26 Jamieson, A.C., Kim, S.H., and Wells, J.A. (1994). In vitro selection of zinc fingers with altered DNA-binding specificity. *Biochemistry* 33 (19): 5689–5695.

27 Wells, J.A., Cunningham, B.C., Fuh, G. et al. (1993). The molecular basis for growth hormone-receptor interactions. *Recent Prog. Horm. Res.* 48: 253–275.

28 Beste, G., Schmidt, F.S., Stibora, T., and Skerra, A. (1999). Small antibody-like proteins with prescribed ligand specificities derived from the lipocalin fold. *Proc. Natl. Acad. Sci. U.S.A.* 96 (5): 1898–1903.

29 Hufton, S.E., van Neer, N., van den Beuken, T. et al. (2000). Development and application of cytotoxic T lymphocyte-associated antigen 4 as a protein scaffold for the generation of novel binding ligands. *FEBS Lett.* 475 (3): 225–231.

30 Piotukh, K., Geltinger, B., Heinrich, N. et al. (2011). Directed evolution of sortase A mutants with altered substrate selectivity profiles. *J. Am. Chem. Soc.* 133 (44): 17536–17539.

31 Schmohl, L., Bierlmeier, J., Gerth, F. et al. (2017). Engineering sortase A by screening a second-generation library using phage display. *J. Pept. Sci.* 23 (7-8): 631–635.

32 Wu, C.H., Liu, I.J., Lu, R.M., and Wu, H.C. (2016). Advancement and applications of peptide phage display technology in biomedical science. *J. Biomed. Sci.* 23: 8.

33 Park, M., Sun, Q., Liu, F. et al. (2014). Positional assembly of enzymes on bacterial outer membrane vesicles for cascade reactions. *PLoS One* 9 (5): 6.

34 Cherf, G.M. and Cochran, J.R. (2015). Applications of yeast surface display for protein engineering. *Methods Mol. Biol.* 1319: 155–175.

35 Mata-Fink, J., Kriegsman, B., Yu, H.X. et al. (2013). Rapid conformational epitope mapping of anti-gp120 antibodies with a designed mutant panel displayed on yeast. *J. Mol. Biol.* 425 (2): 444–456.

36 Boder, E.T., Raeeszadeh-Sarmazdeh, M., and Price, J.V. (2012). Engineering antibodies by yeast display. *Arch. Biochem. Biophys.* 526 (2): 99–106.

37 Ueda, M. (2016). Establishment of cell surface engineering and its development. *Biosci. Biotechnol., Biochem.* 80 (7): 1243–1253.

38 Kuroda, K. and Ueda, M. (2014). Generation of arming yeasts with active proteins and peptides via cell surface display system: cell surface engineering, bio-arming technology. *Methods Mol. Biol.* 1152: 137–155.

39 Pepper, L.R., Cho, Y.K., Boder, E.T., and Shusta, E.V. (2008). A decade of yeast surface display technology: where are we now? *Comb. Chem. High Throughput Screen.* 11 (2): 127–134.

40 Boder, E.T. and Jiang, W. (2011). Engineering antibodies for cancer therapy. *Annu. Rev. Chem. Biomol. Eng.* 2: 53–75.

41 Shusta, E.V., Kieke, M.C., Parke, E. et al. (1999). Yeast polypeptide fusion surface display levels predict thermal stability and soluble secretion efficiency. *J. Mol. Biol.* 292 (5): 949–956.

42 Raeeszadeh-Sarmazdeh, M., Patel, N., Cruise, S. et al. (2018). Identifying stable fragments of arabidopsis thaliana cellulose synthase subunit 3 by yeast display. *Biotechnol. J.*: e1800353.

43 Shusta, E.V., Holler, P.D., Kieke, M.C. et al. (2000). Directed evolution of a stable scaffold for T-cell receptor engineering. *Nat. Biotechnol.* 18 (7): 754–759.

44 Traxlmayr, M.W. and Shusta, E.V. (2017). Directed evolution of protein thermal stability using yeast surface display. *Methods Mol. Biol.* 1575: 45–65.

45 Gera, N., Hussain, M., and Rao, B.M. (2013). Protein selection using yeast surface display. *Methods* 60 (1): 15–26.

46 Konning, D. and Kolmar, H. (2018). Beyond antibody engineering: directed evolution of alternative binding scaffolds and enzymes using yeast surface display. *Microb. Cell Fact.* 17 (1): 32.

47 Bowers, P.M., Horlick, R.A., Kehry, M.R. et al. (2014). Mammalian cell display for the discovery and optimization of antibody therapeutics. *Methods* 65 (1): 44–56.

48 Zhang, J., Zhang, X., Liu, Q. et al. (2014). Mammalian cell display for rapid screening scFv antibody therapy. *Acta Biochim. Biophys. Sin.* 46 (10): 859–866.

49 Ho, M. and Pastan, I. (2009). Mammalian cell display for antibody engineering. *Methods Mol. Biol.* 525: 337, xiv–352.

50 Ho, M. and Pastan, I. (2009). Display and selection of scFv antibodies on HEK-293T cells. *Methods Mol. Biol.* 562: 99–113.

51 Forrer, P., Jung, S., and Pluckthun, A. (1999). Beyond binding: using phage display to select for structure, folding and enzymatic activity in proteins. *Curr. Opin. Struct. Biol.* 9 (4): 514–520.

52 Yang, G. and Withers, S.G. (2009). Ultrahigh-throughput FACS-based screening for directed enzyme evolution. *ChemBioChem* 10 (17): 2704–2715.

53 Olsen, M.J., Stephens, D., Griffiths, D. et al. (2000). Function-based isolation of novel enzymes from a large library. *Nat. Biotechnol.* 18 (10): 1071–1074.

54 Olsen, M.J., Gam, J., Iverson, B.L., and Georgiou, G. (2003). High-throughput FACS method for directed evolution of substrate specificity. *Methods Mol. Biol.* 230: 329–342.

55 Chen, B., Lim, S., Kannan, A. et al. (2016). High-throughput analysis and protein engineering using microcapillary arrays. *Nat. Chem. Biol.* 12 (2): 76–81.

56 Lane, M.D. and Seelig, B. (2014). Advances in the directed evolution of proteins. *Curr. Opin. Chem. Biol.* 22: 129–136.

57 Koryakina, I., Neville, J., Nonaka, K. et al. (2011). A high-throughput screen for directed evolution of the natural product sulfotransferase LipB. *J. Biomol. Screening* 16 (8): 845–851.

58 Yu, J.S., Pertusi, D.A., Adeniran, A.V., and Tyo, K.E. (2017). CellSort: a support vector machine tool for optimizing fluorescence-activated cell sorting and reducing experimental effort. *Bioinformatics* 33 (6): 909–916.

59 Ackerman, M., Levary, D., Tobon, G. et al. (2009). Highly avid magnetic bead capture: an efficient selection method for de novo protein engineering utilizing yeast surface display. *Biotechnol. Progr.* 25 (3): 774–783.

60 Yeung, Y.A. and Wittrup, K.D. (2002). Quantitative screening of yeast surface-displayed polypeptide libraries by magnetic bead capture. *Biotechnol. Progr.* 18 (2): 212–220.

61 Varadarajan, N., Gam, J., Olsen, M.J. et al. (2005). Engineering of protease variants exhibiting high catalytic activity and exquisite substrate selectivity. *Proc. Natl. Acad. Sci. U.S.A.* 102 (19): 6855–6860.

62 Dorr, B.M., Ham, H.O., An, C. et al. (2014). Reprogramming the specificity of sortase enzymes. *Proc. Natl. Acad. Sci. U.S.A.* 111 (37): 13343–13348.

63 Lipovsek, D., Antipov, E., Armstrong, K.A. et al. (2007). Selection of horseradish peroxidase variants with enhanced enantioselectivity by yeast surface display. *Chem. Biol.* 14 (10): 1176–1185.

64 Isticato, R., Cangiano, G., Tran, H.T. et al. (2001). Surface display of recombinant proteins on *Bacillus subtilis* spores. *J. Bacteriol.* 183 (21): 6294–6301.

65 Ostafe, R., Prodanovic, R., Nazor, J., and Fischer, R. (2014). Ultra-high-throughput screening method for the directed evolution of glucose oxidase. *Chem. Biol.* 21 (3): 414–421.

66 Yi, L., Gebhard, M.C., Li, Q. et al. (2013). Engineering of TEV protease variants by yeast ER sequestration screening (YESS) of combinatorial libraries. *Proc. Natl. Acad. Sci. U.S.A.* 110 (18): 7229–7234.

67 Chen, I., Dorr, B.M., and Liu, D.R. (2011). A general strategy for the evolution of bond-forming enzymes using yeast display. *Proc. Natl. Acad. Sci. U.S.A.* 108 (28): 11399–11404.

68 Lim, S., Glasgow, J.E., Filsinger Interrante, M. et al. (2017). Dual display of proteins on the yeast cell surface simplifies quantification of binding interactions and enzymatic bioconjugation reactions. *Biotechnol. J.* 12 (5).

69 Jiang, W. and Boder, E.T. (2010). High-throughput engineering and analysis of peptide binding to class II MHC. *Proc. Natl. Acad. Sci. U.S.A.* 107 (30): 13258–13263.

70 Wang, A.J.A., Mulchandani, A., and Chen, W. (2002). Specific adhesion to cellulose and hydrolysis of organophosphate nerve agents by a genetically engineered *Escherichia coli* strain with a surface-expressed cellulose-binding domain and organophosphorus hydrolase. *Appl. Environ. Microbiol.* 68 (4): 1684–1689.

71 Fujita, Y., Ito, J., Ueda, M. et al. (2004). Synergistic saccharification, and direct fermentation to ethanol, of amorphous cellulose by use of an engineered yeast strain codisplaying three types of cellulolytic enzyme. *Appl. Environ. Microbiol.* 70 (2): 1207–1212.

72 Wang, A.A., Chen, W., and Mulchandani, A. (2005). Detoxification of organophosphate nerve agents by immobilized dual functional biocatalysts in a cellulose hollow fiber bioreactor. *Biotechnol. Bioeng.* 91 (3): 379–386.

73 Yang, C., Zhu, Y.R., Yang, J.J. et al. (2008). Development of an autofluorescent whole-cell biocatalyst by displaying dual functional moieties on *Escherichia coli* cell surfaces and construction of a coculture with organophosphate-mineralizing activity. *Appl. Environ. Microbiol.* 74 (24): 7733–7739.

74 Yang, C., Zhao, Q., Liu, Z. et al. (2008). Cell surface display of functional macromolecule fusions on *Escherichia coli* for development of an autofluorescent whole-cell biocatalyst. *Environ. Sci. Technol.* 42 (16): 6105–6110.

75 Fujita, Y., Takahashi, S., Ueda, M. et al. (2002). Direct and efficient production of ethanol from cellulosic material with a yeast strain displaying cellulolytic enzymes. *Appl. Environ. Microbiol.* 68 (10): 5136–5141.

76 Chen, X.-A., Ishida, N., Todaka, N. et al. (2010). Promotion of efficient saccharification of crystalline cellulose by aspergillus fumigatus Swo1. *Appl. Environ. Microbiol.* 76 (8): 2556–2561.

77 Nakatani, Y., Yamada, R., Ogino, C., and Kondo, A. (2013). Synergetic effect of yeast cell-surface expression of cellulase and expansin-like protein on direct ethanol production from cellulose. *Microb. Cell Fact.* 12: 66.

78 Bayer, E.A., Lamed, R., White, B.A., and Flint, H.J. (2008). From cellulosomes to cellulosomics. *Chem.Rec.* 8 (6): 364–377.

79 Bayer, E.A., Belaich, J.P., Shoham, Y., and Lamed, R. (2004). The cellulosomes: multienzyme machines for degradation of plant cell wall polysaccharides. *Annu. Rev. Microbiol.* 58: 521–554.

80 Lu, Y., Zhang, Y.-H.P., and Lynd, L.R. (2006). Enzyme–microbe synergy during cellulose hydrolysis by Clostridium thermocellum. *Proc. Natl. Acad. Sci. U.S.A.* 103 (44): 16165–16169.
81 Morais, S., Morag, E., Barak, Y. et al. (2012). Deconstruction of lignocellulose into soluble sugars by native and designer cellulosomes. *MBio* 3 (6).
82 Fierobe, H.-P., Mingardon, F., Mechaly, A. et al. (2005). Action of designer cellulosomes on homogeneous versus complex substrates: controlled incorporation of three distinct enzymes into a defined trifunctional scaffoldin. *J. Biol. Chem.* 280 (16): 16325–16334.
83 Tsai, S.L., Oh, J., Singh, S. et al. (2009). Functional assembly of minicellulosomes on the *Saccharomyces cerevisiae* cell surface for cellulose hydrolysis and ethanol production. *Appl. Environ. Microbiol.* 75 (19): 6087–6093.
84 Tsai, S.L., Goyal, G., and Chen, W. (2010). Surface display of a functional minicellulosome by intracellular complementation using a synthetic yeast consortium and its application to cellulose hydrolysis and ethanol production. *Appl. Environ. Microbiol.* 76 (22): 7514–7520.
85 Tsai, S.L., DaSilva, N.A., and Chen, W. (2013). Functional display of complex cellulosomes on the yeast surface via adaptive assembly. *ACS Synth. Biol.* 2 (1): 14–21.
86 Wen, F., Sun, J., and Zhao, H. (2010). Yeast surface display of trifunctional minicellulosomes for simultaneous saccharification and fermentation of cellulose to ethanol. *Appl. Environ. Microbiol.* 76 (4): 1251–1260.
87 Goyal, G., Tsai, S.L., Madan, B. et al. (2011). Simultaneous cell growth and ethanol production from cellulose by an engineered yeast consortium displaying a functional mini-cellulosome. *Microb. Cell Fact.* 10: 8.
88 Fan, L.-H., Zhang, Z.-J., Yu, X.-Y. et al. (2012). Self-surface assembly of cellulosomes with two miniscaffoldins on *Saccharomyces cerevisiae* for cellulosic ethanol production. *Proc. Natl. Acad. Sci. U.S.A.* 109 (33): 13260–13265.
89 You, C., Myung, S., and Zhang, Y.H.P. (2012). Facilitated substrate channeling in a self-assembled trifunctional enzyme complex. *Angew. Chem. Int. Ed.* 51 (35): 8787–8790.
90 Liu, F., Banta, S., and Chen, W. (2013). Functional assembly of a multi-enzyme methanol oxidation cascade on a surface-displayed trifunctional scaffold for enhanced NADH production. *Chem. Commun.* 49 (36): 3766–3768.
91 Han, L., Liang, B., and Song, J. (2018). Rational design of engineered microbial cell surface multi-enzyme co-display system for sustainable NADH regeneration from low-cost biomass. *J. Ind. Microbiol. Biotechnol.* 45 (2): 111–121.
92 You, C., Zhang, X.-Z., Sathitsuksanoh, N. et al. (2012). Enhanced microbial utilization of recalcitrant cellulose by an ex vivo cellulosome-microbe complex. *Appl. Environ. Microbiol.* 78 (5): 1437–1444.
93 Anderson, T.D., Robson, S.A., Jiang, X.W. et al. (2011). Assembly of minicellulosomes on the surface of *Bacillus subtilis*. *Appl. Environ. Microbiol.* 77 (14): 4849–4858.

94 Chen, Q., Rozovsky, S., and Chen, W. (2017). Engineering multi-functional bacterial outer membrane vesicles as modular nanodevices for biosensing and bioimaging. *Chem. Commun.* 53 (54): 7569–7572.

95 Lim, S., Chen, B., Kariolis, M.S. et al. (2017). Engineering high affinity protein-protein interactions using a high-throughput microcapillary array platform. *ACS Chem. Biol.* 12 (2): 336–341.

96 Rouet, R., Jackson, K.J.L., Langley, D.B., and Christ, D. (2018). Next-generation sequencing of antibody display repertoires. *Front. Immunol.* 9: 118.

97 van Rosmalen, M., Janssen, B.M., Hendrikse, N.M. et al. (2017). Affinity maturation of a cyclic peptide handle for therapeutic antibodies using deep mutational scanning. *J. Biol. Chem.* 292 (4): 1477–1489.

98 Harris, D.T., Wang, N., Riley, T.P. et al. (2016). Deep mutational scans as a guide to engineering high affinity T cell receptor interactions with peptide-bound major histocompatibility complex. *J. Biol. Chem.* 291 (47): 24566–24578.

99 Badran, A.H. and Liu, D.R. (2015). In vivo continuous directed evolution. *Curr. Opin. Chem. Biol.* 24: 1–10.

100 Bryson, D.I., Fan, C., Guo, L.T. et al. (2017). Continuous directed evolution of aminoacyl-tRNA synthetases. *Nat. Chem. Biol.* 13 (12): 1253–1260.

101 Esvelt, K.M., Carlson, J.C., and Liu, D.R. (2011). A system for the continuous directed evolution of biomolecules. *Nature* 472 (7344): 499–503.

102 Badran, A.H., Guzov, V.M., Huai, Q. et al. (2016). Continuous evolution of *Bacillus thuringiensis* toxins overcomes insect resistance. *Nature* 533 (7601): 58–63.

5

Iterative Saturation Mutagenesis for Semi-rational Enzyme Design

Ge Qu[1], Zhoutong Sun[1], and Manfred T. Reetz[1,2,3]

[1] *Tianjin Institute of Industrial Biotechnology, Chinese Academy of Sciences, 32 West 7th Avenue, Tianjin Airport Economic Area, Tianjin, 300308, China*
[2] *Philipps-University, Department of Chemistry, Hans-Meerwein-Strasse 4, Marburg, 35032, Germany*
[3] *Max-Planck-Institut für Kohlenforschung, Kaiser-Wilhelm-Platz 1, Mülheim, 45470, Germany*

5.1 Introduction

Chemo-, stereo-, and regioselectivities are central themes in sustainable synthetic organic chemistry, catalytic rather than stoichiometric processes being preferred. When opting for biocatalysis, organic chemists and biotechnologists are generally interested in the selective catalytic transformation of non-natural compounds as substrates, and indeed numerous industrial processes have been reported [1]. For many decades, the major disadvantages of enzyme-based technologies when using non-natural substrates included the following often observed limitations:

- Narrow substrate scope (insufficient activity)
- Poor or wrong enantio- or diastereoselectivity
- Poor or wrong regioselectivity
- Insufficient robustness under operating conditions
- Enzymes cannot catalyze most of the transition metal mediated reaction types that organic chemists have developed over the years.

During the last two decades, directed evolution of enzymes has developed to a point where all of these problems can be addressed and usually solved [2]. Directed evolution consists of recursive cycles of mutagenesis/expression/screening (or selection), the most often used gene mutagenesis techniques being error-prone polymerase chain reaction (epPCR; a shotgun method), saturation mutagenesis (SM, focused randomization), and DNA shuffling (a recombinant technique) [2]. As will be seen in this chapter, it can also be used to manipulate the stereoselectivity of artificial metalloenzymes. The use of epPCR in directed evolution was first reported by Arnold and coworker in successful attempts to enhance the resistance of a protease to a hostile organic solvent [3]. Thereafter, Reetz et al. demonstrated that it is possible to manipulate the enantioselectivity of an enzyme (lipase from *Pseudomonas aeruginosa*) by applying four rounds of epPCR [4]. However, it soon

Protein Engineering: Tools and Applications, First Edition. Edited by Huimin Zhao.
© 2021 WILEY-VCH GmbH. Published 2021 by WILEY-VCH GmbH.

became clear that multiple rounds of epPCR are not optimal, certainly not for evolving enantioselectivity of this enzyme. Nevertheless, today we note that epPCR has been used many times for a variety of different goals by many groups [2], and we do not suggest that it should be ignored. In 2001 we tested combinations of DNA shuffling and SM at residues lining the binding pocket [5], which improved stereoselectivity significantly. Other groups then joined efforts to generalize directed evolution of stereoselectivity in this new and exciting research area [2, 6]. It was clear that the screening step constitutes the labor-intensive part (bottleneck of directed evolution) [7], but limited attention was paid to efficiency. Various display systems can handle much larger libraries of up to $\sim 10^8$ members [7b], but generally they cannot be used to evolve stereoselectivity [7a–c], nor has the droplet-based microfluidic technology solved this problem [7d]. Producing high-quality (smart) libraries that require a minimum of screening emerged as a central challenge [2, 6].

Recalling that SM at a site lining the binding pocket leads to notable enantioselectivity improvement [5], this approach was later generalized and dubbed combinatorial active-site saturation test (CAST) [8]. The term distinguishes it from focused randomization at other (remote) sites for different purposes such as enhancing oxidative robustness [9], thermostabilization [10], or inducing allostery in laboratory evolution [11]. Using an enzyme's crystal structure or homology model, first and second sphere CAST residues lining the binding pocket are identified (Figure 5.1a). Then these can be used as single residue randomization sites, or they can be grouped into multiple-residue randomization sites. In the original CAST/ISM version for controlling stereoselectivity [12], the choice was arbitrary, as was the decision which pathway to explore. Later it was shown that the experimenter can select those residues that appear to be particularly important as suggested by structural and consensus data (Figure 5.1b). Docking the substrate into the binding pocket and performing molecular dynamics (MD) simulations also constitute useful aids, because these techniques generally reveal which amino acid sidechains interact most strongly with the substrate. If single residue sites are chosen, the respective individual SM experiments are called single-site saturation mutagenesis (SSM), as opposed to SM at multi-residue randomization sites designed by appropriate residue-grouping, in which case the term combinatorial saturation mutagenesis (CSM) is appropriate (or simply saturation mutagenesis (SM)) [2].

When the initial mutant libraries, produced either by SSM or CSM, do not provide sufficiently improved mutants, iterative saturation mutagenesis (ISM) can be used for optimizing stereoselectivity and/or activity (Figure 5.1c) [12]. Later the process of ISM was also applied in the manipulation of regioselectivity likewise using CAST residues [14] and to the enhancement of thermostability using SM at remote residues characterized by high *B*-factors [10, 15]. As the size of a randomization site increases, so does oversampling in the screening step for 95% library coverage (or any other percent of library coverage). Using the CASTER computer aid [15] (see Reetz homepage: kofo.mpg.de/en/research/biocatalysis), which is based on the Patrick–Firth algorithm [16], the respective numbers for any percent-coverage are readily accessible, as in the case of NNK codon degeneracy or when applying reduced amino acid alphabets such as NDT (Table 5.1) [15, 17].

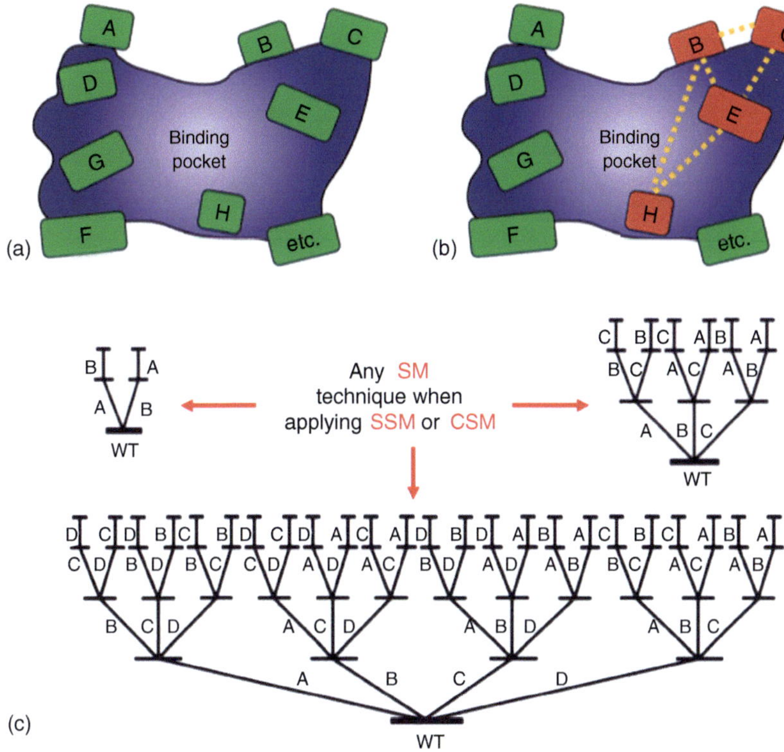

Figure 5.1 Generalization of the CAST/ISM-strategy for manipulating stereoselectivity and substrate scope of an enzyme (activity) [12, 13]. (a) Individual residues lining the binding pocket of an enzyme (CAST residues marked in green); (b) Scheme indicating that it is not necessary to consider all of the identified CAST residues, the red-marked residues being an example of using only a select few; (c) Three cases of iterative saturation mutagenesis (ISM), 2 upward pathways when choosing 2 randomization sites A and B (left), 6 pathways in the case of 3 such sites A, B, and C, and 24 pathways in the case of four such sites, A, B, C, and D (bottom). Source: Acevedo-Rocha et al. [13], Reetz et al. [12a], Reetz [12b].

It is interesting to point out that in an independent study describing enhancement of ligand affinity of a hormone receptor serving as the protein, Zhao and coworkers applied SSM at several rationally chosen residues in a recursive manner, and also combined single mutations and applied epPCR [19]. The chosen randomization residues are not CAST sites because enzyme activity or selectivity is not involved. While not using the term ISM, the authors show an intelligent way to fabricate gene switches. Similarly, ISM has been used to engineer cofactor specificity (NADH versus NADPH) [20].

In order to get acceptable results, it is not absolutely necessary to ensure 95% library coverage [5, 12a]. As already mentioned, reduced amino acid alphabets can be used in SM, which shortens the screening effort considerably [14, 17, 18]. The first example concerned NDT codon degeneracy (12 codons), which encodes 12 amino acids (Phe, Leu, Ile, Val, Tyr, His, Asn, Cys, Arg, Ser, Asp, and Gly), a “cocktail” of amino acids with hydrophobic/hydrophilic, polar/non-polar,

Table 5.1 Oversampling needed in order to ensure 95% library coverage in saturation mutagenesis as a function of NNK versus NDT codon degeneracy [1, 17, 18].

Number of amino acid positions at one site	NNK Codons	NNK Transformants needed	NDT codons	NDT Transformants needed
1	32	94	12	34
2	1 028	3 066	144	430
3	32 768	98 163	1 728	5 175
4	$>1.0\times10^6$	$>3.1\times10^6$	20 736	62 118
5	$>3.3\times10^7$	$>1.0\times10^8$	2 48 832	7 45 433
6	$>1.0\times10^9$	$>3.2\times10^9$	$>2.9\times10^6$	$>8.9\times10^6$
7	$>3.4\times10^{10}$	$>1.0\times10^{11}$	$>3.5\times10^7$	$>1.1\times10^8$
8	$>1.0\times10^{12}$	$>3.3\times10^{12}$	$>4.2\times10^8$	$>1.3\times10^9$
9	$>3.5\times10^{13}$	$>1.0\times10^{14}$	$>5.1\times10^9$	$>1.5\times10^{10}$
10	$>1.1\times10^{15}$	$>3.4\times10^{15}$	$>6.1\times10^{10}$	$>1.9\times10^{11}$

Sources: Drauz et al. [1a], Yang and Withers [7a], Acevedo-Rocha et al. [7b, 18b], Reymond [7c], Fallah-Araghi et al. [7d], Reetz [18a].

aromatic/aliphatic, and small/large sidechains [18, 21]. Here again the screening effort increases as the size of the randomization site increases (Table 5.1), but to a dramatically lesser degree. Another study suggested that it is better to aim for high library coverage at the expense of structural diversity by using a reduced amino acid library, rather than employing NNK codon degeneracy at very low library coverage [17a], which needs to be generalized. The utility of reduced amino acid alphabets in CAST/ISM has emerged as a major pillar in directed evolution of selective enzymes [14, 17, 18]. *Importantly, any molecular biological SM technique can be applied when choosing CAST/ISM* [22], including those based mainly on polymerase chain reaction (PCR) such as QuikChange [22a], MegaPrimer [22b], or overlap-extension PCR [22c]. Therefore, it is not logical to compare such techniques with the concept of CAST/ISM [13]. Finally, by performing deconvolution of multi-mutational variants, it was discovered quite early that CAST/ISM leads to strong cooperative mutational effects (more than traditional additivity) [17c]. The present chapter focuses on the most recent efforts in the quest to increase the efficacy of CAST/ISM.

5.2 Recent Methodology Developments in ISM-Based Directed Evolution

Before focusing on the most recent methodology developments of CAST/ISM, a few remarks concerning library quality are appropriate. The quality of an SM library (on DNA level) can be assessed rapidly but somewhat crudely by the quick quality control (QQC) based on pooling, plasmid extraction and sequencing [23a], or by the more precise but more labor-intensive method of Stewart in which so-called *Q*-values are determined [23b]. In these studies, improvements with respect to enzyme activity or selectivity were also measured (on protein level). A truly general technique for assessing quality on DNA level requires a different procedure utilizing massive DNA sequencing (Sections 5.2.2.1 and 5.2.2.2).

5.2.1 Choosing Reduced Amino Acid Alphabets Properly

In addition to NDT codon degeneracy [18, 21], smaller amino acid alphabets were also tested [14, 17, 18]. Two distinctly different strategies can be chosen: one and the same reduced amino acid alphabet for SM at the entire multi-residue randomization site, or a different amino acid alphabet at each residue of a multi-residue site [24]. Whatever reduced amino acid alphabet is chosen for SM, it can be used in two fundamentally different ways: One option is to apply it to ISM which is composed solely of one-residue randomization sites, which means that amino acid exchanges occur one at a time in the evolutionary upward climb using SSM. The other option is to employ a reduced amino acid alphabet in ISM at multi-residue randomization sites (CSM). This is a superior procedure due to the increased probability of cooperative effects [17c] operating between individual mutations and sets of mutations [2d, 18, 25].

Recently, the choice of three rationally chosen amino acids (in addition to WT), dubbed triple code saturation mutagenesis (TCSM) [26], has been shown to be a viable compromise between structural diversity and library coverage in several studies. How the rational decisions were made is revealed in Sections 5.2.1.1–5.2.1.3 in detail, because this particular system serves as a model for devising other "small" reduced amino acid alphabets for other enzymes.

5.2.1.1 Limonene Epoxide Hydrolase as the Catalyst in Hydrolytic Desymmetrization

In the first TCSM study, limonene epoxide hydrolase (LEH) from *Rhodococcus erythropolis* was chosen as the catalyst in the hydrolytic desymmetrization of cyclohexene oxide (**1**) with formation of (*R*,*R*)- and (*S*,*S*)-**2** (Scheme 5.1) [26]. WT LEH shows 4% ee favoring (*S*,*S*)-**2**. Guided by the crystal structure, **1** was docked in the binding pocket, and 10 residues were chosen for SM. NNK-based randomization would require for 95% library coverage the screening of $>3\times10^{15}$. When opting for TCSM, $\sim10^6$ transformants would have to be screened. Consequently, the 10 residues were grouped into three randomization sites: A (I80/V83/L114/I116), B (L74/M78/L147), and C (M32/L35/L103).

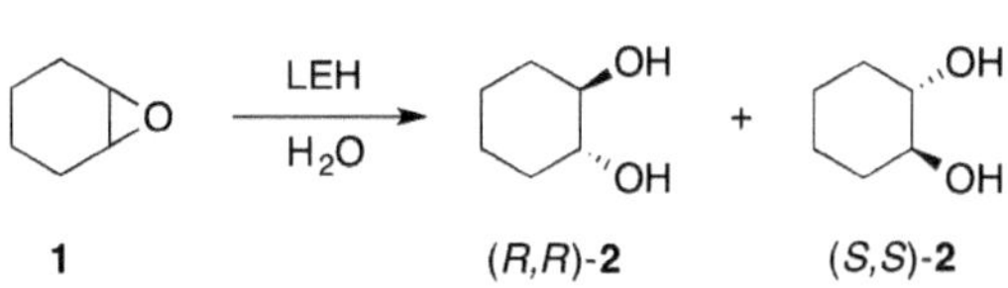

Scheme 5.1 LEH-catalyzed hydrolytic desymmetrization of substrate **1**. Source: Sun et al. [26]. © 2016, American Chemical Society.

Upon docking **1** into the crystal structure of LEH, the CAST residues were found to have hydrophobic character, calling for a triple code that encodes three likewise hydrophobic amino acids. In order to gain additional confidence in such a decision, the consensus approach [27] was tested using 100 related epoxide hydrolases [26]. Consequently, valine, phenylalanine, and tyrosine were chosen as combinatorial building blocks in SM at sites A, B, and C, requiring for 95% library coverage the screening of only 576, 192, and 192 transformants, respectively. This strategy worked better than previous CAST approaches using the same system, (*S*,*S*)-selective variants now being evolved with 97–99% ee, and one ISM step leading to 97% ee in favor of (*R*,*R*)-**2** [26]. Crystal structures of the best (*R*,*R*)- and (*S*,*S*)-selective mutants flanked by docking computations showed how the binding pocket was reshaped and unveiled intriguing mechanistic details [26, 28].

5.2.1.2 Alcohol Dehydrogenase TbSADH as the Catalyst in Asymmetric Transformation of Difficult-to-Reduce Ketones

It was not at all obvious how TCSM can be generalized, since most enzymes do not show a dominance of hydrophobic residues at catalytically active sites. A more difficult enzyme system was therefore chosen comprising the alcohol dehydrogenase (ADH) from *Thermoanaerobacter brockii* (TbSADH) as the catalyst in the enantioselective transformation of "difficult-to-reduce" prochiral ketones **3**, **5**, **7**, and **9** (Scheme 5.2) [29]. Such reductions are synthetically challenging because the best man-made chiral Ru-catalysts fail to deliver acceptable levels of enantioselectivity. In the TCSM study, the reduction of substrate **3** served as the model transformation; WT TbSADH shows slight preference for (*R*)-**4**, but (*S*)-**4** is the desired product due to application in the pharmaceutical industry.

In previous protein engineering of ADHs [2], substrates were employed in which the α- and α′-substituents flanking the carbonyl function differ considerably in size, enabling reasonable choices for mutagenesis, unlike ketone **3**. Docking this

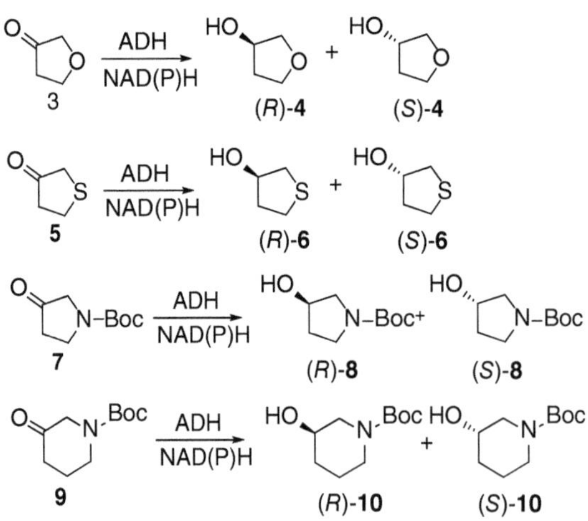

Scheme 5.2 TbSADH-catalyzed stereoselective transformations. Source: Sun et al. [29]. © 2016, American Chemical Society.

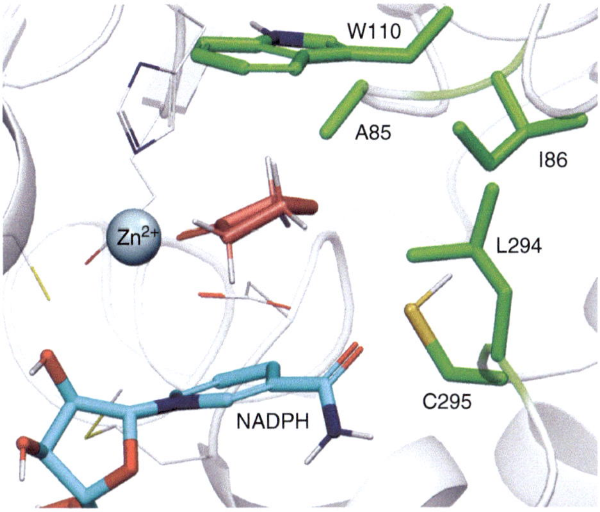

Figure 5.2 Five CAST residues A85, I86, W110, L294, and C295 in contact with substrate **3** in TbSADH. Source: Modified from Sun et al. [29].

substrate in the binding pocket revealed an ensemble of energetically equal, yet differently positioned conformers in which opposite π-faces of the carbonyl function point to the hydride source (NADPH). Figure 5.2 displays one of the conformers, five CAST-residues A85, I86, W110, L294, and C295 being in contact with the substrate.

In order to make an optimal choice of an appropriate triple code, *a straightforward strategy was developed, which can be used in the directed evolution of other enzyme types as well*: Exploratory NNK-based SM was performed individually at each of the five residues, a fast procedure showing that amino acid exchanges lead to the best improvement in (*R*)-selectivity and which ones induce reversed (*S*)-selectivity. At positions 85, 86, and 294, both improved (*R*)-selective and inverted (*S*)-selective single mutants were found, whereas the library generated at position 110 harbored only (*S*)-selective variants, and the library at position 295 contained only (*R*)-selective hits. Therefore, the five single residues were grouped into two different multi-residue randomization sites, the first one being: A (A85, I86, L294, and C295), anticipating upon SM the best (*R*)-selective hits when applying the triple code valine–asparagine–leucine (V–N–L). In exploratory NNK-based SM, these were the amino acids that induced the greatest increase in (R)-selectivity. Analogously, library B (A85, I86, W110, L294) was expected to harbor the best (S)-selective variants when applying triple code valine–glutamine–leucine (V–Q–L). In each case only 576 transformants had to be screened for 95% library coverage. This semi-rational analysis proved to be correct, ISM not being necessary (Figure 5.3). The best variants were also excellent catalysts in the asymmetric reduction of the other ketones **5**, **7**, and **9** without any further mutagenesis [29].

An alternative TCSM strategy would be to generate a single mutant library by applying V–N–L (or V–Q–L) to the large 5-residue SM site A85/I86/W110/L294/

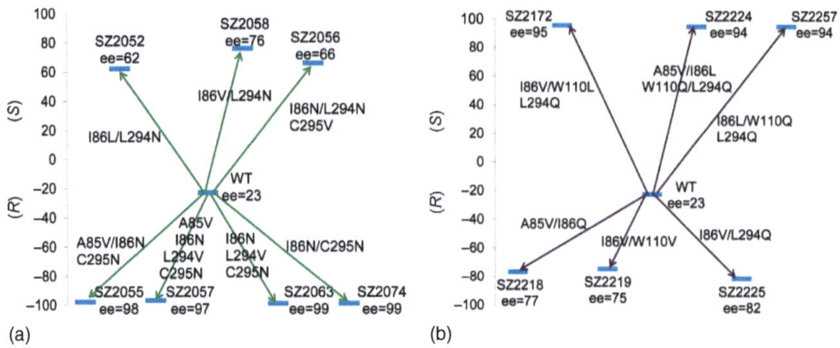

Figure 5.3 Best mutants obtained upon generating and screening (a) TbSADH library A using triple code V–N–L, and (b) TbSADH library B using triple code V–Q–L. Source: Sun et al. [29]. © 2016, American Chemical Society.

C295, but this would require more screening for 95% library coverage (3072 transformants). In view of previous studies indicating that full library coverage is not absolutely mandatory [5, 30] (although advantageous) [17a], screening less than 95% library coverage would be an option.

5.2.1.3 P450-BM3 as the Chemo- and Stereoselective Catalyst in a Whole-Cell Cascade Sequence

Another TCSM-study concerns the successful attempt to create designer cells for one-pot conversion of cyclohexane (**11**) into cyclohexane-1,2-diol in its three stereoisomeric forms (*R*,*R*)-**15**, (*S*,*S*)-**15**, and (*R*,*S*)-**15** (Figure 5.4) [31]. The ambitious plan was to construct *Escherichia coli* cells harboring P450-BM3 and an appropriate ADH. The first three P450-BM3 catalyzed steps leading to the desired acyloins (*R*)- and (*S*)-**14** required directed evolution, whereas the last step was ensured by WT ADHs. Since neither cyclohexane (**11**), cyclohexanol (**12**), nor cyclohexanone (**13**) are essentially accepted by WT P450-BM3, activity and not just regio- and enantioselectivity had to be evolved. We speculated that once active and selective mutants for the conversion of cyclohexanone (**13**) with formation of (*R*)- and (*S*)-**14** had been engineered, then they would also catalyze the first two steps **11** → **12** → **13**. P450-BM3 is a self-sufficient cytochrome P450 monooxygenase, which is active towards long-chain fatty acids and many other sterically large compounds, but it shows no or minimal activity towards small compounds [32].

TCSM was first applied to WT P450-BM3 in the reaction of cyclohexanone (**13**) in hope of evolving two active as well as regio- and stereoselective mutants for accessing (*R*)- and (*S*)-**14**, respectively [31]. Based on size considerations, it was anticipated that such mutants would also accept substrates **11** and **12**. Upon docking ketone **13** into the large binding pocket of P450-BM3 in vicinity of the catalytically active high spin species heme-Fe=O, at least two dozen CAST residues were identified, some of them being shown in Figure 5.5a. Remembering that in CAST/ISM not every residue lining the binding pocket needs to be considered (Figure 5.1b), we chose eight residues that were subjected to exploratory NNK-based SM (Figure 5.5a).

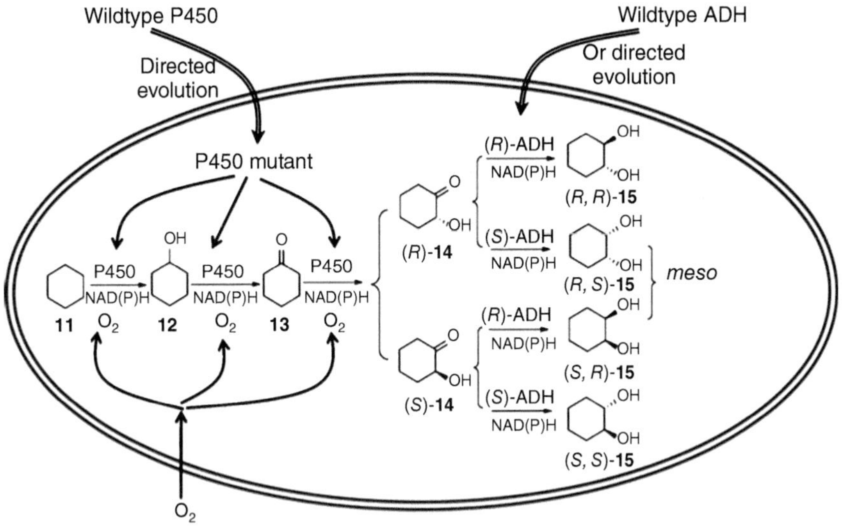

Figure 5.4 Construction of *E. coli* whole cells for producing either (*R*,*R*)-, (*S*,*S*)-, or *meso*-**15**, respectively, in a cascade sequence starting from cyclohexane (**11**). Source: Modified from Li et al. [31].

Following an analysis of the data analogous to the procedure in the ADH-study [29], two randomization sites were considered, A (L75/L181/I263/A264) and B (V78/A82/A328/T438). The exploratory NNK-data also served as a guide in defining a triple code for both sites: asparagine–isoleucine–phenylalanine (N–I–F), entailing upon SM the screening of 552 and 736 transformants, respectively, for 95% library coverage. The results showed that both libraries harbor (*R*)- and (*S*)-selective mutants (Figure 5.4b). In principle, several of them could be chosen for ISM with the aim of enhancing (*R*)- and (*S*)-selectivity, which proved to be the case (Figure 5.5c). The best stereochemically complementary mutants in *E. coli* cells were then shown to be active in the extended cascades leading selectively to the three stereoisomers of **15** (Figure 5.4) [31].

Summarizing all three TCSM-studies [26, 29, 31], a three-step guideline has emerged:

- Perform NNK-based SM at rationally chosen CAST residues for obtaining a limited mutational fingerprint at each position, e.g. identifying positive mutants with improved stereoselectivity, the choices of residues being guided by crystal structures, homology models, MD computations, and/or theoretical techniques that identify hotspots.
- On the basis of improved mutants, make a decision regarding the grouping of the identified single residues into appropriate medium-sized randomization sites for subsequent SM.
- Likewise based on the exploratory NNK-derived data, design an appropriate triple code that defines three amino acids as combinatorial building blocks (in addition to wildtype amino acids).

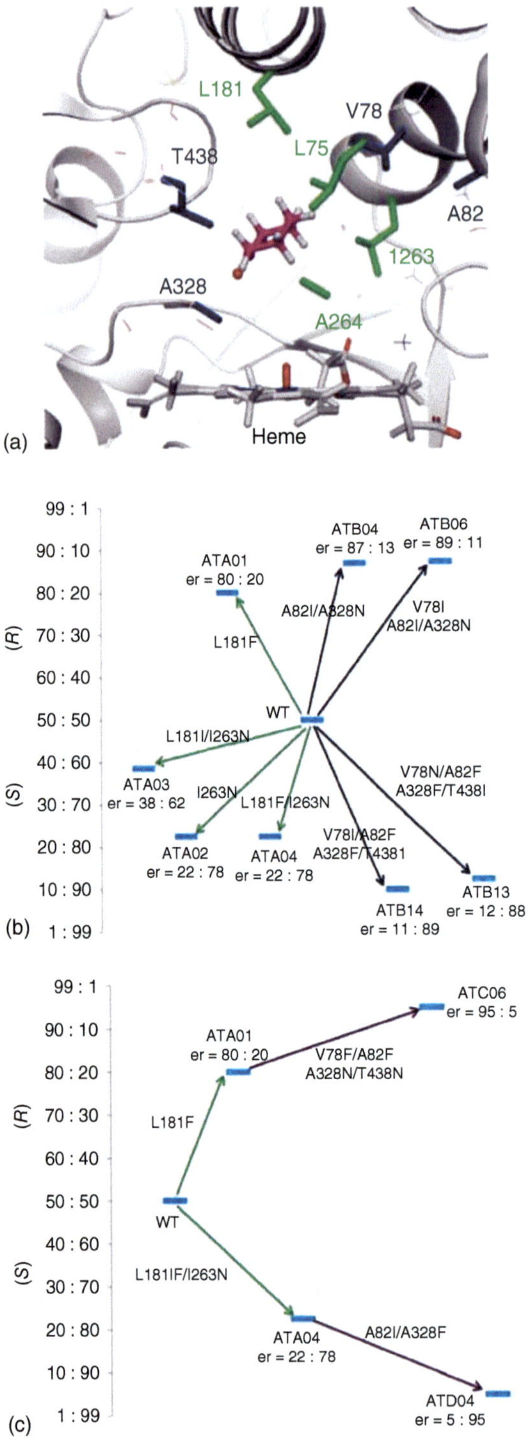

Figure 5.5 Directed evolution of P450-BM3 in the regio- and enantioselective oxidative hydroxylation of cyclohexanone (**13**). (a) Binding pocket of P450-BM3 with **13** docked near the catalytically active heme-Fe=O, featuring two four-residue randomization sites A (green) and B (blue); (b) best active P450-BM3 variants in the reaction providing (*R*)- and (*S*)-**14**; and (c) best ISM pathways for evolving (*R*)- and (*S*)-selective variants; green arrow: variants originating from initial library A; purple arrow: improved (*R*)- and (*S*)-selective variants resulting from ISM at site B. Source: Li et al. [31]. © 2016, John Wiley & Sons.

The three-step guideline, be it for TCSM or other codon choices encoding alternative reduced amino acid alphabets [33], reduces lab work considerably, but it considers only a single defined catalytic property. In the Section 5.2.1.4 this issue is addressed by utilizing mutability landscapes.

5.2.1.4 Multi-parameter Evolution Aided by Mutability Landscaping

The use of mutability landscaping [34] in directed evolution of regio- and stereoselectivity [35] is closely related to the three-step guideline recommended for optimal TCSM (or other reduced amino acid alphabets) [26, 29, 31], but it allows more than one catalytic parameter to be optimized. As before, the first step involves NNK-based SM at each CAST residue. But then a comprehensive fingerprint at each position is generated by extensive DNA sequencing and screening for more than one catalytic property [35]. This procedure requires additional experimental work as well as costs relative to the usual technique outlined earlier. It should therefore be applied primarily in particularly challenging situations, as in the case of *targeted* late-stage hydroxylation of steroids using P450-BM3, which requires activity, regio- and stereoselectivity [35]. In earlier work, we had already reported late-stage P450-BM3 catalyzed oxidation of steroids enabled by ISM, but it was not truly targeted in the strict sense of the word [36]. In that study, the well-known mutant F87A [32] was first tested in the hydroxylation of testosterone, which delivered a 1 : 1 mixture of the 2ß- and 15ß-alcohols. Then ISM was applied to obtain fully 2ß- and 15ß-selective variants, which required labor-intensive screening of 9000 transformants using automated high performance liquid chromatography (HPLC) [36]. While impressive, evolving these mutants was not our goal at the outset of the project, nor are the products of practical significance. Similar scenarios have been observed in other directed evolution studies of late-stage hydroxylation of natural products such as terpenes [37a, b] and alkaloids [37c, d]. Having been asked by pharmaceutical scientists whether we can evolve *highly active* C16α- and C16ß-selective P450-BM3 mutants for steroids in general, because the respective C16-alcohols are interesting corticoids, we set out to solve this difficult problem using testosterone as the model steroid (Scheme 5.3) [35]. The plan was to apply reduced amino acid alphabet in SM at a large CAST randomization site as guided by an appropriately generated mutability landscape.

The first step in mutability landscaping was exploratory NNK-based SM at 10 selected CAST residues, these being chosen on the basis of the P450-BM3 crystal structure. Following extensive DNA sequencing and multi-parameter screening

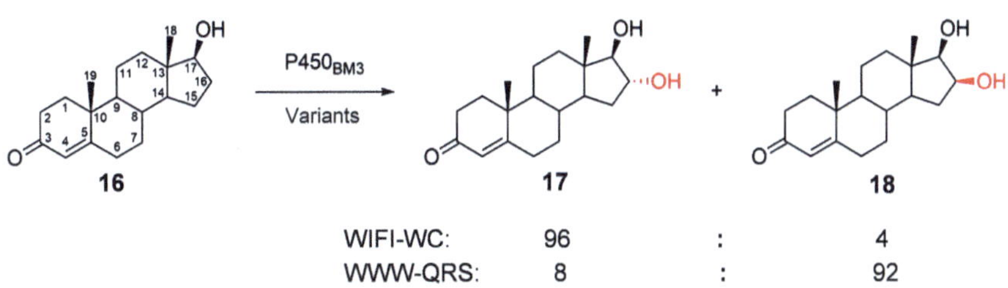

Scheme 5.3 Model reaction for P450-BM3 catalyzed C16-selective hydroxylation of testosterone (**16**). Source: Acevedo-Rocha et al. [35]. © 2018, American Chemical Society.

AA	W	F	M	L	I	V	A	C	G	P	S	T	Y	H	K	R	D	E	Q	N	Parameter (%)
R47	68	20	15	33	20	22	23	15	23	3	11	21	22	11	9	28	8	9	10	10	Testosterone conversion
T49	5	3	19	15	13	16	12	15	9	1	20	28	18	8	4	5	14	88	12	9	
Y51	31	38	20	20	20	17	3	5	1	7	3	7	28	4	7	1	17	19	6	5	
V78	4	14	26	23	8	28	14	29	6	0	7	3	4	0	0	0	2	0	1	2	
A82	19	42	69	76	44	31	28	19	19	2	18	35	4	17	9	1	30	61	57	38	
R47	47	59	55	54	54	59	58	60	58	0	58	49	54	57	56	55	57	58	60	56	2β-hydroxytestosterone
T49	49	48	57	51	57	57	56	60	59	0	62	55	50	55	56	59	62	63	52	53	
Y51	60	69	74	77	82	72	77	76	0	80	77	77	55	62	71	54	60	58	67	71	
V78	48	18	22	38	48	55	30	62	36	0	29	0	0	0	0	0	0	0	0	0	
A82	12	6	16	41	31	41	55	37	60	0	60	43	10	30	26	63	84	63	67	65	
R47	47	37	41	45	43	37	37	36	38	0	40	48	42	38	43	40	41	39	36	42	15β-hydroxytestosterone
T49	43	52	39	45	39	39	40	34	36	0	35	40	45	41	39	39	34	28	44	44	
Y51	39	28	22	18	15	25	0	0	0	13	16	16	40	32	25	43	36	37	26	24	
V78	39	75	71	60	48	40	64	29	60	0	62	0	0	0	0	0	0	0	0	0	
A82	45	87	77	50	63	53	40	57	35	0	35	52	71	65	65	35	13	24	26	30	
R47	3	1	2	1	1	1	1	1	1	0	1	3	2	0	0	3	0	0	1	1	16β-hydroxytestosterone
T49	6	0	2	1	1	1	1	1	0	0	1	3	3	0	0	0	0	0	1	0	
Y51	0	0	0	1	0	1	0	0	0	0	0	0	3	0	0	0	2	2	0	0	
V78	11	5	5	1	0	3	5	2	2	0	5	0	0	0	0	0	0	0	0	1	
A82	41	5	3	1	3	3	3	4	2	0	3	3	19	4	8	0	0	0	0	0	

Figure 5.6 Mutability landscape of P450-BM3 towards testosterone (**16**). The five target residues are indicated on the left side, and the catalytic traits under investigation are shown on the right: Substrate conversion correlating roughly with activity (black/gray), and selectivity towards 2ß-(red), 15ß-(blue), or 16ß-hydroxytestosterone (green). The values of the parent enzyme are shown in squares. Source: Acevedo-Rocha et al. [35]. © 2018, American Chemical Society.

for activity, regio- and diastereoselectivity, the mutability landscape reproduced in Figure 5.6 resulted. It can be seen, inter alia, that some mutations induce high or low activity, others high or low regio- or stereoselectivity. The results from such explorations, coupled with MD data, then allowed semi-rational decisions regarding the nature of the codon degeneracy for CAST/ISM and the way single residues should be grouped into multi-residue randomization sites [35].

On the basis of this information, a four-membered amino acid alphabet was chosen for SM at two multi-residue sites A and B, and only 767 transformants were screened, corresponding to 53% library coverage (instead of 3076 samples for 95% library coverage). For details the interested reader is referred to the original paper [35]. Limited ISM was then performed. Suffice it to say that excellent results were obtained, because in the reaction of **16**, unusually active C16α- and C16ß-selective variants were identified corresponding to 96% and 92% overall selectivity, respectively. The total procedure required the HPLC screening of only 3000 transformants. This demonstrates the remarkable superiority of mutability landscaping in CAST/ISM [35] relative to the earlier CAST/ISM approach [36]. Previous attempts using rational design had provided a lower degree of C16-selectivity at extremely low activity [38]. Some of the best mutants obtained by the new strategy were shown to be excellent catalysts in the oxidation of four other steroids (Scheme 5.4). The X-ray structure of one of the best variants harboring testosterone (**16**), flanked by MD simulations, led to a reasonable model for explaining the source of high activity and selectivity [35].

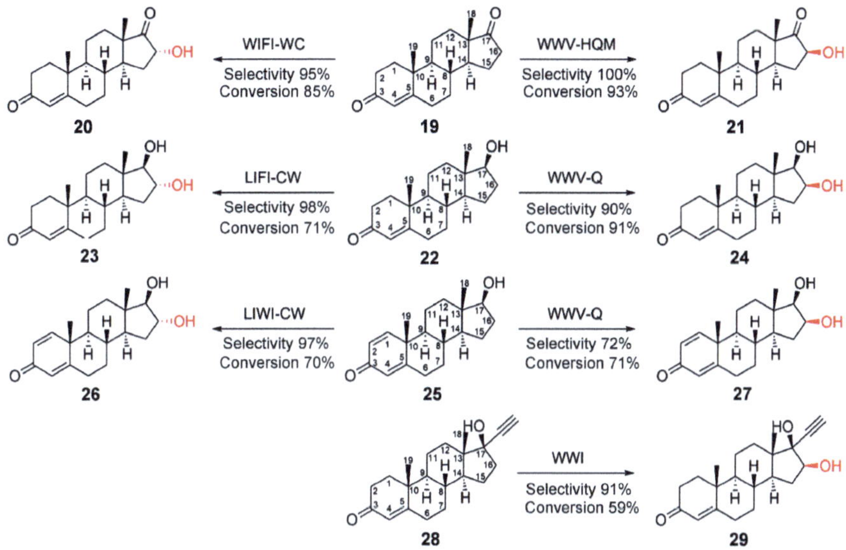

Scheme 5.4 Active mutants for C16-selective hydroxylation of four additional steroids [35]. Substrate **19**: androstenedione; **22**: nandrolone; **25**: boldenone; **28**: norethindrone. WIFI-WC: R47W/S72I/A82F/F87I/Y51W/L181C; WWV-HQM: R47W/A82W/F87V/Y51H/L181Q/A330M; LIFI-CW: R47L/S72I/A82F/F87I/L188C/A330W; WWV-Q: R47W/A82W/F87V/L181Q; LIWI-CW: R47L/S72I/A82W/F87I/L188C/A330W; WWI: R47W/A82W/F87I. Source: Acevedo-Rocha et al. [35]. © 2018, American Chemical Society.

5.2.2 Further Methodology Developments of CAST/ISM

5.2.2.1 Advances Based on Novel Molecular Biological Techniques and Computational Methods

As demonstrated in an early directed evolution study for controlling stereoselectivity, the combination of gene mutagenesis techniques such as epPCR and SM can be successful [5]. As shown later in a CAST/ISM-based study of the glycoside hydrolase from *Streptococcus* pneumonia SP3-BS aimed at increasing activity, several ISM steps were followed by one round of epPCR, the latter inducing only a small improvement [39]. It has also been reported that it may be useful to start with epPCR for identifying positive remote mutations and then to employ CAST/ISM at the binding pocket [6a], as in activity and stability enhancement of the phosphotriesterase PoOPHH-M [40]. However, while the general ideas are worthy of consideration, sequential procedures restrict protein sequence. Relevant is a method in which epPCR and SM are performed simultaneously, which has yet to be applied to engineering stereoselectivity [41]. In yet another publication, a techno-economic analysis of various SM strategies was reported with emphasis on the number, quality, and cost of primers [42]. This issue may influence decision-making when planning directed evolution projects based on SM.

In a noteworthy molecular biological advance, the process of SSM was simplified and improved [43]. Traditionally, SSM involves the introduction of all 19 amino acids at each single residue using a one-step approach based on NNK codon degeneracy (see Introduction, Section 5.1), but in the case of difficult-to- randomize genes (e.g. P450-BM3) [22e], it may not deliver all 19 mutants. A two-step approach was therefore developed in which a mutagenic primer and a non-mutagenic (silent) primer are used to generate a short DNA fragment, which is recovered and then used as a megaprimer to amplify the whole plasmid. The superiority of this new technique was proven by massive sequence data and screening on protein level [43].

A crucial step has been undertaken to explore possible benefits of utilizing CRISPR-Cas9 [44] in directed evolution [45]. An efficient *in vitro* CRISPR/Cas9-mediated mutagenic (ICM) system was developed that enables the construction of designed mutants in a PCR-free manner, thereby avoiding amino acid bias, which is the crucial advantage. The ICM system comprises the following steps: (i) Plasmid digestion by using a complex of Cas9 with specific single guide RNA (sgRNA), followed by degradation with T5 exonuclease leading to the generation of 15 nt homologous region; (ii) Annealing the primers with the desired mutations to form double stranded DNA fragments, which are subsequently ligated into the linearized plasmid. Single and multiple site-directed mutageneses were achieved even in the case of a large size plasmid of up to 9 kb [45]. In further promising experiments, which are of significance in future directed evolution projects, a PCR-free SSM library on single site and two adjacent sites of a green fluorescent protein was also generated [45]. More research is necessary for exploiting CRISPR/Cas9 in directed evolution.

As already emphasized, X-ray data as well as the results of standard docking computations and MD simulations are crucial when planning CAST/ISM experiments with the aim of evolving stereo- and regioselectivity as well as activity. Generally, such aids suffice. Further computational tools can also be used, e.g. those that identify hotspots, which may not only occur as expected at the binding pocket (CAST residues), but possibly also at remote residues (as discovered by epPCR in the directed evolution of a stereoselective lipase [6a]). Useful options include HotSpot Wizard [46], and catalytic selectivity by computational design (CASCO) [47], which is based on RosettaDesign [48]. The portal NewProt [49] features a collection of servers that can be used for computationally designing protein engineering.

5.2.2.2 Advances Based on Solid-Phase Chemical Synthesis of SM Libraries

As already noted (Section 5.1), any genetic technique can be used to generate SM libraries in CAST/ISM. The megaprimer-based approach [22c, d] is commonly applied in directed evolution and can even be used in recalcitrant cases [22e]. Nevertheless, all forms of SM suffer from amino acid bias, but this disadvantage was generally ignored [1, 2]. Indeed, when applying the Patrick–Firth algorithm [16] or the CASTER computer aid for SM design [15], the absence of amino acid bias is assumed. Several factors contribute amino acid bias:

- The degeneracy of the genetic code causes favored amino acid exchange events, while others are disfavored.
- Amplification in PCR steps are imperfect, preventing the occurrence of some planned mutations, while making non-planned insertions and deletions possible.
- Imperfect annealing temperatures cause similar problems.
- Imperfect quality of primers also causes amino acid bias.

Until recently, it was unclear how serious the problem of amino acid bias actually is, because massive DNA sequencing for precise quality control had never been performed. An alternative to standard molecular biological SM techniques is the production of DNA libraries using the known chemical technique of solid-phase gene synthesis, of which several (commercial) variations exist. In our first study, featuring the Sloning method for solid-phase gene synthesis, this chemical approach was found to deliver higher-quality SM libraries, but massive DNA sequencing for reliable quality control was not reported [50]. In contrast, our second study concerning synthetic SM libraries in collaboration with Twist Bioscience led to a breakthrough in this aspect of directed evolution [51a].

The Twist platform based on solid-phase gene synthesis on Si-chips [52] was adapted and extended so that a designed synthetic SM library became available which was compared with the respective library generated by the traditional megaprimer-based SM technique [51a]. The hydrolytic desymmetrization of cyclohexene oxide (**1**) with formation of (*R*,*R*)- and (*S*,*S*)-**2** (Scheme 5.1), catalyzed by the epoxide hydrolase LEH, was used as the model reaction. In the Twist approach, massive numbers of high-fidelity mutagenic primers are rapidly prepared in separate synthetic procedures on Si-chips, which are then extracted. Then primers are mixed (equimolar) and assembled into full gene length by overlap extension PCR with formation of the desired mutants. This procedure provides full-length constructs that are ready for cloning (Figure 5.7) [51a]. Following transformation into *E. coli* BL21 (DE3) and DNA library harvest, massive DNA sequencing was performed.

Using the crystal structure of LEH, a randomization site comprising four CAST residues (Met78, Ile80, Leu114, and Ile116) was considered for TCSM based on valine, phenylalanine, and tyrosine (V-F-Y). In both the traditional PCR-based SM library and the synthetic Twist library, 256 mutants exist theoretically [51a]. The result of massive sequencing on DNA level with oversampling factors (OFs) of 1, 2, and 3 revealed that the traditional PCR-based SM procedure provided only 144 of the expected 256 mutants, corresponding to only 56.3% genetic diversity. In contrast, the synthetic gene LEH library contained 249 different mutants, which means 97.3% genetic diversity, an amazing result (Figure 5.8a) [51a]. The number of WTs observed in the synthetic library is also clearly lower than in the case of the traditional SM library, which likewise contributes to higher quality (Figure 5.8b).

Both libraries were also screened for enantioselectivity on protein level. The synthetic library contained a number of highly selective variants that were not found in the traditional PCR-based SM library [51a]. In final conclusion, the synthetic on-chip Twist method for SM library construction constitutes the presently best way to generate highest-quality SM libraries [51]. The second law of directed evolution emerges: "*You get what you designed.*" Other companies also provide services for

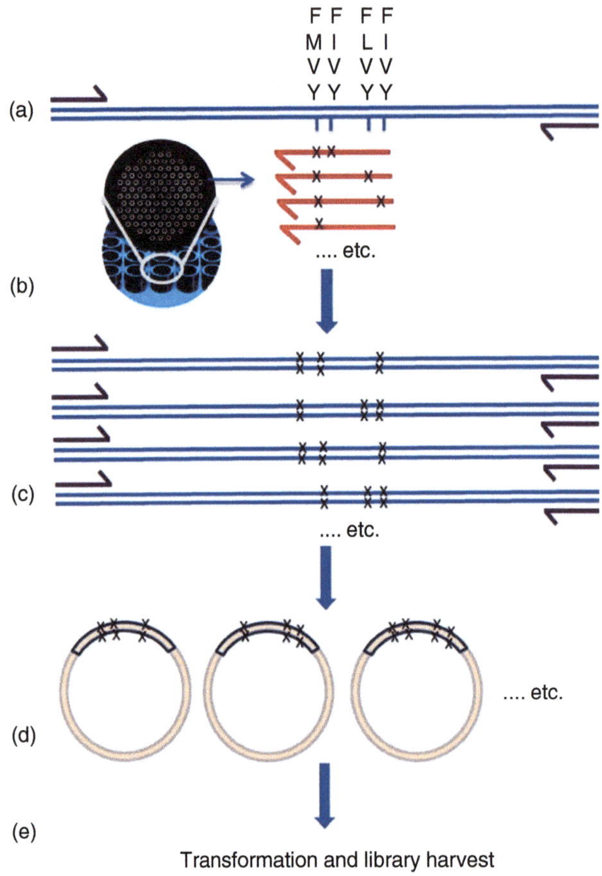

Figure 5.7 The Twist system of precisely controlled combinatorial library fabrication. (a) Design of a mutagenic strategy that allows all variants to be produced in combination. (b) High-quality, high-throughput silicon-based synthesis platform to produce mutagenic primers at each region box. (c) Design and experimental strategy to enable the specific introduction of variants at the DNA level to enable rapid production of high-diversity, full-length gene fragments. (d) Full-length gene fragments were subjected to double restriction enzyme digestion and ligation to plasmid. (e) Transformation into *E. coli* BL21 (DE3) and DNA library harvest. Source: Li et al. [51a]. © 2018, John Wiley & Sons.

synthetic gene production. If the prices of synthetic gene synthesis continue to tumble, and quality control is offered, then the synthetic approach may prove to be the future of directed evolution.

5.3 B-FIT as an ISM Method for Enhancing Protein Thermostability

Although not directly related to CAST/ISM for evolving stereo- and regioselectivity as well as activity, we mention here briefly a technique for enhancing thermostability, in which ISM can also be employed. Many studies have shown

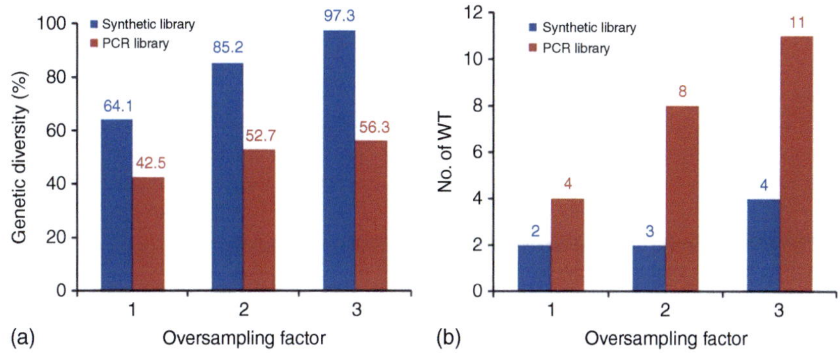

Figure 5.8 Results of massive DNA sequencing as a reliable quality control in the model systems using the enzyme LEH. (a) Genetic diversity of the synthetic (twist) and the combinatorial PCR-based SM library with different oversampling factors. (b) Number of WTs observed in the two libraries. Source: Li et al. [51a]. © 2018, John Wiley & Sons.

that thermostabilization of enzymes by directed evolution, traditionally achieved by multiple rounds of epPCR and/or DNA shuffling [2], is also possible by applying ISM at rationally chosen randomization sites. The respective hot spots can be identified by *B*-factors available from X-ray data, which resulted in the development of the so-called B-FIT method [10, 15]. The B-FITTER computer aid useful in performing B-FIT is available free of charge on the Reetz homepage (kofo.mpg.de/en/research/biocatalysis).

5.4 Learning from CAST/ISM-Based Directed Evolution

Following several CAST/ISM studies and strong cooperative mutational effects [17c] were revealed [53], in contrast to strictly additive influences. *This was achieved by performing complete deconvolution of the multi-mutational variants evolved by CAST/ISM, and utilizing the data to construct fitness pathway landscapes.* Studies of this kind are important in fundamental science, a fascinating research area that has been reviewed [53]. The extensive data from many directed evolution groups allows this protein engineering approach to fuse with rational design based on site-directed mutagenesis. Indeed, in some cases SM can be replaced by a small number of rationally chosen mutations, the hits then being used as a template to perform site-directed mutagenesis at other sites with formation of extremely small libraries, reminiscent of ISM [2].

5.5 Conclusions and Perspectives

This chapter shows that methodology development stands at the heart of directed evolution. CAST/ISM for enhancing activity as well as stereo- and regioselectivity plays a central role in these efforts, Table 5.2 lists additional examples. Since 2016, efficacy of this semi-rational approach has increased further, and more applications

Table 5.2 Selected examples of recent CAST/ISM studies.

Enzyme	Evolved property	Mutagenesis method	Comment	References
Cytochrome P450BM3 monooxygenases	Activity, chemo-selectivity	ISM	Bioorthogonal deprotection of caged compounds	[54]
Cytochrome P450BM3 monooxygenases	Enantioselectivity	CAST/ISM	Asymmetric sulfoxidation	[55]
P450 CYP153A (*Marinobacter aquaeolei*)	Activity	SM, ISM	ω-Hydroxylation of oleic acid, NNK	[56]
Laccase (*Pycnoporus cinnabarinus*)	Activity	ISM	Oxidation of sinapic acid	[57]
Alcohol dehydrogenase (*Lactobacillus kefiri* DSM 20587)	Enantioselectivity	ISM	Inversion of enantioselectivity, NDC/NNK codons	[58]
Ketoreductase *Ch*KRED20 (*Chryseobacterium* sp. CA49)	Substrate scope	ISM	Three rounds of ISM	[59]
L-Aspartate-β-semialdehyde dehydrogenase (*E. coli*)	Cofactor preference NADH acceptance	ISM	L-Homoserine production	[60]
Phenylalanine dehydrogenase (*Rhodococcus* sp. M4)	Substrate scope, enantioselectivity	ISM	Reductive amination	[61]
Monoamine oxidase (*Aspergillus niger*)	Substrate scope, enantioselectivity	ISM	(*R*)-Mexiletine synthesis	[62]
Monoamine oxidase (*Aspergillus niger*)	Stereoselectivity	ISM	ISM on residues lining the tunnel and the binding pocket	[63]
α-1,3-Fucosyl-transferase (*Helicobacter pylori*)	Catalytic efficiency	ISM	NNK degeneracy, oligosaccharides	[64]
Lipase (*Candida rugosa*)	Thermostability	ISM	B-FIT approach and NNK degeneracy	[65]
Xylanase (*Aspergillus oryzae*)	Thermostability, activity	ISM	B-FIT approach also led to higher activity, not just to increased thermostability	[66]
1,3-1,4-β-Glucanase	Thermostability, activity	ISM	Four rounds of ISM	[67]
D-Xylose isomerase (*Thermotoga neapolitana*)	Activity	ISM	Production of d-psicose from sucrose in a cascade reaction	[68]

(Continued)

Table 5.2 (Continued)

Enzyme	Evolved property	Mutagenesis method	Comment	References
Phosphotriesterase (*Po*OPH$_{M2}$)	Activity, thermostability	CAST/ISM DNA shuffling, epPCR	Recombination of multiple strategies	[40]
Limonene epoxide hydrolase	Enantioselectivity	CAST/ISM	Smallest amino acid alphabets	[25]
Limonene epoxide hydrolase	Thermostability, enantioselectivity, activity	CAST/ISM	Multiparameter optimization	[69]
Nitrilase (*Pyrococcus abyssi*)	Activity, enantioselectivity	ISM	Dynamic kinetic resolution	[70]
N-Oligosaccharyl-transferase PglB (*Campylobacter jejuni*)	Activity	ISM	Application in vaccine	[71]
Halohydrin dehalogenase (*Agrobacterium radiobacter* HheC)	Enantioselectivity, activity	CAST/ISM	DC-analyzer applied	[72]
Halohydrin dehalogenase (*Agrobacterium radiobacter* HheC)	Thermostability	epPCR, ISM	NNS, four rounds of ISM	[73]
Nitrile hydratase	Regioselectivity	ISM	In conjunction with mutability landscapes	[74]
Tagatose epimerase (*Pseudomonas cichorii*)	Substrate scope	ISM	C3-epimerization of D-fructose and L-sorbose	[75]
Cyclodextrin glycosyltransferase (*Bacillus stearothermophilus* NO_2)	Activity	ISM	ISM on nine residues in the binding pocket	[76]
ADH-A from *Rhodococcus ruber* DSM 44541	Stereo- and regioselectivity	ISM	ISM combined with activity screening	[77]
β-Glucuronidase from *Aspergillus oryzae*	Substrate specificity	CAST/ISM	Tuning the substrate specificity towards triterpenoid saponins	[78]
Artificial retro-aldolase	Stereoselectivity	CAST/ISM	Design and directed evolution catalysts for the Michael addition	[79]
(+)-γ-Lactamase from *Microbacterium hydrocarbonoxydans*	Enantioselectivity and thermostability	CAST/B-FITTER	Kinetic resolution	[80]

Sources: Lv et al. [78], Garrabou et al. [79], Gao et al. [80].

can be expected in the future. CAST/ISM (sometimes CAST alone) has also emerged as the key technology in controlling activity and stereoselectivity of artificial metalloenzymes [81] as novel catalysts in synthetic organic chemistry, an important concept first illustrated experimentally in 2006 [81b]. Iterative site-directed mutagenesis with formation of smallest libraries (see Section 5.4) has also been applied in the evolution of metalloenzymes [82].

Acknowledgment

Manfred T. Reetz thanks the Max-Planck-Society and the LOEWE Research cluster SynChemBio for generous support. He also acknowledges the support of the Chinese Academy of Sciences (CAS) President's International Fellowship Initiative for 2018 (2018DB0030) as part of the "Distinguished Scientist" award. Zhoutong Sun thanks the CAS Pioneer Hundred Talent Program (Type C) (reference number 2016-053) for initial start support.

References

1 (a) Drauz, K., Gröger, H., and May, O. (eds.) (2012). *Enzyme Catalysis in Organic Synthesis*, 3e. Wiley-VCH: Weinheim. (b) Stewart, J. and Goswami, A. (eds.) (2016). *Organic Synthesis Using Biocatalysis*. Amsterdam: Elsevier.

2 Recent reviews of directed evolution:(a)Hilvert, D. and Zeymer, C. (2018). Directed evolution of protein catalysts. *Annu. Rev. Biochem.* 87: 131–157. (b) Alcalde, M. (2017). *Directed Enzyme Evolution: Advances and Applications*. Stuttgart: Springer. (c) Arnold, F.H. (2019). Innovation by Evolution: Bring New Chemistry to Life (Nobel Lecture). *Angew. Chem. Int. Ed.* 58: 14420–14426. (d) Reetz, M.T. (2016). *Directed Evolution of Selective Enzymes: Catalysts for Organic Chemistry and Biotechnology*. Weinheim: Wiley-VCH.

3 Chen, K. and Arnold, F.H. (1993). Tuning the activity of an enzyme for unusual environments: sequential random mutagenesis of subtilisin E for catalysis in dimethylformamide. *Proc. Natl. Acad. Sci.* 90 (12): 5618–5622.

4 Reetz, M.T., Zonta, A., Schimossek, K. et al. (1997). Creation of enantioselective biocatalysts for organic chemistry by in vitro evolution. *Angew. Chem. Int. Ed.* 36 (24): 2830–2832.

5 Reetz, M.T., Wilensek, S., Zha, D., and Jaeger, K.E. (2001). Directed evolution of an enantioselective enzyme through combinatorial multiple-cassette mutagenesis. *Angew. Chem. Int. Ed.* 40 (19): 3589–3591.

6 (a) Reetz, M.T. (2004). Controlling the enantioselectivity of enzymes by directed evolution: practical and theoretical ramifications. *Proc. Natl. Acad. Sci. U. S. A.* 101 (16): 5716–5722. (b) Parikh, M.R. and Matsumura, I. (2005). Site-saturation mutagenesis is more efficient than DNA shuffling for the directed evolution of ß-fucosidase from ß-galactosidase. *J. Mol. Biol.* 352: 621–628. (c) Lutz, S. and

Patrick, W.M. (2004). Novel methods for directed evolution of enzymes: quality, not quantity. *Curr. Opin. Biotechnol.* 15: 291–297.

7 (a) Yang, G. and Withers, S.G. (2009). Ultrahigh-throughput FACS-based screening for directed evolution. *ChemBioChem* 10: 2704–2715. (b) Acevedo-Rocha, C.G., Agudo, R., and Reetz, M.T. (2014). Directed evolution of stereoselective enzymes based on genetic selection as opposed to screening systems. *J. Biotechnol.* 191: 3–10. (c) Reymond, J.-L. (2006). *Enzyme Assays: High-Throughput Screening, Genetic Selection and Fingerprinting*. Weinheim: Wiley-VCH. (d) Fallah-Araghi, A., Baret, J.C., Ryckelynck, M., and Griffiths, A.D. (2012). A completely in vitro ultrahigh-throughput droplet-based microfluidic screening system for protein engineering and directed evolution. *Lab Chip* 12: 882–891.

8 Reetz, M.T., Bocola, M., Carballeira, J.D. et al. (2005). Expanding the range of substrate acceptance of enzymes: combinatorial active-site saturation test. *Angew. Chem. Int. Ed.* 44: 4192–4196.

9 Estell, D.A., Graycar, T.P., and Wells, J.A. (1985). Engineering an enzyme by site-directed mutagenesis to be resistant to chemical oxidation. *J. Biol. Chem.* 260 (1): 6518–6521.

10 Reetz, M.T., Carballeira, J.D., and Vogel, A. (2006). Iterative saturation mutagenesis on the basis of B factors as a strategy for increasing protein thermostability. *Angew. Chem. Int. Ed.* 45: 7745–7751.

11 Wu, S., Acevedo, J.P., and Reetz, M.T. (2010). Induced allostery in the directed evolution of an enantioselective Baeyer-Villiger monooxygenase. *Proc. Natl. Acad. Sci. U. S. A.* 107: 2775–2780.

12 (a) Reetz, M.T., Wang, L.-W., and Bocola, M. (2006). Directed evolution of enantioselective enzymes: iterative cycles of CASTing for probing protein-sequence space. *Angew. Chem. Int. Ed.* 45: 1236–1241; *Erratum*, 2494. (b) Reetz, M.T. (2005). *Yearbook of the Max-Planck-Society*, 327–331. München: Generalverwaltung der Max-Planck-Gesellschaft.

13 Acevedo-Rocha, C.G., Sun, Z., and Reetz, M.T. (2018). Clarifying the difference between Iterative saturation mutagenesis as a rational guide in directed evolution and OmniChange as a gene mutagenesis technique. *ChemBioChem* 19 (24): 2542–2544.

14 Wang, J., Li, G., and Reetz, M.T. (2017). Enzymatic site-selectivity enabled by structure-guided directed evolution. *Chem. Commun.* 53: 3916–3928.

15 Reetz, M.T. and Carballeira, J.D. (2007). Iterative saturation mutagenesis (ISM) for rapid directed evolution of functional enzymes. *Nat. Protoc.* 2: 891–903.

16 Firth, A.E. and Patrick, W.M. (2008). GLUE-IT and PEDEL-AA: new programmes for analyzing protein diversity in randomized libraries. *Nucleic Acids Res.* 36: 281–285.

17 (a) Reetz, M.T., Kahakeaw, D., and Lohmer, R. (2008). Addressing the numbers problem in directed evolution. *ChemBioChem* 9: 1797–1804. (b) Acevedo-Rocha, C.G. and Reetz, M.T. (2016). Handling the numbers problem in directed evolution. In: *Understanding Enzymes: Function, Design, Engineering and Analysis* (ed. A. Svendsen), 613–642. Singapore: Pan Stanford Publishing. (c) Reetz, M.T.

(2013). The importance of additive and non-additive mutational effects in protein engineering. *Angew. Chem. Int. Ed.* 52: 2658–2666.

18 Early reviews of CAST/ISM with guidelines [17b]:(a)Reetz, M.T. (2011). Laboratory evolution of stereoselective enzymes: a prolific source of catalysts for asymmetric reactions. *Angew. Chem. Int. Ed.* 50: 138–174. (b) Acevedo-Rocha, C.G., Kille, S., and Reetz, M.T. (2014). Iterative saturation mutagenesis: a powerful approach to engineer proteins by simulating darwinian evolution. In: *Directed Evolution Library Creation: Methods and Protocols, Methods in Molecular Biology*, 2e, vol. 1179 (eds. D. Ackerley, J. Copp and E. Gillam), 103–128. Totowa: Humana Press.

19 Chockalingam, K., Chen, Z.L., Katzenellenbogen, J.A., and Zhao, H.M. (2005). Directed evolution of specific receptor-ligand pairs for use in the creation of gene switches. *Proc. Natl. Acad. Sci. U.S.A.* 102 (16): 5691–5696.

20 (a) Liang, L., Zhang, J., and Lin, Z. (2007). Altering coenzyme specificity of Pichia stipitis xylose reductase by the semi-rational approach CASTing. *Microb. Cell Fact.* 6: 36. (b) Cahn, J.K.B., Werlang, C.A., Baumschlager, A. et al. (2017). A general tool for engineering the NAD/NADP cofactor preference of oxidoreductases. *ACS Synth. Biol.* 6: 326–333.

21 Clouthier, C.M., Kayser, M.M., and Reetz, M.T. (2006). Designing new baeyer-villiger monooxygenases using restricted CASTing. *J. Org. Chem.* 71: 8431–8437.

22 (a) Hogrefe, H.H., Cline, J., Youngblood, G.L., and Allen, R.M. (2002). Creating randomized amino acid libraries with the QuikChange® multi site-directed mutagenesis kit. *Biotechniques* 33: 1158–1165. (b) Ho, S.N., Hunt, H.D., Horton, R.M. et al. (1989). Site-directed mutagenesis by overlap extension using the polymerase chain reaction. *Gene* 77: 51–59. (c) Sarkar, G. and Sommer, S.S. (1990). The "megaprimer" method of site-directed mutagenesis. *BioTechniques* 8 (4): 404–407. (d)Miyazaki, K. and Takenouchi, M. (2002). Creating random mutagenesis libraries using megaprimer PCR of whole plasmid. *BioTechniques* 33 (5): 1033–1038. (e) Sanchis, J., Fernández, L., Carballeira, J.D. et al. (2008). Improved PCR method for the creation of saturation mutagenesis libraries in directed evolution: application to difficult-to-amplify templates. *Appl. Microbiol. Biotechnol.* 81 (2): 387–397.

23 (a) Bougioukou, D.J., Kille, S., Taglieber, A., and Reetz, M.T. (2009). Directed evolution of an enantioselective enoate-reductase: testing the utility of iterative saturation mutagenesis. *Adv. Synth. Catal.* 351: 3287–3305. (b) Sullivan, B., Walton, A.Z., and Stewart, J.D. (2013). Library construction and evaluation for site saturation mutagenesis. *Enzyme Microb. Technol.* 53: 70–77.

24 (a) Sun, Z., Wikmark, Y., Bäckvall, J.E., and Reetz, M.T. (2016). New concepts for increasing the efficiency in directed evolution of stereoselective enzymes. *Chem. Eur J.* 22 (15): 5046–5054. (b) Reetz, M.T. and Wu, S. (2008). Greatly reduced amino acid alphabets in directed evolution: making the right choice for saturation mutagenesis at homologous enzyme positions. *Chem. Commun.* 43: 5499–5501. (c) Sandström, A.G., Wikmark, Y., Engström, K. et al. (2012). Combinatorial reshaping of the Candida antarctica lipase A substrate pocket for enantioselectivity using an extremely condensed library. *Proc. Natl. Acad. Sci. U.S.A.* 109 (1): 78–83.

25 Sun, Z., Lonsdale, R., Kong, X.-D. et al. (2015). Reshaping an enzyme binding pocket for enhanced and inverted stereoselectivity: use of smallest amino acid alphabets in directed evolution. *Angew. Chem. Int. Ed.* 54: 12410–12415.

26 Sun, Z., Lonsdale, R., Wu, L. et al. (2016). Structure-guided triple code saturation mutagenesis: efficient tuning of the stereoselectivity of an epoxide hydrolase. *ACS Catal.* 6: 1590–1597.

27 (a) Steipe, B., Schiller, B., Plückthun, A., and Steinbacher, S. (1994). Sequence statistics reliably predict stabilizing mutations in a protein domain. *J. Mol. Biol.* 240: 188–192. (b) Lehmann, M., Pasamontes, L., Lassen, S.F., and Wyss, M. (2000). The consensus concept for thermostability engineering of proteins. *Biochim. Biophys. Acta, Mol. Cell. Res.* 1543: 408–415. (c) García-Guevara, F., Bravo, I., Martínez-Anaya, C., and Segovia, L. (2017). Cofactor specificity switch in Shikimate dehydrogenase by rational design and consensus engineering. *Protein Eng. Des. Sel.* 30: 533–541.

28 Sun, Z., Wu, L., Bocola, M. et al. (2018). Structural and computational insight into the catalytic mechanism of limonene epoxide mutants in stereoselective transformations. *J. Am. Chem. Soc.* 140: 310–318.

29 Sun, Z., Lonsdale, R., Ilie, A. et al. (2016). Catalytic asymmetric reduction of difficult-to-reduce ketones: triple code saturation mutagenesis of an alcohol dehydrogenase. *ACS Catal.* 6: 1598–1605.

30 Examples of SM in which low library coverage sufficed [5]:(a)Parra, L.P., Agudo, R., and Reetz, M.T. (2013). Directed evolution using iterative saturation mutagenesis based on multi-residue sites. *ChemBioChem* 14: 2301–2309. (b) Zhang, Z.-G., Lonsdale, R., Sanchis, J., and Reetz, M.T. (2014). Extreme synergistic mutational effects in the directed evolution of a baeyer-villiger monooxygenase as catalyst for asymmetric sulfoxidation. *J. Am. Chem. Soc.* 136: 17262–17272.

31 Li, A., Ilie, A., Sun, Z. et al. (2016). Whole-cell catalyzed multiple regio- and stereoselective functionalization in cascade reactions enabled by directed evolution. *Angew. Chem. Int. Ed.* 55: 12026–12029.

32 Ortiz de Montellano, P.R. (2010). Hydrocarbon hydroxylation by cytochrome P450 enzymes. *Chem. Rev.* 110: 932–948.

33 (a) Sun, Z., Lonsdale, R., Li, G., and Reetz, M.T. (2016). Comparing different strategies in directed evolution of enzyme stereoselectivity: single versus double code saturation mutagenesis. *ChemBioChem* 17: 1865–1872. (b) Agudo, R., Roiban, G.-D., Lonsdale, R. et al. (2015). Biocatalytic route to chiral acyloins: P450-catalyzed regio- and enantioselective α-hydroxylation of ketones. *J. Org. Chem.* 80: 950–956. (c) Roiban, G.-D., Agudo, R., and Reetz, M.T. (2014). Cytochrome P450 catalyzed oxidative hydroxylation of achiral organic compounds with simultaneous creation of two chirality centers in a single C–H activation step. *Angew. Chem. Int. Ed.* 53: 8659–8663. (d) Agudo, R., Roiban, G.-D., and Reetz, M.T. (2013). Induced axial chirality in biocatalytic asymmetric ketone reduction. *J. Am. Chem. Soc.* 135: 1665–1668.

34 (a) van der Meer, J.-Y., Biewenga, L., and Poelarends, G.J. (2016). The generation and exploitation of protein mutability landscapes for protein engineering. *ChemBioChem* 17: 1792–1799. (b) van der Meer, J.-Y., Poddar, H., Baas, B.-J. et al. (2016). Using mutability landscapes of a promiscuous tautomerase to guide the engineering of enantioselective Michaelases. *Nat. Commun.* 7: 10911.

35 Acevedo-Rocha, C.G., Gamble, C., Lonsdale, R. et al. (2018). P450-catalyzed regio- and diastereoselective steroid hydroxylation: efficient directed evolution enabled by mutability landscaping. *ACS Catal.* 8: 3395–3410.

36 Kille, A., Zilly, F.E., Acevedo, J.P., and Reetz, M.T. (2011). Regio- and stereoselectivity of P450 catalysed hydroxylation of steroids controlled by laboratory evolution. *Nat. Chem.* 3: 738–743.

37 (a) Le-Huu, P., Rekow, D., Krüger, C. et al. (2018). Chemoenzymatic route to oxyfunctionalized cembranoids facilitated by substrate and protein engineering. *Chem. Eur. J.* 24 (46): 12010–12021. (b) Hall, E.A., Sarkar, M.R., Lee, J.H.Z. et al. (2016). Improving the monooxygenase activity and the regio- and stereoselectivity of terpenoid hydroxylation using ester directing groups. *ACS Catal.* 6: 6306–6317. (c) Zhang, K., Shafer, B.M., Demars, M.D. et al. (2012). Controlled oxidation of remote sp_3 C–H bonds in artemisinin via P450 catalysts with fine-tuned regio- and stereoselectivity. *J. Am. Chem. Soc.* 134: 18695–18704. (d) Kolev, J.N., O'Dwyer, K.M., Jordan, C.T., and Fasan, R. (2014). Discovery of potent parthenolide-based antileukemic agents enabled by late-stage P450-mediated C–H functionalization. *ACS Chem. Biol.* 9: 164–173. (e) Dong, J.J., Fernandez-Fueyo, E., Hollmann, F. et al. (2018). Biocatalytic oxidation reactions: a Chemist's perspective. *Angew. Chem. Int. Ed.* 57: 9238–9261.

38 (a) Rea, V., Kolkman, A.J., Vottero, E. et al. (2012). Active site substitution A82W improves the regioselectivity of steroid hydroxylation by cytochrome P450 BM3Mutants as rationalized by spin relaxation nuclear magnetic resonance studies. *Biochemistry* 51: 750–760. (b) Venkataraman, H., de Beer, S.B.A., van Bergen, L.A.H. et al. (2012). A single active site mutation inverts stereoselectivity of 16-hydroxylation of testosterone catalyzed by engineered cytochrome P450 BM3. *ChemBioChem* 13: 520–523.

39 Kwan, D.H., Constantinescu, I., Chapanian, R. et al. (2015). Toward efficient enzymes for the generation of universal blood through structure-guided directed evolution. *J. Am. Chem. Soc.* 137 (17): 5695–5705.

40 Luo, X.J., Zhao, J., Li, C.X. et al. (2016). Combinatorial evolution of phosphotriesterase toward a robust malathion degrader by hierarchical iteration mutagenesis. *Biotechnol. Bioeng.* 113: 2350–2357.

41 Acevedo, J.P., Reetz, M.T., Asenjo, J.A., and Parra, L.P. (2017). One-step combined focused epPCR and saturation mutagenesis for thermostability evolution of a new cold-active xylanase. *Enzyme Microb. Technol.* 100: 60–70.

42 Acevedo-Rocha, C.G., Reetz, M.T., and Nov, Y. (2015). Economical analysis of saturation mutagenesis experiments. *Sci. Rep.* 5: 10654.

43 Li, A., Acevedo-Rocha, C.G., and Reetz, M.T. (2018). Boosting the efficiency of site-saturation mutagenesis for a difficult-to-randomize gene by a two-step PCR strategy. *Appl. Microbiol. Biotechnol.* 102: 6095–6103.

44 (a) Jinek, M., Chylinski, K., Fonfara, I. et al. (2012). A programmable dual-RNA-guided DNA endonuclease in adaptive bacterial immunity. *Science* 337: 816–821. (b) Jiang, F., Zhou, K., Ma, L. et al. (2015). Structural biology: a Cas9-guide RNA complex preorganized for target DNA recognition. *Science* 348: 1477–1481.

45 She, W., Ni, J., Shui, K. et al. (2018). Rapid and error-free site-directed mutagenesis by a PCR-free in vitro CRISPR/Cas9-mediated mutagenic system. *ACS Synth. Biol.* 7: 2236–2244.

46 Sumbalova, L., Stourac, J., Martinek, T. et al. (2018). HotSpot Wizard 3.0: web server for automated design of mutations and smart libraries based on sequence input information. *Nucleic Acids Res.* 46: W356–W362.

47 Wijma, H.J., Floor, R.J., Bjelic, S. et al. (2015). Enantioselective enzymes by computational design and in silico screening. *Angew. Chem. Int. Ed.* 54: 3726–3730.

48 Liu, Y. and Kuhlman, B. (2006). RosettaDesign server for protein design. *Nucleic Acids Res.* 34: W235–W238.

49 Schwarte, A., Genz, M., Skalden, L. et al. (2017). NewProt – a protein engineering portal. *Protein Eng. Des. Sel.* 30: 441–447.

50 Hoebenreich, S., Zilly, F.E., Acevedo-Rocha, C.G. et al. (2015). Speeding up directed evolution: combining the advantages of solid-phase combinatorial gene synthesis with statistically guided reduction of screening effort. *ACS Synth. Biol.* 4 (3): 317–331.

51 (a) Li, A., Acevedo-Rocha, C.G., Sun, Z. et al. (2018). Beating bias in the directed evolution of proteins: combining high-fidelity on-chip solid-phase gene synthesis with efficient gene assembly for combinatorial library construction. *ChemBioChem* 19 (3): 221–228. (b) Li, A., Sun, Z., and Reetz, M.T. (2018). Solid-phase gene synthesis for mutant library construction: the future of directed evolution? *ChemBioChem* 19 (19): 2023–2032.

52 Banyai, W., Peck, B. J., Fernandez, A., Chen, S. and Indermuhle, P. (2015). De novo synthesized gene libraries. Patent WO2015021080 A3.

53 Reviews of lessons learned from directed evolution [17c]:Li, G.Y. and Reetz, M.T. (2016). Learning lessons from directed evolution of stereoselective enzymes. *Org. Chem. Front.* 3 (10): 1350–1358.

54 Ritter, C., Nett, N., Acevedo-Rocha, C.G. et al. (2015). Bioorthogonal enzymatic activation of caged compounds. *Angew. Chem. Int. Ed.* 54: 13440–13443.

55 Wang, J., Ilie, A., and Reetz, M.T. (2017). Chemo- and stereoselective cytochrome P450-BM3 catalyzed sulfoxidation of 1-thiochroman-4-ones enabled by directed evolution. *Adv. Synth. Catal.* 359 (12): 2056–2060.

56 Duan, Y., Ba, L., Gao, J. et al. (2016). Semi-rational engineering of cytochrome CYP153A from *Marinobacter aquaeolei* for improved ω-hydroxylation activity towards oleic acid. *Appl. Microbiol. Biotechnol.* 100 (20): 8779–8788.

57 Pardo, I., Santiago, G., Gentili, P. et al. (2016). Re-designing the substrate binding pocket of laccase for enhanced oxidation of sinapic acid. *Catal. Sci. Technol.* 6 (11): 3900–3910.

58 Wu, K., Wang, H., Chen, L. et al. (2016). Practical two-step synthesis of enantiopure styrene oxide through an optimized chemoenzymatic approach. *Appl. Microbiol. Biotechnol.* 100 (20): 8757–8767.

59 Zhao, F.J., Jin, Y., Liu, Z.C. et al. (2017). Crystal structure and iterative saturation mutagenesis of ChKRED20 for expanded catalytic scope. *Appl. Microbiol. Biotechnol.* 101 (23–24): 8395–8404.

60 Xu, X., Chen, J., Wang, Q. et al. (2015). Mutagenesis of key residues in the binding center of L-aspartate-β-semialdehyde dehydrogenase from *Escherichia coli* enhances utilization of the cofactor NAD(H). *ChemBioChem* 17 (1): 56–64.

61 Ye, L.J., Toh, H.H., Yang, Y. et al. (2015). Engineering of amine dehydrogenase for asymmetric reductive amination of ketone by evolving *Rhodococcus* phenylalanine dehydrogenase. *ACS Catal.* 5 (2): 1119–1122.

62 Chen, Z., Ma, Y., He, M. et al. (2015). Semi-rational directed evolution of monoamine oxidase for kinetic resolution of rac-mexiletine. *Appl. Biochem. Biotechnol.* 176 (8): 2267–2278.

63 Li, G., Yao, P., Gong, R. et al. (2017). Simultaneous engineering of an enzyme's entrance tunnel and active site: the case of monoamine oxidase MAO-N. *Chem. Sci.* 8 (5): 4093–4099.

64 Choi, Y.H., Kim, J.H., Park, B.S., and Kim, B.G. (2016). Solubilization and Iterative saturation mutagenesis of α1,3-fucosyltransferase from *Helicobacter pylori* to enhance its catalytic efficiency. *Biotechnol. Bioeng.* 113 (8): 1666–1675.

65 Zhang, X.-F., Yang, G.-Y., Zhang, Y. et al. (2016). A general and efficient strategy for generating the stable enzymes. *Sci. Rep.* 6 (1): 33797.

66 Li, X.-Q., Wu, Q., Hu, D. et al. (2017). Improving the temperature characteristics and catalytic efficiency of a mesophilic xylanase from *Aspergillus oryzae*, AoXyn11A, by iterative mutagenesis based on *in silico* design. *AMB Express* 7 (1): 97–108.

67 Niu, C., Zhu, L., Xu, X., and Li, Q. (2017). Rational design of thermostability in bacterial 1,3-1,4-β-glucanases through spatial compartmentalization of mutational hotspots. *Appl. Microbiol. Biotechnol.* 101 (3): 1085–1097.

68 Wagner, N., Bosshart, A., Failmezger, J. et al. (2015). A separation-integrated cascade reaction to overcome thermodynamic limitations in rare-sugar synthesis. *Angew. Chem. Int. Ed.* 54 (14): 4182–4186.

69 Li, G., Zhang, H., Sun, Z. et al. (2016). Multiparameter optimization in directed evolution: engineering thermostability, enantioselectivity, and activity of an epoxide hydrolase. *ACS Catal.* 6 (6): 3679–3687.

70 Xue, Y.-P., Shi, C.-C., Xu, Z. et al. (2015). Design of nitrilases with superior activity and enantioselectivity towards sterically hindered nitrile by protein engineering. *Adv. Synth. Catal.* 357 (8): 1741–1750.

71 Ihssen, J., Haas, J., Kowarik, M. et al. (2015). Increased efficiency of *Campylobacter jejuni N*-oligosaccharyltransferase PglB by structure-guided engineering. *Open Biol.* 5 (4): 140227.

72 Guo, C., Chen, Y., Zheng, Y. et al. (2015). Exploring the enantioselective mechanism of halohydrin dehalogenase from *Agrobacterium radiobacter* AD1 by iterative saturation mutagenesis. *Appl. Environ. Microbiol.* 81 (8): 2919–2926.

73 Wu, Z., Deng, W., Tong, Y. et al. (2017). Exploring the thermostable properties of halohydrin dehalogenase from *Agrobacterium radiobacter* AD1 by a combinatorial directed evolution strategy. *Appl. Microbiol. Biotechnol.* 101 (8): 3201–3211.

74 Cheng, Z., Cui, W., Xia, Y. et al. (2018). Modulation of nitrile hydratase regioselectivity towards dinitriles by tailoring the substrate binding pocket residues. *ChemCatChem* 10 (2): 449–458.

75 Bosshart, A., Hee, C.S., Bechtold, M. et al. (2015). Directed divergent evolution of a thermostable D-tagatose epimerase towards improved activity for two hexose substrates. *ChemBioChem* 16 (4): 592–601.

76 Tao, X., Wang, T., Su, L., and Wu, J. (2018). Enhanced 2-*O*-alpha-D-glucopyranosyl-L-ascorbic acid synthesis through Iterative saturation mutagenesis of acceptor subsite residues in *Bacillus stearothermophilus* NO_2 cyclodextrin glycosyltransferase. *J. Agric. Food. Chem.* 66 (34): 9052–9060.

77 Maurer, D., Enugala, T.R., Hamnevik, E. et al. (2018). Stereo- and regioselectivity in catalyzed transformation of a 1,2-disubstituted vicinal diol and the corresponding diketone by wild type and laboratory evolved alcohol dehydrogenases. *ACS Catal.* 8 (8): 7526–7538.

78 Lv, B., Sun, H.L., Huang, S. et al. (2018). Structure-guided engineering of the substrate specificity of a fungal-glucuronidase toward triterpenoid saponins. *J. Biol. Chem.* 293 (2): 433–443.

79 Garrabou, X., Macdonald, D.S., Wicky, B.I.M., and Hilvert, D. (2018). Stereodivergent evolution of artificial enzymes for the michael reaction. *Angew. Chem. Int. Ed.* 57 (19): 5288–5291.

80 Gao, S., Zhu, S., Huang, R. et al. (2018). Engineering the enantioselectivity and thermostability of a (+)-gamma-lactamase from microbacterium hydrocarbonoxydans for kinetic resolution of vince lactam (2-azabicyclo[2.2.1]hept-5-en-3-one). *Appl. Environ. Microbiol.* 84 (1): e01780–e01717.

81 Reviews and selected papers of artificial metalloenzymes in which CAST/ISM was used to increase activity and stereoselectivity (although this acronym was not always used):(a)Ilie, A. and Reetz, M.T. (2015). Directed evolution of artificial metalloenzymes. *Isr. J. Chem.* 55: 51–60. (b) Reetz, M.T., Peyralans, J.J.P., Maichele, A. et al. (2006). Directed evolution of hybrid enzymes: evolving enantioselectivity of an achiral Rh-complex anchored to a protein. *Chem. Commun.* 41: 4318–4320. (c) Creus, M., Pordea, A., Rossel, T. et al. (2008). X-ray structure and designed evolution of an artificial transfer hydrogenase. *Angew. Chem. Int. Ed.* 47: 1400–1404. (d) Durrenberger, M. and Ward, T.R. (2014). Recent achievements in the design and engineering of artificial metalloenzymes. *Curr. Opin. Chem. Biol.* 19: 99–106. (e) Kan, S.B.J., Lewis, R.D., Chen, K., and Arnold, F.H. (2016). Directed evolution of cytochrome c for carbon–silicon bond formation: bringing silicon to life. *Science* 354 (6315): 1048–1051. (f) Prier, C.K., Zhang, R.K., Buller, A.R. et al. (2017). Enantioselective, intermolecular benzylic C–H amination catalysed by an engineered iron-haem enzyme. *Nat. Chem.* 9: 629–634. (g) Chen, K., Huang, X., Kan, S.B.J. et al. (2018). Enzymatic construction of highly strained carbocycles. *Science* 360 (6384): 71–75. (h) Brandenberg, O.F., Prier, C.K., Chen, K. et al. (2018). Stereoselective enzymatic synthesis of heteroatom-substituted cyclopropanes. *ACS Catal.* 8: 2629–2634. (i) Knight, A.M., Kan, S.B.J., Lewis, R.D. et al. (2018). Diverse engineered heme proteins enable stereodivergent cyclopropanation of unactivated alkenes. *ACS Central Sci.* 4 (3): 372–377.

82 (a) Key, H.M., Dydio, P., Clark, D.S., and Hartwig, J.F. (2016). Abiological catalysis by artificial haem proteins containing noble metals in place of iron. *Nature* 534: 534–537. (b) Dydio, P., Key, H.M., Nazarenko, A. et al. (2016). An artificial metalloenzyme with the kinetics of native enzymes. *Science* 6308: 102–106.

Part II

Rational and Semi-Rational Design

6

Data-driven Protein Engineering

Jonathan Greenhalgh[1,], Apoorv Saraogee[1,*], and Philip A. Romero[1,2]*

[1] *University of Wisconsin, Department of Chemical and Biological Engineering, 1415 Engineering Dr, Madison, WI 53706, USA*
[2] *University of Wisconsin, Department of Biochemistry, 433 Babcock Dr, Madison, WI 53706, USA*

6.1 Introduction

A protein's sequence of amino acids encodes its function. This "function" could refer to a protein's natural biological function, or it could also be any other property including binding affinity towards a particular ligand, thermodynamic stability, or catalytic activity. A detailed understanding of how these functions are encoded would allow us to more accurately reconstruct the tree of life and possibly predict future evolutionary events, diagnose genetic diseases before they manifest symptoms, and design new proteins with useful properties. We know that a protein sequence folds into a three-dimensional structure, and this structure positions specific chemical groups to perform a function; however, we are missing the quantitative details of this sequence-structure-function mapping. This mapping is extraordinarily complex because it involves thousands of molecular interactions that are dynamically coupled across multiple length and time scales.

Computational methods can be used to model the mapping from sequence to structure to function. Tools such as molecular dynamics simulations or Rosetta use atomic representations of protein structures and physics-based energy functions to model structures and functions [1–3]. While these models are based on well-founded physical principles, they often fail to capture a protein's overall global behavior and properties. There are numerous challenges associated with physics-based models including consideration of conformational dynamics, the requirement to make energy function approximations for the sake of computational efficiency, and the fact that, for many complex properties such as enzyme catalysis, the molecular basis is simply unknown [4]. In systems composed of thousands of atoms, the propagation of small errors quickly overwhelms any predictive accuracy. Despite tremendous breakthroughs and research progress over the last century, we still lack the key details to reliably predict, simulate, and design protein function.

* These authors contributed equally to this work.

Protein Engineering: Tools and Applications, First Edition. Edited by Huimin Zhao.
© 2021 WILEY-VCH GmbH. Published 2021 by WILEY-VCH GmbH.

Machine learning and artificial intelligence are transforming marketing, finance, healthcare, security, internet search, transportation, and nearly every aspect of our daily lives. These approaches leverage vast amounts of data to find patterns and quickly make optimal decisions. In this chapter, we present how these ideas are starting to impact the field of protein engineering. Instead of physically modeling the relationships between protein sequence, structure, and function, data-driven methods use ideas from statistics and machine learning to infer these complex relationships from data. This top-down modeling approach implicitly captures the numerous and possibly unknown factors that shape the mapping from sequence to function. These statistical models complement physical models and can even be used to improve physics-based models. Statistical models have been used to understand the molecular basis of protein function and provide exceptional predictive accuracy for protein design. We present three key stages in data-driven protein engineering – (i) representation: how to encode protein sequence/structure/function data, (ii) learning: automatic detection of patterns and relationships in data, and (iii) prediction: applying the learned models to design new proteins.

6.2 The Data Revolution in Biology

The volume of biological data has exploded over the last decade. This is being driven by advances in our ability to read and write DNA, which are progressing faster than Moore's law [5]. Simultaneously, we have also gained unprecedented ability to characterize biological systems with advances in automation, miniaturization, multiplex assays, and genome engineering. It is now a routine to perform experiments on thousands to millions of molecules, genes, proteins, and/or cells. The resulting data provides a unique opportunity to study biological systems in a comprehensive and less biased manner.

Protein sequence and structure databases have been growing exponentially for decades (Figure 6.1b,c). Currently, the UniProt database [6] contains over 100 million unique protein sequences and the Protein Data Bank [7] contains over 1 00 000 experimentally determined protein structures. While there is an abundance of protein sequence and structure data, there is still relatively little data mapping sequence to function. ProtaBank is a new effort to build a protein function database [8]. Function data is challenging to standardize because it is highly dependent on experimental conditions and even the particular researcher that performed the experiments. Therefore, statistical modeling approaches are most useful on data that is generated by an individual researcher/research group. This allows for a consistent definition of "function" that is not influenced by uncontrolled experimental factors.

Many sequence-function data sets are generated by protein engineering experiments that involve screening libraries of sequence variants for improved function. These variants may include natural homologs, random mutants, targeted mutants, chimeric proteins generated by homologous recombination, and computationally designed sequences. Each of these sequence diversification methods explores different features of the sequence-function mapping and varies in their information

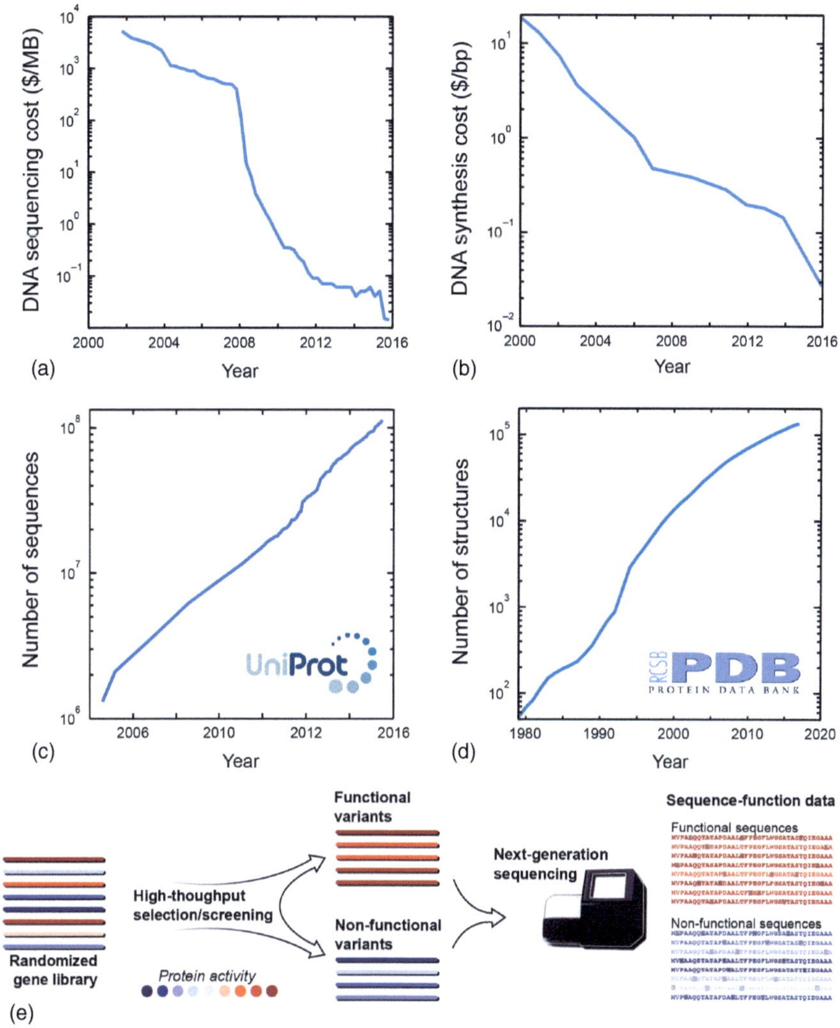

Figure 6.1 The growth of biological data. (a, b) DNA sequencing and synthesis technologies are advancing faster than Moore's law. As a result, costs have decreased exponentially over the last two decades. (c, d) Large-scale genomics, metagenomics, and structural genomics initiatives have resulted in exponential growth of protein sequence and structure databases. (e) Deep mutational scanning experiments combine high-throughput screens/selections with next-generation DNA sequencing to map sequence–function relationships for thousands to millions of protein variants.

content. Important factors include the sequence diversity of a library, the likelihood of functional versus nonfunctional sequences, and the difficulty/cost of building the desired gene sequences.

Recent advances in high-throughput experimentation have enabled researchers to map sequence-function relationships for thousands to millions of protein variants [9, 10]. These "deep mutational scanning" experiments start with a large library of protein variants, and this library is passed through a high-throughput screen/selection to separate variants based on their functional properties (Figure 6.1e). The genes from these variant pools are then extracted and analyzed using next-generation DNA sequencing. Deep mutational scanning experiments generate data containing millions of sequences and how those sequences map to different functional classes (e.g. active/inactive, binds ligand 1/binds ligand 2). The resulting data have been used to study the structure of the protein fitness landscape, discover new functional sites, improve molecular energy functions, and identify beneficial combinations of mutations for protein engineering [9, 11–13].

6.3 Statistical Representations of Protein Sequence, Structure, and Function

The growing trove of biological data can be mined to understand the relationships between protein sequence, structure, and function. This complex and heterogenous protein data needs to be represented in simple, machine-readable formats to leverage advanced tools in pattern recognition and machine learning. There are many possible ways of representing proteins mathematically including simple sequence-based representations or more advanced structure/physics-based representations. In general, a good representation is low dimensional but still captures the system's relevant degrees of freedom.

6.3.1 Representing Protein Sequences

A protein's amino acid sequence contains all the information necessary to specify its structure and function. Each position in this sequence can be modeled as a categorical variable that can take on one of twenty amino acid values. Categorical data can be represented using a one-hot encoding strategy that assigns one bit to each possible category. If a particular observation falls into one of these categories, it is assigned a "1" at that category's bit, otherwise it is assigned a "0." A protein sequence of length l can be represented with a vector of $20l$ bits; 20 bits for each sequence position (Figure 6.2). For example, assuming the amino acid bits are arranged in alphabetical order (A, C, D, E, ..., W, Y), if a protein has alanine (A) at the first position, the first bit would be 1 and the next 19 bits would be 0. If a protein has aspartic acid (D) at the first position, the first two bits would be 0, the third bit 1, and the next 17 bits 0. This encoding strategy can be applied to all amino acid positions in a protein and represent any sequence of length l. One-hot encoding sequence representations are widely used in machine learning because they are simple and flexible. However, they are also very high dimensional ($20l \approx$ thousands of variables for most proteins) and therefore require large quantities of data for learning.

Machine learning is widely used in the fields of text mining and natural language processing to understand sequences of characters and words. The tools word2vec and doc2vec use neural networks to learn vector representations that encode the

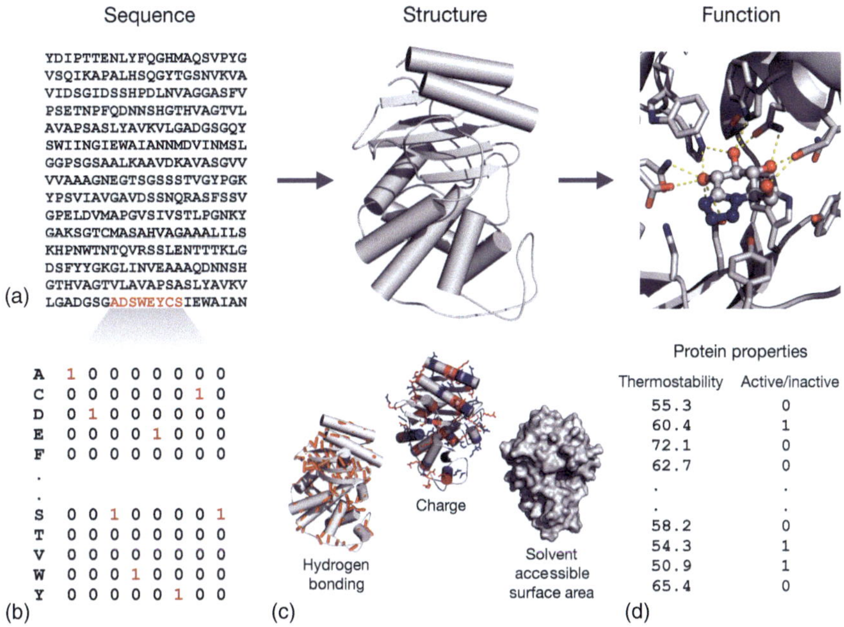

Figure 6.2 Sequence, structure, and function representations. (a) A protein's sequence folds into a three-dimensional structure, and this structure determines its function and properties. (b) Protein sequences can be represented using a one-hot encoding scheme that assigns 20 amino acid bits to each residue position. A bit is assigned a value of "1" if the protein has the corresponding amino acid at a particular residue position. (c) Structure-based representations use modeled protein structures to extract key physiochemical properties such as hydrogen bonds, total charge, or molecular surface areas. (d) Protein functions can be continuous properties such as thermostability or catalytic efficiency, or discrete properties such as active/inactive. Discrete properties can be represented using a binary (0 or 1) encoding.

linguistic context of words and documents [14, 15]. These embeddings attempt to capture word/document "meaning" and are much lower dimensional than the original input space. Similar concepts have recently been applied to learn embedded representations of amino acid sequences [16]. This approach breaks amino acid sequences into all possible subsequences of length k. These subsequences are referred to as k-mers. As an example, the sequence PRFYLA contains the four 3-mers: PRF, RFY, FYL, and YLA. An amino acid sequence's k-mers are treated as "words" and a neural network is used to learn other words that are found before/after a given word (i.e. a word's context). Importantly, words that are found in similar contexts tend to have similar meanings. This concept can be used to build low-dimensional vector spaces that place similar words close together. For an amino acid sequence, this might mean that one amino acid triplet is comparable with another, and therefore, we only need one variable to represent both. This produces a low-dimensional representation or "protein embedding" that captures the entire protein sequence. These protein embeddings can then be used to model specific properties such as thermostability.

6.3.2 Representing Protein Structures

The properties of proteins depend on sequence through their structure, therefore structure-based representations provide a more direct link to function. Experimentally determining a protein's three-dimensional structure (via crystallography, NMR, CryoEM) is significantly more challenging and time consuming than determining sequence or function. Therefore, most sequence-function data sets do not contain experimentally determined protein structures. Instead, this missing structural information can be approximated by taking advantage of the extreme conservation of structures within a family. Homologous proteins with as low as 20% sequence identity still have practically identical three-dimensional structures [17].

A protein's overall fold can be represented by specifying which residues are "contacting" in the three-dimensional structure. These contacting residues could be defined as any pair of residues that has an atom within five angstroms. Other contact definitions could include different distance cutoffs, Cα–Cα distances, or Cβ–Cβ distances. A protein's contact map specifies all pairs of contacting residues and provides a coarse-grained description of the protein's overall fold. Importantly, contact maps are highly conserved within a protein family, and therefore any two evolutionarily related proteins have practically identical contact maps. If we assume a fixed contact map for a protein family, structural information can be represented using a one-hot encoding scheme similar to sequence encoding described earlier. Each pair of contacting residues can take on one of 400 (20^2) possible amino acid combinations, which can be one-hot encoded using 400 bits. Therefore, the structure of a protein with c contacts can be represented with $400c$ bits. In contrast to sequence-based representations, this contact-based representation can capture pairwise interactions between residues. However, this increased flexibility comes at the cost of significantly higher dimensionality.

Three-dimensional protein structures can also be predicted using molecular modeling and simulation software. Most protein sequence-function data sets can take advantage of homology modeling approaches that start with a closely related template structure, mutate differing residues to the target sequence, and run minimization methods to relax the structure into a local energy minimum. State-of-the-art homology modeling methods can reliably predict protein structures with less than 2 Å atomic RMSD [18]. These predicted structures can be analyzed to extract key physiochemical properties such as surface areas, solvent exposure, and physical interactions (Figure 6.2). This approach was recently applied to model the kinetic properties of β-glucosidase point mutants [19]. The substrate was docked into β-glucosidase homology models, and this enzyme–substrate interaction was used to extract 59 physical features such as interface energy, number of intermolecular hydrogen bonds, and change in solvent accessible surface area. A simple linear regression model could relate these physical features to β-glucosidase turnover number, Michaelis constant, and catalytic efficiency. Physics-based representations tend to be lower dimensional than the sequence and contact encodings described earlier. They may also have good generalization within a protein family or even across protein families because they are based on fundamental biophysical principles.

6.4 Learning the Sequence-Function Mapping from Data

Advanced pattern recognition and machine learning techniques can be used to automatically identify key relationships between protein sequence, structure, and function. These tools are used for two primary tasks: supervised learning and unsupervised learning. Supervised methods, such as regression and classification, attempt to learn the mapping between a set of input variables and output variables. The term "supervised learning" arises because the algorithms are given examples of input–output mapping to guide the learning process. In contrast, unsupervised methods are not given information about the output variable, but instead try to learn relationships between the various input variables. Similar concepts have been used extensively in quantitative structure-activity relationship (QSAR) models, which are typically used to predict the chemical and biological properties of small molecules [20]. QSAR models have also been applied to peptide and DNA sequences [21, 22].

6.4.1 Supervised Learning (Regression/Classification)

Regression is a supervised learning technique that is used to model and predict continuous properties. Continuous protein properties could include thermostability, binding affinity, or catalytic efficiency. Regression methods span from simple linear models to advanced, nonlinear models such as neural networks.

Linear regression is the simplest regression technique and applies fixed weights to each input variable. A linear model is described by the following equation:

$$y = X\beta + \varepsilon$$

where y is the vector of continuous output variables, X is the matrix of sequence/structure features (one protein variant per row), β is the weight vector, and ε is the model error. The model parameters (β) can be estimated by minimizing the sum of the squared error. This least-squares parameter estimate has an analytical solution:

$$\hat{\beta} = \left(X^T X\right)^{-1} \left(X^T y\right)$$

Here, $\hat{\beta}$ corresponds to an estimate of the true β. $\hat{\beta}$ can then be applied to new proteins to predict their properties:

$$\hat{y} = X_{\text{new}}\hat{\beta}$$

Linear regression provides a simple framework for relating sequence/structure to function, and predicting the properties of previously uncharacterized proteins.

Linear regression has been used to model chimeric cytochrome P450 thermostability [23]. A library of chimeric P450s was generated by shuffling sequence elements from three related bacterial P450s [24]. The thermostability of 184 randomly chosen chimeric P450s was determined, and a linear regression model was used to relate sequence to thermostability. Each chimeric protein's sequence was one-hot encoded by specifying, which sequence elements were present. This encoding scheme is

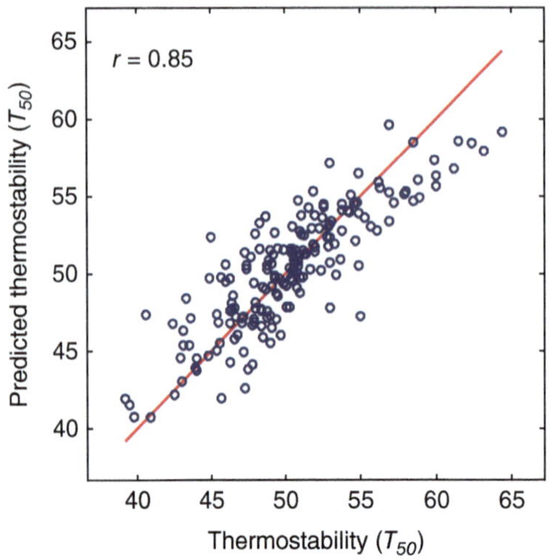

Figure 6.3 A linear regression model for cytochrome P450 thermostability. This model relates sequence blocks of chimeric P450s to their thermostability values. The plot shows the model's cross-validated predictions for 184 chimeric P450s.

similar to the sequence-based one-hot encoding described earlier, but sequence "blocks" are used rather than individual amino acids. This simple regression model revealed a strong correlation between the predicted and observed thermostability (Figure 6.3). The model was applied to predict the thermostabilities of all 6351 possible sequences in the chimeric P450 library, and the most stable predicted sequences were validated experimentally.

Supervised learning methods, including linear regression, are highly susceptible to overfitting data. A linear model must have at least as many data points as model parameters to avoid overfitting. More complex nonlinear models require even more data. Overfitting occurs when there is not sufficient data and the model fits spurious correlations or noise, rather than the true underlying signal. An overfit model will display very small error on the training data, but large prediction error on new data points.

All statistical models must be evaluated for overfitting and their ability to generalize to new, unseen data points. One method for model validation involves training the model on some fraction of the data and using the remainder to evaluate the model's predictive ability. For example, one could train a model on 60% of the data and test the model on the remaining 40%. This holdout method is simple to implement, but also throws out valuable information because the model is not learning from the entire data set. Cross-validation is another method for model evaluation that more effectively utilizes the available data. Cross-validation is similar to the holdout method, but rotates through multiple training set-test set combinations. For example, 10-fold cross-validation breaks the data into ten subsets; a model is trained on nine of these subsets and used to predict the tenth subset. This process is repeated over all 10 data folds (i.e. testing on all 10 subsets) and the results are averaged. Cross-validation allows all data points to be used in model training and evaluation.

Overfitting can be reduced using regularization methods that favor simpler models. Regularized parameter estimation involves minimizing the model's squared error in addition to the magnitude of the model parameters. This can be achieved by including a penalty term on the norm of the parameter vector:

$$\min_{\beta} (X\beta - y)^2 + \lambda ||\beta||_n$$

Here, the first term corresponds to the model's squared error, the second term is the magnitude of the model parameters, and λ tunes the relative influence of these two terms. n determines the type of vector norm and is typically equal to 0, 1, or 2. L0 regularization ($n = 0$) penalizes the total number of non-zero parameters in the model, L1 regularization ($n = 1$) penalizes the sum of the parameter absolute values, and L2 regularization ($n = 2$) penalizes the sum of the squared parameters. This minimization problem can be solved analytically if $n = 2$ or using convex optimization if $n = 1$. The hyperparameter λ can be determined using cross-validation. Combinations of these penalties can also be used, such as elastic net regression, which utilizes both L1 and L2 norms.

While regression methods model continuous properties, classification methods are used to model discrete protein properties such as folded/unfolded or active/inactive. Classifiers are especially important for modeling data generated by high-throughput methods such as deep mutational scanning because these methods often bin proteins into broad functional classes. Classification methods try to relate input feature vectors to functional classes (e.g. active/inactive or folded/unfolded). Like the regression models discussed earlier, classification models can be evaluated using cross-validation, and regularization can be used to prevent overfitting.

Logistic regression is simple classification method that transforms a linear model through the logistic (sigmoid) function to produce binary outputs. The name "logistic regression" is a misnomer because it actually performs classification rather than regression. Logistic regression parameters can be identified using iterative methods or convex optimization. Logistic regression was recently used to refine molecular energy functions for designing de novo miniproteins [25]. Thousands of miniproteins were designed using Rosetta protein design software, and these designs were screened for folding using a high-throughput yeast display assay. Each protein's structure was modeled and used to generate physical input features such as number of H-bonds, Lennard-Jones energies, and net charge. Logistic regression was then used to map these physical features to whether a design was successful or unsuccessful. The statistical model revealed that a protein's buried nonpolar surface area was a dominant factor in determining design success. The logistic regression model was used to rank designs and drastically improved the rate of successful designs.

Kernel methods are another modeling approach that is widely used in machine learning and bioinformatics. In contrast to the parametric regression/classification methods described earlier, kernel methods do not require input feature vectors, but instead a user defined similarity function (or kernel function) is used to compute the "implicit features" by comparing pairs of data points. Kernel methods are more effective at dealing with high dimensional problems than parametric models because they do not have to store large parameter matrices. The similarity function could be as simple as an inner product between feature vectors, or it can represent more complex, potentially infinite dimensional, relationships between data points [26]. This flexibility allows them to learn from unstructured objects such as biological systems. Popular kernel methods include support vector machines (SVMs) and Gaussian process (GP) regression/classification.

GPs use kernel functions to define a prior probability distribution over a function space. This allows predictions of both the function mean and its confidence intervals. GPs have been used to model stability and activity of cytochrome P450s [27]. A structure-based kernel function was developed to define structural similarity between pairs of proteins. GP regression using this kernel function explained 30% more of the variation in P450 thermostability in comparison with linear regression and sequence-based kernels. The structure-based kernel was also used to model enzyme activity and binding affinity for several P450 substrates.

6.4.2 Unsupervised/Semisupervised Learning

Unlike supervised learning, where the data is labeled or categorized, in unsupervised learning there are no labels associated with each data point. Unsupervised learning can be used to find patterns such as clusters or correlations within data. The main drawback of unsupervised techniques is that the outputs are unknown, i.e. there is no mapping to protein function. However, these techniques still provide valuable information about proteins because of the massive amount of protein sequence data that is currently available. Examples of unsupervised methods include clustering, where data points are grouped based on similarity, and principal component analysis (PCA). PCA is a projection of data onto lower dimensional space in a way that maximizes the variance of the projection. This converts high dimensional input variables into a set of uncorrelated principle components that are ranked based on their variance. These principle components can be used to reduce the dimensionality of a problem and identify important relationships among variables [28].

Unsupervised methods can be used to identify patterns in multiple sequence alignments (MSAs) of evolutionarily related proteins. Statistical coupling analysis (SCA) analyzes residue coevolution by performing PCA on a protein family's MSA [29]. The dominant principle components consist of positions that coevolve and can reveal networks of spatially connected amino acids called protein sectors (Figure 6.4). Protein sectors have been demonstrated to play roles in protein dynamics and allostery and may represent functional modules [30, 31]. EVmutation is another unsupervised method that models natural sequence variation and simultaneously considers epistasis (non-independence of mutational effects) [32]. Although EVmutation is only parameterized on an MSA (i.e. it is unsupervised), it is capable of predicting the functional effects of amino acid substitutions and residue interdependencies.

Semisupervised methods learn from data sets that contain both unlabeled and labeled data points. Semisupervised approaches can be used in protein engineering to transfer knowledge across protein families. A semisupervised approach was recently developed that trained an unsupervised embedding model (doc2vec) on a large protein sequence database [16]. These embeddings were then used as the inputs for supervised GP regression. This approach was used to model

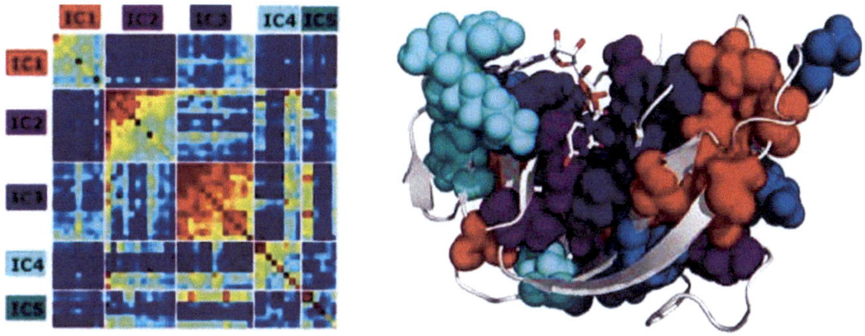

(a) (b)

Figure 6.4 Unsupervised learning from protein sequences. (a) Statistical coupling analysis of the RNase superfamily reveals five independent components (ICs) that correspond to groups of coevolving residues. (b) These five ICs form contiguous "sectors" in the three-dimensional protein structure. Source: Adapted from Narayanan et al. [31].

channelrhodopsin membrane localization, P450 thermostability, and epoxide hydrolase enantioselectivity.

6.5 Applying Statistical Models to Engineer Proteins

Statistical modeling approaches provide unprecedented predictive accuracy for a wide variety of complex protein functions/properties. These models can be used to understand protein function and design new proteins. In addition, many classes of statistical models can provide confidence intervals for their predictions. These confidence intervals can be used to gauge whether a prediction is valid or if it contains too much uncertainty to be useful. We discuss several protein engineering strategies that leverage the predictive power of statistical models.

The most straightforward data-driven protein engineering approach involves training a model on a data set and then extrapolating that model to design best predicted sequences. This method was applied to engineer thermostable fungal cellobiohydrolase class II (CBHII) cellulases [33]. A panel of 33 chimeric CBHIIs was characterized for their thermal inactivation half-lives at elevated temperatures. This data was used to train a linear regression model that related sequence blocks to thermal tolerance. This model was then used to design 18 chimeras that were predicted to have enhanced stability relative to the parent enzymes. Most of these designed CBHII chimeras could hydrolyze cellulose at higher temperatures than most stable parent. A key feature of this extrapolation-based design approach is a relatively small training set (<1% of possible chimeras) can be used to make predictions over a massive combinatorial sequence space. The CBHII regression model also pointed to a single sequence block that contributed over 8 °C of thermostability [34]. Further analysis revealed that a single amino acid substitution in that block (C313S) was responsible for the elevated thermostability. This example highlights how statistical models can be used to uncover molecular mechanisms contributing to protein function.

It is important to consider the space of sequences that a statistical model can make valid predictions on. This prediction domain is highly dependent on the model's sequence/structure representation. For example, consider a model that uses one-hot encoding to represent protein sequences. This model can only learn the effect of amino acids that are observed in the training set, and therefore can only make predictions about sequences composed of combinations of these observed amino acids. Representations that include information about amino acid properties and/or protein structure can broaden a model's prediction domain. Representations that use three-dimensional structural models to extract key physiochemical properties have potential to generalize well within a protein family and even across protein families.

Statistical models can be incorporated into an iterative directed evolution framework. ProSAR uses a statistical model to guide the search for beneficial mutations [35]. This model consists of a one-hot encoded sequence representation and a partial least squares linear regression model to relate sequence to function. A mutational library is screened, and the model classifies each amino acid substitution as deleterious, neutral, beneficial, or underdetermined (i.e. needing more information). Substitutions that are beneficial or underdetermined are combined with new substitutions in the next round, and this screen-and-learn process is repeated over multiple rounds. The ProSAR method was used to engineer bacterial halohydrin dehalogenases (HHDH) to perform a cyanation reaction important for the synthesis of the cholesterol-lowering drug Lipitor [35]. Eighteen rounds of ProSAR yielded HHDH variants with over 35 mutations and increased the volumetric productivity of target reaction by ~4000-fold. More recently, ProSAR-driven evolution was used to evolve ultra-stable carbonic anhydrase variants (107 °C thermostability at pH 10 in 4.2 M solvent) that enhanced the rate of CO_2 capture by 25-fold over the natural enzyme [36].

Statistical models can also be used in an active learning setting that very efficiently explores protein sequence space. Active learning involves sequentially designing an informative experiment, performing that experiment, learning from the resulting data, and repeating the process over multiple cycles (Figure 6.5a). For protein engineering, the active learning algorithm must first learn the sequence-function mapping and then apply this knowledge to design optimized sequences. The primary challenge is how to allocate experimental resources towards understanding the sequence-function mapping versus designing optimized sequences. This trade-off is referred to as the "exploration-exploitation dilemma," and the objective is to minimize the amount of exploration that is needed to predict optimized sequences. Upper confidence bound (UCB) algorithms provide a principled framework for trading off between exploration and exploitation modes [37]. The UCB algorithm iteratively selects the point with the largest UCB (predicted mean plus confidence interval) and therefore encourages sampling of points that are simultaneously optimized and uncertain (Figure 6.5b). A UCB search algorithm was combined with a GP regression model to optimize cytochrome P450 thermostability [27]. Eight rounds of UCB optimization identified thermostable P450s that were more stable than variants made by rational design, recombination, or directed evolution.

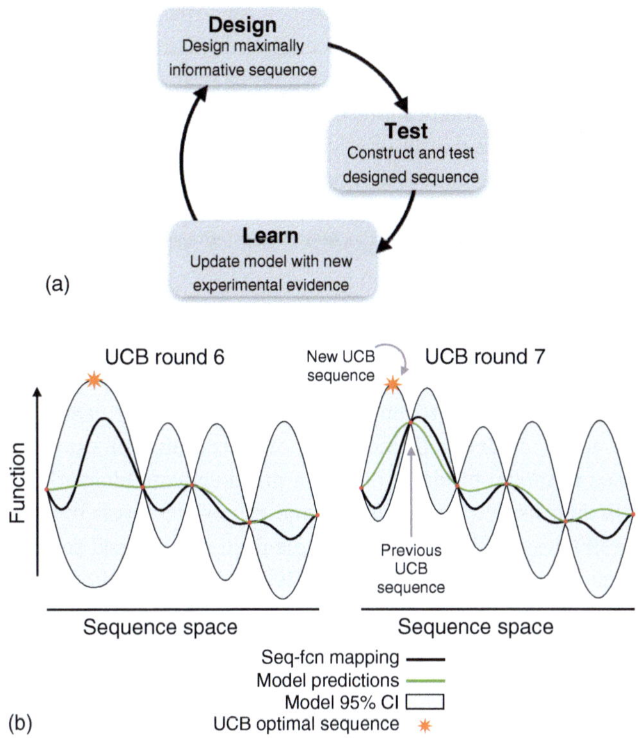

Figure 6.5 Active machine learning. (a) Active learning involves designing maximally informative sequences, experimentally characterizing these sequences, learning from the resulting data, and repeating this process over multiple iterations. (b) Upper-confidence bound (UCB) optimization involves iteratively selecting the sequence with the largest upper confidence bound (mean + confidence interval). The schematic illustrates sequence space in one dimension and the true mapping from sequence to function as a black line. Characterized sequences (small red dots) have accurate model predictions and small confidence intervals. The first panel shows five characterized sequences, which cause the model to propose one UCB optimal sequence (marked with a star). The second panel shows the results after this UCB optimal sequence is characterized – this causes a new UCB sequence to be proposed. This iterative process is guaranteed to efficiently converge to the optimal point.

6.6 Conclusions and Future Outlook

The protein sequence-structure-function mapping involves thousands of interacting atoms, a practically infinite number of dynamic conformational states, and physical processes that span multiple length and time scales. This mapping is extremely difficult to model from a physical perspective. In contrast, statistical methods are able to learn complex interrelationships directly from experimental data. This top-down understanding of complex systems allows discovery of new functional mechanisms and provides exceptional predictive accuracy.

This chapter provides an overview of emerging data-driven approaches to model and engineer proteins. We have described statistical representations of proteins, how these representations can be used to learn from data, and practical protein engineering applications of these models. As a relatively new field, there is still significant room for improving these methods, especially in the area of sequence and structure representations. Ideal representations would be sparse, but still have a broad prediction domain. These representations may integrate different sources of information (evolutionary, biochemical, and physical) into a single unified model. Advanced machine learning methods such as dictionary learning and deep learning attempt to learn new representations directly from data and could play an important role in protein modeling. Another key challenge for the field is data access and sharing. While there are many interesting sequence-structure-function data sets, they are often buried in a publication's supplemental information and very difficult to parse/organize. Efforts to share data on public repositories and databases such as ProtaBank will greatly accelerate progress in the field.

In addition to proteins, statistical approaches can be used to model genotype-phenotype relationships across all levels of biological organization. For example, linear regression was used to model product titers in a multi-enzyme biosynthetic pathway; this model was then used to optimize enzyme expression levels to maximize overall product production [38]. Another example used compressed sensing methods to model a protein's DNA-binding specificity [39]. Statistical methods have been widely used in genetics relate phenotypes to genetic loci using quantitative trait locus (QTL) mapping [40].

Data-driven approaches are transforming every field of science and engineering. This revolution has been triggered by the confluence of advances in data generation, data access, and data analysis/interpretation. Advanced experimental technologies are allowing us to analyze biological systems on an unprecedented scale and resolution. The resulting data is also becoming readily accessible through large, public biological databases and repositories. At the same time, there have been tremendous advances in artificial intelligence and pattern recognition. Widespread interest in machine learning has also driven improvements in software packages such as the Scikit-learn and Keras deep learning Python libraries. Data-driven approaches leverage the continuously expanding sea of data and will play an increasingly important role in biological discovery and engineering.

References

1 Lazaridis, T. and Karplus, M. (2000). Effective energy functions for protein structure prediction. *Curr. Opin. Struct. Biol.* 10 (2): 139–145.
2 Li, Z., Yang, Y., Zhan, J. et al. (2013). Energy functions in de novo protein design: current challenges and future prospects. *Annu. Rev. Biophys.* 42: 315–335.
3 Bradley, P., Misura, K.M., and Baker, D. (2005). Toward high-resolution de novo structure prediction for small proteins. *Science* 309 (5742): 1868–1871.
4 Baker, D. (2010). An exciting but challenging road ahead for computational enzyme design. *Protein Sci.* 19 (10): 1817–1819.

5 Kosuri, S. and Church, G.M. (2014). Large-scale de novo DNA synthesis: technologies and applications. *Nat. Methods* 11 (5): 499–507.

6 Apweiler, R., Bairoch, A., Wu, C.H. et al. (2017). UniProt: the universal protein knowledgebase. *Nucleic Acids Res.* 45: D158–D169.

7 Rose, P.W., Prlic, A., Altunkaya, A. et al. (2017). The RCSB protein data bank: integrative view of protein, gene and 3D structural information. *Nucleic Acids Res.* 45: D271–D281.

8 Wang, C.Y., Chang, P.M., Ary, M.L. et al. (2018). ProtaBank: a repository for protein design and engineering data. *Protein Sci.* https://doi.org/10.1002/pro.3406.

9 Hietpas, R.T., Jensen, J.D., and Bolon, D.N.A. (2011). Experimental illumination of a fitness landscape. *Proc. Natl. Acad. Sci. U.S.A.* 108 (19): 7896–7901.

10 Araya, C.L. and Fowler, D.M. (2011). Deep mutational scanning: assessing protein function on a massive scale. *Trends Biotechnol.* 29 (9): 435–442.

11 Romero, P.A., Tran, T.M., and Abate, A.R. (2015). Dissecting enzyme function with microfluidic-based deep mutational scanning. *Proc. Natl. Acad. Sci. U.S.A.* 112 (23): 7159–7164.

12 Whitehead, T.A., Chevalier, A., Song, Y. et al. (2012). Optimization of affinity, specificity and function of designed influenza inhibitors using deep sequencing. *Nat. Biotechnol.* 30 (May): 1–9.

13 Bloom, J.D. (2014). An experimentally determined evolutionary model dramatically improves phylogenetic fit. *Mol. Biol. Evol.* 31 (8): 1956–1978.

14 Mikolov, T., Chen, K., Corrado, G., and Dean, J. (2013). Efficient estimation of word representations in vector space. *CoRR* abs/1301.3. doi:https://doi.org/10.1162/153244303322533223.

15 Le, Q.V. and Mikolov, T. (2014). Distributed representations of sentences and documents. *CoRR* abs/1405.4. doi:https://doi.org/10.1145/2740908.2742760.

16 Yang, K.K., Wu, Z., Bedbrook, C.N., and Arnold, F.H. (2018). Learned protein embeddings for machine learning. *Bioinformatics* 34 (15): 2642–2648.

17 Chothia, C. and Lesk, A.M. (1986). The relation between the divergence of sequence and structure in proteins. *EMBO J.* 5 (4): 823–826.

18 Misura, K.M., Chivian, D., Rohl, C.A. et al. (2006). Physically realistic homology models built with ROSETTA can be more accurate than their templates. *Proc. Natl. Acad. Sci.* 103 (14): 5361–5366.

19 Carlin, D.A., Caster, R.W., Wang, X. et al. (2016). Kinetic characterization of 100 glycoside hydrolase mutants enables the discovery of structural features correlated with kinetic constants. *PLoS One* https://doi.org/10.1371/journal.pone.0147596.

20 Dudek, A., Arodz, T., and Galvez, J. (2006). Computational methods in developing quantitative structure-activity relationships (QSAR): a review. *Comb. Chem. High Throughput Screening* 9 (3): 213–228.

21 Hellberg, S., Sjöström, M., Skagerberg, B., and Wold, S. (1987). Peptide quantitative structure–activity relationships, a multivariate approach. *J. Med. Chem.* 30 (7): 1126–1135.

22 Jonsson, J., Norberg, T., Carlsson, L. et al. (1993). Quantitative sequence-activity models (QSAM) – tools for sequence design. *Nucleic Acids Res.* 21 (3): 733–739.

23 Li, Y., Drummond, D.A., Sawayama, A.M. et al. (2007). A diverse family of thermostable cytochrome P450s created by recombination of stabilizing fragments. *Nat. Biotechnol.* 25 (9): 1051–1056.

24 Otey, C.R., Landwehr, M., Endelman, J.B. et al. (2006). Structure-guided recombination creates an artificial family of cytochromes P450. *PLoS Biol.* 4 (5): e112.

25 Rocklin, G.J., Chidyausiku, T.M., Goreshnik, I. et al. (2017). Global analysis of protein folding using massively parallel design, synthesis, and testing. *Science* 357 (6347): 168–175.

26 Rasmussen, C.E. and Williams, C. (2006). *Gaussian Processes for Machine Learning*. Cambridge, MA: MIT Press.

27 Romero, P.A., Krause, A., and Arnold, F.H. (2013). Navigating the protein fitness landscape with Gaussian processes. *Proc. Natl. Acad. Sci. U.S.A.* 110 (3): E193–E201.

28 Bishop, C.M. (2006). *Pattern Recognition and Machine Learning*, 1e. New York, NY: Springer.

29 Lockless, S.W. and Ranganathan, R. (1999). Evolutionarily conserved pathways of energetic connectivity in protein families. *Science* 286 (5438): 295–299.

30 Halabi, N., Rivoire, O., Leibler, S., and Ranganathan, R. (2009). Protein sectors: evolutionary units of three-dimensional structure. *Cell* 138 (4): 774–786.

31 Narayanan, C., Gagné, D., Reynolds, K.A., and Doucet, N. (2017). Conserved amino acid networks modulate discrete functional properties in an enzyme superfamily. *Sci. Rep.* 7 (1) https://doi.org/10.1038/s41598-017-03298-4.

32 Hopf, T.A., Ingraham, J.B., Poelwijk, F.J. et al. (2017). Mutation effects predicted from sequence co-variation. *Nat. Biotechnol.* 35 (2) https://doi.org/10.1038/nbt.3769.

33 Heinzelman, P., Snow, C.D., Wu, I. et al. (2009). A family of thermostable fungal cellulases created by structure-guided recombination. *Proc. Natl. Acad. Sci. U.S.A.* 106 (14): 5610–5615.

34 Heinzelman, P., Snow, C.D., Smith, M.A. et al. (2009). SCHEMA recombination of a fungal cellulase uncovers a single mutation that contributes markedly to stability. *J. Biol. Chem.* 284 (39): 26229–26233.

35 Fox, R.J., Davis, S.C., Mundorff, E.C. et al. (2007). Improving catalytic function by ProSAR-driven enzyme evolution. *Nat. Biotechnol.* 25 (3): 338–344.

36 Alvizo, O., Nguyen, L.J., Savile, C.K. et al. (2014). Directed evolution of an ultrastable carbonic anhydrase for highly efficient carbon capture from flue gas. *Proc. Natl. Acad. Sci.* 111 (46): 16436–16441.

37 Auer, P. (2002). Using confidence bounds for exploitation-exploration trade-offs. *J. Mach .Learn Res.* 3 (3): 397–422.

38 Lee, M.E., Aswani, A., Han, A.S. et al. (2013). Expression-level optimization of a multi-enzyme pathway in the absence of a high-throughput assay. *Nucleic Acids Res.* 41 (22): 10668–10678.

39 AlQuraishi, M. and McAdams, H.H. (2011). Direct inference of protein–DNA interactions using compressed sensing methods. *Proc. Natl. Acad. Sci.* 108 (36): 14819–14824.

40 Doerge, R.W. (2002). Mapping and analysis of quantitative trait loci in experimental populations. *Nat. Rev. Genet.* 3 (1): 43–52.

7

Protein Engineering by Efficient Sequence Space Exploration Through Combination of Directed Evolution and Computational Design Methodologies

Subrata Pramanik[1], *Francisca Contreras*[1], *Mehdi D. Davari*[1], *and Ulrich Schwaneberg*[1,2]

[1] *RWTH Aachen University, Institute of Biotechnology, Worringerweg 3, 52074 Aachen, Germany*
[2] *DWI – Leibniz Institute for Interactive Materials, Forckenbeckstraße 50, 52074 Aachen, Germany*

7.1 Introduction

Everybody uses an enzyme or enzyme product on a daily base, e.g. in food products (e.g. cheese), laundry, biofuels, pharmaceuticals, etc. [1–3]. Enzymes are by their natural design often not made to efficiently operate under conditions required for cost-effective production of chemicals and pharmaceuticals. Directed evolution advanced in the last decades to become a robust methodology for tailoring enzyme's properties to application demands in chemical synthesis and production of antibodies. The key methodologies in directed evolution are diversity generation and high-throughput screening [2, 4–7]. The importance of directed evolution for our society was acknowledged by awarding the Noble Prize in Chemistry in 2018 to Frances H. Arnold, George P. Smith, and Sir Gregory P. Winter for their pioneering work on the directed evolution of enzymes and the phage display of peptides and antibodies.

The beauty of directed evolution is that enzyme properties within physical boundaries can be improved in the absence of any structural knowledge or molecular understanding of the target property; prerequisite is that the selected property can be reflected in the screening system [2, 8]. Still today, a computational analysis of improved enzyme variants can lead to the discovery of fundamental design principles [4, 9–11], for example, disulfide bond engineering to improve thermostability [12], understanding polymers and protease interactions to boost the protease activity [13, 14], optimization of the pH profile of proteases through surface charge engineering [15], ionic liquid and deep eutectic solvent-activated cellulose CelA2 [16], and ionic liquid tolerance of laccase [17].

In this chapter, we highlighted protein engineering strategies that combine directed evolution and computational analysis to efficiently reengineer enzymes.

Protein Engineering: Tools and Applications, First Edition. Edited by Huimin Zhao.
© 2021 WILEY-VCH GmbH. Published 2021 by WILEY-VCH GmbH.

These combined strategies partially contribute to a molecular understanding of sequence–structure–function relationship, which can likely be transferred from enzymes to another. In the first part, we provided recent advancement of current computational methods, including framework for rapid enzyme stabilization by computational libraries (FRESCO) [18, 19], FoldX [20–22], constraint network analysis (CNA) [23], protein repair one-stop shop (PROSS) [24], protein sequence activity relationship (ProSAR) [25, 26], and their successful application to improve a broad range of enzyme properties. In the second part, we delineated different knowledge-based directed evolution methods, which combine directed evolution with computational analysis to explore the protein structure and function relationship. In this respect, an emphasis was given to the "KnowVolution" (knowledge gaining directed evolution; [4]) strategy, which is generally applicable, minimizes experimental efforts, generates a molecular understanding on each position/amino acid substitution, and was successfully applied to a broad range of enzymes and properties (e.g. laccase [17], sortase [27], sulfotransferase [ST] [28], and phytase [29]). Finally, we drew attention to a possible future of robust computational design by "*in silico*" methodologies through development of quantum computing and machine-learning approaches that hold promise to fully explore the protein sequence and thereby to efficiently make use of the vast design space that nature offers.

7.2 Protein Engineering Strategies

The combination of directed evolution and computer-aided rational design holds particular promise for optimizing enzyme properties, which are not yet exploited by natural selection. Since it is impossible to explore natural protein sequence space through directed evolution based on random mutagenesis methods, computer-aided "small, but smart" mutant libraries derived from sophisticated analysis promise a high chance to have variants with desired enzymatic properties [4, 10, 30, 31]. The present knowledge about directed evolution has been extensively reviewed elsewhere (see, e.g. [2, 4, 6, 32, 33]), and will not be further discussed in this chapter. So far, thermostability, resistance/tolerance in cosolvents (e.g. ionic liquids), catalytic efficiency, and product selectivities (e.g. enantioselectivity, regioselectivity) are the properties successfully improved by computer-aided rational design methods. In Section 7.2.1, we briefly review the computer-aided rational design methods (FRESCO [18, 19], FoldX [20–22], CNA [23], PROSS [24], ProSAR [25, 26]) and their successful applications to the broad range of properties and enzymes. In section 7.2.2, we briefly review the knowledge based directed evolution methods (iterative saturation mutagenesis [ISM] [34], Mutagenic Organized Recombination Process by Homologous IN vivo Grouping [MORPHING] [35], KnowVolution [4]) and their successful applications to the broad range of properties and enzymes.

7.2.1 Computer-Aided Rational Design

7.2.1.1 FRESCO

FRESCO is a computational library design method based on scanning of the entire protein structure to identify stabilizing disulfide bonds and point mutations, explores their effect by molecular dynamics (MD) simulations, and provides mutant libraries with variants that have a good chance to enhance stability [18]. FRESCO has been successful in engineering thermostability of haloalkane dehalogenase from *Sphingomonas paucimobilis* UT26s [36], functionalization (e.g. to improve thermostability and solvent-compatibility) of peptide amidase from *Stenotrophomonas maltophilia* [19, 37, 38].

FRESCO requires 3D protein structure as input to scan the entire protein structure to identify stabilizing disulfide bonds, point mutations, and explores their effect by MD simulations (details are shown in Figure 7.1a and Table 7.1). Three steps were

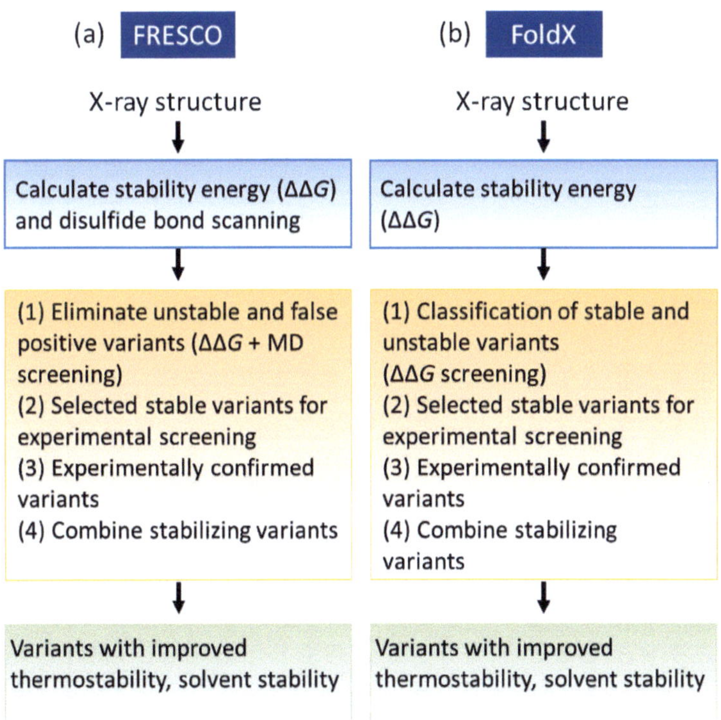

Figure 7.1 Framework of the principle strategy of (a) FRESCO [18, 19], (b) FoldX [20–22] methods (details are shown in Table 7.1). Calculation of the relative changes in folding free energy $\Delta\Delta G$ due to point substitutions based on the X-ray structure is the primary step to screen *in silico* variants in the both methods. Additionally, disulfide bond scanning is implemented in FRESCO [18, 19]. Source: (a) Wijma et al. [18, 19], (b) Delgado et al. [20], Schymkowitz et al. [22].

Table 7.1 Summary of computational methods on rational protein design approaches (workflows are shown in Figures 7.1 and 7.2).

Method	Input	Description	Improved property	Case study
FRESCO	X-ray structure	(i) Stability energy ($\Delta\Delta G$) calculation using Rosetta and FoldX, (ii) disulfide bond scanning, (iii) elimination of unreasonable and false positive variants based on $\Delta\Delta G$ and MD simulations screening, (iv) experimentally confirmed variants, (v) combine stabilizing variants	Thermostability, solvent stability	Thermostability of haloalkane dehalogenase from *Sphingomonas paucimobilis* UT26s [36], thermostability and solvent stability of peptide amidase from *Stenotrophomonas maltophilia* [37]
FoldX	X-ray structure	(i) Stability energy ($\Delta\Delta G$) calculation, (ii) classification of stabilizing and destabilizing variants, (iii) combine stabilizing variants	Thermostability, solvent stability	Thermostability of phosphotriesterase from *Pseudomonas oleovorans*, [39] and haloalkane dehalogenase from *Sphingomonas paucimobilis* [36], ionic liquid resistance of laccase lcc2 from *Trametes versicolor* [17]
CNA	X-ray structure	(i) Thermal unfolding simulation, (ii) network analysis, (iii) calculation of global and local indices, (iv) detection of structural weak spots	Thermostability	Thermostability of lipase A from *Bacillus subtilis* [40]
PROSS	X-ray structure	(i) Sequence conservation analysis, (ii) identification of likelihood position, (iii) stability energy (Rosetta $\Delta\Delta G$) calculation, (iv) combinatorial design within the reduced sequence space	Thermostability, solubility, solvent stability, catalytic efficiency	Thermostability of malaria invasion protein RH5 [41], hydrolytic activity of human butyrylcholinesterase [42], and catalytic efficiency of phosphotriesterase [43]
ProSAR	Experimental data set	(i) Generate features using sequence and 3D structure, (ii) model training using machine learning and artificial intelligence, (iii) test prediction including single and combine variants, (iv) repeat for next set of positions	Thermostability, solubility, solvent stability, catalytic efficiency, product selectivity	Enantioselectivity of epoxide hydrolase from *Aspergillus niger* [44], and nitric oxide dioxygenase (NOD) from *Rhodothermus marinus* [45]

It provides required input, methodology description, improved property, and case studies for successful application of each method.
Sources: Wallraf et al. [17], Floor et al. [36], Wu et al. [37].

applied in FRESCO methods. Firstly, point mutations were selected using input 3D protein structure, if they significantly improved the predicted free energy of protein folding. Secondly, disulfide bonds were designed using sampling of backbone conformational space, which tripled the number of experimentally stabilizing disulfide bridges. Thirdly, orthogonal *in silico* screening steps were used to remove chemically unreasonable substitutions and substitutions that are predicted to increase protein flexibility [18]. Fourthly, experimental screening is used to validate predicted mutant libraries. Finally, most stabilizing variants are combined based on stability energy and rigidity analysis using MD simulations (for a recent review, see [19]).

Recent studies have shown that FRESCO can likely be used to enhance thermostability, substrate specificity, and solvent tolerance [18, 19]. In contrast, properties such as stereoselectivity, regioselectivity, enantioselectivity, and pH profile might not be addressable since the stabilization energy is not crucial for such properties. However, implementation of mechanism-based geometry optimizations of active site with FRESCO approach could successfully redesign aspartase from *Bacillus* sp. YM55-1 for regio- and enantioselective hydroamination production [38, 46]. In summary, FRESCO is from our point of view best used to enhance thermostability of enzymes.

7.2.1.2 FoldX

FoldX is a computational protein design method based on an empirical force field, which was developed for the rapid evaluation of the effect of mutations on the stability, folding, and dynamics of proteins and nucleic acids [20, 22, 47]. The key applications of FoldX include energy minimization, *in silico* mutagenesis, calculation of free energy (ΔG), prediction of free energy changes between alternative structures ($\Delta\Delta G$), and alanine scanning of a protein based on its high-resolution 3D structure [22]. It is user-friendly and can be easily used with a graphical user interface as plugin in the protein structure visualizer yet another scientific artificial reality application (YASARA) [47]. Substitute residue function calculates the folding free energy difference ($\Delta\Delta G$) between the wild type and a variant of the enzyme ($\Delta\Delta G$ (kcal mol^{-1}) = $\Delta G_{\text{wild type}} - \Delta G_{\text{variant}}$). FoldX is also able to calculate energy difference ($\Delta\Delta G$) between the wild type and a variant with multiple residue substitutions.

FoldX has been successfully applied for different stability tests, especially when protein design was performed to predict whether distinct mutations are destabilizing. It has been instrumental in engineering the thermostability of various enzymes, including phosphotriesterase from *Pseudomonas oleovorans* [39], haloalkane dehalogenase from *S. paucimobilis* [36], limonene-1,2-epoxidehydrolase from *Rhodococcus erythropolis* [21], amine transaminase from *Aspergillus terreus* [48], and many more. Similarly, it was also successfully applied to predict destabilizing variants of various enzymes, such as thermoalkalophilic lipase from *Bacillus thermocatenulatus* [49], cellobiohydrolase Cel6A from *Humicola insolens* [50], and glycoside-hydrolase from *Neisseria polysaccharea* [51]. Detailed applications of different types of enzymes have been extensively reviewed by Buss et al. [21]. Moreover, our group also applied FoldX based stability calculations to engineer laccases with improved resistance in ionic liquids for using hardly soluble lignin lysates and increased tolerance toward high molarity of salts [17, 52].

FoldX requires 3D protein structure as input to calculate stability energy on empirical force field in which crystal structure quality is crucial for the accurate calculations (details are shown in Figure 7.1b and Table 7.1). In this regard, study carried out by Christensen and Kepp on fungal laccases *Trametes versicolor* demonstrated that FoldX is very structure-sensitive [53]. Likewise, another comparative study suggested to use only structures resolved in scales of near-atomic-resolutions [54]. It contains different subroutines such as RepairPDB, BuildModel, PrintNetworks, AnalyseComplex, stability, mutate residue/DNA, multiple residues, and so on [21, 47].

As FoldX is an energy-based method, it can firmly be used to enhance thermostability [17, 21, 52], and most likely other types of stability such as solvent stability [17, 52]. However, properties that are independent of stability energy such as substrate specificity, stereoselectivity, regioselectivity, enantioselectivity, and pH profile might not be achievable using FoldX based calculations. In summary, FoldX is from our point of view best used to enhance stability of enzymes.

7.2.1.3 CNA

CNA is a graph theory-based rigidity analysis method, which analyzes global and local flexibility, rigidity, and stability characteristics of enzymes by carrying out thermal unfolding simulations [23]. CNA is a user-friendly web server-based tool that predicts thermostability based on calculation of a set of global and local indices from temperature dependence unfolding simulations. Global indices, including floppy mode density, mean rigid cluster size identify phase transition points in which the structure switches from a rigid to a floppy state; these phase transition points can be related to the thermostability of enzymes [40, 55–57]. In order to identify residue-wise local stability, several local indices such as flexibility index, percolation index, and rigidity index are widely applied to identify potential weak spot sensitive thermostability. Rigidity index of zero indicates that region is always in a flexible region of the enzyme. Oppositely, the lower rigidity index indicates residue part of a rigid cluster during the thermal unfolding simulation. Similarly, lowest percolation index indicates most stable region of enzyme, whereas value zero indicates region is always in a flexible region of the enzyme [56].

Recently, Rathi et al. applied CNA approach to correlate the structural rigidity of a well-characterized set of 16 mutants from *Bacillus subtilis* lipase A (BSLA) with their thermodynamic thermostability [40]. This study suggested that improving structural stability is a common way to obtain a higher thermostability [40]. So far, CNA approach has not been widely applied yet on various enzymes to enhance thermostability and thus further studies should emphasize on implication of CNA to improve thermostability.

CNA only requires 3D protein structure as input to perform thermal unfolding simulations (details are shown in Figure 7.2a and Table 7.1). As it is network-based analysis from 3D structure, it is supposed to be sensitive toward crystal structure quality [23]. Although this method is not widely applied yet on various enzymes,

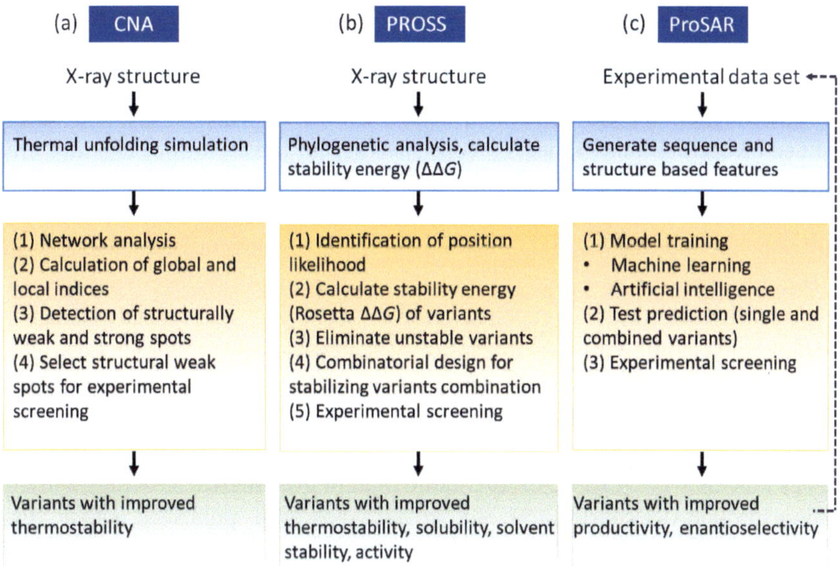

Figure 7.2 Framework of the principle strategy of (a) CNA [23], (b) PROSS [24], and (c) ProSAR [25, 26] methods (details are shown in Table 7.1). CNA detects structural weak spots based on thermal unfolding simulations using X-ray structure. Calculation of the relative changes in folding free energy $\Delta\Delta G$ of the point substitutions and phylogenetic analysis are the steps to screen *in silico* variants in PROSS [24] method. In the ProSAR method, experimental data of desired property (dependent variable) are used as input, which is usually split into the training and test sets. Independent variables are the generated features based on sequence and/or 3D structure of enzymes. Training set are used for model development by various algorithms such as regression analysis and machine learning, whereas trained models are used to predict the relevant properties for test set. Source: (a) Based on Kruger et al. [23], (b) Based on Goldenzweig and Fleishman [24], (c) Based on Fox [25].

it is very useful to enhance enzyme thermostability and solvent stability from our viewpoint.

7.2.1.4 PROSS

PROSS is a hybrid method in which Rosetta design and phylogenetic analysis are implemented to the entire protein except for the active or binding site [24, 58]. This method has been successfully applied to improve several protein properties, including thermostability of malaria invasion protein RH5 [41], hydrolytic activity of human butyrylcholinesterase [42], and catalytic efficiency of phosphotriesterase [43]. Details about these have been extensively reviewed elsewhere (see, e.g. [24]).

It requires 3D protein structure as input for multiple sequence alignment and stability energy calculations (details are shown in Figure 7.2b and Table 7.1). Diversity calculation can be performed up to 15 positions [58]. As it is network-based analysis from 3D structure, it is supposed to be sensitive toward crystal structure quality. It starts by two-step filtering of all amino acid identities that are likely to be destabilizing as single-point mutations relative to the wild-type native state. Firstly, it retains only mutations with at least a reasonable possibility of occurrence in the natural diversity according to a multiple sequence alignment of homologs. Secondly, Rosetta modeling is applied to scan all identities that pass the first filter and to eliminate single-point mutations that destabilize the native state [24, 58]. Taken together, PROSS combines aspects of consensus design and atomistic modeling, and thus may introduce both positive- and negative-design elements, including elements that address thermal stability and expressibility [24]. From our point of view, PROSS is best used for improvement of solubility, catalytic efficiency, and resistance to high temperature, denaturant, proteases, and non-physiological pH. Properties that are independent of stability energy such as substrate specificity, stereoselectivity, regioselectivity, and enantioselectivity might not be achievable using this approach.

7.2.1.5 ProSAR

ProSAR methodology is a computational multivariate library design and screening strategy [25, 26, 44, 45] based on the foundation of quantitative structure–activity relationship (QSAR) concept [59, 60]. In 2007, Fox et al. published a first comprehensive study on ProSAR approach in protein engineering by application of machine learning strategies based on partial least-squares (PLSs) regression [26]. ProSAR model was developed on halohydrin dehalogenase with the potential for use in the manufacture of ethyl (*R*)-4-cyano-3-hydroxybutyrate, which is a starting material for the production of the cholesterol-lowering drug atorvastatin [26]. Thereafter, advancement of ProSAR in protein engineering was quite slow in the community. Recently, an innovative sequence activity relationship (iSAR) model based on the combination of fast Fourier transform and PLS regression was developed using physicochemical properties of the amino acids to predict thermostability of cytochrome P450 and enterotoxins, binding affinity of tumor necrosis factor (TNF-α) [61]. Model showed a good predictive quality; however, novel mutations appeared in the test set that have not been featured in the training set, remain to be discovered [61]. In further work, machine-learning approach was applied to predict amino acid interactions, which was eventually applied to improve the enantioselectivity of an epoxide hydrolase from *Aspergillus niger* [44]. To minimize the expense of experimentally testing a large number of protein variants, Arnold and co-workers integrated machine learning with directed evolution [45]. Machine learning-guided directed evolution showed enrichment on the libraries and efficiently navigates large regions of sequence space to identify improved enantiomeric property of nitric oxide dioxygenase (NOD) from *Rhodothermus marinus*. As described in numerous reports, machine learning can be expected to serve as an assistive guide in directed evolution as depicted in the excellent reviews by Reetz and coworkers [62].

It requires screening and sequencing data from previous experimental study to generate training and test sets, in which training set are used for model development by various algorithms such as regression analysis and machine learning (details are shown in Figure 7.2c and Table 7.1). Model quality can be assessed by fitness parameter R^2 (at least ≥0.5) [63]. The trained model on the entire training set was then used to predict the relevant properties for test set sequences. Predictive quality on test set can be assessed by external validation parameter $R^2_{\text{prediction}}$ (at least ≥0.5)

[64]. Afterward, applicability and predictive quality of model can be evaluated by implementation on true external data set, which might be a property/function of similar enzymes.

In summary, ProSAR is a predictive modeling approach based on features/descriptors of variants and experimental data set. Thus, it might be applicable to predict broad range of enzyme properties from our point of view, including stability, stereo-, regio-, enantio-, and chemo-selectivities of enzymes reflecting reliable and sensitive features/descriptors and experimental data set.

7.2.2 Knowledge Based Directed Evolution

In order to improve the sampling of the vast universe of a protein sequence, directed evolution needs to meet rational design. Different approaches were developed in combination of directed evolution and the accumulated knowledge regarding the structure–function relationship between different properties and the enzymes. This knowledge can be used to generate smaller and more efficient libraries, where the molecular understanding of the addressed property could improve the overall design of new catalysts. In this section, we use three methods (ISM [34, 65], MORPHING [35], and KnowVolution [4]) to illustrate the combination of directed evolution and knowledge-based protein engineering and how this concludes in success stories of enzyme improvement.

7.2.2.1 Iterative Saturation Mutagenesis (ISM)

ISM is a technique where amino acid residues are selected and saturated in a sequential manner by multi- or single-site saturation mutagenesis. The selection of the mutation sites is aided by computational approaches and is used to generate more efficient libraries and enables to determine synergistic effects in mutations, thus decreasing the screening efforts [34]. The criteria for the multi-site amino acid selection will be dependent on the property that is addressed in combination with structural analysis. Engineering properties regarding stability, such as thermostability or solvent resistance, will require analysis on the protein structure, and a B-factor iterative test (B-FIT) can be applied [65]. On the other hand, improvement of the activity, stereoselectivity, and substrate scope will require analysis in the binding pocket of the protein; hence, a structure-based combinatorial active-site saturation test (CAST) can be applied [66]. ISM was used to improve the thermal resistance of the epoxide hydrolase (*A. niger* epoxide hydrolase [ANEH]) from *A. niger*, where the performance of ISM led to the improvement of $\Delta T_{50}{}^{60}$ values in the range of 14–21 °C in several variants [67]. ISM has also been used to change the enantioselectivity of lipase A of *Candida antarctica* toward α-substituted esters, where a 30-fold increase in activity was observed for most substrates and the developed variants reversed the (*S*)-selectivity of the wild type for most substrates [68].

7.2.2.2 Mutagenic Organized Recombination Process by Homologous *In Vivo* Grouping (MORPHING)

Combination of structural analysis and random mutagenesis can be used to decrease the screening efforts and a targeted library can be created. MORPHING is a method by which random mutations are introduced in the selected areas of the protein. Therefore, knowledge about the structure–function relationship of different protein areas is needed beforehand. After the mutagenesis process, the recombination machinery of yeast is utilized, and all the fragments are cloned *in vivo* in a one-pot reaction. Two case studies have been used to validate it: the oxidative stability of a versatile peroxidase (VP) was improved in high concentrations of H_2O_2 and the signal peptide of an unspecific peroxygenase (UPO) was evolved toward functional expression in yeast [35]. Another example is the UPO from *Agrocybe aegerita*, where a focused library in specific loops of the protein led to the identification of key positions that produced a shift in the peroxygenative:peroxidative activity ratio [69]. From our point of view, this method could be employed to improve properties of enzymes expressed in yeast, with a known structure–function relationship.

7.2.2.3 Knowledge Gaining Directed Evolution (KnowVolution)

The KnowVolution is an integrative protein engineering strategy by combining directed evolution with computational analysis for improving enzymes as a standard operation in process development (summarized in Figure 7.3). Over the last few years, KnowVolution has been successfully applied to engineer different properties of enzymes/proteins such as chiral separation [70], pH profile [71], surface binding [72], and solvent resistance [27]. In order to achieve a successfully engineered enzyme in a cost and time efficient manner, a process-optimized strategy is urgently essential to fulfil the demand of engineered enzymes/proteins. In this regard, KnowVolution has been designed into four phases (phase I–IV) (Figure 7.3). In phase I, a standard directed evolution round is performed for diversity generation. Depending on the gene length and the mutation frequency, approximately 20 of the most beneficial variants are sequenced to identify the substituted positions. Based on improvement of beneficial variants, ~12 positions can be identified that potentially contribute to the improved properties. In general, beneficial positions are thereby often clustered in close proximity within two to three regions of the targeted protein. In phase II, site-saturation mutagenesis (SSM) is performed at ~12 selected positions obtained from phase I. It was often observed that less than 50% of the saturated positions contribute to the improvement of the targeted properties. In general, it is expected that SSM at the targeted position should nearly always result in more improved variants compared with random mutagenesis as random mutagenesis methods usually generate 0–4 amino acid substitutions per residue [73]. Usually, 2–3 out of 5 SSM positions yield further improvements. Subsequently, sequencing is again performed at each SSM position. If beneficial variants are found, ~16 clones are sequenced or only a few are sequenced in case no beneficial variant is found in order to ensure that the SSM experiment worked properly (no wild-type background or faulty primer). Eventually, 4–6 positions with known substitutions should be identified from SSM results that contribute to an improvement of the targeted properties. Subsequently, further analysis of the obtained substitutions will provide detailed insights into the type of chemical interaction (charge, size, H-bond, and hydrophobicity) that contributes to the improvement of the targeted properties. In phase III, detailed computational analysis from the 3D

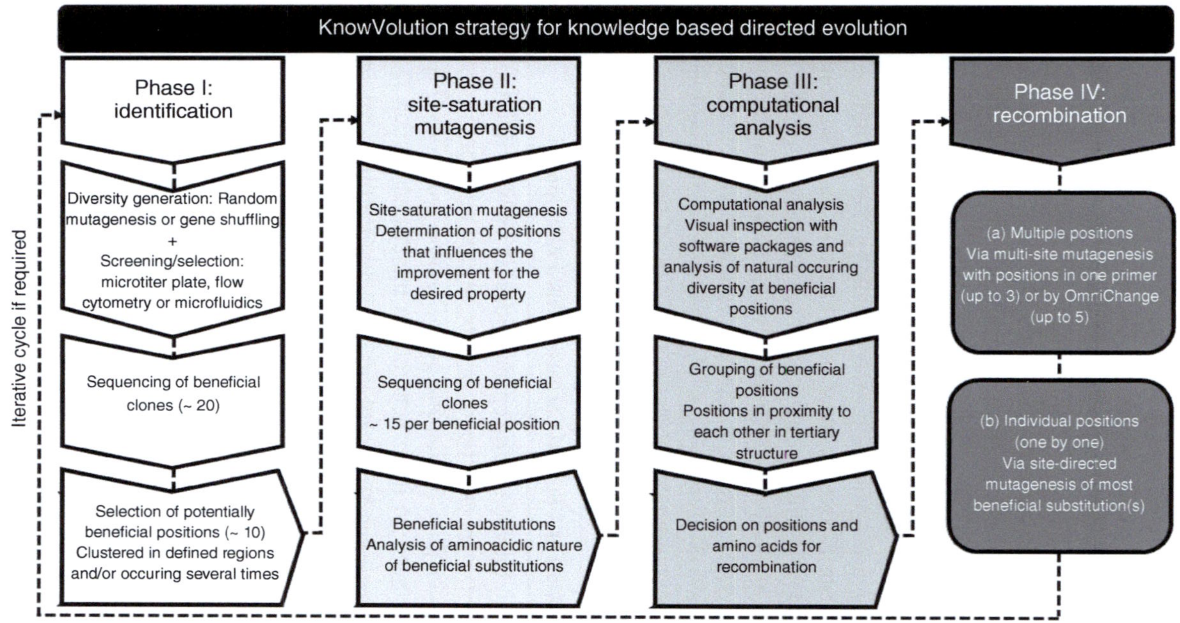

Figure 7.3 Overview of the KnowVolution strategy. Four phases are (I) identification of potentially beneficial amino acid positions, (II) determination of beneficial amino acid positions and substitutions, (III) computational analysis and a group of amino acid substitutions which might interact with each other, and (IV) recombination of beneficial substitutions in a simultaneous or iterative manner. Source: Cheng et al. [4]. © 2015, Royal Society of Chemistry.

structural perspective is performed to explore whether amino acid substitutions are in the close proximity or might even interact with each other. Computer-assisted structural analysis will provide detailed insights to decide how the beneficial substitutions should be combined, how these positions should be grouped, and which amino acid substitutions should be generated at the selected positions. In phase IV, the recombination is performed depending on the outcome of phase III. If several positions are in close proximity to each other, multi-SSM experiments should be performed preferentially. The number of targeted positions and the diversity of generated amino acid substitutions should be limited in order to ensure an efficient screening (for a recent review, see [4]). The beauty of KnowVolution is that it combines experimental directed evolution with computational strategies to determine regions in a protein that are responsible for the improvement in the targeted property and thus lead to a molecular understanding.

Over the last few years, KnowVolution was successfully applied to engineer several enzymes, including glucose oxidase, phytase, alkaline protease, arginine deiminase, and protease subtilisin E. Details about these enzymes have been reviewed elsewhere (see review [4] and will not be further discussed here). In Table 7.2, we have summarized recent applications (2015–2019) of KnowVolution and related strategies to engineer several enzymes and peptides such as laccase, sortase, aryl sulfotransferases, phytase, hyaluronic acid (HA) synthase from *Pasteurella multocida* (pmHAS), and liquid chromatography peak I (LCI). In the following, we provide some insights into the KnowVolution strategy to evolve these enzymes.

Case 1 KnowVolution of a Fungal Laccase Toward Alkaline pH

Laccases belong to the blue multicopper oxidases family, having the potential ability of oxidation. They have been used for several purposes in many industries including paper, pulp, textile, and petrochemical industries [71]. Recently, fungal laccases played an important role in lignin degradation [71]. So far, commercial laccase preparations are used at mild pH in the food, textile, and paper and pulp industries, whereas application of laccases in the synthesis of dye molecules or lignin degradation requires a high pH ($\geq$8.0) stability [71]. By employing the KnowVolution strategy, the fungal laccase from *Melanocarpus albomyces* (MaL) was successfully engineered for increased activity toward the substrate 2,6-dimethoxyphenol at pH $\geq$ 9.0 [71]. Two key positions L365 and L513 were identified to potentially be important for alkaline tolerance from phase I–III. 3D structural analysis revealed that both positions are located in close proximity of the T1Cu site. Further, computational analysis showed that both substitutions act in a synergic way to stabilize and improve laccase activity at higher pH [71]. Synergic effects of the both positions were investigated by simultaneous saturation mutagenesis and a variant MaL-M1 (L365E/L513M) was identified with an improvement of 3.1 ± 0.1 (variant/wild-type [V/WT] ratio, unpurified). Kinetic characterization showed a threefold k_{cat} improvement for the MaL-M1 ($k_{cat} = 6.0 \pm 0.2\, s^{-1}$) compared with wild-type MaL laccase ($k_{cat} = 2.11 \pm 0.07\, s^{-1}$) at pH 9.8 [71].

Table 7.2 Successful application of Know Volution strategy to improve enzymes properties.

Targeted enzyme	Screening system	P. I directed evolution (number of screened clones)	P. I number of potentially beneficial amino acid positions (number of sequenced clones)	P. II SSM at identified beneficial amino acid positions	P. IV recombination of identified beneficial substitutions	The final variant/ obtained improvement	Reasoning of improvement	References
Laccase MaL (*M. albomyces*)	2,6-Dimeth oxyphenol screening assay at pH 8.0	SeSaM-Tv (1200)	9 (7)	9 positions (N29, Q34, F141, L319, L365, N422, F427, L513, and P548)	L365E and L513 (NNK for both positions 800 clones)	MaL-M1 (L365E/L513M) at least twofold increase of specific activity for the phenolic substrate 2,6-DMP in the range of pH 4.0–10.2. Highest specific activity of a fungal laccase at pH 9.8	Salt bridge formation with neighboring residues leading to the stabilization of the loop at the entrance toward the T1Cu catalytic site. Small shift of the α-helix leading to conformational re-arrangement	[71]
Sortase (Sa-SrtA)	Fluorescence for eGFP as reporter	SeSaM (1680)	8 (12)	8 positions (94, 159, 160, 165, 170, 172, 186, and 196)	SSM in D186 and K196 (NNK for both positions). Later, SDM in variant M2 (D186G/K196V)	The variant M3 (D165Q/D186G/ K196V) showed a 5.5-fold improved activity in the transpeptidation of Abz-LPETGK-Dnp and tri-glycine in 45% (v/v) DMSO and a 17-fold increased specific activity in the absence of DMSO	Amino acid substitutions in close proximity to the Leu-Pro-x-Thr-Gly (LPxTG) binding pocket can partially retain essential water molecules in 45% (v/v) DMSO	[27]

(Continued)

Table 7.2 (Continued)

Targeted enzyme	Screening system	P. I directed evolution (number of screened clones)	P. I number of potentially beneficial amino acid positions (number of sequenced clones)	P. II SSM at identified beneficial amino acid positions	P. IV recombination of identified beneficial substitutions	The final variant/ obtained improvement	Reasoning of improvement	References
Aryl sulfotransferases (ASTB) (*Desulfitobacterium hafniense*)	pNPS (*para*-nitrophenyl sulfate) screening system for cellobiose as acceptor molecule for ASTB	SeSaM-Tv (1716)	7 (4)	7 positions (E111, T144, L446, V579, G608, S615, and V624)	SDM (iterative recombination of L446P, V579K, S608D, and S615G)	ASTB-M5 (L446P, V579K) showed 7.6-fold improvement of specific activity toward cellobiose and 2.6-fold improvement toward resorcinol	Substitution L446P facilitates more flexibility to the substrate-binding site and substitution V579K change the protein conformational motion leading to improved sulfation activity	[28]
Phytase (YmPh)	4-MUP screening system	Error-prone PCR (3600)	12 (8)	2 positions (T44 and K45)	SSM in T44 and K45 (NNK)	YmPhM10 (T44I/K45E) and YmPh-M16 (T44V/K45E) showed twofold increase in k_{cat} and ~sevenfold increase in the specific activity at pH 6.6	Substitution T44I/V could probably lead to a stronger interaction with the $-CH_3$ group at the aromatic ring of the 4-MUP substrate. Substitution K45E might interact upon hydrogen bond formation with the putative catalytic residues (H38, D327, and T328) and interact with the oxygen at the aromatic ring of the 4-MUP substrate	[29]

pmHAS	Cetyl trimethyl ammonium bromide (CTAB) turbidimetric assay	Error-prone PCR (1392)	8	8 positions (T40, V59, D90, T104, I175, H186, E313 and T480 with NNK)	SDM in 4 positions (T40, V59, T104, I175)	T40L/V59M/T104A has a 2.2-fold improvement in the mass-based turnover number and production of longer chains of HA	Substitutions (T40L/V59M/T104A) in the N-terminal part of pmHAS-VF lead to increased flexibility	[74]
LCI	Assay for better polymer binding anchor peptides (ABBA) screening system in the presence surfactant	Error-prone PCR (1044)	11	11 positions (K3, P8, D19, I24, S27, Y29, D31, G35, I40, E42, and D45 with NNK)	Simultaneous SSM of Y29 and G35 with MVW and NNK codons, respectively	EGFP-LCI KR-2 (Y29R/G35R) presented an improved binding capacity of 8.8 pmol cm^{-2} in the presence of 1 mM Triton X-100	The existence of a positive charge at the interface may de-structure water molecules induced in hydrophobic hydration and consequently increase hydrophobic interactions	[75, 76]

Four phases (Ps), including (I) identification, (II) determination, (III) selection (not shown in this table), and (IV) recombination were applied for each enzyme.
Structure guided computational analyses were performed for selection of amino acid positions to recombine in phase III for all six enzymes.
Sources: Zou et al. [27], Islam et al. [28], Körfer et al. [29].

Case 2 KnowVolution of Sortase A for Efficient Site-Specific Bioconjugations in Organic Co-solvents

Sortase mediated ligations have emerged as an attractive tool for the site-specific ligation of peptides and proteins in material science and biocatalysis [27, 77]. Recently, Zou et al. applied KnowVolution strategy to evolve sortase A from *Staphylococcus aureus* (Sa-SrtA) to improve specific activity and/or resistance against dimethylsulfoxide (DMSO) [27]. In phase I, 8 positions (P94, R159, D160, D165, D186, D170, Q172, and K196) were identified to increase specific activity when compared with wild type Sa-SrtA. In phase II, SSM at each position and screening revealed that 7 positions (P94, R159, D160, D165, D170, D186, and K196) improved the activity and 2 positions (R159 and Q172) improved the DMSO resistance [27]. In phase III, computational analysis indicated that two substitutions D186G and K196V are close proximity to each other and in close proximity to the active sites, C184 and R197. Finally, two variants R159G and D165Q/D186G/K196V were identified from phase IV. R159G showed highest improved resistance (2.2-fold) in 45% DMSO when compared with wild type Sa-SrtA and D165Q/D186G/K196V showed 6.3-fold improvement of catalytic efficiency in 45% (v/v) DMSO when compared with wild type. Computational analysis suggested that conformational mobility is important to improve DMSO resistance [27].

Case 3 KnowVolution of an Aryl Sulfotransferase to Increase Activity Toward Cellobiose

Sulfotransferases (STs) catalyze sulfation of many molecules, including polysaccharides, phenols, lipids, proteins, hormones, and metabolites [28]. Enzymatic degradability of sulfated cellulose offers a green chemistry route over chemically sulfated cellulose to selective sulfation of oligosaccharides in aqueous solution, at ambient temperature, and high chemoselectivity [28, 78]. A KnowVolution campaign was employed on an aryl sulfotransferase B (ASTB) from *Desulfitobacterium hafniense* to increase its specific activity. 7 positions E111, T144, L446, V579, G608, S615, and V624 were identified from phase I. 4 positions (L446, V579, G608, and S615) out of 7 positions showed improved volumetric activity for cellobiose sulfation from phase II. Subsequently, in phase II, computational analysis indicated that all positions are diversely located in which position L446 is localized in the close proximity to the catalytic residue H350 and positions V579, G608, and S615 are located in a loop of the C-terminal flexible region. Due to the diverse nature of identified positions, individual recombinations were performed for the identified substitutions to obtain double substituted variants in phase IV. Finally, a recombination variant comprised of two amino acid substitutions (L446P and V579K) resulted in up to 7.6-fold increase in specific activity (6.15 U mg^{-1}) compared with wild type [28]. Nuclear magnetic resonance (NMR)-based structure elucidation confirmed the sulfation at the positions C-3/C-4 of GlcNAc-linker-tBoc. Further, computational analysis indicated that L446P might influence substrate-binding and V579K might result in the indirect and far-reaching structural effects [28].

Case 4 KnowVolution of Phytase from *Yersinia mollaretii* (YmPh) for Broadened Activity at Neutral pH

Phytases hydrolyze phosphate from phytate, which is the main source of phosphorous in various plant seeds, specifically in cereals and grains [29]. Microbial phytases are widely used in plant-based (e.g. soy) animal feed to improve nutrition by enhanced phosphorus as phytate is indigestible for most monogastric animals, such as poultry, swine, and fish [29]. In order to use phytases for animal feed industrially, it requires high thermal resistance to withstand the processing, pelleting and storage, low production costs, and activity over a broad pH range (due to pH variation in the gastrointestinal tract of animals) [29]. By employing the KnowVolution, Körfer et al. engineered a phytase from *Y. mollaretii* (YmPh) toward a broader pH spectrum [29]. In phase I of diversity generation, 3600 clones produced by error-prone polymerase chain reaction (epPCR) were screened with the 4-methylumbelliferyl phosphate (4-MUP) substrate at pH 6.6 and 5.6 [29]. In phase II, eight variants with increased activity at pH 5.6 and 6.6 were selected and sequenced. In phase III, a homology model was generated to guide the recombination of beneficial positions found in phase II. Positions T44 and K45 were selected for recombination, due to their strategic positions within the phytase. They were close to the active center and the substrate access. The selection of recombination strategy in phase IV depends on the location of the substitutions in the protein. In this case, as the positions are next to each other, simultaneous SSM was selected as the recombination strategy. After phase IV, variants YmPh-M10 (T44I/K45E) and -M16 (T44V/K45E) showed improved activity at pH 6.6 and 5.6. After purification and characterization with the natural substrate, phytate, the variants YmPh-M10 and -M16 improved their activity between pH 6.6 and 7.6 when compared with wild type YmPh. To understand the molecular basis for the improvement, a homology model was built, which revealed that the positions located in the flexible loop that control the substrate access were exchanged for a hydrophobic or negatively charged residue, which may interact strongly with the substrate.

Case 5 KnowVolution of Hyaluronic Acid Synthase from *Pasteurella multocida* (pmHAS) Toward High-Molecular-Weight Hyaluronic Acid

Hyaluronan synthases (HASs) catalyze uridine diphosphate (UDP)-*N*-acetylglucosamine to produce HA, which has diverse cosmetic and medical applications [79]. In addition to metabolic and process engineering,

(Continued)

Case 5 (Continued)

engineered HASs offer a potential solution for industrial production of HA [74]. In this study, a class II hyaluronic acid synthase (pmHAS) was engineered for improved catalytic activity and reduced dispersion in the HA chain-length [74]. In phase I, a library with 1392 variants was generated by epPCR. Seven improved variants were identified and sequenced, revealing eight potentially beneficial positions. In phase II, 8 positions identified in the previous phase were determined as relevant and further subjected to SSM. In phase III, 3 positions exhibited a ~twofold improvement, representing suitable candidates for recombination. The selection of these positions was corroborated in the homology model. A substitution cluster was identified in the N-terminal domain, which was previously reported as dispensable for functional *N*-acetylglucosamine-transferase and polymerizing activity. In phase IV, the recombination was performed by site-directed mutagenesis (SDM), utilizing the wild type pmHAS and variant pmHAS-P6.H8 as parental genes. The best recombination variant, pmHAS-VF, showed a 2.1-fold improvement in HA titers and longer chains. From the homology model and MD simulations of the pmHAS-VF variant, it could be observed that the substitutions are located in a flexible loop, which may influence the length in HA production.

Case 6 KnowVolution of Polypropylene and Polystyrene Binding Peptides

Surface functionalization of polymer surfaces by polymer-binding peptides opens up numerous opportunities for immobilization of enzymes, bioactive peptides, and antigens [75]. The application of polymer-binding peptides requires stable and efficient binding under application conditions [75, 76]. The interaction between the polymer binding peptide (PBP) LCI and polypropylene (PP) was studied by a single KnowVolution campaign [75]. The peptide LCI was improved for PP-binding strength in the presence of the non-ionic surfactant Triton X-100, leading to an understanding of the PBP-surface interaction. In phase I, 11 beneficial positions were identified from an epPCR library of 1044 LCI variants. In phase II, SSM was performed in each beneficial position identified in phase I. In this phase, 8 positions were determined as influential toward the binding strength in the presence of Triton X-100. The identified position was studied from the crystal structure (protein data bank [PDB] ID: 2B9K) and a hot spot was located in turn 2. Positions identified in the best variants and in this area were subjected to recombination. In phase IV, simultaneous SSM at positions Y29 and G35 was performed, resulting in the variant enhanced green fluorescent protein (EGFP)-LCI Y29R/G35R with a 5.4-fold improvement in the binding strength in the presence of Triton-X100. As a trend, detected substitutions were positively charged amino acids; these charges may disrupt the water layer over the hydrophobic surface and consequently increase the hydrophobic interactions.

7.3 Conclusions and Future Perspectives

The combination of computational analysis and experimental mutagenesis represents a time-efficient and powerful approach to optimize various enzyme properties, especially localized properties (e.g. stability, activity, selectivity, catalytic promiscuity), for biocatatylic applications [80–82]. Here, the methods FRESCO and FoldX do an excellent job in improving thermostability and resistance/tolerance in cosolvents (e.g. ionic liquids) [19, 21]. Machine learning based on ProSAR method showed excellent results in enhancing enantioselectivity, stereoselectivity, and activity [45, 62]. However, especially in directed evolution campaigns with random mutagenesis methods, we scratch only at the surface of the potential that nature offers (e.g. ~20% positions identified in traditional directed evolution campaigns [73, 83, 84]). The latter is attributed to the complexity of the protein sequence space [2]. Generating the knowledge to enable molecular understanding of property improvements that can be transferred between protein folds or can be used within a protein fold should be the focus in directed evolution and computational protein design. In this respect, generally applicable protein engineering strategies such as KnowVolution, ISM, MORPHING, and variant libraries that offer the full natural diversity (e.g. BSLA studies; [73, 84]) can provide the data sets and molecular knowledge for machine learning and artificial intelligence research to enable on the long run a "pure" computational design. With the development of quantum computing in contrast to experimental screening technologies, computational design by "*in silico*" methodologies holds the promise to fully explore the protein sequence and thereby to efficiently use the vast design space that nature offers.

Acknowledgments

This work was supported through funds from RWTH Aachen University, DWI-Leibniz Institute for Interactive Materials. We gratefully acknowledge the German Federal Ministry of Education and Research (BMBF) (FKZ: 031B0506) for the financial support in the framework of Bioeconomy International. We thank Dr. Daniel Friedrich Sauer for help in preparing Figure 7.3.

References

1 Turner, N.J. (2009). Directed evolution drives the next generation of biocatalysts. *Nat. Chem. Biol.* 5 (8): 567.

2 Bornscheuer, U.T., Hauer, B., Jaeger, K.E. et al. (2019). Directed evolution empowered redesign of natural proteins for the sustainable production of chemicals and pharmaceuticals. *Angew. Chem. Int. Ed.* 58 (1): 36–40.

3 Otten, L.G. and Quax, W.J. (2005). Directed evolution: selecting today's biocatalysts. *Biomol. Eng.* 22 (1–3): 1–9.

4 Cheng, F., Zhu, L., and Schwaneberg, U. (2015). Directed evolution 2.0: improving and deciphering enzyme properties. *Chem. Commun.* 51 (48): 9760–9772.

5 Markel, U., Sauer, D.F., Schiffels, J. et al. (2019). Towards the evolution of artificial metalloenzymes – a protein engineer's perspective. *Angew. Chem. Int. Ed.* 58: 4454–4464.

6 Shivange, A.V., Marienhagen, J., Mundhada, H. et al. (2009). Advances in generating functional diversity for directed protein evolution. *Curr. Opin. Chem. Biol.* 13 (1): 19–25.

7 Zeymer, C. and Hilvert, D. (2018). Directed evolution of protein catalysts. *Annu. Rev. Biochem.* 87: 131–157.

8 Dougherty, M.J. and Arnold, F.H. (2009). Directed evolution: new parts and optimized function. *Curr. Opin. Biotechnol.* 20 (4): 486–491.

9 Barrozo, A., Borstnar, R., Marloie, G. et al. (2012). Computational protein engineering: bridging the gap between rational design and laboratory evolution. *Int. J. Mol. Sci.* 13 (10): 12428–12460.

10 Lutz, S. (2010). Beyond directed evolution – semi-rational protein engineering and design. *Curr. Opin. Biotechnol.* 21 (6): 734–743.

11 Chen, R. (2001). Enzyme engineering: rational redesign versus directed evolution. *Trends Biotechnol.* 19 (1): 13–14.

12 Bashirova, A., Pramanik, S., Volkov, P. et al. (2019). Disulfide bond engineering of an endoglucanase from *Penicillium verruculosum* to improve its thermostability. *Int. J. Mol. Sci.* 20 (7): 1–14.

13 Thiele, M.J., Davari, M.D., König, M. et al. (2018). Enzyme–polyelectrolyte complexes boost the catalytic performance of enzymes. *ACS Catal.* 8 (11): 10876–10887.

14 Thiele, M.J., Davari, M.D., Hofmann, I. et al. (2018). Enzyme-compatible dynamic nanoreactors from electrostatically bridged like – charged surfactants and polyelectrolytes. *Angew. Chem. Int. Ed.* 57 (30): 9402–9407.

15 Jakob, F., Martinez, R., Mandawe, J. et al. (2013). Surface charge engineering of a *Bacillus gibsonii* subtilisin protease. *Appl. Microbiol. Biotechnol.* 97 (15): 6793–6802.

16 Lehmann, C., Bocola, M., Streit., W.R. et al. (2014). Ionic liquid and deep eutectic solvent-activated CelA2 variants generated by directed evolution. *Appl. Microbiol. Biotechnol.* 98 (12): 5775–5785.

17 Wallraf, A.-M., Liu, H., Zhu, L. et al. (2018). A loop engineering strategy improves laccase lcc2 activity in ionic liquid and aqueous solution. *Green Chem.* 20 (12): 2801–2812.

18 Wijma, H.J., Floor, R.J., Jekel, P.A. et al. (2014). Computationally designed libraries for rapid enzyme stabilization. *Protein Eng. Des. Sel.* 27 (2): 49–58.

19 Wijma, H.J., Furst, M., and Janssen, D.B. (2018). A computational library design protocol for rapid improvement of protein stability: FRESCO. *Methods Mol. Biol.* 1685: 69–85.

20 Delgado, J., Radusky L.G., Cianferoni, D. et al. (2019). FoldX 5.0: working with RNA, small molecules and a new graphical interface. *Bioinformatics* 35 (20): 4168–4169.

21 Buss, O., Rudat, J., and Ochsenreither, K. (2018). FoldX as protein engineering tool: better than random based approaches? *Comput. Struct. Biotechnol. J.* 16: 25–33.

22 Schymkowitz, J., Borg J., Stricher, F. et al. (2005). The FoldX web server: an online force field. *Nucleic Acids Res.* 33 (Web Server issue): W382–W388.

23 Kruger, D.M., Rathi, P.C., Pfleger, C. et al. (2013). CNA web server: rigidity theory-based thermal unfolding simulations of proteins for linking structure, (thermo-)stability, and function. *Nucleic Acids Res.* 41 (Web Server issue): W340–W348.

24 Goldenzweig, A. and Fleishman, S.J. (2018). Principles of protein stability and their application in computational design. *Annu. Rev. Biochem.* 87: 105–129.

25 Fox, R. (2005). Directed molecular evolution by machine learning and the influence of nonlinear interactions. *J. Theor. Biol.* 234 (2): 187–199.

26 Fox, R.J., Davis S.C., Mundorff, E.C. et al. (2007). Improving catalytic function by ProSAR-driven enzyme evolution. *Nat. Biotechnol.* 25 (3): 338–344.

27 Zou, Z., Alibiglou, H., Mate, D.M. et al. (2018). Directed sortase A evolution for efficient site-specific bioconjugations in organic co-solvents. *Chem. Commun.* 54 (81): 11467–11470.

28 Islam, S., Laaf, D., Infanzón, B. et al. (2018). KnowVolution campaign of an aryl sulfotransferase increases activity toward cellobiose. *Chem. Eur. J.* 24 (64): 17117–17124.

29 Körfer, G., Novoa, C., Kern, J. et al. (2018). Directed evolution of an acid *Yersinia mollaretii* phytase for broadened activity at neutral pH. *Appl. Microbiol. Biotechnol.* 102 (22): 9607–9620.

30 Verma, R., Schwaneberg, U., and Roccatano, D. (2012). Computer-aided protein directed evolution: a review of web servers, databases and other computational tools for protein engineering. *Comput. Struct. Biotechnol. J.* 2 (3): e201209008.

31 Romero-Rivera, A., Garcia-Borras, M., and Osuna, S. (2016). Computational tools for the evaluation of laboratory-engineered biocatalysts. *Chem. Commun.* 53 (2): 284–297.

32 Packer, M.S. and Liu, D.R. (2015). Methods for the directed evolution of proteins. *Nat. Rev. Genet.* 16 (7): 379.

33 Lane, M.D. and Seelig, B. (2014). Advances in the directed evolution of proteins. *Curr. Opin. Chem. Biol.* 22: 129–136.

34 Reetz, M.T. and Carballeira, J.D. (2007). Iterative saturation mutagenesis (ISM) for rapid directed evolution of functional enzymes. *Nat. Protoc.* 2 (4): 891.

35 Gonzalez-Perez, D., Molina-Espeja, P., Garcia-Ruiz, E. et al. (2014). Mutagenic organized recombination process by homologous *in vivo* grouping (MORPHING) for directed enzyme evolution. *PLoS One* 9 (3): e90919.

36 Floor, R.J., Wijma, H.J., Colpa, D.I. et al. (2014). Computational library design for increasing haloalkane dehalogenase stability. *ChemBioChem* 15 (11): 1660–1672.

37 Wu, B., Wijma, H.J., Song, L. et al. (2016). Versatile peptide C-terminal functionalization via a computationally engineered peptide amidase. *ACS Catal.* 6 (8): 5405–5414.

38 Wijma, H.J., Floor, R.J., Bjelic, S. et al. (2015). Enantioselective enzymes by computational design and in silico screening. *Angew. Chem. Int. Ed.* 54 (12): 3726–3730.

39 Luo, X.J., Zhao, J., Li., C.-X. et al. (2016). Combinatorial evolution of phosphotriesterase toward a robust malathion degrader by hierarchical iteration mutagenesis. *Biotechnol. Bioeng.* 113 (11): 2350–2357.

40 Rathi, P.C., Jaeger, K.E., and Gohlke, H. (2015). Structural rigidity and protein thermostability in variants of lipase A from *Bacillus subtilis*. *PLoS One* 10 (7): e0130289.

41 Campeotto, I., Goldenzweig, A., Davey, J. et al. (2017). One-step design of a stable variant of the malaria invasion protein RH5 for use as a vaccine immunogen. *Proc. Natl. Acad. Sci. U.S.A.* 114 (5): 998–1002.

42 Brazzolotto, X., Igert., A., Guillon, V. et al. (2017). Bacterial expression of human butyrylcholinesterase as a tool for nerve agent bioscavengers development. *Molecules* 22 (11): 1–16.

43 Goldsmith, M., Aggarwal, N., Ashani, Y. et al. (2017). Overcoming an optimization plateau in the directed evolution of highly efficient nerve agent bioscavengers. *Protein Eng. Des. Sel.* 30 (4): 333–345.

44 Cadet, F., Fontaine, N., Li, G. et al. (2018). A machine learning approach for reliable prediction of amino acid interactions and its application in the directed evolution of enantioselective enzymes. *Sci. Rep.* 8 (1): 16757.

45 Wu, Z., Kan, S.B.J., Lewis R.D. et al. (2019). Machine learning-assisted directed protein evolution with combinatorial libraries. *Proc. Natl. Acad. Sci. U.S.A.* 116 (18): 8852–8858.

46 Li, R., Wijma, H.J., Song, L. et al. (2018). Computational redesign of enzymes for regio-and enantioselective hydroamination. *Nat. Chem. Biol.* 14 (7): 664.

47 Van Durme, J., Delgado J., Stricher, F. et al. (2011). A graphical interface for the FoldX forcefield. *Bioinformatics* 27 (12): 1711–1712.

48 Huang, J., Xie, D.-F., and Feng, Y. (2017). Engineering thermostable (R)-selective amine transaminase from *Aspergillus terreus* through in silico design employing B-factor and folding free energy calculations. *Biochem. Biophys. Res. Commun.* 483 (1): 397–402.

49 Timucin, E. and Sezerman, O.U. (2013). The conserved lid tryptophan, W211, potentiates thermostability and thermoactivity in bacterial thermoalkalophilic lipases. *PLoS One* 8 (12): e85186.

50 Wu, I. and Arnold, F.H. (2013). Engineered thermostable fungal Cel6A and Cel7A cellobiohydrolases hydrolyze cellulose efficiently at elevated temperatures. *Biotechnol. Bioeng.* 110 (7): 1874–1883.

51 Daudé, D., Topham, C.M., Remaud-Siméon, M. et al. (2013). Probing impact of active site residue mutations on stability and activity of *Neisseria polysaccharea* amylosucrase. *Protein Sci.* 22 (12): 1754–1765.

52 Liu, H., Zhu, L., Bocola, M. et al. (2013). Directed laccase evolution for improved ionic liquid resistance. *Green Chem.* 15 (5): 1348–1355.

53 Christensen, N.J. and Kepp, K.P. (2012). Accurate stabilities of laccase mutants predicted with a modified FoldX protocol. *J. Chem. Inf. Model.* 52 (11): 3028–3042.

54 Kepp, K.P. (2015). Towards a "Golden Standard" for computing globin stability: stability and structure sensitivity of myoglobin mutants. *Biochim. Biophys. Acta* 1854 (10): 1239–1248.

55 Pfleger, C. et al. (2013). Constraint network analysis (CNA): a Python software package for efficiently linking biomacromolecular structure, flexibility, (thermo-)stability, and function. *J. Chem. Inf. Model.* 53 (4): 1007–1015.

56 Pfleger, C., Rathi, P.C., Klein, D.L. et al. (2013). Global and local indices for characterizing biomolecular flexibility and rigidity. *J. Comput. Chem.* 34 (3): 220–233.

57 Rathi, P.C., Radestock, S., and Gohlke, H. (2012). Thermostabilizing mutations preferentially occur at structural weak spots with a high mutation ratio. *J. Biotechnol.* 159 (3): 135–144.

58 Khersonsky, O., Lipsh, R., Avizemer, Z. et al. (2018). Automated design of efficient and functionally diverse enzyme repertoires. *Mol. Cell* 72 (1): 178–186.e5.

59 Cumming, J.G., Davis, A.M., Muresan, S. et al. (2013). Chemical predictive modelling to improve compound quality. *Nat. Rev. Drug Discovery* 12 (12): 948.

60 Rekker, R.F. (1992). The history of drug research: from Overton to Hansch. *Quant. Struct.-Act. Relat.* 11 (2): 195–199.

61 Cadet, F., Fontaine, N., Vetrivel, I. et al. (2018). Application of Fourier transform and proteochemometrics principles to protein engineering. *BMC Bioinf.* 19 (1): 382.

62 Li, G., Dong, Y., and Reetz, M. (2019). Can machine learning revolutionize directed evolution of selective enzymes? *Adv. Synth. Catal.* 361: 2377–2386.

63 Alexander, D.L., Tropsha, A., and Winkler, D.A. (2015). Beware of R(2): simple, unambiguous assessment of the prediction accuracy of QSAR and QSPR models. *J. Chem. Inf. Model.* 55 (7): 1316–1322.

64 Kubinyi, H., Hamprecht, F.A., and Mietzner, T. (1998). Three-dimensional quantitative similarity-activity relationships (3D QSiAR) from SEAL similarity matrices. *J. Med. Chem.* 41 (14): 2553–2564.

65 Reetz, M.T., Carballeira, J.D., and Vogel, A. (2006). Iterative saturation mutagenesis on the basis of B factors as a strategy for increasing protein thermostability. *Angew. Chem. Int. Ed.* 45 (46): 7745–7751.

66 Reetz, M.T., Bocola, M., Carballeira, J.D. et al. (2005). Expanding the range of substrate acceptance of enzymes: combinatorial active-site saturation test. *Angew. Chem. Int. Ed.* 44 (27): 4192–4196.

67 Gumulya, Y. and Reetz, M.T. (2011). Enhancing the thermal robustness of an enzyme by directed evolution: least favorable starting points and inferior mutants can map superior evolutionary pathways. *ChemBioChem* 12 (16): 2502–2510.

68 Engström, K., Nyhlén, J., Sandström, A.G. et al. (2010). Directed evolution of an enantioselective lipase with broad substrate scope for hydrolysis of α-substituted esters. *J. Am. Chem. Soc.* 132 (20): 7038–7042.

69 Mate, D.M., Palomino, M.A., Molina-Espeja, P. et al. (2017). Modification of the peroxygenative: peroxidative activity ratio in the unspecific peroxygenase from *Agrocybe aegerita* by structure-guided evolution. *Protein Eng. Des. Sel.* 30 (3): 191–198.

70 Anand, D., Dhoke, G.V., Gehrmann, J. et al. (2019). Chiral separation of D/L-arginine with whole cells through an engineered FhuA nanochannel. *Chem. Commun.* 55 (38): 5431–5434.

71 Novoa, C., Dhoke, G.V., Mate, D.M. et al. (2019). KnowVolution of a fungal laccase toward alkaline pH. *ChemBioChem* 20: 1458–1466.

72 Islam, S., Apitius, L., Jakob, F. et al. (2019). Targeting microplastic particles in the void of diluted suspensions. *Environ. Int.* 123: 428–435.

73 Zhao, J., Kardashliev, T., Ruff, A.J. et al. (2014). Lessons from diversity of directed evolution experiments by an analysis of 3,000 mutations. *Biotechnol. Bioeng.* 111 (12): 2380–2389.

74 Mandawe, J., Infanzon, B., Eisele, A. et al. (2018). Directed evolution of hyaluronic acid synthase from pasteurella multocida towards high-molecular-weight hyaluronic acid. *ChemBioChem* 19 (13): 1414–1423.

75 Rübsam, K., Davari, M.D., Jakob, F. et al. (2018). KnowVolution of the polymer-binding peptide LCI for improved polypropylene binding. *Polymers* 10 (4): 423.

76 Rübsam, K., Weber, L., Jakob, F. et al. (2018). Directed evolution of polypropylene and polystyrene binding peptides. *Biotechnol. Bioeng.* 115 (2): 321–330.

77 Zou, Z., Mate, D.M., Rübsam, K. et al. (2018). Sortase-mediated high-throughput screening platform for directed enzyme evolution. *ACS Comb. Sci.* 20 (4): 203–211.

78 Yamamoto, I., Takayama, K., Honma, K. et al. (1990). Synthesis, structure and antiviral activity of sulfates of cellulose and its branched derivatives. *Carbohydr. Polym.* 14 (1): 53–63.

79 Sze, J.H., Brownlie, J.C., and Love, C.A. (2016). Biotechnological production of hyaluronic acid: a mini review. *3 Biotech* 6 (1): 67.

80 Monza, E., Acebes, S., Lucas, M.F. et al. (2017). Molecular modeling in enzyme design, toward in silico guided directed evolution. In: *Directed Enzyme Evolution: Advances and Applications* (ed. M. Alcalde), 257–284. Springer.

81 Amrein, B.A., Steffen-Munsberg, F., Szeler, I. et al. (2017). CADEE: computer-aided directed evolution of enzymes. *IUCrJ* 4 (1): 50–64.

82 Damborsky, J. and Brezovsky, J. (2014). Computational tools for designing and engineering enzymes. *Curr. Opin. Chem. Biol.* 19: 8–16.

83 Zhao, J., Frauenkron-Machedjou, V.J., Kardashliev, T. et al. (2017). Amino acid substitutions in random mutagenesis libraries: lessons from analyzing 3000 mutations. *Appl. Microbiol. Biotechnol.* 101 (8): 3177–3187.

84 Frauenkron-Machedjou, V.J., Fulton, A., Anker, C. et al. (2015). Towards understanding directed evolution: more than half of all amino acid positions contribute to ionic liquid resistance of *Bacillus subtilis* lipase A. *ChemBioChem* 16 (6): 937–945.

8

Engineering Artificial Metalloenzymes

Kevin A. Harnden[1], Yajie Wang[2*], Lam Vo[1,2], Huimin Zhao[1,2#], and Yi Lu[1#]*

[1] *University of Illinois at Urbana-Champaign, Department of Chemistry, Urbana, Illinois 61801, United States*
[2] *University of Illinois at Urbana-Champaign, Department of Chemical and Biomolecular Engineering, Urbana, Illinois 61801, United States*

8.1 Introduction

Roughly half of all characterized proteins contain a metal cofactor, often used as an integral part of the proteins' function [1]. Many of these metalloproteins catalyze some of the most challenging yet industrially useful reactions, such as nitrogen fixation, oxygen reduction, and regio- and stereoselective organic transformations [2]. There is a problem, however, in that these enzymes are not always stable under the conditions for industrial applications and do not work with a desired substrate or produce the intended product. Engineering will allow us to increase the stability, change the substrate specificity, while gaining a deeper understanding of metalloenzymes structures and reaction mechanisms. There are two main ways to engineer enzymes, through rational design and directed evolution. Rational design involves the use of prediction, either manually or with the aid of computation, to engineer new or different functionality in enzymes. Directed evolution takes advantage of random mutations or gene shuffling to augment enzymes' function in ways that cannot be predicted through rational means.

8.2 Rational Design

8.2.1 Rational Design of Metalloenzymes Using *De Novo* Designed Scaffolds

De novo protein design has always been connected to the challenge of understanding protein folding. If the process of protein folding is truly understood, then a novel protein with a defined secondary and tertiary structure can be designed from scratch

These are corresponding authors
* These authors contributed equally

Protein Engineering: Tools and Applications, First Edition. Edited by Huimin Zhao.
© 2021 WILEY-VCH GmbH. Published 2021 by WILEY-VCH GmbH.

purely based off the amino acid sequence. Early work in this field focused on using structural interactions observed in native proteins to design simple, short α-helical peptides, which later developed into designing α-helical bundles [3]. More recent work has seen the development of computational tools to search through a larger number of sequence space and predict, using what has been learned from protein folding research, the structure of a protein based off of its sequence [4]. These *de novo* designed scaffolds have been used to design novel metalloenzymes with various activities.

Phenol oxidation using a diiron cofactor. A nonheme diiron center called Due Ferri (DF) has been designed into a *de novo* designed four-helical bundle to display phenol oxidation activity [5]. The diiron center is coordinated by four glutamate and two histidine residues that allows the diferrous form to react with O_2 to yield a diferric μ-oxo species. The initial design (DF1) did not include a substrate-binding site so further work went into adding a phenol binding site, thus creating an artificial metalloenzyme for phenol oxidation. In order to create a substrate-binding site, four leucine residues were mutated into glycines, which subsequently reduced the stability of the four-helical bundle. Compensatory mutations in the α_R-α_L-β turn to improve stability were found by searching the Protein Data Bank for sequences that frequently appear in such turns. Val24-Lys25-Leu26 of DF1 was replaced with Thr24-His25-Asn26 and this new protein was designated DF3 (Figure 8.1a). DF3 with two equivalents of iron bound was shown to undergo a reversible two electron redox cycle and catalyze oxidation of 4-aminophenol to the corresponding benzoquinone mono-imine. The Michaelis–Menten kinetics parameters of 1.97 ± 0.27 mM for K_M and $2.72 \pm 0.19\ \text{min}^{-1}$ for k_{cat} were obtained, and the total number of turnovers was estimated to be 50.

Hydroquinone amine oxidation using a diiron cofactor. Another *de novo* nonheme diiron protein based on the DF family called DFsc was designed to catalyze the hydroxylation of aryl amines, a reaction that is natively catalyzed by *p*-aminobenzoate *N*-oxygenase (AurF) [6]. The original design of this scaffold contained four alanines along the active site channel. Following the success of previous models, these four alanines were mutated into glycine to allow for an

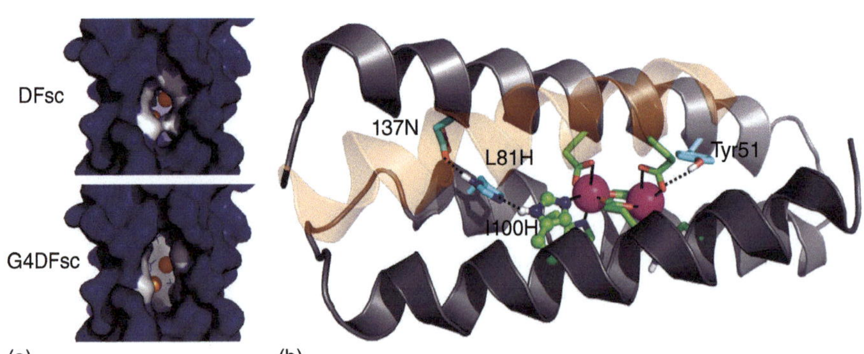

Figure 8.1 (a) Model of DFsc (top) and G4DFsc (bottom) from computational design of DFsc. (b) Structure of 3His-G4DFsc showing the primary, secondary, and tertiary coordination sphere interactions. Source: Reig et al. [6]. © 2012, Springer Nature.

increase in the size of the active site cavity (designated G4DFsc). Comparison of the diiron site of the G4DFsc and the diiron site of AurF reveals high similarity with the exception of an additional histidine ligand in AurF, with Ile100 at the position in G4DFsc (Figure 8.1b). Using the software package Molecular Software Library, compensatory mutations to allow for the Ile100His mutation were scanned and two mutations were tested experimentally: Tyr18Phe and Tyr18Leu. Additional mutations were tested to help stabilize the extra histidines that were added and a new designation was given (3His-G4DFsc). The diferrous center reacts with O_2 to give a μ-oxo diferric center and, in the presence of *p*-anisidine, affords the product 4-nitroso-4′-methoxydiphenylamine, the product of *N*-hydroxylation followed by nucleophilic aromatic substitution.

Ester hydrolysis using a zinc cofactor. Enzymatic ester hydrolysis is a reaction often catalyzed by metals acting as Lewis acids that have a hydroxyl ligand. An artificial metalloenzyme with zinc and mercury cofactors was designed based on the TRI peptide scaffold, containing three cysteine residues to coordinate Hg(II) and three histidine residues to coordinate Zn(II) ([Hg(II)$_S$[Zn(II)(OH$^-$)$_N$(TRIL9L23H)$_3$) (where *S* denotes coordination by sulfur of cysteine and *N* denotes coordination by nitrogen of histidine) [7]. Mercury is used for structural stabilization while zinc is the site for catalysis. Zinc is coordinated by three histidine residues and one aquo/hydroxo ligand, similar to the active site of carbonic anhydrase (CAII). The catalytic performance was evaluated using the hydrolysis of *p*-nitrophenyl acetate. The highest activity was observed at pH 9.5 with a k_{cat} of $40 \pm 12 \times 10^{-3}\ s^{-1}$ and a K_M of 1.7 ± 0.5 mM. The activity for CO_2 hydration was also investigated and compared with the fastest reported synthetic complexes and native CAII. The fastest synthetic complex Zn(II)-tris(4,5-di-*n*-propyl-2-imidazolyl)phosphine had a rate of $2\,480\ M^{-1}\ s^{-1}$ while CAII features a turnover frequency of 8.2×10^5 and a catalytic efficiency of $9.2 \times 10^7\ M^{-1}\ s^{-1}$. [Hg(II)$_S$[Zn(II)(OH$^-$)$_N$(TRIL9L23H)$_3$ was found to have a k_{cat} of $1.8 \pm 0.4 \times 10^3\ s^{-1}$, a k_{cat}/K_M of $1.8 \pm 0.3 \times 10^5\ M^{-1}\ s^{-1}$, and a K_M of 10.0 ± 2.4 mM.

Nitrite reduction using a copper cofactor. Another example of a *de novo* designed metalloenzyme with the three-helical bundle contains a mononuclear copper site coordinated by three histidines, designated Cu(**TRI**L23H)$_3^{n+}$, and is a structural and functional model of nitrite reductase [8, 9]. The binding affinity for copper was very high, with a K_d in the nano- to micromolar range for Cu(II) and picomolar for Cu(I). Using ascorbic acid as a reductant, Cu(**TRI**L23H)$_3^+$ catalytically reduces nitrite to nitric oxide with at least five turnovers. By modifying the secondary coordination sphere, the catalytic activity can be greatly improved. Replacing Leu residues at position 19 and 26 with Ala opens the active site cavity while Asp in the same positions provides hydrogen bonding that can assist in catalysis. All sets of these mutations increased the activity of nitrite reduction, from $4.6 \times 10^{-4}\ s^{-1}$ for the starting model to $3.5 \times 10^{-2}\ s^{-1}$ for L19A or $2.2 \times 10^{-2}\ s^{-1}$ for L26A and $2.4 \times 10^{-2}\ s^{-1}$ for L19D or $3.5 \times 10^{-2}\ s^{-1}$ for L26D.

8.2.2 Rational Design of Metalloenzymes Using Native Scaffolds

8.2.2.1 Redesign of Native Proteins

Using native proteins as scaffolds to design artificial metalloenzymes carries several benefits. First, through many years of evolution, native proteins are not only stable under physiological conditions, but also maintain the stability despite many mutations. By choosing a known scaffold with specific properties, engineering an enzyme with those desired properties becomes easier. Another reason for using native proteins as scaffold is that many proteins share common tertiary folds. This allows for native enzymes to be modeled in smaller, more stable proteins that have similar folds that can be used to both study the native enzyme and develop industrial biocatalysts.

Oxygen reduction using a heme–nonheme heterobimetallic system. Heme-copper oxidases (HCOs) are responsible for the four-electron, four-proton reduction of O_2 to H_2O, the oxygen reduction reaction (ORR) in fuel cells that are often catalyzed by precious metal ion such as platinum. Designing artificial enzymes to mimic HCOs will offer deeper insight into how nature uses earth abundant metal ions such as Fe and Cu to catalyze ORR. Sperm whale myoglobin (Mb) has been used as a structural and functional model of HCOs through mutagenesis guided by active site comparison. The HCO model contains the native Mb heme cofactor and a copper center coordinated by three histidines at positions 29, 43, and 64. Nitric oxide reductase (NOR), a related class of proteins that has a similar active site and cross reactivity, can also be modeled in Mb by adding an additional glutamate residue at position 68, which allows for higher affinity with different metals such as manganese, iron, cobalt, nickel, and zinc [10]. Coordination of a metal ion in the nonheme site turns Mb into a heterobimetallic system with heme and nonheme metal cofactors, allowing for the rate of O_2 reduction with different metals to be compared. When the nonheme site is empty or when zinc is bound, the rate of O_2 reduction is low ($0.21 \pm 0.03\ \mu M\ s^{-1}$ for empty and $0.22 \pm 0.02\ \mu M\ s^{-1}$ for zinc), but when copper is bound in the nonheme site, the rate of O_2 reduction is highest ($2.72 \pm 0.1\ \mu M\ s^{-1}$) [11]. Manganese, iron, or cobalt in the nonheme site gives moderate O_2 reduction rates of 1.14, 1.15, and 1.01 $\mu M\ s^{-1}$, respectively. More importantly, when a conserved tyrosine is introduced into the Mb model at a similar position as in the native HCO [12], combined with design of interface between the Mb model and its redox partner (cytochrome b5) to increase the electron transfer rate, the resulting Mb model display catalytic ORR activity that rivals native HCO [13]. When placed onto the surface of an electrode, the ORR activity is even better than native HCOs [14].

C–H functionalization using a heme cofactor. Enzymatic functionalization of C—H bonds is a useful way to add functional groups or modifications to natural products or pharmaceuticals in a regio-, stereo-, and chemoselective manor. Sperm whale myoglobin (Mb) containing its native heme cofactor was used as a scaffold for C—C bond formation between unprotected indoles and ethyl α-diazoacetate [15]. Residues surrounding the active site were mutated in order to augment activity and selectivity, with H64V/V68A as the best result with conversion of 85% turnover number (TON) of 106 at pH 9.0 for the reaction between indole and ethyl α-diazoacetate. The substrate scope was examined using a variety of substituted indoles, with methyl and chloro-substituted indoles having >99% conversion. In order to demonstrate the applicability of this artificial metalloenzyme, indomethacin, a non-steroidal anti-inflammatory drug, was synthesized using Mb (L29F,H64V).

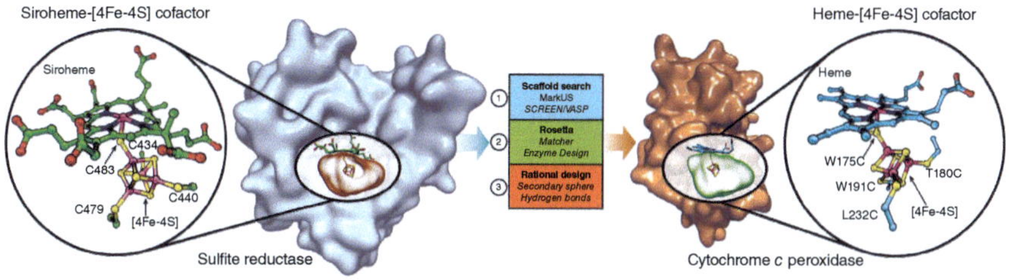

Figure 8.2 Design of SiRC*c*P starting from scaffold search followed by design of the binding site for the 4Fe–4S cluster. Source: Mirts et al. [16]. Reprinted with permission from AAAS.

Sulfite reduction using a heme-[4Fe–4S] system. Sulfite reductase (SiR) is a metalloenzyme that contains a siroheme and [4Fe–4S] cluster bridged by a cysteine residue and catalyzes the six-electron, seven-proton reduction of sulfite to hydrogen sulfide. It is among the most complex metalloenzymes that contain heteronuclear metal centers, which are required to catalyze a highly demanding multiple-step reaction in a single enzyme. To meet this challenge, a model of SiR was engineered using cytochrome *c* peroxidase (C*c*P) as a scaffold by engineering a binding site for a [4Fe–4S] cluster beneath the native heme of C*c*P (Figure 8.2) [16]. The engineered C*c*P (SiRC*c*P.1) was chosen as a scaffold after a scaffold search was performed using MarkUS and the specific mutations for the FeS cluster were selected using Rosetta Matcher and Enzyme Designs with further mutations in the secondary sphere were found by inspection. The engineered model protein was reconstituted with ferrous sulfate and sodium sulfide, followed by heme. The reconstituted model was characterized using UV–Vis, X-ray, and electron paramagnetic resonance (EPR) spectroscopies and was found to be consistent with the formation of an [4Fe–4S] cluster. The rate of sulfite reduction of the first design (SiRC*c*P.1) was very low at $0.348 \pm 0.15\ \text{min}^{-1}$. Native SiR contains multiple Lys and Arg residues in the substrate binding pocket above the heme that help bind and stabilize the −2 charge of sulfite. Mimicking these secondary coordination sphere residues increases the activity by 5.3-fold, to $1.26 \pm 0.20\ \text{min}^{-1}$. Next, the residues surrounding the [4Fe–4S] cluster were examined for potential mutations to increase the activity further. Native [4Fe–4S] proteins contain an asparagine or a cysteine, which is oriented toward the [4Fe–4S] cluster, at the same position as residue Asp235 in C*c*P. The mutations Asp235Asn (SiRC*c*P.2) and Asp235Cys (SiRC*c*P.3) increase the activity to 5.91 ± 1.6 and $21.8 \pm 2.4\ \text{min}^{-1}$, respectively, which is ~18% of the native SiR's activity.

8.2.2.2 Cofactor Replacement in Native Proteins

Metalloenzymes use a variety of cofactors, such as metal ions and metal complexes, to perform their functions. By replacing the cofactor in an enzyme with biological or abiological metal ions and metal complexes, the activity of that enzyme can be altered, enhanced, or completely changed. Heme and heme-like cofactors are very common in biology, and many heme proteins can be reconstituted with cofactors that are structurally similar to heme, which makes them great scaffolds for designing artificial metalloenzymes using heme-like cofactors.

Cyclopropanation using an Iron–Chlorin e6 cofactor. Chlorin e6 (Ce6) is a derivative of the tetrapyrrole chromophore of chlorophyll, similar to protoporphyrin IX (ppIX) found in myoglobin. Ce6 contains three carboxylate groups and a saturated C—C bond in one of the pyrrole rings, making Ce6 a more electron deficient ligand compared with protoporphyrin IX. A Ce6 cofactor containing an iron was postulated to increase the activity toward olefin cyclopropanation by increasing the reactivity of the iron–carbene intermediate found in carbene transfer reactions [17]. Ce6 containing iron ([Fe(Ce6)]) was used to replace the native heme cofactor in myoglobin in order to create an artificial metalloenzyme for stereoselective olefin cyclopropanation reactions. Fe(Ce6) was prepared by refluxing ferrous chloride and ascorbic acid with Ce6 in acetone, yielding [Fe(Cl)(Ce6)]. ApoMb was purified and reconstituted with [Fe(Cl)(Ce6)], resulting in a green-colored Mb[Fe(Ce6)] complex. Following previous results that demonstrate His64Val/Val68Ala mutations increased activity, the variant Mb(H64V,V68A)[Fe(Ce6)] was prepared and its activity toward olefin cyclopropanation was examined. The cyclopropanation of styrene using ethyl α-diazoacetate was used to evaluate the catalytic performance under different reaction conditions. Mb(H64V,V68A)[Fe(Ce6)] was found to have high yield (>99%), TON (>990) compared with Mb(H64V,V68A)[Fe(ppIX)], which had a yield of 43% and TON of 434. Diastereo- and enantioselectivities were also high, at 99.6% de and 98.5% ee. Substrate scope was shown to include a variety of ortho-, meta-, and para-substituted styrene derivatives.

Cyclopropanation using an iron porphycene. Porphycene is an aromatic macrocycle and an isomer of porphyrin that has different electronic effects compared to porphyrin. The effect a porphycene ligand would have on catalysis was examined by reconstituting an iron–porphycene into myoglobin (rMb) (Figure 8.3) [18]. Compared with native myoglobin (nMb) for the cyclopropanation reaction between styrene and ethyl α-diazoacetate, rMb had a turnover frequency of $2.2\,s^{-1}$, 35 times higher than nMb, and the Michaelis–Menten parameters were also more favorable, with a k_{cat} of $1.7 \pm 0.3\,s^{-1}$ versus $0.06 \pm 0.01\,s^{-1}$ and a K_M of 1.3 ± 0.4 mM versus 1.2 ± 0.5 mM. In order to further understand the effect of the different macrocycle on the reactivity, density-functional theory (DFT) calculations were performed at the B97D/6-31g* level. The potential energy diagrams for the reaction between the metal complex and ethyl α-diazoacetate reveal that the porphycene derivative begins in a triple state and passes an intersystem crossing point to end in a single state upon Fe–carbene formation. In contrast, the native porphyrin cofactor starts in a quintet state, passes a first intersystem crossing point to a triplet state, and then passes a second intersystem crossing point to a single state, causing the carbene formation to be less efficient than the iron–porphycene complex.

C–H hydroxylation using a manganese porphycene. Heme-containing proteins, such as cytochrome P450, form high valent Fe-oxo species that are capable

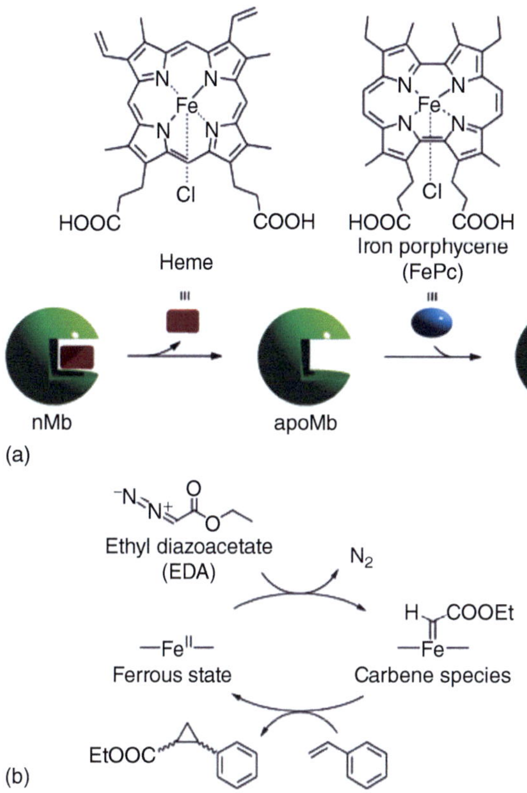

Figure 8.3 (a) Structure of heme and iron porphycene (top) and cartoon of the procedure for replacing the native heme cofactor of myoglobin with iron porphycene. (b) Reaction catalyzed by rMb. Source: Oohora et al. [18]. © 2017, American Chemical Society.

of hydroxylating inert C—H bonds [19]. Manganese-containing macrocycles, including porphycene complexes, are also capable of forming high valent Mn-oxo species that are capable of oxidizing hydrocarbons [20]. A manganese porphycene (MnPc) was incorporated into myoglobin (rMb(Mn^{III}Pc)) and its activity toward C—H bond hydroxylation was evaluated [21]. Oxidation of Mn^{III}Pc by H_2O_2 affords Mn^{V}(O)Pc, which can then hydroxylate ethylbenzene. The pH was optimized and found to give the best catalytic results at pH 8.5. The formation of a Mn^{V}(O) species was confirmed by electron paramagnetic resonance in perpendicular and parallel modes.

Cyclopropanation and C–H insertion using heme and cobalt porphyrin. Carbene transfer chemistry can be used in a variety of chemical reactions, such as cyclopropanation and Y–H insertion, which makes designing chemoselective catalysts challenging [22]. Protoporphyrin IX (ppIX) can be metallated with a wide scope of transition metals, which can be subsequently incorporated into a protein scaffold, thus creating a gamut of artificial metalloenzymes. Protoporphyrin IX containing iron, manganese, cobalt, and rhodium was prepared and incorporated into myoglobin so that the catalytic performance and selectivity could be evaluated [23].

The native heme-coordinating residue found in Mb, His93, was also mutated to other Lewis basic residues (Cys, Ser, Tyr) and non-Lewis basic residues (Ala, Phe) to further expand the number of enzymes being tested. The activity and selectivity were evaluated using 4-(dimethylsilyl)styrene as a substrate with both an olefin and a silane functional group, where cyclopropanation can occur at the olefin and insertion into the Si–H can occur at the silane. Each catalyst showed varying levels for cyclopropanation and Si–H insertion products with most favoring Si–H insertion. Mb(H64V,V68A)–[Co(ppIX)] variants, on the other hand, favored cyclopropanation at much higher levels compared with the other metal-ppIXs, with Mb(H64V,V68A,H93S)[Co(ppIX)] having the highest yield of cyclopropanation over Si–H insertion. Selectivity using 4-amino-styrene was also evaluated and the results were similar to those with 4-(dimethylsilyl)styrene with the exception of Mb(H64V,V68A)[Rh(ppIX)] also exhibiting higher amounts of cyclopropanation albeit with lower overall yield.

8.2.2.3 Covalent Anchoring in Native Protein

There are a wide variety of metal complexes used in organic and organometallic chemistry to catalyze reactions. Unlike enzymes, however, these catalysts tend to have limited enantio-, regio-, and chemoselectivities and poor water solubility. In order to expand the activities of native enzymes, these catalysts can be incorporated into proteins. Unlike replacing native hemes with structurally similar planer cofactors, incorporating these complexes require covalent anchoring in order to control the stereo- and enantioselectivity. There are several factors of covalently anchored catalysts that can be changed in order to alter their performance, such as the position of anchoring into a protein, the length of the linker between the anchor and the catalyst, and the residues surrounding the catalyst.

Hydrogen evolution using a nickel catalyst. Catalysts for hydrogen evolution are useful in addressing the need for renewable, alternative energy and for new technologies, such as hydrogen fuel cells. The DuBois-type nickel-diphosphine catalysts are some of best performing first-row transition metal catalysts for hydrogen evolution, and therefore make an excellent candidate for biocatalytic hydrogen production [24]. In conjunction with a ruthenium photosensitizer, a DuBois-type catalyst was used to create an artificial metalloenzyme capable of light-driven hydrogen evolution (Figure 8.4) [25]. The protein scaffold used was a flavodoxin (Fld) from *Synechococcus lividus*, an electron transfer protein that natively holds a flavin mononucleotide cofactor. The flavin mononucleotide (FMN) cofactor of Fld can be removed by trichloroacetic acid precipitation, allowing for reconstitution with the DuBois catalyst in place of the flavin. The Ru photosensitizer $[Ru(4\text{-}CH_2Br\text{-}4'\text{-}bpy)(bpy)_2]\cdot 2PF_6$ (RuPS) was conjugated to the protein via Cys54. Analysis by inductively coupled plasma–atomic emission spectroscopy (ICP–AES) indicates that the binding stoichiometry of 1.2 ± 0.3 for Ru and 1.0 ± 0.2 for Ni per Fld. Photocatalysis was performed using a 300 W Xe lamp and, in pH 6.2 MES buffer, the turnover frequency was found to be 410 ± 30 mol H_2 (mol Fld)$^{-1}$ h^{-1} with a TON of 620 ± 80.

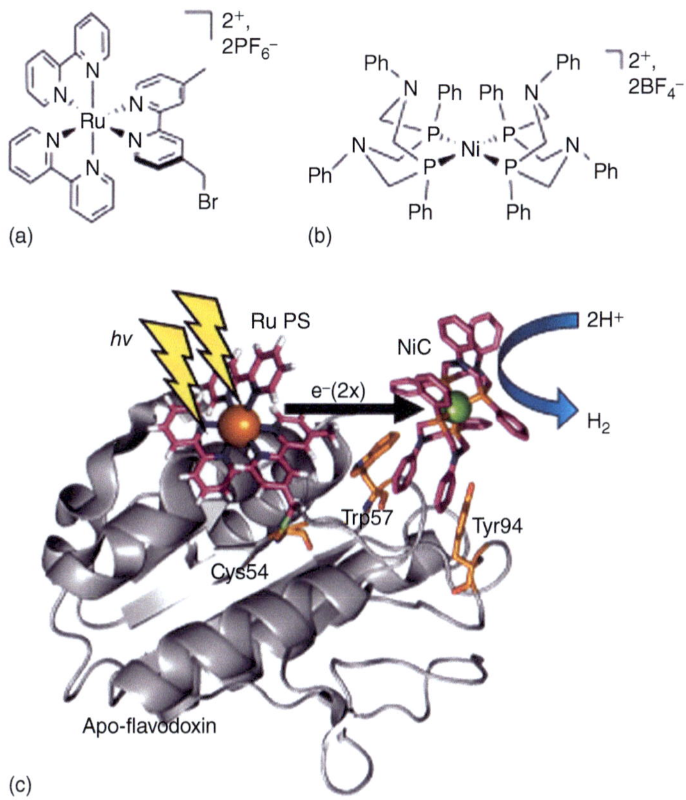

Figure 8.4 (a) Structure of ruthenium photosensitizer, (b) structure of DuBois-type catalyst, (c) and model of apo-flavodoxin conjugated with the ruthenium photosensitizer and DuBois-type catalyst. Source: Soltau et al. [25]. © 2017, American Chemical Society.

Hydroformylation using a rhodium catalyst. Hydroformylation reactions convert alkenes into aldehydes using carbon monoxide and H_2 and usually require a metal-based catalyst. There are no native proteins that catalyze hydroformylation reactions; however there are a large number of metal complexes that are catalysts for this reaction [26]. Therefore, an artificial metalloenzyme can be made by conjugating a hydroformylation catalyst to a protein. Since rhodium–phosphine complexes are known to catalyze hydroformylation, a rhodium system was chosen as the cofactor to create a metalloenzyme for hydroformylation [27]. Steroid carrier protein type 2 (SPC-2L) was chosen as the scaffold and different phosphines (P1, P2, or P3) were conjugated to a cysteine residues that were added via mutagenesis (V83C or A100C). Rhodium-bound proteins were prepared by adding $Rh(acac)(CO)_2$, which was found to selectively bind to the phosphine. The catalytic activity was tested using the hydroformylation of 1-octene and SCP-2L A100C conjugated to the P3 phosphine was found to have the highest TON (408.7 ± 57.79) and highest product yield ($78.8 \pm 4.86\%$). X-ray absorption spectroscopy was used to characterize the rhodium center in the protein.

Comparison of the X-ray absorption near edge structure (XANES) of SCP-2L A100C-1-P3-Rh to other Rh complexes reveals loss of carbonyl ligands while extended X-ray absorption fine structure (EXAFS) suggests that Rh is coordinated by the phosphine and Met105.

Phenylacetylene polymerization using a rhodium catalyst. The majority of industrial biocatalysis use whole-cell systems based on natural enzymes. Transition metal catalysts in whole-cell systems are poisoned by thiols and other components from cellular debris. In order to circumvent this problem, a microbial cell surface display (CSD) system was used in conjunction with a rhodium-containing engineered metalloenzyme [28]. The nitrobindin variant NB4 was chosen as the protein scaffold for its hydrophobic internal cavity, where the metal catalyst can be anchored. An inactivated esterase autotransporter (EstA) was used the CSD system and an Rh(cp)(cod) catalyst was covalently anchored to the NB4 displayed on the cell surface. The catalytic activity was evaluated using the polymerization reaction of phenylacetylene. The TON of the reaction was found to be very high, at 39×10^6, and is in fact much higher than the free catalyst (5×10^3). The cis:trans ratio of the polymerization product was altered compared with the free catalyst, with the Rh-protein conjugate giving 20 : 80 and the free catalyst giving 95 : 5.

Olefin metathesis using a ruthenium catalyst. Nitrobindin has also been used as a scaffold to anchor other transition metal catalysts [29]. Grubbs–Hoveyda (GH)-type ruthenium catalysts, with varying lengths in the alkyl spacer (C1–C3), were conjugated to Cys96 residue inside the hydrophobic cavity of NB4. In an effort to enlarge the cavity of NB4, the YASARA structure software suite was used to duplicate two β-strands (referred to as NB4exp). Estimating the cavity volume using the HotSpot Wizard server v2.0 shows that the cavity increased from 855 to 1389 $Å^3$. Ring-closing metathesis, ring-opening metathesis polymerization, and cross-metathesis reactions were used to assess the catalytic function of NB4 and NB4exp. The best results for both ring-closing metathesis and cross-metathesis reactions were >99% conversion and TON of 100, obtained with the NB4exp protein and C2 or C3 catalysts, and 98% conversion and TON of 296, obtained with the NB4exp protein and C2 catalyst. For ring-opening metathesis polymerization, the best results were obtained with the NB4exp protein with the C2 catalyst giving a 81% conversion, a TON of 10 000, an average molecular weight of 750×10^3 g mol^{-1}, and a polydispersity index of 1.29.

Carbon dioxide reduction using a nickel catalyst. Efficient carbon dioxide reduction is an important target for fixing carbons in the atmosphere and development of environmentally benign catalysts still remains a challenge. In an effort to develop such catalysts, an artificial metalloenzyme containing $[Ni(II)(cyclam)]^{2+}$ and a ruthenium photosensitizer were engineered [30]. $[Ni(II)(cyclam)]^{2+}$ is a CO_2 reduction catalyst that is well-known for its ability to selectively reduce CO_2 to CO under aqueous conditions. $[Ni(II)(cyclam)]^{2+}$ was attached to azurin (Az) through coordination to native histidine residue His83 and the ruthenium photosensitizer ($[Ru(bpy)_2$(epoxy-phen)]) was covalently attached to one of three surface cysteine residues placed by mutagenesis. In addition to the two non-natural cofactors, Az contains a native type 1 Cu center, which can be reconstituted with Zn. The effect of the identity of the metal in the type 1 site was examined. The activity for light-driven CO evolution of the three variants, S66C, S78C, and S100C, was tested. The activity was found to be inversely related to the distance between the Ru and Ni centers, with S78C-RuCuAz-[Ni(II)(cyclam)] being the most active. As a comparison, the activity of the free Ni complex with the photosensitizer was also evaluated and found to be higher but with worse selectivity, producing ten times the amount of H_2 while RuAz-[Ni(II)(cyclam)] produced CO_2 exclusively.

8.2.2.4 Supramolecular Anchoring in Native Protein

In addition to covalent anchoring in a single protein, it is possible to take advantage of supramolecular assembly of two or more subunits of native proteins to anchor inorganic catalysts. In this way, one can use the interface of proteins for tuning both reactivity and selectivity of the artificial enzymes.

Styrene oxidation using an iron catalyst. While natural enzymes are homogeneous catalysts, there are some examples of artificial enzymes that have been engineered or modified to be heterogenous catalysts [31]. One approach in designing heterogeneous enzymes is cross-linked enzyme crystals (CLECs) allows for the benefits of enzymes (chemo-, region, and enantioselectivity) with a high density of active sites and increased stability. Crystals of NikA, a nickel-binding protein, were cross-linked between lysine residues using glutaraldehyde [31]. For catalysis, iron-ethylenediaminetetraacetic acid (EDTA)-like complexes (Fe-L0, Fe-L1, and Fe-L2) that have been shown to bind to NikA via noncovalent interactions were chosen for oxidative transformations of organic compounds. Catalytic activity was evaluated using the oxidation of the carbon–carbon double bond in β-methoxystyrene in the presence of O_2 and dithiothreitol (DTT), producing benzaldehyde and 2-methoxyacetophenone. The best results were obtained with NikA/Fe-L2 CLEC, with yields as high as 74% and an accumulative TON of 28 000 after 50 successive runs. Additionally, substrate chlorination was observed when the reaction was performed in the presence of sodium chloride and potassium persulfate, with both chlorohydrin and dechlorinated products produced. Reaction with NikA/Fe-L2 CLEC and 4-methoxystyrene exclusively produces the chlorohydrin product with a yield of 96% and accumulated TON over 5900.

Cyclopropanation using a heme cofactor. Using the concepts of supramolecular assembly, artificial metalloenzymes can be designed to incorporate metal complexes and impart the selectivity of natural proteins through mutations around the active site. One such system was designed using LmrR, a homodimeric protein with an unusually large hydrophobic binding pocket, and a heme cofactor (Figure 8.5) [32]. Two mutations, K55D and K59Q, were made to improve expression and purification and an additional mutation, W96A, was made to assess the effect of tryptophan on the access to the hydrophobic pocket. UV–Vis characterization of the protein-heme assembly shows that is monomeric and there is one heme per protein dimer.

Diels–Alder reaction using a copper catalyst. Due to their biocompatibility, large substrate scope, and catalytic specificity, artificial metalloenzymes have great potential for therapeutic applications. A vital aspect of therapeutic applications is the

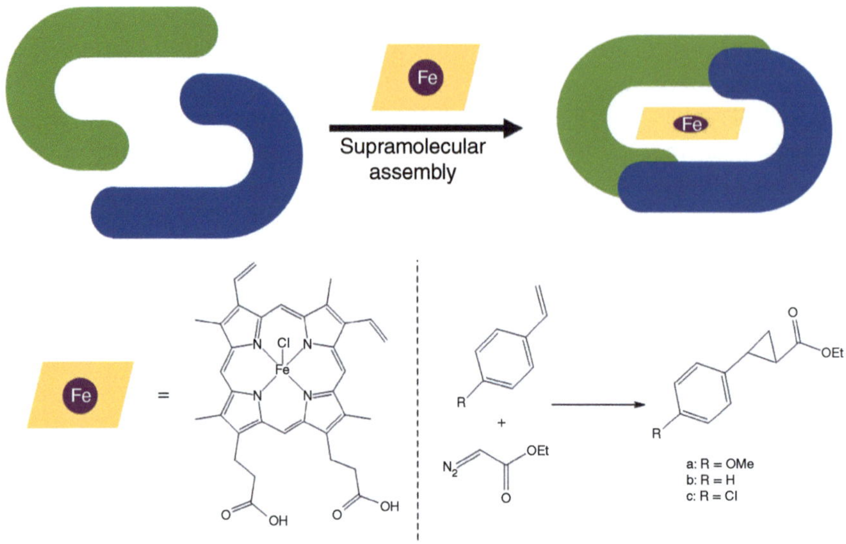

Figure 8.5 Schematic representation of the LmrR-heme assembly (top), structure of heme (bottom left), and reaction catalyzed by the LmrR-heme system (bottom left).

ability to work in living human cells. Toward this goal, an artificial metalloenzyme was engineered using the guanine nucleotide-binding protein-coupled receptor subtype A_{2A} [33]. By connecting a 1,10-phenanthroline moiety capable of coordinating Cu(II) to antagonists of A_{2A}, a catalytically active metal complex can be incorporated into A_{2A}. Catalytic activity was evaluated with A_{2A} on the surface of living human embryonic kidney (HEK) cells for Diels–Alder cycloaddition of cyclopentadiene and various alkenes. The protein-bound catalyst was found to have a moderate ee of $28 \pm 0.5\%$ was observed with a yield of $22 \pm 2\%$ compared with no ee for the free catalyst.

8.3 Engineering Artificial Metalloenzyme by Directed Evolution in Combination with Rational Design

Since methods for engineering Cytochromes P450 have been reviewed thoroughly in Chapter 9 of this book, this chapter will focus on the use of site-saturation mutagenesis, directed evolution, and related strategies to construct artificial metalloenzymes possessing new reactivities from other protein scaffolds. The strategies are categorized by the protein scaffolds being utilized.

8.3.1 Directed Evolution of Metalloenzymes Using *De Novo* Designed Scaffolds

Growing computational modeling power and more sophisticated design methodologies have been used to generate several metalloenzymes using *de novo* designed scaffolds that could catalyze different reactions. Like *de novo* artificial enzymes, artificial metalloenzymes generally remain suboptimal with limited reactivity and selectivity and require directed evolution to achieve high activity [34, 35]. For example, the binary code strategy was used to construct α-helical proteins consisting of "binary patterning" of polar and non-polar amino acids. Those α-helical bundles could bind the heme cofactor and the resulted novel heme proteins catalyzed peroxidation with very low activity [36]. Random mutagenesis of two previously characterized binary-patterned proteins S-824 and S-836 were performed and the resulted library was screened either by a 96-well plate-based assay or colony-based assay. Several rounds of screening and selection led to the identification of two evolved mutants with nearly threefold higher activity than the parental enzyme S836. This example demonstrated that the *de novo* designed artificial metalloenzymes, although not related to any natural protein and not specifically designed for any catalytic activity, were useful starting templates for directed evolution. In another study, a *de novo* designed bifunctional enzyme Syn-IF was used to rescue two different auxotrophic strains of *E. coli*: Δ*ilvA* and Δ*fes* that catalyzed isoleucine biosynthesis and iron assimilation respectively. Directed evolution of Syn-IF for the faster rescue of either Δ*ilvA* or Δ*fes* resulted in two different proteins that were capable of rescuing the selected function much faster than Syn-IF, but lost the ability to rescue the unselected function at the same time. This study successfully demonstrated the divergent evolutionary trajectories that were hard to reconstruct in natural systems [37].

8.3.2 Directed Evolution of Metalloenzymes Using Native Scaffolds

8.3.2.1 Cofactor Replacement in Native Proteins

As discussed in Section 8.2.2.2, replacing the native co-factor of metalloenzyme with an abiotic one is one of the most essential strategies to create artificial metalloenzyme with new reactivity. However, in many cases, the activity and selectivity need to be further improved by directed evolution strategies [38].

Hydrogenation and hydroformylation using hCAII scaffold. In nature, nicotinamide cofactors like NAD(P)H are normally needed to hydrogenate organic substrates. However, this is not an economically viable strategy due to the high cost of cofactors. Anchoring metal complexes into protein scaffolds by linking them with a binding ligand is one strategy to introduce new-to-nature reactivities. Another approach is to link the metal to the protein scaffold directly, but this is limited by nonspecific binding. This barrier was overcome by Kazlauskas and coworkers in 2009, when they found a way to link Rh(I) to carbonic anhydrase [39]. Nonspecific binding was minimized by either site-directed mutagenesis of the His residues on the protein surface or modifying them chemically. Though this catalyst had a slower rate of hydrogenation than the isolated metal complex, it was proved to be superior in diastereoselectivity with a cis:trans ratio of approximately 20 : 1. The same strategy of replacing the zinc in carbonic anhydrase's active site with

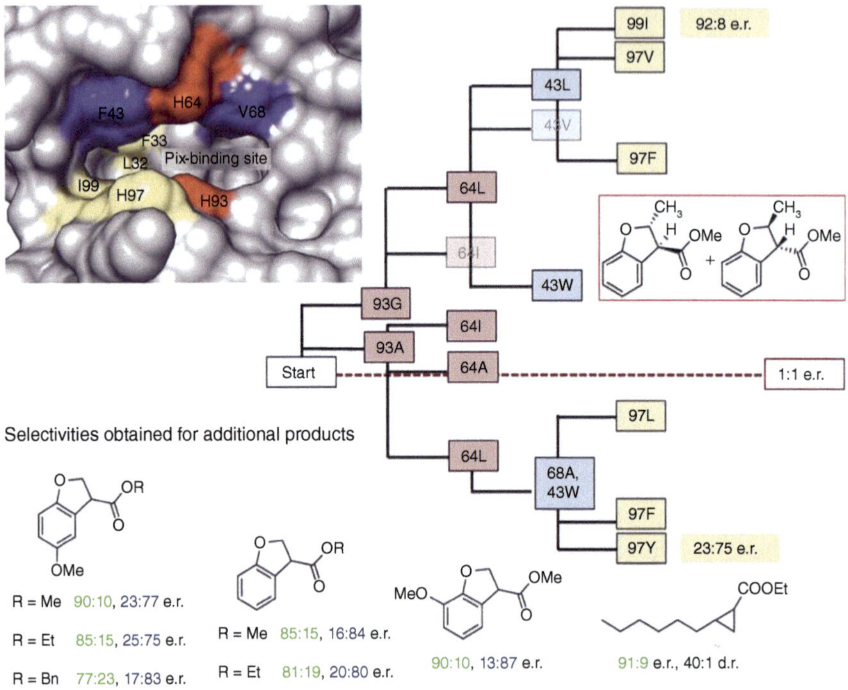

Figure 8.6 Directed evolution strategy used to obtain Ir(Me)-mOCR-Myo mutants capable of producing either enantiomer of the products of C–H insertion reactions of varied substrates. Source: Key et al. [42]. © 2016, Springer Nature.

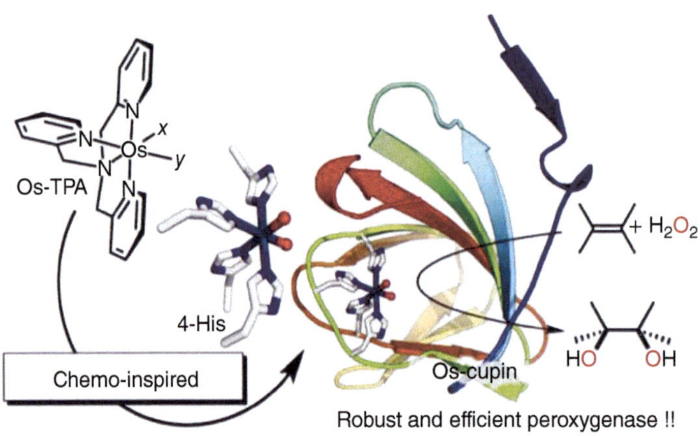

Figure 8.7 Artificial osmium peroxygenase from Osmium–Cupin complex. Source: Fujieda et al. [46]. © 2017, American Chemical Society.

rhodium yielded a metalloenzyme that could perform cofactor-free hydroformylation of styrene with syngas [40]. The new metalloenzyme carried the transformation with 40-fold higher regioselectivity than the isolated metal catalyst, with the ratio of linear over branched aldehyde product being 8.4 : 1.

Carbene insertion. Aside from the organometallic anchoring approach used by Ward and Lewis [2, 41], engineering natural enzyme scaffolds to endow them with new reactivities is another approach that has recently seen several breakthroughs. In a study published in *Nature* in 2016 [42], the iron in Fe–porphyrin IX in myoglobin porphyrin core was replaced by abiological metal with reactivity different from iron. This was achieved by first identifying suitable expressing conditions for the apo- form of heme proteins in *E. coli*, followed by reconstitution with [M]-PIX cofactors with M being a new metal. Scanning a library of metalloenzymes generated with this method using Fe, Co, Cu, Mn, Rh, Ir, Ru, and Ag, Ir(Me)-mOCR-Myo was identified as the most active in performing carbene insertion into C—H bond to form C—C bonds. The versatility of this method was demonstrated by engineering [M]-mOCR–Myo that can carry out cyclopropanation on β-substituted vinyl arenes and aliphatic α-olefins that could not be catalyzed by using native Fe-porphyrin myoglobin at that time. Furthermore, through directed evolution, the authors identified mutants of this enzyme that effectively control the enantioselectivity of these transformations to form either enantiomer of the products (Figure 8.6) [42]. Meanwhile, variants of the P450 enzyme CYP119 with iridium in place of iron were created and site-directed evolution was applied to engineer its substrate binding sites. The resulted quadruple mutant CYP119-max (C317G, T213G, L69V, V254L) can catalyze intramolecular carbene insertion with up to 98% enantiomeric excess, 35 000 turnovers, and 2550 h^{-1} turnover frequency. This study demonstrated that directed evolution could be applied to evolve artificial metalloenzyme to catalyze abiotic transformation with the kinetics comparable to the native enzymes [43]. Additionally, site-directed evolution further broadened the substrate scope of abiotic transformations and resulted in Ir(Me)-PXI CYP119 mutants catalyzing the cyclopropanation of terminal and internal, activated and inactivated, electron-rich and electron-deficient, conjugated and nonconjugated alkenes; and C–H functionalization of phthalan derivatives with high yield and good chemo-, stereo-, and enantioselectivities [44, 45].

Peroxidation using osmium–cupin complex. Metal substitution of natural metalloenzymes based on their metal-binding promiscuity is a well-recognized strategy to create novel activity and improve the thermal stability of enzymes. An artificial peroxygenase was created by substituting the native metal manganese in the TM1459 cupin superfamily protein from *Thermotoga maritima* with osmium (Figure 8.7) [46]. The Os center endowed the protein with new catalytic activity and protein folding thermodynamic stability. This artificial metalloenzyme was shown to catalyze cis-dihydroxylation of a range of alkenes with high TON under low catalyst loading (up to 9100 TONs under 0.01% mol catalyst when the alkene was 2-methody-6-vinyl-naphthalene). Furthermore, the regioselectivity of the enzyme was also demonstrated by the fact that terminal octene isomers form products with higher yield than internal octene isomers. Having the protein scaffold around the Os center also limited disproportionation of H_2O_2 by Os and enhanced the efficiency of the dihydroxylation. Through site-directed mutagenesis of this artificial metalloenzyme at position 106, the authors also identified C106S to improve the catalytic activity by threefold, suggesting that further genetic modification could allow discovery of better performing variants of this enzyme.

8.3.2.2 Covalent Anchoring in Native Protein

Among different methods to prepare artificial metalloenzymes, covalent anchoring has key advances in flexibility on the nature of the metal catalysts and scaffold protein. Unlike non-covalent methods for artificial metalloenzyme formation, the covalent linkage in principle allows the use of any desired protein as a scaffold, including those without native co-factors. A good cross-linkage should have high specificity and biorthogonality and can easily locate the metal complex to the desired location without changing the overall structure of the scaffold protein. Like artificial metalloenzymes prepared from other methods, the covalent linkage may also result in metalloenzymes with low or even no activity that is mainly because the absence of the entrance point or defined binding site [47, 48]. Directed evolution is a promising strategy to resolve those issues and improve the activity and selectivity of the artificial metalloenzymes created by covalent linkage.

Carbene formation and insertion. A new approach for the construction of artificial metalloenzymes using strain-promoted azide–alkyne cycloaddition (SPAAC) was reported by the Lewis group in 2014 [49]. Unlike the biotin-streptavidin technology, which relies on a nature binding using strain-promoted azide-interaction to help anchor the organometallic moiety, this new method introduces a *p*-azido-L-phenylalanine (Az) residue into the scaffold protein and uses click chemistry to covalently link it with the bicyclononyne-substituted metal complex (Figure 8.8). The high efficiency of this reaction allows the linkage of metal complex within the protein barriers and the biorthogonality of SPAAC eliminates the need to remove residues such as cysteine that might react with the electrophiles used in conventional bioconjugation methods [34]. Mutating the scaffold protein allows tuning of the second coordination sphere of the metal complex and controlling its activity. Dirhodium artificial metalloenzymes constructed using this method was shown to catalyze decomposition of diazocompounds into carbene precursors as well as their insertion into Si—H bonds and olefins. Though the yield and enantioselectivity of these reactions did not exceed that of the metal cofactor alone, the study served as a proof-of-concept of a promising new method for artificial metalloenzyme construction and engineering.

Cyclopropanation. Dirhodium complexes can catalyze a broad range of reactions including cyclopropanation and X–H insertion (X = C, N, O, and so on) [51]. However, the low selectivity remains as a big challenge for many of these reactions [52]. To resolve this issue, a dirhodium artificial metalloenzyme was created to link Esp-based dirhodium cofactor **1** to a prolyl oligopeptidase

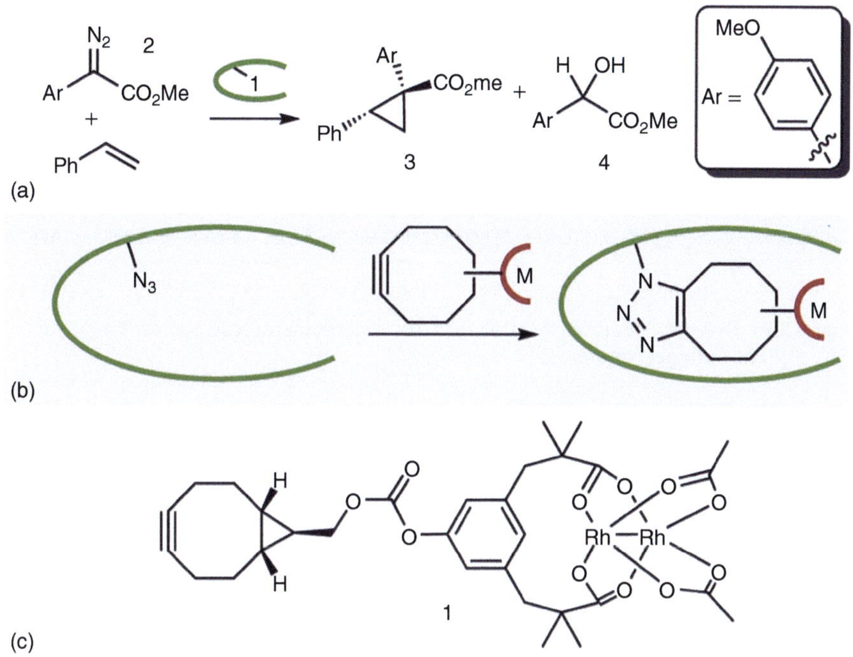

Figure 8.8 Model reaction of SPAAC and artificial metalloenzyme structure. Source: Yang et al. [50]. © 2018, Springer Nature.

(POP) scaffold possessing an L-4-azidophenylalanine residue by SPAAC reaction (Figure 8.8) [53]. The performance of this artificial metalloenzyme as a catalyst for olefin cyclopropanation was improved via scaffold mutagenesis. POP mutant H328-F99-F354 achieved a very good ee of 92%. Furthermore, it also accepted a range of styrenes and donor–acceptor carbene precursors. Notably, the artificial metalloenzyme was water tolerant, which is remarkable for metal-catalyzed reactions in water.

A significant step forward was reported three years later, when a platform for the directed evolution of artificial metalloenzymes generated via the SPAAC-based anchoring method was developed by Lewis and coworkers (Figure 8.9) [50]. Error-prone PCR and combinatorial codon mutagenesis were employed to create libraries of artificial metalloenzymes that can be screened for improved performance. This approach allowed the exploration and discovery of beneficial mutations distal to the metal-binding site. Using this platform, new artificial cyclopropanases with high enantioselectivity for each enantiomer of the cyclopropane product as well as ones with higher activity than artificial metalloenzymes previously created via targeted mutagenesis were created. These artificial metalloenzymes were also found to catalyze N–H, S–H, and Si–H carbene insertion, which means they can serve as starting points for further rounds of directed evolution toward these reactions. This strategy of forming artificial metalloenzymes via SPAAC-based metal complex-host protein linking followed by directed evolution via random

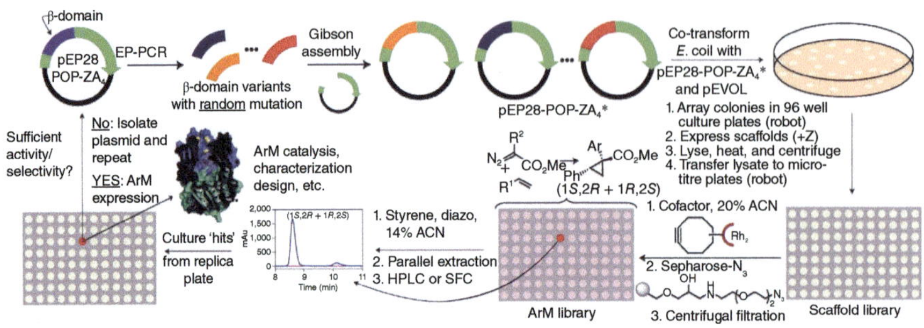

Figure 8.9 Overview of artificial metalloenzyme evolution protocol. Source: Yang et al. [50]. © 2018, Springer Nature.

mutagenesis of the host proteins is a potentially versatile and useful method for engineering new artificial metalloenzymes with novel activities in the future.

8.3.2.3 Non-covalent Anchoring in Native Proteins

Hydrogenation. Biotin-streptavidin technology might be one of the most popular techniques used to incorporate an achiral metal catalyst within the protein noncovalently due to the extraordinary affinity of biotin for either avidin or streptavidin ($K_a \sim 1 \times 10^{14}$ M^{-1}) [41]. More importantly, the derivatization of the valeric acid side chain of biotin does not affect biotin–avidin binding affinity, which warrants the integrity of the diverse organometallic species; and allows the use of orthogonal diversity-generating procedures: evolution for protein optimization and organometallic fragment diversification to optimize the artificial metalloenzymes from both perspectives. An artificial hydrogenase was created by incorporating rhodium (I) complex within streptavidin and the enantioselectivity in the asymmetric hydrogenation of two dehydroamino acid derivatives was optimized using both chemical and genetic methods (Figure 8.10) [54, 55]. On the metal complex side, the chelating biphosphine ligands were tested as the aspects of both flexible and rigid scaffolds as well as spacers between the biotin and the metal complex with different topologies. On the host protein side, site-directed mutagenesis at the S112 position was performed to generate a library of 20 isoforms. Notably, it was observed that mutation at the S112 position could invert the enantioselectivity of the reaction: while wild type (WT) streptavidin produced mostly the (*R*) product (~93% ee), the S112Q mutant produced mostly the (*S*) product (~90% ee). These studies not only demonstrated artificial metalloenzymes constructed from the biotin-streptavidin as highly enantioselective catalysts for hydrogenation, but also showed that genetic optimization allows control of the reaction's enantioselectivity.

Transfer hydrogenation. The biotin–streptavidin technology was further adapted to carry out artificial transfer hydrogenation for a wide range of substrates, including ketones, enones, imines, $NAD(P)^+$, and its analogs. For each substrate group, a metal complex was carefully selected.

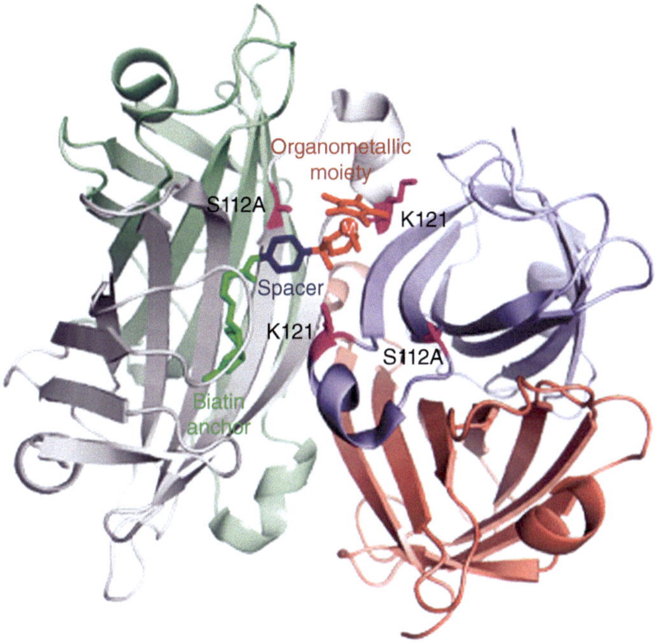

Figure 8.10 Chemoenzymatic optimization of artificial metalloenzymes based on the biotin-streptavidin technology. Source: Heinisch and Ward [41]. © 2016, American Chemical Society.

Transfer hydrogenation: Ketones. An artificial metalloenzyme was created by conjugating biotinylated Noyori's homogeneous d^6 piano-stool complexes of Ru (II) with (strept)avidin. The resulted metalloenzyme can catalyze transfer hydrogenation of ketones by using boric acid-formate mixture as a hydrogen source (Figure 8.11a) [56]. Protein crystal structure analysis revealed that positions K121L and L124 along with position S112 on the streptavidin scaffold have the most significant effects on the enantioselectivity of the reaction. Thus, 80 double and 40 singles streptavidin mutants at these positions were chosen to combine with the biotinylated d^6-pianostool complexes for screening. The best performers achieved up to 96% ee (*R*) and 92% ee (*S*) for acetophenone derivatives and up to 90% ee (*R*) for model dialkyl alcohols [57].

Transfer hydrogenation: Imines. A similar strategy was used by the Ward group to construct an artificial imine reductase for cyclic imines. By anchoring a biotinylated iridium pianostool complex into streptavidin scaffolds mutated at positions S112 and K121 for screening, the authors achieved 96% ee (*R*) and 78% ee (*S*) for the model reaction (Figure 8.11b) [58]. A later structural, kinetic, and docking study on this system revealed mechanistic insights into the preference of 2 streptavidin mutants S112A and S112K for *S* and *R* form of the Ir cofactor, leading to opposite stereochemistry in the reduced amine [59]. A dual anchoring strategy, in which a His residue was introduced into the streptavidin scaffold at the S112 position and helped anchoring the organometallic moiety through the formation of a

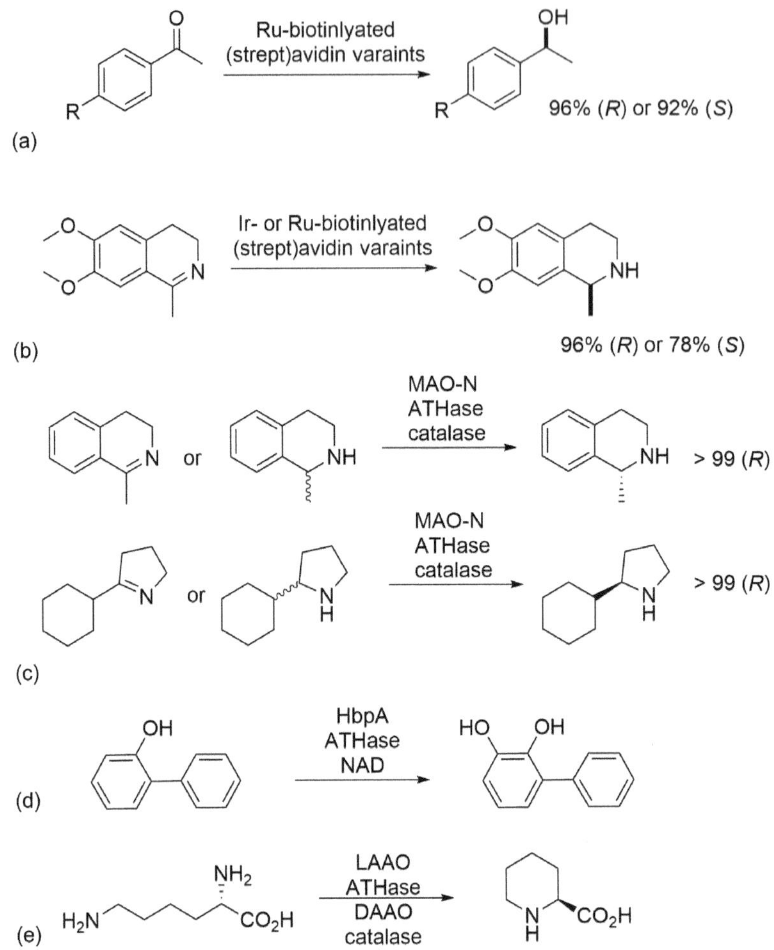

Figure 8.11 Representative transfer hydrogenation reactions catalyzed by artificial metalloenzyme based on biotin-streptavidin technology (a) transfer hydrogenation on ketones, (b) transfer hydrogenation on imines, (c) redox cascade for pure secondary amines formation, (d) ATHase for NADH regeneration, (e) redox cascade for L-pipecolic acid formation.

dative bond, was also explored [60]. While anchoring the Ir(II) and Rh(II) complexes into WT streptavidin did not offer any advantage over the "naked" metal complexes, anchoring them into streptavidin S112H resulted in the formation of the (*S*) product with 55% ee and into streptavidin K121H resulted in the formation of the (*R*) product with 79% ee at pH 5 and 55 °C.

Notably, to facilitate the directed evolution of artificial imine reductase, a high-throughput screening system was established [61]. After good glutathione scavengers that can prevent glutathione from coordinating with the metal complex in artificial metalloenzymes were identified, a screening system utilizing 24-deep well plates for a library of streptavidin mutants carrying mutations at 28 positions with 12 possible amino acids at each position was developed [15]. This system is versatile and could be used to screen and evolve other artificial metalloenzymes constructed based on the biotin-streptavidin technology.

Transfer hydrogenation: Redox cascades. Creating artificial metalloenzymes is an important strategy to create concurrent cascade reactions that combine an organometallic catalyst with additional enzymes, which has been proven challenging due to mutual inactivation of both catalysts. For example, an Ir-based artificial transfer hydrogenase (ATHases) can participate in orthogonal redox cascades with many NADH, FAD, and heme-dependent enzymes [62]. Coupling ATHase with evolved monoamine oxidases (MAO-N), a double stereoselective amine deracemization was established to prepare pure secondary amines (Figure 8.11c). Furthermore, ATHase catalyzed NADH regeneration for monooxygenase (HbpA)-catalyzed hydroxylation of 2-hydroxybiphenl (Figure 8.11d). Additionally, ATHases worked concurrent with L-amino acid oxidase (LAAO) and D-selective amino acid oxidase (DAAO) to prepare L-pipecolic acid from L-lysine (Figure 8.11e). Most notably, a horseradish peroxidase-coupled colorimetric assay for ATHase activity was also developed, which facilitates the directed evolution of the enzyme for future new activities.

C–H activation. Pentamethylcyclopentadienyl Rh catalyst $[Cp^{*}RhCl_2]_2$ was a versatile catalyst for electrophilic aromatic C–H activation reaction. However, it was difficult to introduce an asymmetric ligand around Cp*Rh moiety for enantioselective synthesis [63]. An engineered streptavidin-biotinylated Rh(III) complex was reported for the first time by Ward and Rovis collaboratively for catalytic asymmetric carbon–hydrogen (C–H) activation (Figure 8.12) [64]. The mechanism study of $[Cp^{*}RhCl_2]_2$ catalyzed aromatic C–H activation occurred through a concerted metalation–deprotonation mechanism and the presence of a base significantly lowed the activation energy of this step. Active-site tailoring was conducted by introducing basic residue at $S112_A$ and $K121_B$ that were closet to the metal center by site-directed mutagenesis. The double streptavidin mutant S112Y-K121E was discovered to be the best performer in the selective coupling of benzamides and alkenes to dihydroisoquinoline, achieving 100-fold faster rate than when the Rh complex acted alone as well as e.r. up to 93 : 7.

Olefin metathesis. A breakthrough in improving the throughput of the genetic optimization process for artificial metalloenzymes constructed based on the biotin–streptavidin technology was made in a 2016 study [65]. By expressing streptavidin mutants, which were formed by saturation mutagenesis of the 20 residues closest to the metal (Ru in this case), inside *E. coli* cells and secret them into the cells' periplasm through fusion with the OmpA signal peptide, the need for protein purification and inhibitors (e.g. glutathione) scavenging was eliminated. This allowed rapid screening and directed evolution of the artificial metalloenzymes for new reactivities, which was demonstrated by the construction and evolution of artificial olefin metathesis by anchoring a biotinylated Hoveyda–Grubbs second-generation catalyst into Sav mutants to form artificial metalloenzymes and screen them for the formation of umbelliferone, a fluorescent product, from its precursor via ring-closing metathesis (Figure 8.13). It should also be noted

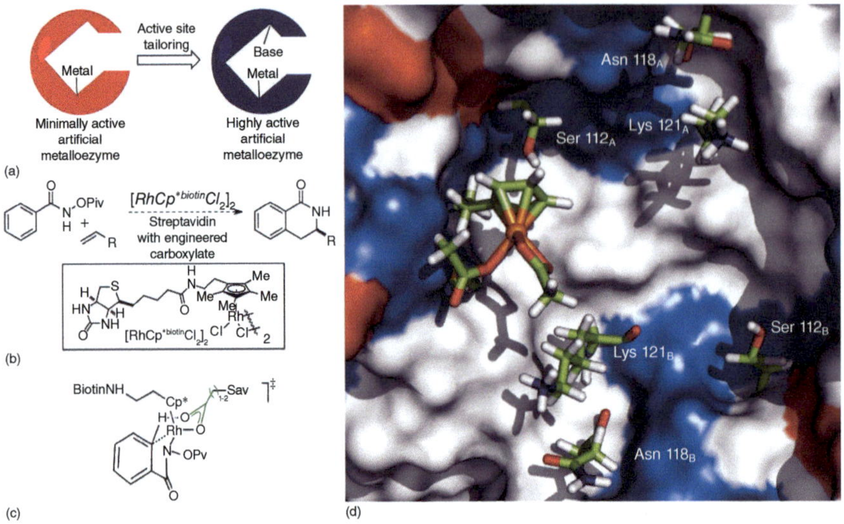

Figure 8.12 Synergistic effect of a basic residue introduced by site-directed mutagenesis and a biotinylated $RhCp^{*biotin}Cl_2$ moiety acting as a catalyst for selective benzannulation of benzamides and alkenes. (a) Improve activity of artificial metalloenzyme by active site evolution, (b) artificial metalloenzyme catalyzed benzannulation for enantionselective synthesis of dihydroisoquinolones, (c) postulated transition state for the C–H activation steps, (d) Auto-Dock model of biotinylated $RhCp^{biotin}(OAc)_2$ complex in the active site of streptavidin. Source: Hyster et al. [64]. Reprinted with permission from AAAS.

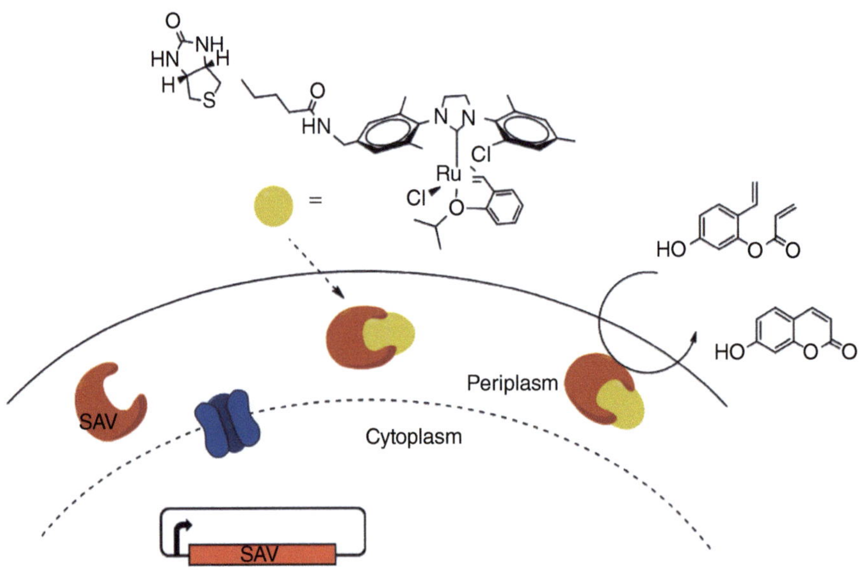

Figure 8.13 Streptavidin-based artificial metalloenzymes for *in vivo* metathesis and directed evolution. Source: Based on Jeschek et al. [65].

that the biotinylated Ru complex did not show any activity *in cellulo*. This work showed that the biotin-streptavidin could be applied for the directed evolution of artificial metalloenzymes that can perform non-natural reactions *in vivo*, which has many potential applications in bio-orthogonal catalysis.

Except for using (strept)avidin as a protein scaffold, human carbonic anhydrase II was used to hold a Hoveyda–Grubbs catalyst through an arylsulfonamide anchor. The resulted metalloenzyme could catalyze ring-closing metathesis [66]. Mutants of this metalloenzyme at residues I91, F131, L198, and K170 were created via site-directed mutagenesis. The L198H mutant turned out to achieve up to 28 TONs, which is a significant improvement in catalytic performance. More notably, this enzyme worked under aerobic physiological conditions and low concentration, which are remarkable advantages.

Genetic switch. After the success of performing streptavidin-based artificial metalloenzyme-catalyzed olefin metathesis in *E. coli*, a biotinylated Ru complex and a biotinylated cell-penetrating poly(disulfide) was anchored into the streptavidin scaffold to form cell-penetrating artificial metalloenzymes [67]. The artificial metalloenzymes could be taken up into HEK-293T cells and catalyze the uncaging of the allyl-carbamate-protected thyroid hormone triiodothyronine, leading to the expression of a secreted NanoLuc-luciferase via upregulating the gene circuit that controls this process (Figure 8.14). The artificial metalloenzymes' catalytic performance could be improved by controlling the ratio of the

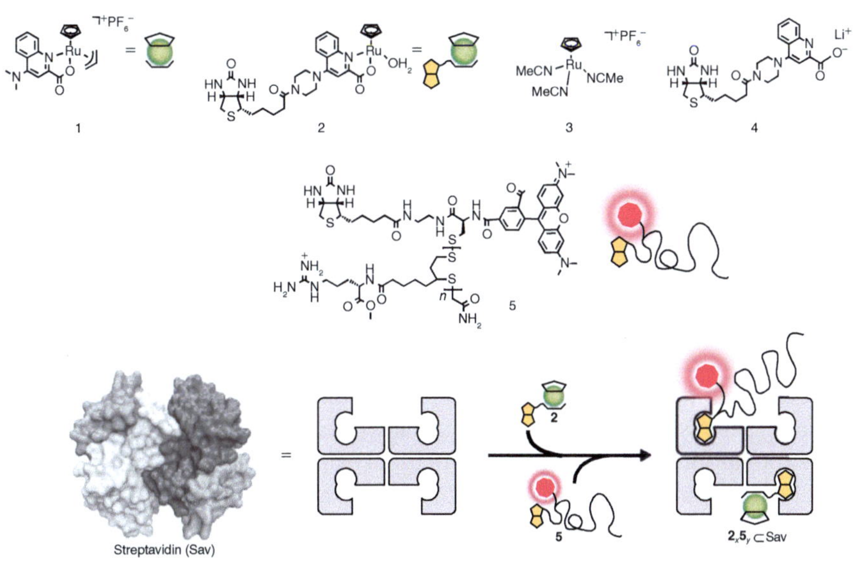

Figure 8.14 Assembly of cell-penetrating ArMs. Ruthenium complexes **1** and **2** catalyze a bioorthogonal uncaging reaction. The biotinylated cell-penetrating poly (disulfide) (CPD) bears a fluorescent TAMRA moiety **5** allowing the monitoring of cellular uptake. Incorporation of both biotinylated moieties **2** and **5** in various ratios (*x* and *y*) in tetrameric Sav affords a cell-permeable ArM for the uncaging of allyl carbamate-containing substrates within cell. Source: Okamoto et al. [67]. CC BY 4.0.

two biotinylated components and modifying the streptavidin scaffold by site-directed mutagenesis. The study signified a major step forward in introducing new-to-nature reactions intracellularly by using artificial metalloenzymes armed with cell-penetrating modules.

Suzuki–Miyaura cross-coupling. Except for Ru and Ir metal complex, biotinylated monophosphine palladium complex was also conjugated within streptavidin to generate an enantioselective artificial Suzukiase [68]. Site-directed mutagenesis was conducted to optimize the activity and enantioselectivity of artificial Suzukiase and the authors demonstrated the activity of the optimized artificial Suzukiase in the asymmetric synthesis of 2-methoxy-binaphthyl, achieving up to 90% ee and 50 TONs for this transformation.

8.4 Summary and Outlook

The variety of protein scaffolds and strategies to empower them with new reactivities presented in this chapter shows our current abilities to engineer enzymes to carry out transformations that nature has not yet explored. Furthermore, some methods have been demonstrated to work *in vivo*, opening the door to introducing artificial metalloenzymes into, and modifying biological processes inside living systems. *De novo* rational design methods have become powerful enough to yield not just structurally correct but functionally active enzymes. In other cases, they provide good starting points for later mutagenesis and directed evolution campaigns to improve the scaffolds toward desired reactivities.

A prevalent theme that is observed in many studies presented here is the introduction of novel factors into native protein scaffolds, such as new metal cofactors, through either covalent or non-covalent linkage. In many cases, this is facilitated by strategic point mutations on the scaffolds, most often in or near the active sites. Advances have been made to adapt random mutagenesis techniques to perform truly directed evolution on artificial metalloenzymes. This is a challenging direction due to the limitations in current methods to incorporate new metal cofactors. The work from the Lewis group on developing a platform for directed evolution of artificial metalloenzymes generated via the SPAAC-based anchoring method was a notable step in this direction [50].

Though both rational design and directed evolution have led to significant innovations in engineering new artificial metalloenzymes in recent years, many exciting discoveries lie ahead at the intersection of these two strategies. New computational modelling techniques have been getting better and better at simulating and designing not just scaffolds based on natural proteins, but also completely new ones. Advances in automation, high-throughput screening, and smart library-design have allowed directed evolution to explore various new-to-nature transformations. Protein engineers now possess an extensive and powerful arsenal of tools that they can utilize to tackle any challenges they might face in tailoring biomolecules for desirable properties. Much work lies ahead, but the progress that has been made in recent years is both encouraging and inspiring.

Acknowledgment

The authors acknowledge the funding of the DOE Center for Advanced Bioenergy and Bioproducts Innovation (US Department of Energy, Office of Science, Office of Biological, and Environmental Research under Award Number DE-SC0018420). Any opinions, findings, and conclusions or recommendations expressed in this publication are those of the author(s) and do not necessarily reflect the views of the US Department of Energy.

References

1 Thomson, A.J. and Gray, H.B. (1998). Bio-inorganic chemistry. *Curr. Opin. Chem. Biol.* 2 (2): 155–158.
2 Schwizer, F., Okamoto, Y., Heinisch, T. et al. (2018). Artificial metalloenzymes: reaction scope and optimization strategies. *Chem. Rev.* 118 (1): 142–231.
3 Lombardi, A., Bryson, J.W., and DeGrado, W.F. (1996). *De novo* design of heterotrimeric coiled coils. *Pept. Sci.* 40 (5): 495–504.
4 Lombardi, A., Pirro, F., Maglio, O. et al. (2019). *De novo* design of four-helix bundle metalloproteins: one scaffold, diverse reactivities. *Acc. Chem. Res.* 52 (5): 1148–1159.
5 Faiella, M., Andreozzi, C., de Rosales, R.T.M. et al. (2009). An artificial di-iron oxo-protein with phenol oxidase activity. *Nat. Chem. Biol.* 5: 882–884.
6 Reig, A.J., Pires, M.M., Snyder, R.A. et al. (2012). Alteration of the oxygen-dependent reactivity of *de novo* due Ferri proteins. *Nat. Chem.* 4: 900–906.
7 Zastrow, M.L., Peacock, A.F.A., Stuckey, J.A., and Pecoraro, V.L. (2012). Hydrolytic catalysis and structural stabilization in a designed metalloprotein. *Nat. Chem.* 4 (2): 118–123.
8 Tegoni, M., Yu, F., Bersellini, M. et al. (2012). Designing a functional type 2 copper center that has nitrite reductase activity within α-helical coiled coils. *Proc. Natl. Acad. Sci. U.S.A.* 109 (52): 21234.
9 Koebke, K.J., Yu, F., Salerno, E. et al. (2018). Modifying the steric properties in the second coordination sphere of designed peptides leads to enhancement of nitrite reductase activity. *Angew. Chem. Int. Ed.* 57 (15): 3954–3957.
10 Reed, J.H., Shi, Y., Zhu, Q. et al. (2017). Manganese and cobalt in the nonheme-metal-binding site of a biosynthetic model of heme-copper oxidase superfamily confer oxidase activity through redox-inactive mechanism. *J. Am. Chem. Soc.* 139 (35): 12209–12218.
11 Bhagi-Damodaran, A., Michael, M.A., Zhu, Q. et al. (2016). Why copper is preferred over iron for oxygen activation and reduction in haem–copper oxidases. *Nat. Chem.* 9: 257–263.
12 Miner, K.D., Mukherjee, A., Gao, Y.-G. et al. (2012). A designed functional metalloenzyme that reduces O_2 to H_2O with over one thousand turnovers. *Angew. Chem. Int. Ed.* 51 (23): 5589–5592.

13 Yu, Y., Cui, C., Liu, X. et al. (2015). A designed metalloenzyme achieving the catalytic rate of a native enzyme. *J. Am. Chem. Soc.* 137 (36): 11570–11573.

14 Mukherjee, S., Mukherjee, A., Bhagi-Damodaran, A. et al. (2015). A biosynthetic model of cytochrome c oxidase as an electrocatalyst for oxygen reduction. *Nat. Commun.* 6: 8467.

15 Vargas, D.A., Tinoco, A., Tyagi, V., and Fasan, R. (2018). Myoglobin-catalyzed C–H functionalization of unprotected indoles. *Angew. Chem. Int. Ed.* 57 (31): 9911–9915.

16 Mirts, E.N., Petrik, I.D., Hosseinzadeh, P. et al. (2018). A designed heme-[4Fe–4S] metalloenzyme catalyzes sulfite reduction like the native enzyme. *Science* 361 (6407): 1098–1101.

17 Sreenilayam, G., Moore, E.J., Steck, V., and Fasan, R. (2017). Stereoselective olefin cyclopropanation under aerobic conditions with an artificial enzyme incorporating an iron-chlorin E6 cofactor. *ACS Catal.* 7 (11): 7629–7633.

18 Oohora, K., Meichin, H., Zhao, L. et al. (2017). Catalytic cyclopropanation by myoglobin reconstituted with iron porphycene: acceleration of catalysis due to rapid formation of the carbene species. *J. Am. Chem. Soc.* 139 (48): 17265–17268.

19 Denisov, I.G., Makris, T.M., Sligar, S.G., and Schlichting, I. (2005). Structure and chemistry of cytochrome P450. *Chem. Rev.* 105 (6): 2253–2278.

20 Guo, M., Lee, Y.-M., Gupta, R. et al. (2017). Dioxygen activation and O–O bond formation reactions by manganese corroles. *J. Am. Chem. Soc.* 139 (44): 15858–15867.

21 Oohora, K., Meichin, H., Kihira, Y. et al. (2017). Manganese(V) porphycene complex responsible for inert C–H bond hydroxylation in a myoglobin matrix. *J. Am. Chem. Soc.* 139 (51): 18460–18463.

22 Baumann, L.K., Mbuvi, H.M., Du, G., and Woo, L.K. (2007). Iron porphyrin catalyzed N–H insertion reactions with ethyl diazoacetate. *Organometallics* 26 (16): 3995–4002.

23 Moore, E.J., Steck, V., Bajaj, P., and Fasan, R. (2018). Chemoselective cyclopropanation over carbene Y–H insertion catalyzed by an engineered carbene transferase. *J. Org. Chem.* 83 (14): 7480–7490.

24 Rakowski Dubois, M. and Dubois, D.L. (2009). Development of molecular electrocatalysts for CO_2 reduction and H_2 production/oxidation. *Acc. Chem. Res.* 42 (12): 1974–1982.

25 Soltau, S.R., Niklas, J., Dahlberg, P.D. et al. (2017). Charge separation related to photocatalytic H_2 production from a Ru–apoflavodoxin–Ni biohybrid. *ACS Energy Lett.* 2 (1): 230–237.

26 Cornils, B. and Herrmann, W.A. (2004). Aqueous-phase organometallic catalysis: concepts and applications. In: *Green Chemistry*, 2e, Completely Revised and Enlarged Edition. Weinheim: Wiley-VCH.

27 Jarvis, A.G., Obrecht, L., Deuss, P.J. et al. (2017). Enzyme activity by design: an artificial rhodium hydroformylase for linear aldehydes. *Angew. Chem. Int. Ed.* 56 (44): 13596–13600.

28 Grimm, A.R., Sauer, D.F., Polen, T. et al. (2018). A whole cell *E. coli* display platform for artificial metalloenzymes: poly(phenylacetylene) production with a rhodium–nitrobindin metalloprotein. *ACS Catal.* 8 (3): 2611–2614.

29 Grimm, A.R., Sauer, D.F., Davari, M.D. et al. (2018). Cavity size engineering of a β-barrel protein generates efficient biohybrid catalysts for olefin metathesis. *ACS Catal.* 8 (4): 3358–3364.

30 Schneider, C.R., Manesis, A.C., Stevenson, M.J., and Shafaat, H.S. (2018). A photoactive semisynthetic metalloenzyme exhibits complete selectivity for CO_2 reduction in water. *Chem. Commun.* 54 (37): 4681–4684.

31 Lopez, S., Rondot, L., Leprêtre, C. et al. (2017). Cross-linked artificial enzyme crystals as heterogeneous catalysts for oxidation reactions. *J. Am. Chem. Soc.* 139 (49): 17994–18002.

32 Villarino, L., Splan, K.E., Reddem, E. et al. (2018). An artificial heme enzyme for cyclopropanation reactions. *Angew. Chem. Int. Ed.* 57 (26): 7785–7789.

33 Ghattas, W., Dubosclard, V., Wick, A. et al. (2018). Receptor-based artificial metalloenzymes on living human cells. *J. Am. Chem. Soc.* 140 (28): 8756–8762.

34 Lewis, J.C. (2013). Artificial metalloenzymes and metallopeptide catalysts for organic synthesis. *ACS Catal.* 3 (12): 2954–2975.

35 Nastri, F., Chino, M., Maglio, O. et al. (2016). Design and engineering of artificial oxygen-activating metalloenzymes. *Chem. Soc. Rev.* 45 (18): 5020–5054.

36 Patel, S.C. and Hecht, M.H. (2012). Directed evolution of the peroxidase activity of a de novo-designed protein. *Protein Eng. Des. Sel.* 25 (9): 445–452.

37 Smith, B.A., Mularz, A.E., and Hecht, M.H. (2015). Divergent evolution of a bifunctional de novo protein. *Protein Sci.* 24 (2): 246–252.

38 Reetz, M.T. (2019). Directed evolution of artificial metalloenzymes: a universal means to tune the selectivity of transition metal catalysts? *Acc. Chem. Res.* 52 (2): 336–344.

39 Jing, Q., Okrasa, K., and Kazlauskas, R.J. (2009). Stereoselective hydrogenation of olefins using rhodium-substituted carbonic anhydrase – a new reductase. *Chem. Eur. J.* 15 (6): 1370–1376.

40 Jing, Q. and Kazlauskas, R.J. (2010). Regioselective hydroformylation of styrene using rhodium-substituted carbonic anhydrase. *ChemCatChem* 2 (8): 953–957.

41 Heinisch, T. and Ward, T.R. (2016). Artificial metalloenzymes based on the biotin–streptavidin technology: challenges and opportunities. *Acc. Chem. Res.* 49 (9): 1711–1721.

42 Key, H.M., Dydio, P., Clark, D.S., and Hartwig, J.F. (2016). Abiological catalysis by artificial haem proteins containing noble metals in place of iron. *Nature* 534 (7608): 534–537.

43 Dydio, P., Key, H.M., Nazarenko, A. et al. (2016). An artificial metalloenzyme with the kinetics of native enzymes. *Science* 354 (6308): 102–106.

44 Key, H.M., Dydio, P., Liu, Z. et al. (2017). Beyond iron: iridium-containing P450 enzymes for selective cyclopropanations of structurally diverse alkenes. *ACS Cent. Sci.* 3 (4): 302–308.

45 Gu, Y., Natoli, S.N., Liu, Z. et al. (2019). Site-selective functionalization of (sp^3)C–H bonds catalyzed by artificial metalloenzymes containing an iridium-porphyrin cofactor. *Angew. Chem. Int. Ed.* 58 (39): 13954–13960.

46 Fujieda, N., Nakano, T., Taniguchi, Y. et al. (2017). A well-defined osmium–cupin complex: hyperstable artificial osmium peroxygenase. *J. Am. Chem. Soc.* 139 (14): 5149–5155.

47 Reetz, M.T., Rentzsch, M., Pletsch, A. et al. (2008). A robust protein host for anchoring chelating ligands and organocatalysts. *ChemBioChem* 9 (4): 552–564.

48 Köhler, V., Wilson, Y.M., Lo, C. et al. (2010). Protein-based hybrid catalysts – design and evolution. *Curr. Opin. Biotechnol.* 21 (6): 744–752.

49 Yang, H., Srivastava, P., Zhang, C., and Lewis, J.C. (2014). A general method for artificial metalloenzyme formation via strain-promoted azide-alkyne cycloaddition. *Chembiochem Eur. J. Chem. Biol.* 15 (2): 223–227.

50 Yang, H., Swartz, A.M., Park, H.J. et al. (2018). Evolving artificial metalloenzymes via random mutagenesis. *Nat. Chem.* 10 (3): 318–324.

51 Albert, F.C., Carlos, A.M., and Richard, A.W. (2005). *Multiple Bonds Between Metal Atoms*. Springer.

52 Nicolas, I., Le Maux, P., and Simonneaux, G. (2008). Asymmetric catalytic cyclopropanation reactions in water. *Coord. Chem. Rev.* 252 (5): 727–735.

53 Srivastava, P., Yang, H., Ellis-Guardiola, K., and Lewis, J.C. (2015). Engineering a dirhodium artificial metalloenzyme for selective olefin cyclopropanation. *Nat. Commun.* 6: 7789.

54 Skander, M., Humbert, N., Collot, J. et al. (2004). Artificial metalloenzymes: (strept)avidin as host for enantioselective hydrogenation by achiral biotinylated rhodium–diphosphine complexes. *J. Am. Chem. Soc.* 126 (44): 14411–14418.

55 Klein, G., Humbert, N., Gradinaru, J. et al. (2005). Tailoring the active site of chemzymes by using a chemogenetic-optimization procedure: towards substrate-specific artificial hydrogenases based on the biotin–avidin technology. *Angew. Chem. Int. Ed.* 44 (47): 7764–7767.

56 Letondor, C., Humbert, N., and Ward, T.R. (2005). Artificial metalloenzymes based on biotin-avidin technology for the enantioselective reduction of ketones by transfer hydrogenation. *Proc. Natl. Acad. Sci. U.S.A.* 102 (13): 4683–4687.

57 Creus, M., Pordea, A., Rossel, T. et al. (2008). X-ray structure and designed evolution of an artificial transfer hydrogenase. *Angew. Chem. Int. Ed.* 47 (8): 1400–1404.

58 Dürrenberger, M., Heinisch, T., Wilson, Y.M. et al. (2011). Artificial transfer hydrogenases for the enantioselective reduction of cyclic imines. *Angew. Chem. Int. Ed.* 50 (13): 3026–3029.

59 Robles, V.M., Dürrenberger, M., Heinisch, T. et al. (2014). Structural, kinetic, and docking studies of artificial imine reductases based on biotin–streptavidin technology: an induced lock-and-key hypothesis. *J. Am. Chem. Soc.* 136 (44): 15676–15683.

60 Zimbron, J.M., Heinisch, T., Schmid, M. et al. (2013). A dual anchoring strategy for the localization and activation of artificial metalloenzymes based on the biotin–streptavidin technology. *J. Am. Chem. Soc.* 135 (14): 5384–5388.

61 Hestericová, M., Heinisch, T., Alonso-Cotchico, L. et al. (2018). Directed evolution of an artificial imine reductase. *Angew. Chem. Int. Ed.* 57 (7): 1863–1868.

62 Köhler, V., Wilson, Y.M., Dürrenberger, M. et al. (2013). Synthetic cascades are enabled by combining biocatalysts with artificial metalloenzymes. *Nat. Chem.* 5 (2): 93–99.

63 Satoh, T. and Miura, M. (2010). Oxidative coupling of aromatic substrates with alkynes and alkenes under rhodium catalysis. *Chem. Eur. J.* 16 (37): 11212–11222.

64 Hyster, T.K., Knoerr, L., Ward, T.R., and Rovis, T. (2012). Biotinylated Rh(III) complexes in engineered streptavidin for accelerated asymmetric C–H activation. *Science* 338 (6106): 500–503.

65 Jeschek, M., Reuter, R., Heinisch, T. et al. (2016). Directed evolution of artificial metalloenzymes for in vivo metathesis. *Nature* 537 (7622): 661–665.

66 Zhao, J., Kajetanowicz, A., and Ward, T.R. (2015). Carbonic anhydrase II as host protein for the creation of a biocompatible artificial metathesase. *Org. Biomol. Chem.* 13 (20): 5652–5655.

67 Okamoto, Y., Kojima, R., Schwizer, F. et al. (2018). A cell-penetrating artificial metalloenzyme regulates a gene switch in a designer mammalian cell. *Nat. Commun.* 9 (1): 1943.

68 Chatterjee, A., Mallin, H., Klehr, J. et al. (2015). An enantioselective artificial suzukiase based on the biotin–streptavidin technology. *Chem. Sci.* 7 (1): 673–677.

9

Engineered Cytochromes P450 for Biocatalysis

Hanan Alwaseem and Rudi Fasan

University of Rochester, Department of Chemistry, Rochester, NY 120 Trustee Rd, 14627, USA

9.1 Cytochrome P450 Monooxygenases

Cytochromes P450 are a large superfamily of *b*-heme containing monooxygenases (>40 000) [1] widespread in all kingdoms of life including bacteria, yeast, mammals, plants, and even viruses. These enzymes are involved in a wide range of oxidative processes including the biosynthesis of many natural products such as terpenes, alkaloids, steroids, and polyketides [2–7]. In humans, P450s play a key role in the metabolism of drugs and xenobiotics and in the biosynthesis of vitamins, cholesterol, and steroid hormones. P450s derive their name from a characteristic electron absorption (Soret) band observed at 450 nm for their ferrous (Fe^{II}) carbon monoxide-bound state [8]. This characteristic spectral feature derives from the heme cofactor (iron-protoporphyrin IX), which is bound to the core of protein via non-covalent interaction and coordination of the heme iron via the thiolate group of a conserved cysteine residue [9, 10].

Catalytically, P450s utilize their heme group and reducing equivalents derived from NAD(P)H, to bind and reductively activate molecular oxygen (O_2), resulting in the formation of a highly oxidizing species, called Compound (I), which mediates the insertion of a single oxygen atom into the substrate molecule [8, 11]. The catalytic mechanism of P450 enzymes is now quite well understood and is illustrated in Figure 9.1 [8, 11–13]. Briefly, the catalytic cycle is typically initiated by binding of the substrate into the active site of the ferric enzyme. This process results in displacement of the distal water ligand and an increase of the redox potential of the heme, which enables reduction of the heme iron to the ferrous state (Fe^{II}) via an electron transfer from NAD(P)H through the redox partner proteins. The reduced heme binds molecular oxygen leading to an Fe^{III}-superoxide complex, which is reduced by a second electron and rapidly protonated to yield a ferric hydroperoxy intermediate (Compound 0). Protonation and cleavage of the O—O bond liberates water and generates a high-valent iron-oxo species formally known as Compound I. Compound I oxidizes the substrate resulting in the formation of the mono-oxygenated product and restoring the heme to its ferric resting state. Depending on the substrate and type

Protein Engineering: Tools and Applications, First Edition. Edited by Huimin Zhao.
© 2021 WILEY-VCH GmbH. Published 2021 by WILEY-VCH GmbH.

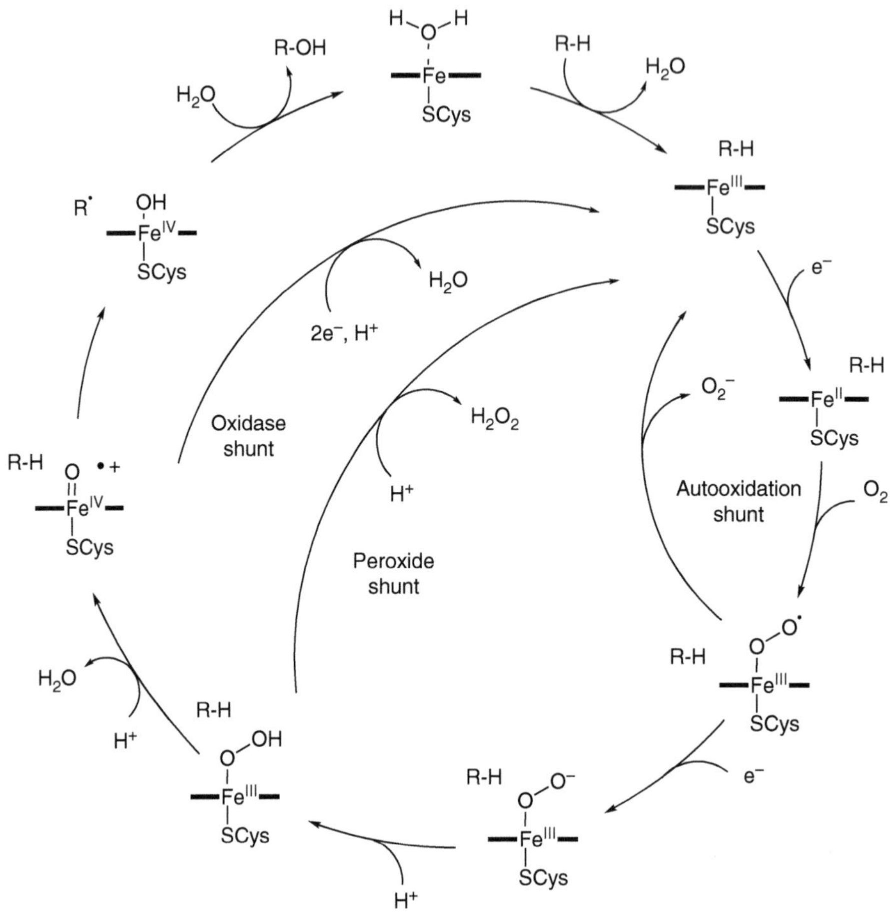

Figure 9.1 The catalytic cycle of P450 enzymes. Sources: Denisov et al. [8], Ortiz de Montellano et al. [11], Sono et al. [12].

of P450, unproductive pathways can also occur resulting in the release of superoxide (O_2^-), H_2O_2, or H_2O from the various intermediates generated during the major catalytic cycle. Furthermore, certain P450s natively utilize H_2O_2 as the source of oxygen and reducing equivalents to promote the formation of Compound I via the peroxide shunt pathway and subsequent oxidation of the substrate.

The general mechanism outlined earlier is responsible for the diverse range of P450-catalyzed monooxygenation reactions characterized by natural and engineered P450s, which include aliphatic and aromatic hydroxylation, olefin epoxidation, heteroatom oxidation, and heteroatom dealkylation reactions [11]. In addition, these enzymes are capable of catalyzing a variety of less canonical oxidative processes, including aromatic dehalogenation, C–C and C–O cross-coupling, and rearrangement reactions [14]. From a biocatalysis standpoint, these enzymes have attracted significant attention for their ability to perform synthetically challenging oxidation reactions, including the hydroxylation of unactivated C—H bonds, under

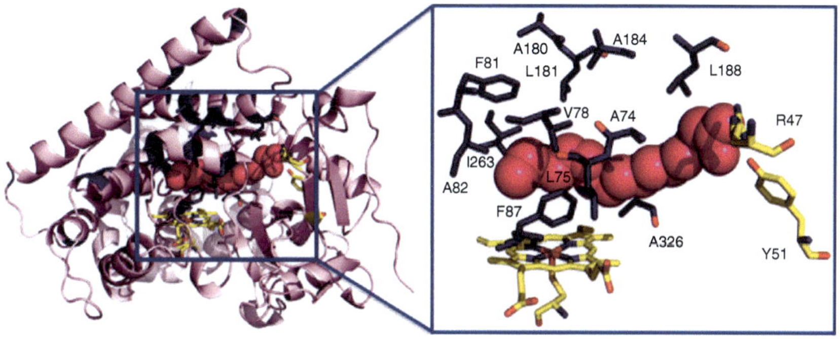

Figure 9.2 Model of heme domain of $P450_{BM3}$ (left) in complex with *N*-palmitoylglycine (pink, sphere model) PDB code 1JPZ. The heme prosthetic group is displayed in yellow (stick model). Close-up view of the enzyme active site (right).

very mild reaction conditions and using readily available or inexpensive oxidants (O_2 or H_2O_2) [15].

Despite displaying a sequence homology as low as 20–30%, P450s share a common three-dimensional fold as exemplified by the structure of the fatty acid hydroxylase $P450_{BM3}$ from *Bacillus megaterium* shown in Figure 9.2. The heme cofactor is embedded within the core of these enzymes and anchored to the protein scaffold via coordination of the heme iron to a conserved cysteine residue along with non-covalent interactions. A cavity above the heme cofactor defines the active site where the substrate binds and undergoes oxidation. Structural comparison of the substrate-free and substrate-bound form of P450s has revealed a significant degree of conformational flexibility of certain elements of the P450 structure upon substrate binding. In particular, the so-called F/G helices have been found to act as a "lid" that permits access of the substrate to the heme pocket and then seals the active site in a "close" substrate-bound form [16].

With the exception of P450s functioning as peroxygenases, redox protein partners are required for delivering the electrons from the NAD(P)H cofactors to the heme of the P450 and thus support its monooxygenase activity. Natural P450 systems exhibit a range of different arrangements with respect to the structure and organization of the respective redox partners, occurring as three-, two-, or single component systems (Figure 9.3) [17, 18]. Class I P450 systems, which comprise most of bacterial P450 systems, comprise an NAD(P)H dependent FAD-containing reductase and a ferredoxin [2Fe-2S] or FMN-dependent flavodoxin for transferring the electrons from the reductase to the P450 (Figure 9.3a). Class II P450 systems, on the other hand, are the most common systems found in eukaryotes (e.g. liver and plant P450s) and consist of a P450 component and a diflavin (FAD/FMN)-containing cytochrome P450 reductase (CPR), both of which are typically bound to the membrane of the endoplasmic reticulum (Figure 9.3b). In addition to these predominant classes, there are also so-called "catalytically self-sufficient" P450s, in which the heme domain is naturally fused to a reductase domain (Figure 9.3c,d). A prototypical example of this type of P450 systems is the prokaryotic $P450_{BM3}$ (CYP102A1) isolated from the soil bacterium *B. megaterium*, in which a CPR-like reductase domain

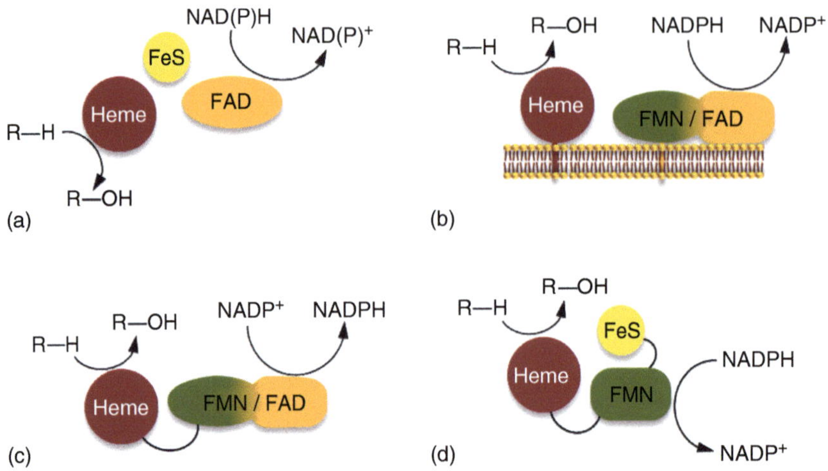

Figure 9.3 P450 classes: (a) bacterial, class I P450 system (e.g. P450cam); (b) eukaryotic, membrane-bound class II P450 systems (e.g. human liver P450s); (c) P450-CPR fusion system (e.g. P450$_{BM3}$); (d) P450RhF-type fusion system. FeS, iron–sulfur cluster; FMN, flavin mononucleotide; FAD, flavin adenine dinucleotide. Sources: Munro et al. [17], Hannemann et al. [15].

is fused to the heme-containing monooxygenase domain [19]. In 2002, a different type of self-sufficient P450 system was isolated from *Rhodococcus* sp., which is exemplified by P450$_{RhF}$ (CYP116B2) and features a heme domain fused to a reductase domain comprising an FMN-binding domain, a nicotinamide adenine dinucleotide (NADH)-binding domain and a [2Fe-2S] ferredoxin domain [20–23].

9.2 Engineered Bacterial P450s for Biocatalytic Applications

Because of their soluble nature, bacterial P450 systems, and in particular the catalytically self-sufficient bacterial P450$_{BM3}$, have represented major targets for protein engineering efforts directed at developing biocatalysts for the oxidation of a broad spectrum of substrates, ranging from simple alkanes to drug molecules and complex natural products [24–26]. In this section, representative contributions made in this area over the course of the past two decades are discussed with an emphasis on the different protein engineering strategies, involving directed evolution, rational design, or semi-rational approaches, applied for the development of engineered P450 catalysts for biotechnological applications. These contributions have been organized according to the class of target substrates, which illustrates the diverse range of applications of engineered P450s in the context of the synthesis of bulk chemicals and chiral building blocks, the production of drug metabolites, and the late-stage functionalization of natural products.

9.2.1 Oxyfunctionalization of Small Organic Substrates

The selective oxidation of petrochemical derived compounds (e.g. linear and cyclic alkanes/alkenes, alkanoic/alkenoic acids) and other small molecular weight (MW) organic molecules finds important applications for the synthesis of polymers, detergents, and fine chemicals. Among the earliest examples of P450 engineering applied to the oxidation of non-native organic substrates was the development of engineered $P450_{BM3}$ variants for selective oxidation of alkanes, which are notoriously inert compounds. $P450_{BM3}$ catalyzes the subterminal hydroxylation of long-chain (C_{12}–C_{20}) fatty acids [27, 28] and it was found to have some basal oxidation activity on medium chain alkanes such as octane. Using a directed evolution approach in combination with a plate-based high-throughput screen involving octane surrogate substrates (*p*-nitrophenyl octyl ether [29] or hexyl methyl ether [30]), Arnold and coworkers evolved a $P450_{BM3}$ variant (called 9-10A) with significantly enhanced oxidation activity on octane (30 → 3000 total turnovers [TTN]) and other medium-chain alkanes [30, 31]. Out of the 13 heme domain mutations accumulated in 9-10A, only two were found to be located in the active site (V78A, A184V), illustrating the power of directed evolution toward identifying beneficial mutations outside of the active site. Site-saturation mutagenesis of active site positions in 9-10A then enabled the development of engineered $P450_{BM3}$ variants with improved selectivity for the subterminal (ω-1) hydroxylation of C_6–C_{10} alkanes (76–86% regioselect.; 39–83% ee; Figure 9.4a) [30], the α-hydroxylation of 2-aryl-acetic acid esters (57–99% regioselect.; 74–94% ee; 105–1640 TTN; Figure 9.4c) [32], and the epoxidation of terminal C_5–C_8 alkenes 92–94% regioselect.; 65–83% ee; 560–1370 TTN) [33] (Figure 9.4b).

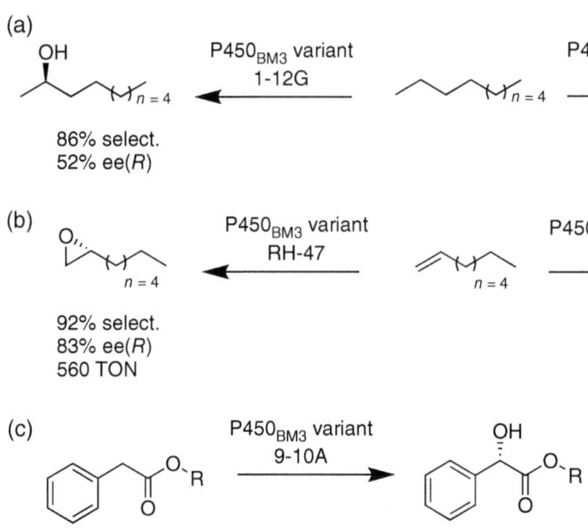

R	% ee	% select.
Me	77	57
Et	52	63
n-Pr	85	41
n-Bu	95	25

Figure 9.4 Oxyfunctionalization of non-native substrates by engineered variants of $P450_{BM3}$: (a) alkanes, (b) alkenes, and (c) aryl-esters. TON, turnover number; ee, enantiomeric excess; select., selectivity. Sources: (a) Based on Peters et al. [30], (b) Based on Kubo et al. [33], (c) Based on Landwehr et al. [32].

The increase in octane hydroxylation activity achieved in 9-10A was found to be accompanied by detectable hydroxylation activity on smaller alkane substrates such as propane (1100 TTN), which is not oxidized by wild-type $P450_{BM3}$ [34]. To obtain a more efficient biocatalyst for the selective oxidation of propane to 2-propanol, variant 9-10A was further evolved using a "domain engineering strategy," where beneficial mutations within the heme-, FMN-, and FAD-domain of this P450 were first identified via random and active site mutagenesis and then combined in a final step [35]. This process resulted in the development of a proficient propane monooxygenase, named $P450_{PMO}$, exhibiting high catalytic efficiency (k_{cat}/k_M: $4.4 \times 10^4\ M^{-1}\ s^{-1}$; 45 800 TTN) and coupling efficiency (98%) for this reaction, i.e. catalytic properties that are comparable with that of wild-type $P450_{BM3}$ with its preferred substrates [36, 37]. Characterization of $P450_{PMO}$ along with its evolutionary precursors showed a significant remodeling of the active site of the enzyme during the overall directed evolution and "substrate walking" process, revealing a significant reduction of the active site cavity above the heme (590 → 315 Å) to better accommodate the small gaseous substrate [38]. This work also illustrated the importance of addressing the activity/stability trade-off often observed in directed evolution experiments to permit the accumulation of beneficial yet destabilizing mutations [38], reiterating the connection between protein thermostability and evolvability [39]. The laboratory-evolved $P450_{PMO}$ could be then applied for the efficient conversion of propane to 2-propanol in a whole cell system involving a metabolically engineered *Escherichia coli* strain [40]. Starting from $P450_{BM3}$, the same group later explored a computationally-guided combinatorial site-saturation mutagenesis strategy for the development of a propane- and ethane hydroxylating biocatalyst [41]. This approach yielded a $P450_{BM3}$ variant that harbors 7 active site mutations (E32) and exhibits significantly improved propane oxidation activity (16 800 TTN; 36% coupling), albeit the catalytic performance of this enzyme in propane oxidation did not exceed that of $P450_{PMO}$. More recently, directed evolution strategies have been applied also for the development of $P450_{BM3}$ variants for the terminal hydroxylation of palmitic acid (74% regioselect.; 400–600 TTN) [42] and the hydroxylation of polycyclic aromatics [43].

Engineered P450s evolved for a particular non-native substrate often exhibit a broadened substrate profile [38], enabling their application to the oxidation of a broader range of target molecules. This concept was illustrated by Fasan and coworkers through the application of a panel of ~100 engineered $P450_{BM3}$ variants, previously evolved for alkane substrates, for the selective hydroxylation of a series of small molecule building blocks and drugs with regiodivergent selectivity [44]. P450-catalyzed hydroxylation could be coupled to chemical deoxyfluorination to rapidly access fluorinated derivatives of the target molecules, including the nonsteroidal anti-inflammatory drug (NSAID) drug ibuprofen, on a preparative scale. By coupling chemoenzymatic C—H fluorination using two regiodivergent P450s, a difluorinated derivative of this drug could be also obtained (Figure 9.5).

Complementary to directed evolution, structure-based rational mutagenesis has also represented an effective strategy for developing $P450_{BM3}$ based biocatalysts for the oxidation of various non-native substrates, including linear and cyclic

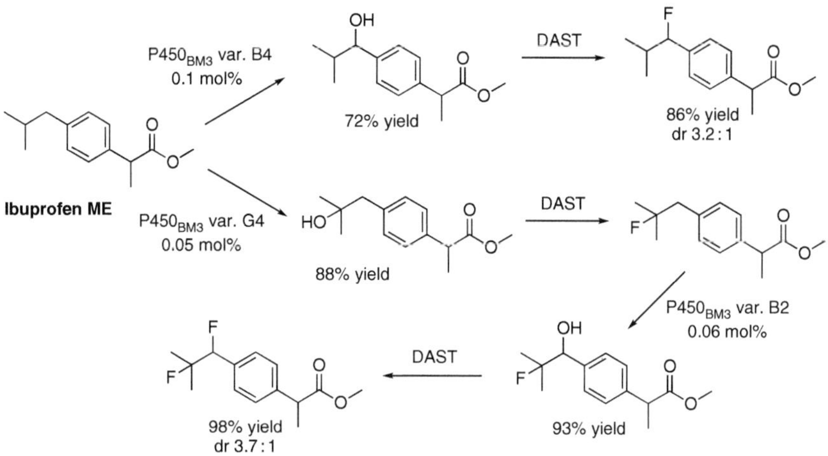

Figure 9.5 Selective chemoenzymatic mono- and difluorination of ibuprofen methyl ester via P450-catalyzed hydroxylation/deoxofluorination. var, variant; dr, diastereomeric ratio. Source: Based on Rentmeister et al. [44].

alkanes and alkenes. Mutagenesis of amino acid residues most proximal to the heme (e.g. F87, A328) or implicated in fatty acid recognition (R47, Y51) have provided a particularly effective means to extend the substrate scope of this P450. Using this approach in combination with an indole-based screen, the Wong and Schmid groups have reported the development of two promiscuous P450 catalysts, $P450_{BM3}$(R47L/Y51F/F87A) [45] and $P450_{BM3}$(A74G/F87V/L188Q) [46], that are able to hydroxylate large polycyclic aromatic hydrocarbons [47, 48], heteroarenes [49], branched fatty acids [50], and cyclohexane [51]. Related active site variants of $P450_{BM3}$, along with engineered variants of CYP101A2, were reported by Bell and coworkers to catalyze the oxidation of a broad range of small aryl alkanes with varying degree of regioselectivity [52, 53].

By screening a small library of $P450_{BM3}$ variants created by mutating F87 and A328 with five apolar residues, Urlacher and coworkers identified engineered P450 variants useful for the selective epoxidation of limonene (Figure 9.6a) [54] and the hydroxylation of C_8–C_{12} cycloalkanes [55]. One of the variants identified in these studies was then further engineered via iterative rounds of molecular modeling, mutagenesis, and screening, to obtain a triple mutant (A264V/A328V/L437F) with high regioselectivity for the conversion of limonene to perillyl alcohol (Figure 9.6a) [56]. Using a similar approach based on the generation of a small library (65) of active site variants, Pietruszka and coworkers identified $P450_{BM3}$(A74G/L188Q) as a highly selective biocatalyst for the regio- (90%) and stereoselective (93–99% ee) allylic hydroxylation of C_8–C_{11} ω-alkenoic esters (Figure 9.6b) [57]. This enzyme could be applied to the transformation of 10-undecanoic acid on a preparative scale enabling the isolation of the corresponding *S*-configured allylic alcohol in high yield (80%). By focusing mutagenesis on a single "hot spot" residue in $P450_{BM3}$ active site (A328), Reetz and coworkers were able to develop engineered variants, such as $P450_{BM3}$(A328F), that catalyze the benzylic hydroxylation of 1-tetralones

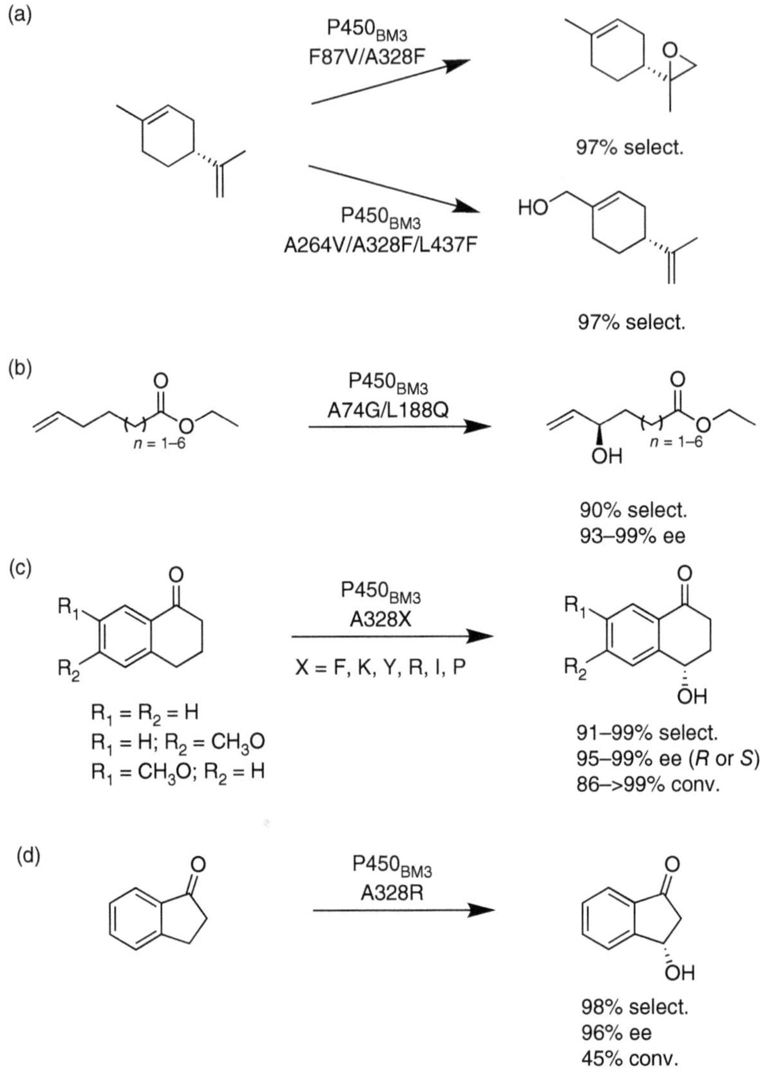

Figure 9.6 Oxyfunctionalization of non-native substrates by engineered variants of $P450_{BM3}$: (a) limonene, (b) alkenoic esters, (c) tetralones, and (d) 1-indalone. Sources: (a) Based on Seifert et al. [54, 56], (b) Based on Neufeld et al. [57], (c, d) Based on Roiban et al. [58].

and 1-indanone with high regioselectivity (91–95%) and stereoselectivity (95–99% ee) (Figure 9.6c,d) [58].

In a series of recent studies, the Reetz group has addressed the important challenge of developing P450 oxidation catalysts with high stereoselectivity for the formation of both enantiomers of a target oxidation product. In these cases, a semi-rational protein engineering strategy based on iterative saturation mutagenesis [59–61], in combination with mutagenesis with reduced amino acid alphabets [62], was shown to provide an effective solution to this problem. Using this approach and chiral gas chromatography (GC) for library screening, Agudo et al. developed a panel of $P450_{BM3}$ variants featuring high *S*- or *R*-stereoselectivity (97–98% ee) for the allylic hydroxylation of a cyclopentene derivative in whole cells (Figure 9.7a) [63].

Notably, some of these P450 variants could be also applied to the stereoselective hydroxylation of a structurally related cyclopentene and cyclohexadiene compound (84–94% ee). Using a similar strategy, engineered $P450_{BM3}$ variants for the biocatalytic synthesis of chiral acyloins were obtained (Figure 9.7b,c) [64]. Upon screening a series of single and double site libraries, highly regio- and stereoselective variants for both the *S*- and *R*-selective α-hydroxylation of ketones (Figure 9.7b,c) were identified (91–99 regioselect.; 76–99% ee; 355–1964 TTN) [64]. More recently, the same group reported the development of stereocomplementary $P450_{BM3}$ variants for the asymmetric sulfoxidation of 1-thiochroman-4-one and its derivatives (Figure 9.7d) [65]. In this case, a single codon saturation mutagenesis (SCSM) approach [66, 67] entailing the combinatorial incorporation of valine at 7 active site positions led to the discovery of a P450 variant with improved *R*-stereoselectivity for this reaction (WAJ-1, 70% ee). On the other hand, a second Phe-based SCSM library yielded an *S*-stereoselective mutant (WAJ-2, 80% ee). Further engineering of these enzymes led to the identification of $P450_{BM3}$ variants with improved stereoselectivity for formation of both enantiomers of the sulfoxidation product of 1-thiochroman-4-one (86% ee; 294–550 TTN) [65].

Class I P450s have also constituted promising scaffolds for the development of oxidation biocatalysts. Initial efforts in this area have involved the engineering of P450cam (CYP101A1), a well characterized class I P450 that catalyzes the regio- and stereoselective oxidation of (+)-camphor to 5-exo-hydroxycamphor with a high turnover rate (>2000 min^{-1}) and coupling efficiency (95%) in the presence of saturating concentrations of the redox partners putidaredoxin (Pdx) and putidaredoxin reductase (PdR) [68, 69]. Engineered variants of $P450_{cam}$, primarily obtained via rational structure-guided mutagenesis, have been reported to oxidize a variety of non-native substrates, including other terpenes (e.g. α-pinene, valencene) [70–72], alkanes [73, 74], styrene [75], and aromatic compounds [76, 77]. Along with $P450_{cam}$, class I $P450_{pyr}$ has also been recently engineered for biotechnological applications. Isolated from *Sphingomonas* sp. HXN-200 [78], $P450_{pyr}$ was initially found to catalyze the regioselective hydroxylation of *N*-heterocycles including pyrrolidines, azetidines, and pyrrolidinones [79–81]. Via iterative rounds of site saturation mutagenesis and library screening using a high-throughput enantioselectivity assay (*vide infra*), Li and coworkers developed a P450pyr variant featuring inverted stereoselectivity for the β-hydroxylation of *N*-benzyl-pyrrolidine compared with the wild-type enzyme (83% (*R*) versus 43% (*S*)) [82]. Leveraging structural information derived from solving the crystal structure of this P450 along with an MS-based assay for enantioselectivity determination [83], Li and coworkers were later able to obtain a P450pyr variant exhibiting high stereoselectivity for the formation of the *S*-configured product (98% ee; Figure 9.8a) [84]. Using a similar protein engineering strategy (i.e. iterative saturation mutagenesis), the Li group also reported the development of a highly regio- and stereoselective P450pyr variant for

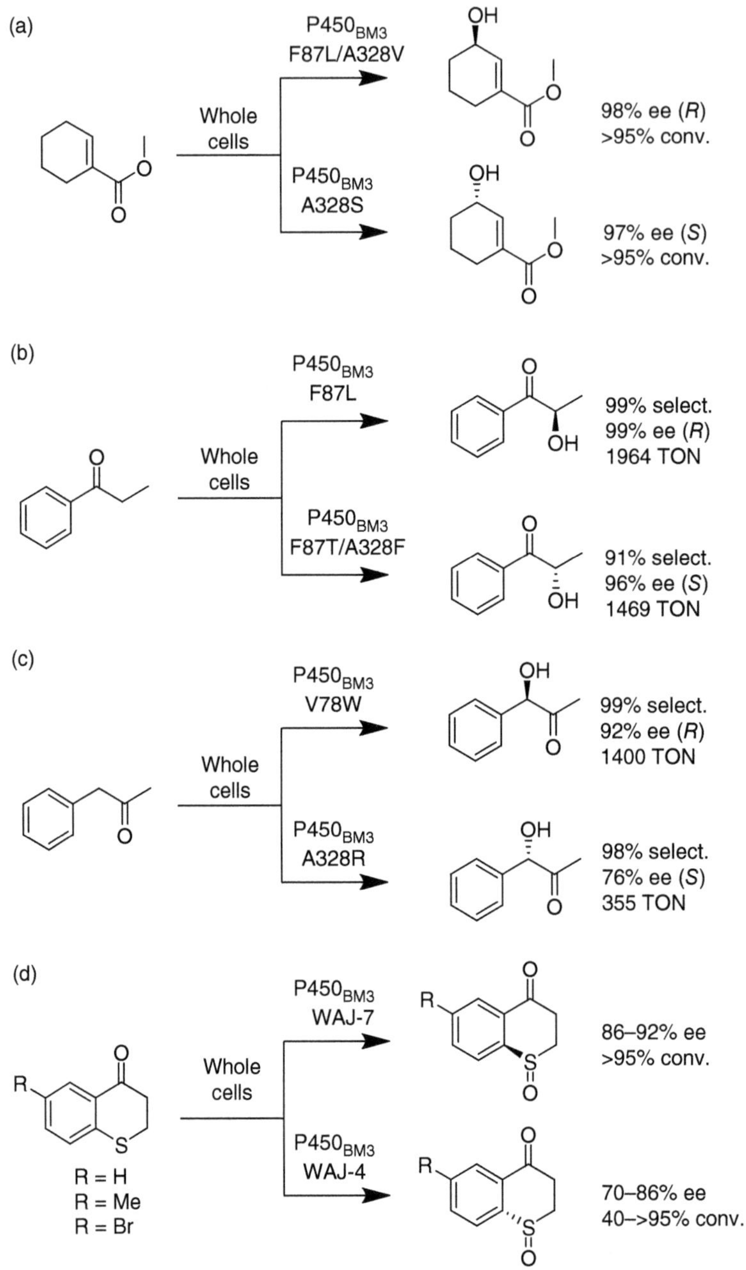

Figure 9.7 Oxyfunctionalization of non-native substrates by engineered variants of $P450_{BM3}$ in whole cells: (a) cyclopentane derivatives, (b, c) ketones, (d) 1-thiochroman-4-one. TON, number of turnovers. Sources: (a) Based on Agudo et al. [63], (b, c) Based on Agudo et al. [64]. (d) Based on Jian-bo et al. [65].

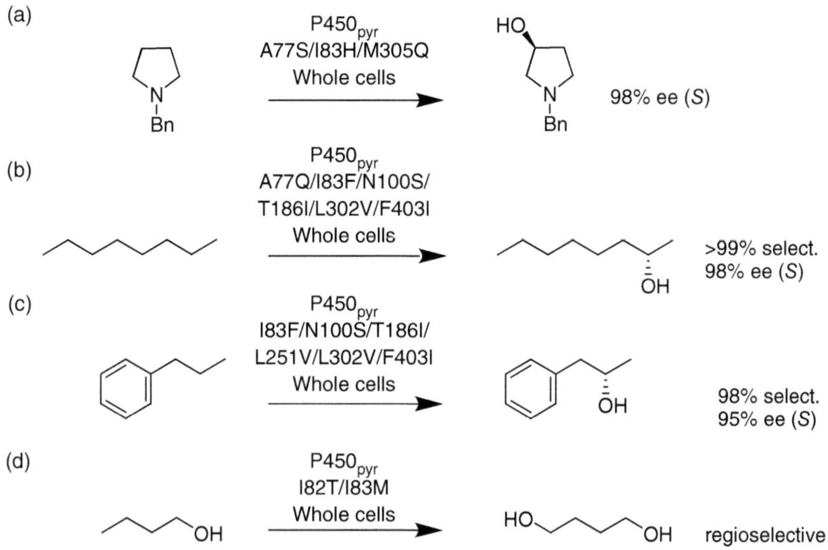

Figure 9.8 Oxyfunctionalization of non-native substrates by engineered variants of $P450_{pyr}$ in whole cells: (a) *N*-benzyl-pyrrolidine, (b) octanol, (c) propylbenzene, and (d) 1,4-butanediol. Sources: Based on (a–c) Yang et al. [85]. (d) Based on Yang et al. [86].

the conversion of octanol and propylbenzene to (*S*)-2-octanol (99% regioselect.; 98% ee) and (*S*)-1-phenyl-propanol (98% regioselect.; 95% ee), respectively (Figure 9.8b,c) [85]. The latter is a valuable intermediate for the synthesis of various pharmacologically active molecules, including the monoamine oxidase (MAO) inhibitor drug selegiline. In another study, $P450_{pyr}$ was evolved via iterative saturation mutagenesis to obtain a $P450_{pyr}$ variant (I83M/I82T) with improved activity toward the terminal hydroxylation of *n*-butanol to give 1,4-butanediol, a reaction not catalyzed by the wild type enzyme (Figure 9.8d) [86].

Recently, P450 enzymes have been engineered also for promoting non-canonical reactions, including abiotic carbene and nitrene transfer reactions. While the latter are discussed elsewhere [87], two examples of non-canonical P450-catalyzed reactions are worth highlighting here. In a first example, the Arnold recently reported the directed evolution of $P450_{LaMO}$ from the rhodobacterium *Labrenzia aggregata* for the anti-Markovnikov oxidation of styrene derivatives (Figure 9.9) [88]. This P450 belongs to the family of self-sufficient CYP116 enzymes and was previously found to catalyze the oxidation of styrene to give styrene epoxide along with an anti-Markovnikov aldehyde as a minor product (90 : 10 ratio; ~1000 TTN) [89]. After multiple rounds [10] of directed evolution, first involving random mutagenesis followed by site-saturation mutagenesis on putative active site residues, Arnold and coworkers were able to greatly increase the activity and chemoselectivity of the P450 for the desired anti-Markovnikov reactions (81% selectivity; 3800 TTN). The evolved enzyme, called aMOx (13 mutations), could be applied in combination with an alcohol dehydrogenase in a single pot reaction to convert various styrene derivatives to the corresponding anti-Markovnikov alcohol products with high selectivity (69–94%) and turnover numbers (750–4500 TTN) (Figure 9.9) [88].

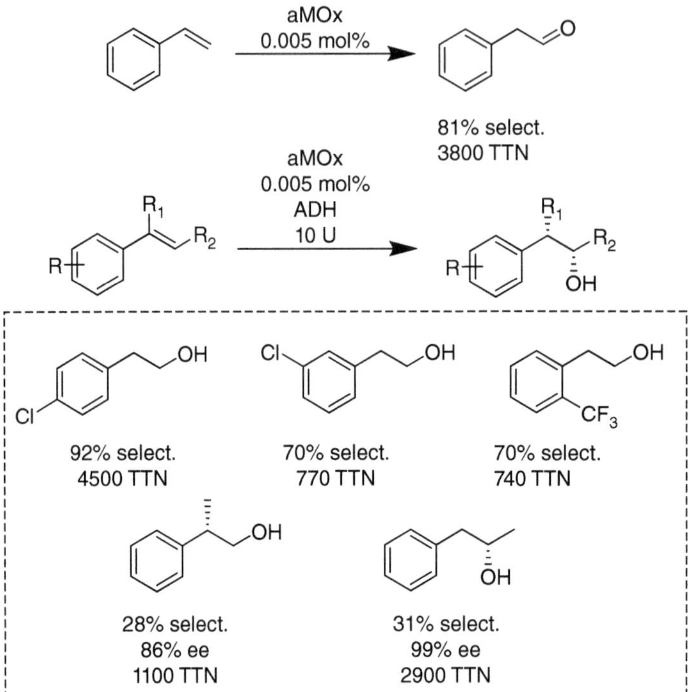

Figure 9.9 Application of the engineered P450 variant aMOx for the anti-Markovnikov hydration of styrenes. TTN, total turnover numbers. Source: Based on Hammer et al. [88].

In a separate study, Wong and coworkers serendipitously discovered the ability of engineered $P450_{BM3}$ variants to catalyze the oxidative cyclization of the drug lidocaine to give an imidazolidin-4-one product [90]. The latter likely arises from oxidation of the tertiary amine to give an iminium intermediate, which then is attacked intramolecularly by the proximal amide group to form a new C—N bond. Following this observation, the Wong group later reported the screening and identification of additional $P450_{BM3}$ variants capable of transforming a series 2-aminoacetamides to the corresponding imidazolidin-4-ones in high yield (55–97%; 290–1980 TTN) and selectivity (72–100%) [91]. Notably, two $P450_{BM3}$ variants featuring regiodivergent selectivity could be applied to generate two different cyclized products (Figure 9.10a).

Finally, unnatural amino acid (UAA) mutagenesis has been recently applied for altering the regioselectivity and improving the catalytic efficiency of engineered P450s [92]. Four UAAs with diverse aromatic side chains were incorporated at 11 active site positions of a substrate-promiscuous $P450_{BM3}$ variant. The resulting UAA-containing P450s, and in particular those harboring single substitutions with the UAA *para*-acetyl-phenylalanine (Figure 9.10b), were found to exhibit large shifts in regioselectivity in hydroxylation of a small molecule drug (ibuprofen) and a terpene natural product (nootkatone, Figure 9.10c) [92]. In addition, mutagenesis with the UAA *para*-amino-phenylalanine (Figure 9.10b) was found to boost the catalytic activity of the P450, resulting in an engineered P450 catalyst capable

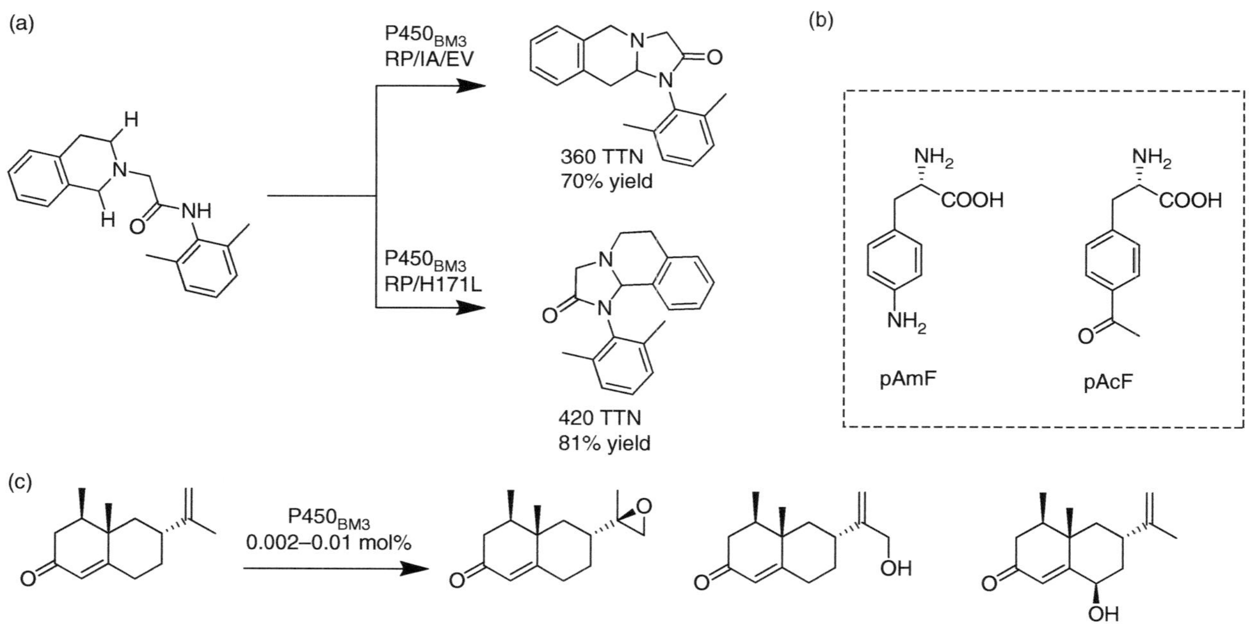

P450				TTN
139-3 (parent)	96%	4%	0%	6,980
/L188pAmF	97%	3%	0%	15,470
/A82pAcF	38%	62%	0%	720
/A78pAcF	27%	0%	73%	1,760

Figure 9.10 (a) P450$_{BM3}$-catalyzed intramolecular C–H amination of 2-aminoacetamides to imidazolidin-4-ones; (b) The unnatural amino acids (UAA) pAmF: *para*-amino-phenylalanine; pAcF: *para*-acetyl-phenylalanine; (c) Oxidation products obtained from the P450-catalyzed transformations of (+)-nootkatone with P450$_{BM3}$ 139-3 and 139-3 UAA variants. Sources: (a) Based on Ren et al. [91], (c) Based on Kolev et al. [92].

of supporting the highest TTN reported to date on a complex molecule (34 650 TON). The functional changes induced by the UAAs could not be reproduced by any of the 20 natural amino acids, highlighting the potential advantage of unnatural mutagenesis toward tuning the catalytic activity and regioselectivity of P450 oxidation catalysts.

9.2.2 Late-Stage Functionalization of Natural Products

Natural products constitute a major source of bioactive compounds and late-stage modification of these molecules is often a critical step toward their development into pharmaceuticals. While conventional strategies have involved the chemical manipulation of pre-existing reactive functional groups (e.g. artemisinin-derived antimalarial drugs artesunate and artemether), the Fasan group has demonstrated the potential of engineered P450s and P450-mediated chemoenzymatic synthesis for the late-stage functionalization of unactivated C—H bonds in complex natural products. In a first example, engineered $P450_{BM3}$ variants were developed for performing the highly regio- and stereoselective hydroxylation of three distinct aliphatic positions (C6a, C7) in the upper hemisphere of the antimalarial natural product artemisinin (Figure 9.11a) [93]. Starting from a promiscuous $P450_{BM3}$ variant (called FL#62) that produces a mixture of the three hydroxylated products (83 : 10 : 7 ratio; 340 TTN), three regio- and diastereodivergent P450 catalysts for the selective synthesis of each of these products (92–99% select.; 380–440 TTN) were obtained via a multi-tier strategy involving (i) mutagenesis of first-sphere active site residues, (ii) high-throughput P450 fingerprinting (*vide infra*) [94], and (ii) fingerprint-driven P450 reactivity predictions [93]. The resulting P450 enzymes could be applied to the hydroxylation of artemisinin on a preparative scale (0.2–0.4 g; >90% yield) and enabled the chemoenzymatic synthesis of two analogs of the antimalarials, artemether, and artesunate, in which a metabolically labile C–H site is protected via a C–F substitution.

In a subsequent study, the same group applied a similar strategy based on active site mutagenesis and P450 fingerprint-based predictions, for the development of three regio- and stereoselective $P450_{BM3}$-based catalysts for the hydroxylation of allylic position C9 and C14 and epoxidation of the C1,C10 double bond in parthenolide, a sesquiterpene lactone with antileukemic activity (Figure 9.11b) [95]. Also in this case, the starting point was an unselective parthenolide oxidizing $P450_{BM3}$ variant (FL#62; 77 : 13 : 10 product ratio) and refinement of its regio- and stereoselectivity for the formation of the three oxyfunctionalized products was achieved by means of three to seven active site mutations (81–90% select.; 420–4980 TTN). The evolved P450 enzymes could be then applied to the chemoenzymatic synthesis of a range of C9- and C14-functionalized parthenolide derivatives, resulting in the discovery of parthenolide analogs with significantly improved activity against leukemia and other types of cancers [36, 96].

In a separate study, Urlacher and coworkers reported the development of engineered $P450_{BM3}$ variants for the late-stage oxidation of the macrocyclic diterpenoid β-cembrenediol, a flavor ingredient in tobacco [97]. Following enzyme optimization via first-sphere active site mutagenesis using a reduced amino acid alphabet

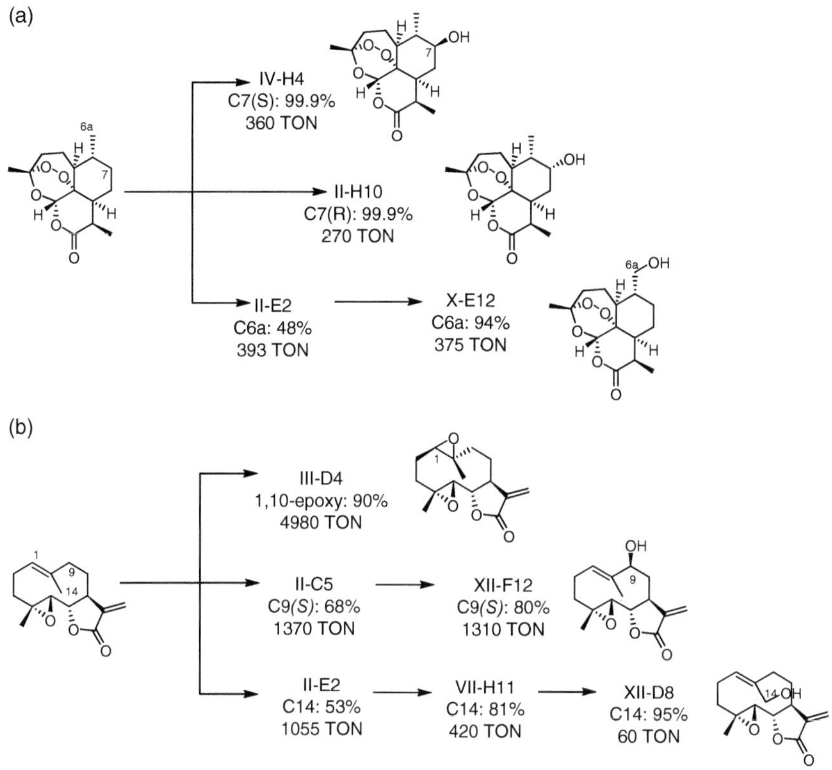

Figure 9.11 Development of regio- and stereoselective $P450_{BM3}$-based catalysts for the late-stage oxyfunctionalization of the antimalarial natural product artemisinin (a) and the antileukemic natural product parthenolide (b). Sources: (a) Based on Zhang et al. [93]. (b) Based on Kolev et al. [95].

(Ala, Val, Gly, Leu), the authors obtained a $P450_{BM3}$ variant (F87A/I263L) capable of producing 9-hydroxy-cembrendiol with 100% regioselectivity and 78% diastereoselectivity and a second variant (L75A/V78A/F87G) capable of hydroxylating position C10 in β-cembrenediol with 97% regioselectivity and 48% diastereoselectivity (Figure 9.12). The enzymatic hydroxylation reactions could be performed on a preparative scale resulting in 45–57% yields. Later, the same authors carried out sequential hydroxylations (one-pot) of the terpene natural product using multiple $P450_{BM3}$ variants to produce novel 9(*S*),10(*R*/*S*)-β-cembrenetetraols with a diastereomeric ratio from 48 : 52 to 10 : 90 (Figure 9.12) [98].

The P450-catalyzed oxidation of steroids has a long history and represents one of the (few) biocatalytic processes involving P450 enzymes that are currently performed at the industrial scale [99]. Targeting the selective hydroxylation of testosterone, the Reetz group identified in $P450_{BM3}$(F87A) an unselective catalyst that produces 2β- and 15β-hydroxy-testosterone in approximately 1 : 1 ratio. Optimization of the regioselectivity of this initial P450 variant was then achieved using iterative site-saturation mutagenesis at selected active site positions using a reduced amino acid alphabet (NDC codon degeneracy) to minimize screening

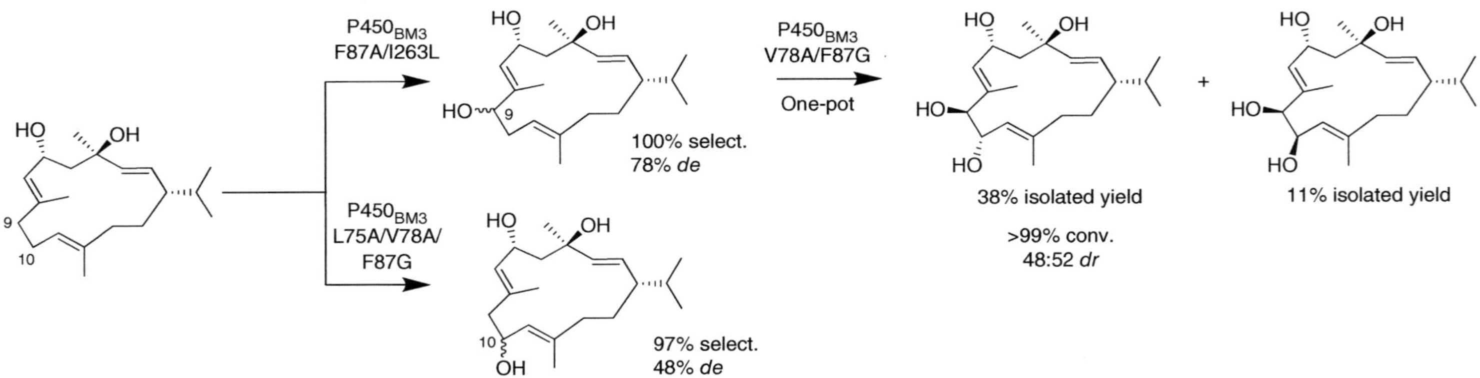

Figure 9.12 Chemo-and regioselective oxidation of the monocyclic diterpenoid β-cembrenediol using engineered $P450_{BM3}$ variants. de, diastereomeric excess; dr, diastereomeric ratio. Source: Based on Le-Huu et al. [98].

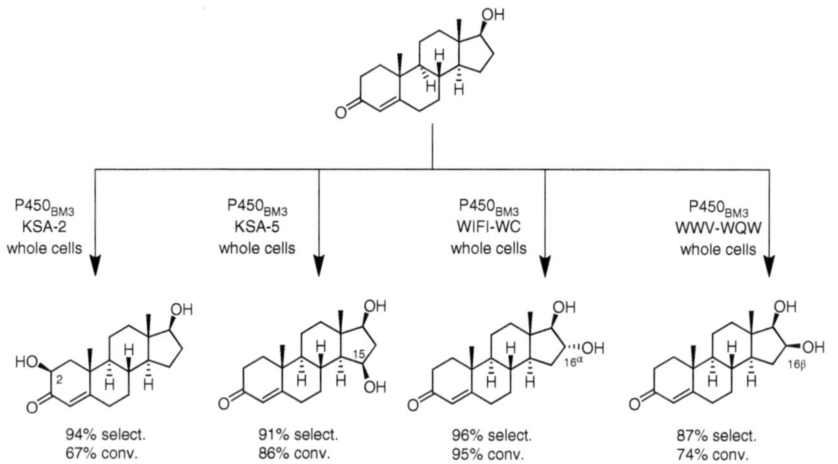

Figure 9.13 Regio- and stereoselective oxidation of testosterone with engineered P450$_{BM3}$ variants expressed in whole cells. Source: Based on Acevedo-Rocha et al. [37].

efforts [100]. Upon screening ∼9000 enzyme variants using a high performance liquid chromatography (HPLC)-based method, the authors successfully identified two optimized P450$_{BM3}$-based biocatalysts (KSA-2 and KSA-5) capable of producing either hydroxylated product with high selectivity (91–94%) in whole cell systems (Figure 9.13). Engineered P450s developed in this process could be also applied to the transformation of a structurally related steroid, progesterone, to obtain the corresponding 2β- and 16β-hydroxylated derivatives with high regioselectivity (91–100%) and excellent diastereoselectivity. Using a similar strategy but in combination with a mutability landscape analysis approach (=systematic analysis of all 19 amino acid substitutions at 20 active site positions), the same group more recently reported the development of additional P450$_{BM3}$ variants for the highly regio- and diastereoselective hydroxylation of testosterone at the C16 position with complementary diastereoselectivity (Figure 9.13) [37]. The same or related P450 variants could be applied to catalyze diastereocomplementary hydroxylation reactions at the C16 position of a variety of related steroid molecules, including androstenedione, nandrolone, and boldenone. The C16 alcohols are of practical value for the synthesis of bioactive glucocorticoids. Prior to this work, the Commandeur group had reported the engineering of P450$_{BM3}$ variants capable of 16α/β hydroxylation in testosterone with 60–85% selectivity [101, 102].

Starting from the bacterial class I P450 CYP106A2, which converts progesterone to multiple (15β-, 11α-, 9α-, 6β-) hydroxylated products with low activity [103], the Bernhardt group applied semi-rational mutagenesis of selected active site residues to obtain a CYP106A2 variant with significantly improved activity (15-fold higher k_{cat}/K_M) and regioselectivity (28% → 81%) for the production of 11α-hydroxyprogesterone [104]. Unlike for the P450$_{BM3}$ derived variants described earlier, 2β- and 16β-hydroxylation of progesterone was not observed with the CYP106A2 variants, indicating a complementary scope of these P450s for steroid hydroxylation. The Arnold group found that the P450$_{BM3}$ variant "F1" could

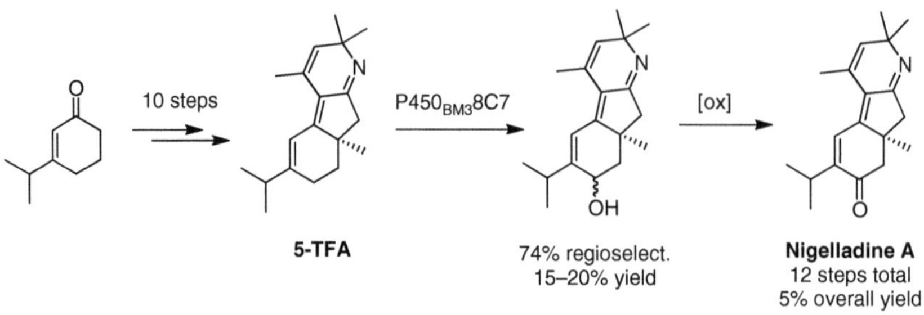

Figure 9.14 Total synthesis of nigelladine A involving a C–H oxidation step catalyzed by an engineered P450$_{BM3}$ variant. Source: Based on Loskot et al. [110].

hydroxylate 11α-hydroxyprogesterone to 2,11-α-dihydroxyprogesterone (82% regioselectivity, 20% yield) [105].

Another important industrial process involving P450-catalyzed oxidation is the conversion of compactin to pravastatin, a major cholesterol-lowering drug. The current process involves a dual-step fermentation with a *Penicillium* sp. to produce compactin, followed by biohydroxylation with an oxidizing bacterial strain (*Streptomyces carbophilus*). Recently, a remarkable example of P450 engineering and metabolic engineering was reported toward the development of simplified, single-strain platform for pravastatin production [106]. Specifically, Munro and coworkers identified a class I P450 CYP105AS1 (*Amycolatopsis orientalis*) capable of hydroxylating the C6 position of compactin to give the undesired epimer 6-epi-pravastatin, along with pravastatin in a 9 : 1 ratio. To generate CYP105AS1 into a self-sufficient P450, the P450 was fused to the reductase domain of P450 RhF, following a strategy previously applied to other class I P450s [107–109].The CYP105AS1 gene was then evolved using multiple rounds of random mutagenesis and screening to obtain a variant, called P450$_{Prava}$, capable of producing pravastatin with 96% diastereoselectivity over the undesired epimer as a result of five amino acid mutations, three of which are located in the active site as determined by solving the structure of the P450 [106]. The optimized P450 could be integrated into an engineered compactin-producing *Penicillium* strain to produce pravastatin in high titers (>6 g l^{-1}), which represents a two- to threefold improvement compared to the classical two-step process [106].

Finally, the application of a P450-catalyzed oxidation for the total synthesis of a complex natural product (nigelladine A) was recently reported by the Stoltz and Arnold groups [110]. While chemical oxidation of compound 5-TFA at the allylic C1 carbon (Figure 9.14) was found to exhibit poor selectivity leading a mixture of products, an engineered P450 useful for this reaction (8C7; ~60% selectivity) was identified upon screening a panel of P450$_{BM3}$ variants previously engineered for the oxidation of large substrates [105]. 8C7-catalyzed oxidation of intermediate 5-TFA followed by chemical oxidation of the resulting alcohol with Dess–Martin periodinane gave nigelladine A in 21% yield over two steps.

9.2.3 Synthesis of Drug Metabolites

Investigating the metabolic fate of drug molecules is integral part of the drug development process and the characterization of the pharmacological activity and toxicity of drug metabolites is a critical step toward the approval of new pharmaceutical agents for use in humans. Since the human liver P450s account for the majority (>90%) of Phase I metabolic transformations of drugs *in vivo*, the synthesis of the corresponding metabolites is particularly important [111], yet often challenged by the need to design *de novo* synthetic routes to access these compounds. While biotransformations with recombinant human P450s have provided a means to produce authentic drug metabolites [112–114], an alternative approach has involved the application of bacterial P450s and engineered variants thereof for this purpose in view of their higher potential for scalability as a result of their higher stability, expression levels, and catalytic activity compared to their mammalian counterparts. Early studies showed that wild-type $P450_{BM3}$ as well as engineered variants thereof are able to accept and oxidize various drug molecules, including acetaminophen and propranolol, resulting in the formation of authentic human metabolites of these drugs [115]. Starting from a promiscuous drug-metabolizing $P450_{BM3}$ variant (R47L/F87V/L188Q) [115], Commandeur and coworkers applied directed evolution to generate a P450 variant (M11) with improved activity for the oxidation of dextromethorphan [102]. In a later study, the same group reported the identification of a small panel of $P450_{BM3}$ variants capable of metabolizing 41 out of 43 drugs tested with >20% conversion [116]. Previously, Arnold and coworkers applied a mini library (~120) of engineered $P450_{BM3}$ variants for producing 12 out of 13 mammalian metabolites for two marketed drugs, verapamil and astemizole, and a drug candidate [117].

Engineered $P450_{BM3}$ variants have been successfully applied toward the transformation of NSAIDs into their respective metabolites via hydroxylation or dealkylation chemistries. For example, the aforementioned variant "M11" proved useful for the synthesis of hydroxylated metabolites of fenamic acid-based NSAIDs [118]. Meclofenamic acid was converted into three metabolites hydroxylated at the aromatic positions C4′ and C5 and benzylic position C3′ with high turnovers (>2000 TTN) [118]. Furthermore, the activity and selectivity of this P450 variant toward the benzylic oxidation reaction could be improved through the introduction of two additional active site mutations (V87I/L437N; 85% select.; >5000 TTN) [118]. In another study, Wong and coworkers identified a highly active $P450_{BM3}$ variant "RP/FV/EV/FW" ($P450_{BM3}$ R47L/Y51F/F81W/F87V/E267V/I401P) capable of hydroxylating diclofenac at the C4′ (45%) and C5 (55%) sites with quantitative conversion, whereas the $P450_{BM3}$ variant "RT2/AP/PV," which contains 9 amino acid mutations, is capable of converting naproxen to desmethylnaproxen in 57% isolated yields (Figure 9.15a) [90]. Interestingly, the R47L and F87V mutations of the "RP/FV/EV/FW" variant are shared by the M11 and MT35 variants reported by Commandeur for the oxidation of other drug molecules [116]. Arnold and coworkers reported the directed evolution a $P450_{BM3}$ variant for the oxidation of acidic drugs, which are typically metabolized by human CYP2C9 [119]. The laboratory evolved P450, called "X3H1," bears 23 amino acid mutations in comparison with the wild type enzyme and could be applied to the demethylation of naproxen in 43% isolated yields (Figure 9.15a) [119]. A L75R mutation was found to be critical for the enhanced activity of the P450 on this acidic NSAID despite the fact that Arg75 was not found to establish direct interactions with the substrate via crystallographic studies.

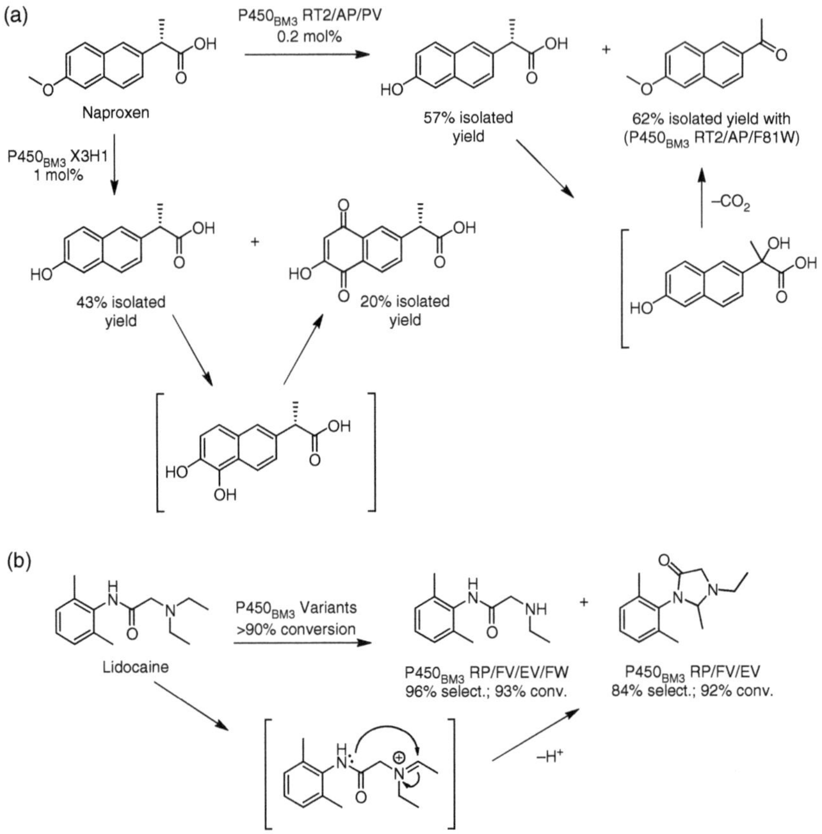

Figure 9.15 Synthesis of drug metabolites using engineered $P450_{BM3}$ variants. Sources: (a) Based on Ren et al. [90], Rentmeister et al. [119], (b) Based on Ren et al. [91].

$P450_{BM3}$ variants have also been evolved to target cationic substrates such as amitriptyline and lidocaine. The aforementioned M11 P450 variant was reported to oxidize amitriptyline producing six uncharacterized metabolites (70% conversion), three of which were unique to $P450_{BM3}$ and not detected upon incubation of amitriptyline with human liver microsomes [116]. The RP/FV/EV/FW variant reported by the Wong group, on the other hand, was found to be highly active toward lidocaine (93% conversion) producing the dealkylation product (Figure 9.15b) [90], a known human metabolite of CYP3A4 [120]. The RP/FV/EV variant, which does not contain the F81W mutation, was found to favor the formation of the *N*,*N*-acetal compound (Figure 9.15b, 84% select., 92% conversion).

Yun and coworkers reported the use of engineered $P450_{BM3}$ variants for targeting the cholesterol-lowering drugs simvastatin and lovastatin [121]. The authors found that $P450_{BM3}$ variant M#16 could oxidize the C6 site of simvastatin and lovastatin to produce the 6′β-OH statin products as the major metabolite as well as the dehydrogenation product, 6′-exomethylene statin (~800 TONs). In a subsequent study, the same group swapped the reductase domain of variant M#16 with the reductase domain of a natural variant of CYP102A1, called CYP102Al.2 (V2) [122].

Using random mutagenesis, the "M16V2 chimera" was further evolved to show improved oxidation activity (up to 10-fold faster turnover rates = 100–200 min^{-1}) on simvastatin and lovastatin compared to M#16 [122].

9.3 High-throughput Methods for Screening Engineered P450s

The success of P450 engineering campaigns heavily relies on the availability of high-throughput methods for library screening, which often represents the bottleneck in the development of engineered P450s with altered or improved function. Over the past two decades, a variety of colorimetric and fluorescence-based assays have been developed for the high-throughput screening (HTS) of P450 libraries expressed in multi-well plates. These HTS assays have proven useful for measurement of P450-dependent indigo formation [46], C–H hydroxylation in *p*-nitrophenol-containing substrates [123], aromatic C–H oxidation [124], alkene epoxidation [125], and O-demethylation [30]. In addition, the O-dealkylation of resorufin- and coumarin-based substrates has provided a sensitive fluorescence-based assay for the screening of both bacterial and mammalian P450s [126–129]. The latter assay was recently integrated with flow cytometry as a means to facilitate the screening of engineered P450 libraries expressed in whole cells [130].

Whereas the aforementioned methods have proven useful for the analysis of "generic" P450-dependent oxidation activity or P450 activity against a surrogate substrate, high-throughput methods useful for the development of P450s with refined regio- and stereoselectivity have also recently emerged. The latter represents indeed a challenging task which typically requires the screening of large collections of enzymes (>10^4–10^5) using laborious and time-consuming (chiral) GC or HPLC methods [15]. To facilitate the identification of stereoselective P450s for a target substrate, the Li group has reported a tandem enzymatic assay in which P450-catalyzed hydroxylation is coupled to colorimetric conversion of the hydroxylated product by a dehydrogenase enzyme [82].Using two dehydrogenases with high specificity for either enantiomer of the hydroxylate product, it is possible to measure the % ee of the P450 enzyme (Figure 9.16a). This method was successfully applied to the development of highly stereoselective $P450_{pyr}$ variants for the hydroxylation of *N*-benzyl-pyrrolidine [82] and the subterminal hydroxylation of octane [85], although its scope remains substrate-specific and is contingent upon the availability of dehydrogenases with high enantioselectivity for the product of the P450 reaction.

As an alternative and potentially more general strategy for the development of engineered P450s with fine-tuned regio- and stereoselectivity, the Fasan group has recently introduced a "P450 fingerprinting" method, in which a panel of chromogenic methoxy-functionalized probes based on different molecular scaffolds is used to map the active site geometry of P450 enzymes in a high-throughput manner (Figure 9.16b) [94]. This approach was shown to provide a convenient means for both (i) rapidly identifying functional P450s with divergent regio- and stereoselectivity properties and (ii) predicting the reactivity of these P450s against a target

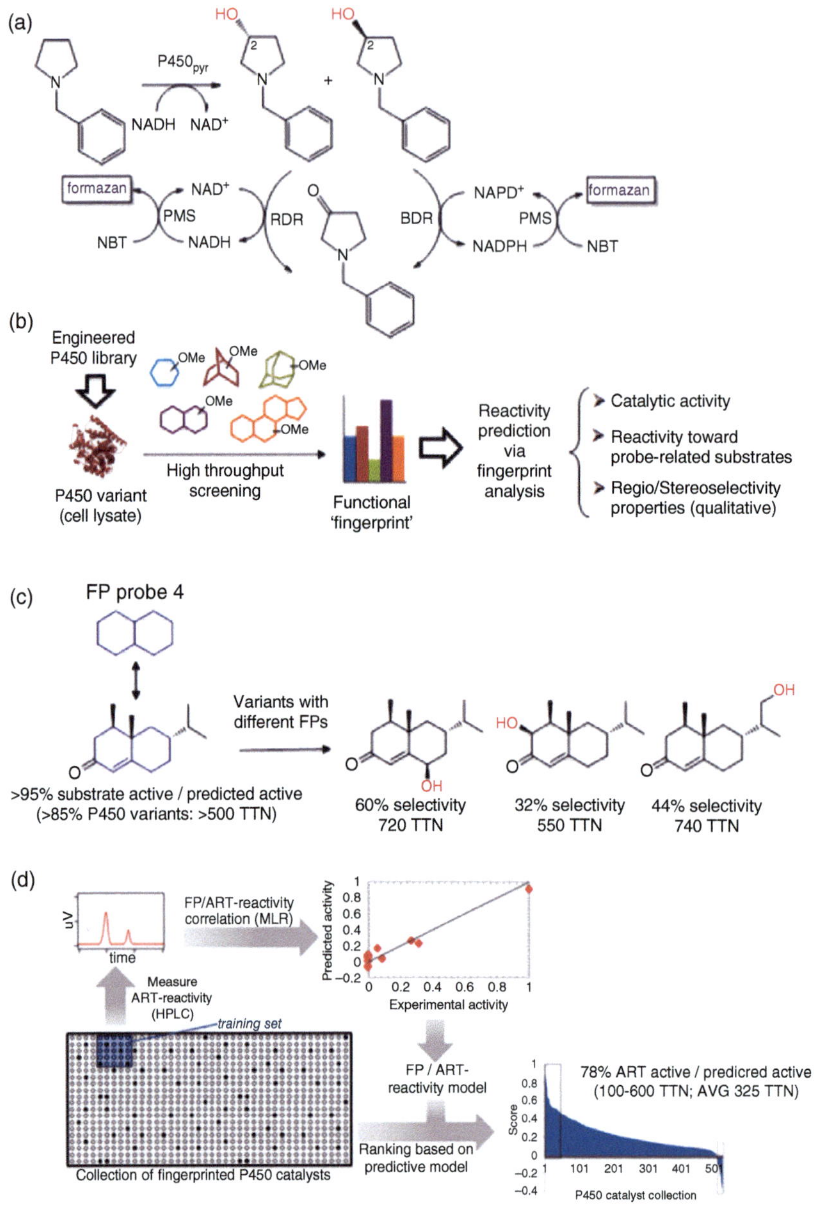

Figure 9.16 High-throughput methods for engineering of regio- and stereoselective P450 enzymes. (a) High-throughput colorimetric assay for screening for P450 variants with *R/S*-stereoselectivity toward *N*-benzyl-pyrrolidine C2 hydroxylation; (b) High-throughput fingerprinting method for discovery of functionally diverse P450 catalysts; (c) Fingerprint-driven identification of terpene-hydroxylating P450 catalysts via fingerprint single component analysis; (d) Fingerprint-driven identification of artemisinin-hydroxylating P450 catalysts via fingerprint multicomponent analysis. Sources: (a) Based on Tang et al. [82], (b-c) Based on Zhang et al. [93], (d) Based on Zhang et al. [94] copyright 2011, American Chemical Society.

substrate structurally related [94] or structurally unrelated [93, 95] to the fingerprint probes via single-component (Figure 9.16c) or multi-component fingerprint analysis (Figure 9.16d), respectively. Combined with active site mutagenesis, these fingerprint-based methods could be applied to enable the rapid development of high regio- and stereoselective P450 catalysts for the hydroxylation of multiple C–H sites in complex natural products such as artemisinin and parthenolide (Figure 9.16d). More recently, a similar approach, but involving a broader set of chromogenic probes [26], was applied to examine the substrate profile of three members of the CYP116B family, namely, $P450_{RpMO}$, $P450_{CtMO}$, and $P450_{ArMO}$, from which useful insights could be gained about the preferential reactivity of these P450s for aromatic substrates over alkane or steroid substrates [131].

9.4 Engineering of Hybrid P450 Systems

Electron transfer from the redox partner(s) to the heme can be a limiting factor in reactions catalyzed by multicomponent P450 systems (e.g. class I P450 systems). While careful optimization of the relative ratio of the P450 and cognate redox partners is required for optimal function of these systems, genetic fusion of the P450 to a non-cognate reductase domain from a self-sufficient P450 has provided a promising strategy for improving the catalytic performance of class I P450 monooxygenases. A particularly useful reductase domain for this purpose has turned out to be the FMN-/[2Fe-2S]-containing reductase domain found in $P450_{RHF}$ (CYP116B2) from *Rhodococcus* sp. (RhFRed), which has been successfully fused to an increasing number of class I P450s, including $P450_{bzo}$ [107], $P450_{balk}$ [132], $P450_{PikC}$ [108], $P450_{EryF}$ [108], $P450_{cam}$ [109], and $P450_{Cin}$ [133]. For example, an efficient $P450_{cam}$–RhFRed chimera was prepared by Flitsch and coworkers through the fusion of $P450_{cam}$ to RhFRed and optimization of the intervening linker [109]. The best variant, which incorporated a 9-amino acid linker in addition to RhFRed native linker, was found to maintain a similar affinity for camphor as wild-type $P450_{cam}$ (1.4 versus 1.6 μM) and to enable the conversion of this substrate to 5-exo-hydroxycamphor in yields (>80% with 30 mM camphor) comparable to those obtained in whole-cell systems expressing $P450_{CAM}$ along with its cognate putidaredoxin reductase (PdR) and putidaredoxin (Pdx).

More recently, Hauer and coworkers investigated several fusion constructs for a class I P450 from *Marinobacter aquaeolei* VT8 ($CYP153A_{M.aq}$), which catalyzes the highly selective terminal hydroxylation (>95%) of medium-chain fatty acids [134]. In this case, the reductase domain from CYP116B3 (PFOR), CYP102A1 (CPR), and CYP116B2 (RhFRed) was fused to the C-terminus of $CYP153A_{M.aq}$ and the performance of the resulting hybrid P450s was analyzed in the context of dodecanoic acid hydroxylation. The fusion constructs containing CPR and PFOR were found to exhibit a threefold faster rate in comparison with that containing RhFRed. Further optimization of the linker in the $CYP153A_{M.aq}$-PFOR fusion resulted in a variant showing 94% coupling efficiency in the fatty acid hydroxylation reaction, which represents a two- to three-fold improvement compared with the fusion constructs harboring other linker sequences.

Evolutionary approaches have also been applied for improving the interactions between the P450 and the non-cognate reductase domain toward improving the catalytic performance of engineered hybrid P450 systems. As an example, Lin and coworkers applied this strategy for optimizing a fusion between $P450_{sca\text{-}2}$ (*S. carbophilus*), a class I P450 that stereoselectively hydroxylates mevastatin to produce pravastatin, and the redox partners Pdx and PdR [135]. To this end, five sites located in the substrate binding pocket, substrate access entrance, and putative P450/Pdx interaction interface as predicted based on homology modeling were subjected to site-saturation mutagenesis, followed by library screening. The best variant, which incorporates two mutations at the putative P450/Pdx interface, was found to produce 377 mg of pravastatin/liter in whole cell biotransformations, which represents a seven-fold improvement compared to parent fusion construct and a 29-fold improvement as compared with the wild type P450 system.

9.5 Engineered P450s with Improved Thermostability and Solubility

Improving the thermal stability of P450 systems have represented an important protein engineering goal in view of the advantages of this feature for the application of P450s in chemical processes as well as toward enhancing their robustness to mutagenesis for evolving new or improved functions [38, 39]. In addition to the mining of naturally occurring thermophilic P450 systems [136], P450s with enhanced thermostability have been obtained via directed evolution using random mutagenesis [137] or, alternatively, through structure-guided recombination using the SCHEMA algorithm [138, 139]. Both studies focused on the isolated heme domain of $P450_{BM3}$, which can function as a peroxygenase through a single active site mutation (F87A) [140]. Using directed evolution, a thermostable peroxygenase variant was obtained (5H6) that exhibits a 15 °C higher T_{50} (46 °C → 61 °C) compared with the parent P450 as a result of 8 mutations [137]. The stabilizing mutations identified through this process could be later used to increase the thermal stability of other engineered $P450_{BM3}$ variants evolved for other functions [35, 141]. In another study, SCHEMA-guided recombination of the heme domain of $P450_{BM3}$ with those of other homologous P450s (CYP102A2 and A3) enabled the identification of thermostable peroxygenase variants with up to 10 °C higher T_{50} compared to the most thermostable parent protein (64.4 versus 55 °C) [139]. To obtain more stable $P450_{BM3}$ variants, Urlacher and coworkers constructed a chimera by fusing the heme domain of $P450_{BM3}$ to the more stable reductase domain from the homologous CYP102A3 from *Bacillus subtilis* [142]. The chimeric enzyme was shown to exhibit a 10-fold longer half-life (8 → 100 minutes) at 50 °C, although its catalytic activity was significantly reduced (30%) relative to that of $P450_{BM3}$, presumably due to suboptimal electron transfer between these domains. More recently, Saab-Rincon et al. applied a consensus-guided mutagenesis approach to enhance the thermal stability of the reductase domain of $P450_{BM3}$ [143]. Through the combination of consensus residues identified via phylogenetic analysis of a set of distantly related P450 reductase sequences, several variants were identified that feature a 10-fold longer half-life at 50 °C as well as increased catalytic performance at elevated temperatures compared with the wild-type enzyme. Characterization of the engineered $P450_{BM3}$ variants indicated that the introduced mutations increased the thermal stability of the FAD-binding domain without compromising the catalytic activity of the enzyme or its substrate selectivity profile. This work highlighted the effectiveness of consensus-guided mutagenesis for enhancing the thermal stability of the reductase component of a multidomain P450 enzyme.

Human liver P450s have also attracted significant interest as biocatalysts in particular in the context of the synthesis of authentic drug metabolites. Protein engineering efforts toward improving the stability of these eukaryotic P450s have involved both rational design and evolutionary methods. In an initial study by Halpert and coworkers, 11 variants of human CYP2B6 were rationally engineered based on sequence comparison with three other mammalian CYP2B enzymes with higher stability (i.e. 3–10 °C higher T_m), resulting in the identification of a human CYP2B6 variant with improved thermostability ($\Delta T_m = +7\,°C$) [144, 145]. In another study, the same group applied directed evolution via random mutagenesis to successfully evolve a CYP2B6 variant with higher thermostability ($\Delta T_{50} = +6\,°C$), along with increased tolerance to higher concentrations of the organic solvent dimethylsulfoxide (DMSO) [146].

The poor solubility and limited expression levels in heterologous hosts (e.g. *E. coli*) typically exhibited by membrane-bound eukaryotic P450s represents a major obstacle toward the exploitation of these systems for biocatalytic applications. Toward addressing this challenge, Gillam and coworkers applied a gene shuffling approach to generate a more stable variant of the mammalian cytochrome $P450_{2F1}$, which is implicated in drug/xenobiotic metabolism in the human respiratory tract [147]. Through deoxyribonucleic acid (DNA) shuffling with an homologous P450 (i.e. $P450_{2F3}$; 84% sequence identity), the authors were able to obtain an engineered $P450_{2F1}$ variant that can be more efficiently expressed in *E. coli*, making possible the characterization of the drug metabolism activity of this poorly characterized P450. In another recent study, a chimeragenesis strategy for improving the expression of membrane-bound plant P450s in *E. coli* was developed by screening a library of N-terminal expression tag chimeras in combination with HTS platform based on C-terminal green fluorescent protein (GFP) fusions [148]. Upon identification of an optimal N-terminal tag using the plant P450 CYP79A1, this chimeragenesis strategy was applied to 49 different P450s from medicinal plants. Notably, nearly half of these P450s showed improved levels of heterologous expression in *E. coli* (2- to nearly 40-fold improvement), thereby facilitating the recombinant production of these plant P450s for further characterization and potential application in biocatalysis [148].

9.6 Conclusions

P450 enzymes represent an important class of biological oxidation catalysts and the contributions highlighted in this chapter highlight the versatility of this class of enzymes toward engineering for improved catalytic activity, relaxed substrate specificity, and/or improved stability for a broad range of applications. Both rational design and evolutionary methods have proven effective toward expanding the substrate scope of these enzymes to include a broad spectrum of non-native substrates, whereas structure-guided semi-rational mutagenesis has provided a means to improve the regio- and stereoselectivity of these enzymes for a desired synthetic transformation. Domain engineering and chimeragenesis, on the other hand, have represented promising strategies toward the development of catalytically self-sufficient P450 systems for biocatalytic applications and/or for improving the accessibility of eukaryotic P450s. These protein engineering strategies have further benefited from the introduction of "smarter" mutagenesis schemes to reduce the burden associated with library screening, along with novel high-throughput methods and platform for analysis of P450 function and selectivity. While a few industrial processes already exist for P450-catalyzed transformations (e.g. steroid derivatization, pravastatin production), these enzymes have historically been challenging to implement in large-scale biocatalytic processes due to their molecular complexity, limited stability, and cofactor/cosubstrate requirements [99]. We anticipate P450 engineering will provide a way to overcome these limitations and enable a broader implementation and exploitation of engineered P450s in both academia and industry, opening the way to the development of efficient and sustainable P450-catalyzed processes for the synthesis and manufacturing of pharmaceuticals, fine chemicals, and other high value compounds.

Acknowledgments

This work was supported by the US National Science Foundation grant CHE-1609550.

References

1 Nelson, D.R. (2018). Cytochrome P450 diversity in the tree of life. *Biochim. Biophys. Acta, Proteins Proteomics* 1866: 141–154.

2 Jennewein, S. and Croteau, R. (2001). Taxol: biosynthesis, molecular genetics, and biotechnological applications. *Appl. Environ. Microbiol.* 57: 13–19.

3 Jennewein, S., Rithner, C.D., Williams, R.M., and Croteau, R.B. (2001). Taxol biosynthesis: taxane 13α-hydroxylase is a cytochrome P450-dependent monooxygenase. *Proc. Natl. Acad. Sci. U.S.A.* 98: 13595–13600.

4 Morrone, D., Chen, X., Coates, R.M., and Peters, R.J. (2010). Characterization of the kaurene oxidase CYP701A3, a multifunctional cytochrome P450 from gibberellin biosynthesis. *Biochem. J.* 431: 337–344.

5 Anzai, Y., Li, S.Y., Chaulagain, M.R. et al. (2008). Functional analysis of MycCI and MycG, cytochrome P450 enzymes involved in biosynthesis of mycinamicin macrolide antibiotics. *Chem. Biol.* 15: 950–959.

6 Song, L.J., Laureti, L., Corre, C. et al. (2014). Cytochrome P450-mediated hydroxylation is required for polyketide macrolactonization in stambomycin biosynthesis. *J. Antibiot.* 67: 71–76.

7 Rudolf, J.D., Dong, L.B., Zhang, X. et al. (2018). Cytochrome P450-catalyzed hydroxylation initiating ether formation in platensimycin biosynthesis. *J. Am. Chem. Soc.* 140: 12349–12353.

8 Denisov, I.G., Makris, T.M., Sligar, S.G., and Schlichting, I. (2005). Structure and chemistry of cytochrome P450. *Chem. Rev.* 105: 2253–2277.

9 Pylypenko, O. and Schlichting, I. (2004). Structural aspects of ligand binding to and electron transfer in bacterial and fungal P450s. *Annu. Rev. Biochem.* 73: 991–1018.

10 Unger, B.P., Gunsalus, I.C., and Sligar, S.G. (1986). Nucleotide-sequence of the *Pseudomonas-putida* cytochrome P-450cam gene and its expression in *Escherichia-coli*. *J. Biol. Chem.* 261: 1158–1163.

11 Ortiz de Montellano, P.R. (2010). Hydrocarbon hydroxylation by cytochrome P450 enzymes. *Chem. Rev.* 110: 932–948.

12 Sono, M., Roach, M.P., Coulter, E.D., and Dawson, J.H. (1996). Heme-containing oxygenases. *Chem. Rev.* 96: 2841–2888.

13 Whitehouse, C.J., Bell, S.G., and Wong, L.L. (2012). P450(BM3) (CYP102A1): connecting the dots. *Chem. Soc. Rev.* 41: 1218–1260.

14 Guengerich, F.P. and Yoshimoto, F.K. (2018). Formation and cleavage of C-C bonds by enzymatic oxidation reduction reactions. *Chem. Rev.* 118: 6573–6655.

15 Fasan, R. (2012). Tuning P450 enzymes as oxidation catalysts. *ACS Catal.* 2: 647–666.

16 Pochapsky, T.C., Kazanis, S., and Dang, M. (2010). Conformational plasticity and structure/function relationships in cytochromes P450. *Antioxid. Redox Signaling* 13: 1273–1296.

17 Munro, A.W., Girvan, H.M., and McLean, K.J. (2007). Variations on a (t)heme – novel mechanisms, redox partners and catalytic functions in the cytochrome P450 superfamily. *Nat. Prod. Rep.* 24: 585–609.

18 Hannemann, F., Bichet, A., Ewen, K.M., and Bernhardt, R. (2007). Cytochrome P450 systems – biological variations of electron transport chains. *Biochim. Biophys. Acta* 1770: 330–344.

19 Miura, Y. and Fulco, A.J. (1974). (Omega - 2) hydroxylation of fatty-acids by a soluble system from *Bacillus-megaterium*. *J. Biol. Chem.* 249: 1880–1888.

20 Roberts, G.A., Grogan, G., Greter, A. et al. (2002). Identification of a new class of cytochrome P450 from a *Rhodococcus* sp. *J. Bacteriol.* 184: 3898–3908.

21 Hunter, D.J., Roberts, G.A., Ost, T.W. et al. (2005). Analysis of the domain properties of the novel cytochrome P450 RhF. *FEBS Lett.* 579: 2215–2220.

22 De Mot, R. and Parret, A.H. (2002). A novel class of self-sufficient cytochrome P450 monooxygenases in prokaryotes. *Trends Microbiol.* 10: 502–508.

23 Liu, L., Schmid, R.D., and Urlacher, V.B. (2006). Cloning, expression, and characterization of a self-sufficient cytochrome P450 monooxygenase from *Rhodococcus ruber* DSM 44319. *Appl. Environ. Microbiol.* 72: 876–882.

24 Bernhardt, R. and Urlacher, V.B. (2014). Cytochromes P450 as promising catalysts for biotechnological application: chances and limitations. *Appl. Microbiol. Biotechnol.* 98: 6185–6203.

25 Urlacher, V.B. and Girhard, M. (2012). Cytochrome P450 monooxygenases: an update on perspectives for synthetic application. *Trends Biotechnol.* 30: 26–36.

26 Behrendorff, J., Huang, W.L., and Gillam, E.M.J. (2015). Directed evolution of cytochrome P450 enzymes for biocatalysis: exploiting the catalytic versatility of enzymes with relaxed substrate specificity. *Biochem. J* 467: 1–15.

27 Ravichandran, K.G., Boddupalli, S.S., Hasermann, C.A. et al. (1993). Crystal structure of hemoprotein domain of P450BM-3, a prototype for microsomal P450's. *Science* 261: 731–736.

28 Li, H. and Poulos, T.L. (1997). The structure of the cytochrome p450BM-3 haem domain complexed with the fatty acid substrate, palmitoleic acid. *Nat. Struct. Biol.* 4: 140–146.

29 Farinas, E.T., Schwaneberg, U., Glieder, A., and Arnold, F.H. (2001). Directed evolution of a cytochrome P450 monooxygenase for alkane oxidation. *Adv. Synth. Catal.* 343: 601–606.

30 Peters, M.W., Meinhold, P., Glieder, A., and Arnold, F.H. (2003). Regio- and enantioselective alkane hydroxylation with engineered cytochromes P450 BM-3. *J. Am. Chem. Soc.* 125: 13442–13450.

31 Glieder, A., Farinas, E.T., and Arnold, F.H. (2002). Laboratory evolution of a soluble, self-sufficient, highly active alkane hydroxylase. *Nat. Biotechnol.* 20: 1135–1139.

32 Landwehr, M., Hochrein, L., Otey, C.R. et al. (2006). Enantioselective alpha-hydroxylation of 2-arylacetic acid derivatives and buspirone catalyzed by engineered cytochrome P450 BM-3. *J. Am. Chem. Soc.* 128: 6058–6059.

33 Kubo, T., Peters, M.W., Meinhold, P., and Arnold, F.H. (2006). Enantioselective epoxidation of terminal alkenes to (R)- and (S)-epoxides by engineered cytochromes P450 BM-3. *Chemistry* 12: 1216–1220.

34 Meinhold, P., Peters, M.W., Chen, M.M. et al. (2005). Direct conversion of ethane to ethanol by engineered cytochrome P450 BM3. *ChemBioChem* 6: 1765–1768.

35 Fasan, R., Chen, M.M., Crook, N.C., and Arnold, F.H. (2007). Engineered alkane-hydroxylating cytochrome P450(BM3) exhibiting nativelike catalytic properties. *Angew. Chem. Int. Ed.* 46: 8414–8418.

36 Tyagi, V., Alwaseem, H., O'Dwyer, K.M. et al. (2016). Chemoenzymatic synthesis and antileukemic activity of novel C9-and C14-functionalized parthenolide analogs. *Bioorg. Med. Chem.* 24: 3876–3886.

37 Acevedo-Rocha, C.G., Gamble, C.G., Lonsdale, R. et al. (2018). P450-catalyzed regio- and diastereoselective steroid hydroxylation: efficient directed evolution enabled by mutability landscaping. *ACS Catal.* 8: 3395–3410.

38 Fasan, R., Meharenna, Y.T., Snow, C.D. et al. (2008). Evolutionary history of a specialized P450 propane monooxygenase. *J. Mol. Biol.* 383: 1069–1080.

39 Bloom, J.D., Labthavikul, S.T., Otey, C.R., and Arnold, F.H. (2006). Protein stability promotes evolvability. *Proc. Natl. Acad. Sci. U.S.A.* 103: 5869–5874.

40 Fasan, R., Crook, N.C., Peters, M.W. et al. (2011). Improved product-per-glucose yields in P450-dependent propane biotransformations using engineered *Escherichia coli*. *Biotechnol. Bioeng.* 108: 500–510.

41 Chen, M.M.Y., Snow, C.D., Vizcarra, C.L. et al. (2012). Comparison of random mutagenesis and semi-rational designed libraries for improved cytochrome P450 BM3-catalyzed hydroxylation of small alkanes. *Protein Eng. Des. Sel.* 25: 171–178.

42 Bruhlmann, F., Fourage, L., Ullmann, C. et al. (2014). Engineering cytochrome P450 BM3 of *Bacillus megaterium* for terminal oxidation of palmitic acid. *J. Biotechnol.* 184: 17–26.

43 Sideri, A., Goyal, A., Di Nardo, G. et al. (2013). Hydroxylation of non-substituted polycyclic aromatic hydrocarbons by cytochrome P450 BM3 engineered by directed evolution. *J. Inorg. Biochem.* 120: 1–7.

44 Rentmeister, A., Arnold, F.H., and Fasan, R. (2009). Chemo-enzymatic fluorination of unactivated organic compounds. *Nat. Chem. Biol.* 5: 26–28.

45 Carmichael, A.B. and Wong, L.L. (2001). Protein engineering of *Bacillus megaterium* CYP102. The oxidation of polycyclic aromatic hydrocarbons. *Eur. J. Biochem.* 268: 3117–3125.

46 Li, Q.S., Schwaneberg, U., Fischer, P., and Schmid, R.D. (2000). Directed evolution of the fatty-acid hydroxylase P450 BM-3 into an indole-hydroxylating catalyst. *Chemistry* 6: 1531–1536.

47 Li, Q.S., Ogawa, J., Schmid, R.D., and Shimizu, S. (2001). Engineering cytochrome P450 BM-3 for oxidation of polycyclic aromatic hydrocarbons. *Appl. Environ. Microbiol.* 67: 5735–5739.

48 Whitehouse, C.J., Bell, S.G., Tufton, H.G. et al. (2008). Evolved CYP102A1 (P450BM3) variants oxidise a range of non-natural substrates and offer new selectivity options. *Chem. Commun.* 8: 966–968.

49 Appel, D., Lutz-Wahl, S., Fischer, P. et al. (2001). A P450 BM-3 mutant hydroxylates alkanes, cycloalkanes, arenes and heteroarenes. *J. Biotechnol.* 88: 167–171.

50 Budde, M., Morr, M., Schmid, R.D., and Urlacher, V.B. (2006). Selective hydroxylation of highly branched fatty acids and their derivatives by CYP102A1 from *Bacillus megaterium*. *ChemBioChem* 7: 789–794.

51 Maurer, S.C., Kuhnel, K., Kaysser, L.A. et al. (2005). Catalytic hydroxylation in biphasic systems using CYP102A1 variants. *Adv. Synth. Catal.* 347: 802–810.

52 Munday, S.D., Dezvarei, S., Lau, I.C.K., and Bell, S.G. (2017). Examination of selectivity in the oxidation of ortho- and meta-disubstituted benzenes by CYP102A1 (P450Bm3) variants. *ChemCatChem* 9: 2512–2522.

53 Sarkar, M.R., Lee, J.H.Z., and Bell, S.G. (2017). The oxidation of hydrophobic aromatic substrates by using a variant of the P450 monooxygenase CYP101B1. *ChemBioChem* 18: 2119–2128.

54 Seifert, A., Vomund, S., Grohmann, K. et al. (2009). Rational design of a minimal and highly enriched CYP102A1 mutant library with improved regio-, stereo- and chemoselectivity. *ChemBioChem* 10: 853–861.

55 Weber, E., Seifert, A., Antonovici, M. et al. (2011). Screening of a minimal enriched P450 BM3 mutant library for hydroxylation of cyclic and acyclic alkanes. *Chem. Commun.* 47: 944–946.

56 Seifert, A., Antonovici, M., Hauer, B., and Pleiss, J. (2011). An efficient route to selective bio-oxidation catalysts: an iterative approach comprising modeling, diversification, and screening, based on CYP102A1. *ChemBioChem* 12: 1346–1351.

57 Neufeld, K., Henssen, B., and Pietruszka, J. (2014). Enantioselective allylic hydroxylation of omega-alkenoic acids and esters by P450 BM3 monooxygenase. *Angew. Chem. Int. Ed.* 53: 13253–13257.

58 Roiban, G.D., Agudo, R., Ilie, A. et al. (2014). CH-activating oxidative hydroxylation of 1-tetralones and related compounds with high regio- and stereoselectivity. *Chem. Commun.* 50: 14310–14313.

59 Reetz, M.T. and Carballeira, J.D. (2007). Iterative saturation mutagenesis (ISM) for rapid directed evolution of functional enzymes. *Nat. Protoc.* 2: 891–903.

60 Reetz, M.T. (2011). Laboratory evolution of stereoselective enzymes: a prolific source of catalysts for asymmetric reactions. *Angew. Chem. Int. Ed.* 50: 138–174.

61 Reetz, M.T., Wang, L.W., and Bocola, M. (2006). Directed evolution of enantioselective enzymes: iterative cycles of CASTing for probing protein-sequence space. *Angew. Chem. Int. Ed.* 45: 1236–1241.

62 Reetz, M.T., Kahakeaw, D., and Lohmer, R. (2008). Addressing the numbers problem in directed evolution. *ChemBioChem* 9: 1797–1804.

63 Agudo, R., Roiban, G.D., and Reetz, M.T. (2012). Achieving regio- and enantioselectivity of P450-catalyzed oxidative CH activation of small functionalized molecules by structure-guided directed evolution. *ChemBioChem* 13: 1465–1473.

64 Agudo, R., Roiban, G.D., Lonsdale, R. et al. (2015). Biocatalytic route to chiral acyloins: P450-catalyzed regio- and enantioselective alpha-hydroxylation of ketones. *J. Org. Chem.* 80: 950–956.

65 Jian-bo, W., Adriana, I., and Reetz, M.T. (2017). Chemo- and stereoselective cytochrome P450-BM3-catalyzed sulfoxidation of 1-thiochroman-4-ones enabled by directed evolution. *Adv. Synth. Catal.* 359: 2056–2060.

66 Sun, Z.T., Lonsdale, R., Li, G.Y., and Reetz, M.T. (2016). Comparing different strategies in directed evolution of enzyme stereoselectivity: single- versus double-code saturation mutagenesis. *ChemBioChem* 17: 1865–1872.

67 Sun, Z.T., Lonsdale, R., Kong, X.D. et al. (2015). Reshaping an enzyme binding pocket for enhanced and inverted stereoselectivity: use of smallest amino acid alphabets in directed evolution. *Angew. Chem. Int. Ed.* 54: 12410–12415.

68 Gelb, M.H., Heimbrook, D.C., Malkonen, P., and Sligar, S.G. (1982). Stereochemistry and deuterium isotope effects in camphor hydroxylation by the cytochrome P450cam monoxygenase system. *Biochemistry* 21: 370–377.

69 Kadkhodayan, S., Coulter, E.D., Maryniak, D.M. et al. (1995). Uncoupling oxygen transfer and electron transfer in the oxygenation of camphor analogues by cytochrome P450-CAM. Direct observation of an intermolecular isotope effect for substrate C-H activation. *J. Biol. Chem.* 270: 28042–28048.

70 Sowden, R.J., Yasmin, S., Rees, N.H. et al. (2005). Biotransformation of the sesquiterpene (+)-valencene by cytochrome P450(cam) and P450(BM-3). *Org. Biomol. Chem.* 3: 57–64.

71 Bell, S.G., Chen, X.H., Sowden, R.J. et al. (2003). Molecular recognition in (+)-alpha-pinene oxidation by cytochrome P450(cam). *J. Am. Chem. Soc.* 125: 705–714.

72 Bell, S.G., Sowden, R.J., and Wong, L.L. (2001). Engineering the haem monooxygenase cytochrome P450(cam) for monoterpene oxidation. *Chem. Commun.*: 635–636.

73 Bell, S.G., Stevenson, J.A., Boyd, H.D. et al. (2002). Butane and propane oxidation by engineered cytochrome P450cam. *Chem. Commun.* 5: 490–491.

74 Xu, F., Bell, S.G., Lednik, J. et al. (2005). The heme monooxygenase cytochrome P450cam can be engineered to oxidize ethane to ethanol. *Angew. Chem. Int. Ed.* 44: 4029–4032.

75 Nickerson, D.P., Harford-Cross, C.F., Fulcher, S.R., and Wong, L.L. (1997). The catalytic activity of cytochrome P450cam towards styrene oxidation is increased by site-specific mutagenesis. *FEBS Lett.* 405: 153–156.

76 Jones, J.P., O'Hare, E.J., and Wong, L.L. (2001). Oxidation of polychlorinated benzenes by genetically engineered CYP101 (cytochrome P450(cam)). *Eur. J. Biochem.* 268: 1460–1467.

77 Harford-Cross, C.F., Carmichael, A.B., Allan, F.K. et al. (2000). Protein engineering of cytochrome p450(cam) (CYP101) for the oxidation of polycyclic aromatic hydrocarbons. *Protein Eng.* 13: 121–128.

78 Li, Z., Feiten, H.J., van Beilen, J.B. et al. (1999). Preparation of optically active N-benzyl-3-hydroxypyrrolidine by enzymatic hydroxylation. *Tetrahedron: Asymmetry* 10: 1323–1333.

79 Chang, D.L., Feiten, H.J., Engesser, K.H. et al. (2002). Practical syntheses of N-substituted 3-hydroxyazetidines and 4-hydroxypiperidines by hydroxylation with *Sphingomonas* sp HXN-200. *Org. Lett.* 4: 1859–1862.

80 Chang, D.L., Witholt, B., and Li, Z. (2000). Preparation of (S)-N-substituted 4-hydroxy-pyrrolidin-2-ones by regio-and stereoselective hydroxylation with *Sphingomonas* sp HXN-200. *Org. Lett.* 2: 3949–3952.

81 Li, Z., Feiten, H.J., Chang, D.L. et al. (2001). Preparation of (R)- and (S)-N-protected 3-hydroxypyrrolidines by hydroxylation with *Sphingomonas* sp HXN-200, a highly active, regio- and stereoselective, and easy to handle biocatalyst. *J. Org. Chem.* 66: 8424–8430.

82 Tang, W.L., Li, Z., and Zhao, H. (2010). Inverting the enantioselectivity of P450pyr monooxygenase by directed evolution. *Chem. Commun.* 46: 5461–5463.

83 Chen, Y.Z., Tang, W.L., Mou, J., and Li, Z. (2010). High-throughput method for determining the enantioselectivity of enzyme-catalyzed hydroxylations based on mass spectrometry. *Angew. Chem. Int. Ed.* 49: 5278–5283.

84 Pham, S.Q., Pompidor, G., Liu, J. et al. (2012). Evolving P450pyr hydroxylase for highly enantioselective hydroxylation at non-activated carbon atom. *Chem. Commun.* 48: 4618–4620.

85 Yang, Y., Liu, J., and Li, Z. (2014). Engineering of P450pyr hydroxylase for the highly regio- and enantioselective subterminal hydroxylation of alkanes. *Angew. Chem. Int. Ed.* 53: 3120–3124.

86 Yang, Y., Chi, Y.T., Toh, H.H., and Li, Z. (2015). Evolving P450pyr monooxygenase for highly regioselective terminal hydroxylation of n-butanol to 1,4-butanediol. *Chem. Commun.* 51: 914–917.

87 Brandenberg, O.F., Fasan, R., and Arnold, F.H. (2017). Exploiting and engineering hemoproteins for abiological carbene and nitrene transfer reactions. *Curr. Opin. Biotechnol.* 47: 102–111.

88 Hammer, S.C., Kubik, G., Watkins, E. et al. (2017). Anti-Markovnikov alkene oxidation by metal-oxo-mediated enzyme catalysis. *Science* 358: 215.

89 Yin, Y.C., Yu, H.L., Luan, Z.J. et al. (2014). Unusually broad substrate profile of self-sufficient cytochrome P450 monooxygenase CYP116B4 from *Labrenzia aggregata*. *ChemBioChem* 15: 2443–2449.

90 Ren, X.K., Yorke, J.A., Taylor, E. et al. (2015). Drug oxidation by cytochrome P450(BM3): metabolite synthesis and discovering new P450 reaction types. *Chem. Eur. J.* 21: 15039–15047.

91 Ren, X.K., O'Hanlon, J.A., Morris, M. et al. (2016). Synthesis of imidazolidin-4-ones via a cytochrome P450-catalyzed intramolecular C-H amination. *ACS Catal.* 6: 6833–6837.

92 Kolev, J.N., Zaengle, J.M., Ravikumar, R., and Fasan, R. (2014). Enhancing the efficiency and regioselectivity of P450 oxidation catalysts by unnatural amino acid mutagenesis. *ChemBioChem* 15: 1001–1010.

93 Zhang, K.D., Shafer, B.M., Demars, M.D. et al. (2012). Controlled oxidation of remote sp(3) C-H bonds in artemisinin via P450 catalysts with fine-tuned regio- and stereoselectivity. *J. Am. Chem. Soc.* 134: 18695–18704.

94 Zhang, K., El Damaty, S., and Fasan, R. (2011). P450 fingerprinting method for rapid discovery of terpene hydroxylating P450 catalysts with diversified regioselectivity. *J. Am. Chem. Soc.* 133: 3242–3245.

95 Kolev, J.N., O'Dwyer, K.M., Jordan, C.T., and Fasan, R. (2014). Discovery of potent parthenolide-based antileukemic agents enabled by late-stage P450-mediated C-H functionalization. *ACS Chem. Biol.* 9: 164–173.

96 Alwaseem, H., Frisch, B.J., and Fasan, R. (2018). Anticancer activity profiling of parthenolide analogs generated via P450-mediated chemoenzymatic synthesis. *Bioorg. Med. Chem.* 26: 1365–1373.

97 Le-Huu, P., Heidt, T., Claasen, B. et al. (2015). Chemo-, regio-, and stereoselective oxidation of the monocyclic diterpenoid beta-cembrenediol by P450 BM3. *ACS Catal.* 5: 1772–1780.

98 Le-Huu, P., Petrovic, D., Strodel, B., and Urlacher, V.B. (2016). One-pot, two-step hydroxylation of the macrocyclic diterpenoid beta-cembrenediol catalyzed by P450 BM3 mutants. *ChemCatChem* 8: 3755–3761.

99 Julsing, M.K., Cornelissen, S., Buhler, B., and Schmid, A. (2008). Heme-iron oxygenases: powerful industrial biocatalysts? *Curr. Opin. Chem. Biol.* 12: 177–186.

100 Kille, S., Zilly, F.E., Acevedo, J.P., and Reetz, M.T. (2011). Regio- and stereoselectivity of P450-catalysed hydroxylation of steroids controlled by laboratory evolution. *Nat. Chem.* 3: 738–743.

101 de Beer, S.B., van Bergen, L.A., Keijzer, K. et al. (2012). The role of protein plasticity in computational rationalization studies on regioselectivity in testosterone hydroxylation by cytochrome P450 BM3 mutants. *Curr. Drug Metab.* 13: 155–166.

102 van Vugt-Lussenburg, B.M.A., Stjernschantz, E., Lastdrager, J. et al. (2007). Identification of critical residues in novel drug metabolizing mutants of cytochrome P450BM3 using random mutagenesis. *J. Med. Chem.* 50: 455–461.

103 Bleif, S., Hannemann, F., Lisurek, M. et al. (2011). Identification of CYP106A2 as a regioselective allylic bacterial diterpene hydroxylase. *ChemBioChem* 12: 576–582.

104 Nguyen, K.T., Virus, C., Gunnewich, N. et al. (2012). Changing the regioselectivity of a P450 from C15 to C11 hydroxylation of progesterone. *ChemBioChem* 13: 1161–1166.

105 Lewis, J.C., Mantovani, S.M., Fu, Y. et al. (2010). Combinatorial alanine substitution enables rapid optimization of cytochrome P450BM3 for selective hydroxylation of large substrates. *ChemBioChem* 11: 2502–2505.

106 McLean, K.J., Hans, M., Meijrink, B. et al. (2015). Single-step fermentative production of the cholesterol-lowering drug pravastatin via reprogramming of *Penicillium chrysogenum*. *Proc. Natl. Acad. Sci. U.S.A.* 112: 2847–2852.

107 Nodate, M., Kubota, M., and Misawa, N. (2006). Functional expression system for cytochrome P450 genes using the reductase domain of self-sufficient P450RhF from *Rhodococcus* sp. NCIMB 9784. *Appl. Microbiol. Biotechnol.* 71: 455–462.

108 Li, S.Y., Podust, L.M., and Sherman, D.H. (2007). Engineering and analysis of a self-sufficient biosynthetic cytochrome P450 PikC fused to the RhFRED reductase domain. *J. Am. Chem. Soc.* 129: 12940–12941.

109 Robin, A., Roberts, G.A., Kisch, J. et al. (2009). Engineering and improvement of the efficiency of a chimeric [P450cam-RhFRed reductase domain] enzyme. *Chem. Commun.*: 2478–2480.

110 Loskot, S.A., Romney, D.K., Arnold, F.H., and Stoltz, B.M. (2017). Enantioselective total synthesis of nigelladine A via late-stage C-H oxidation enabled by an engineered P450 enzyme. *J. Am. Chem. Soc.* 139: 10196–10199.

111 Guengerich, F.P. (2002). Cytochrome P450 enzymes in the generation of commercial products. *Nat. Rev. Drug Discovery* 1: 359–366.

112 Schroer, K., Kittelmann, M., and Lutz, S. (2010). Recombinant human cytochrome P450 monooxygenases for drug metabolite synthesis. *Biotechnol. Adv.* 106: 699–706.

113 Orhan, H. and Vermeulen, N.P.E. (2011). Conventional and novel approaches in generating and characterization of reactive intermediates from drugs/drug candidates. *Curr. Drug Metab.* 12: 383–394.

114 Caswell, J.M., O'Neill, M., Taylor, S.J.C., and Moody, T.S. (2013). Engineering and application of P450 monooxygenases in pharmaceutical and metabolite synthesis. *Curr. Opin. Chem. Biol.* 17: 271–275.

115 van Vugt-Lussenburg, B.M.A., Damsten, M.C., Maasdijk, D.M. et al. (2006). Heterotropic and homotropic cooperativity by a drug-metabolising mutant of cytochrome P450BM3. *Biochem. Biophys. Res. Commun.* 346: 810–818.

116 Reinen, J., van Leeuwen, J.S., Li, Y.M. et al. (2011). Efficient screening of cytochrome P450 BM3 mutants for their metabolic activity and diversity toward a wide set of drug-like molecules in chemical space. *Drug Metab. Dispos.* 39: 1568–1576.

117 Sawayama, A.M., Chen, M.M., Kulanthaivel, P. et al. (2009). A panel of cytochrome P450 BM3 variants to produce drug metabolites and diversify lead compounds. *Chemistry* 15: 11723–11729.

118 Venkataraman, H., Verkade-Vreeker, M.C.A., Capoferri, L. et al. (2014). Application of engineered cytochrome P450 mutants as biocatalysts for the synthesis of benzylic and aromatic metabolites of fenamic acid NSAIDs. *Bioorg. Med. Chem.* 22: 5613–5620.

119 Rentmeister, A., Brown, T.R., Snow, C.D. et al. (2011). Engineered bacterial mimics of human drug metabolizing enzyme CYP2C9. *ChemCatChem* 3: 1065–1071.

120 Wang, J.S., Backman, J.T., Taavitsainen, P. et al. (2000). Involvement of CYP1A2 and CYP3A4 in lidocaine N-deethylation and 3-hydroxylation in humans. *Drug Metab. Dispos.* 28: 959–965.

121 Kim, K.H., Kang, J.Y., Kim, D.H. et al. (2011). Generation of human chiral metabolites of simvastatin and lovastatin by bacterial CYP102A1 mutants. *Drug Metab. Dispos.* 39: 140–150.

122 Kang, J.Y., Ryu, S.H., Park, S.H. et al. (2014). Chimeric cytochromes P450 engineered by domain swapping and random mutagenesis for producing human metabolites of drugs. *Biotechnol. Bioeng.* 111: 1313–1322.

123 Schwaneberg, U., Schmidt-Dannert, C., Schmitt, J., and Schmid, R.D. (1999). A continuous spectrophotometric assay for P450 BM-3, a fatty acid hydroxylating enzyme, and its mutant F87A. *Anal. Bioanal.Chem.* 269: 359–366.

124 Wong, T.S., Wu, N., Roccatano, D. et al. (2005). Sensitive assay for laboratory evolution of hydroxylases toward aromatic and heterocyclic compounds. *J. Biomol. Screening* 10: 246–252.

125 Alcalde, M., Farinas, E.T., and Arnold, F.H. (2004). Colorimetric high-throughput assay for alkene epoxidation catalyzed by cytochrome P450 BM-3 variant 139-3. *J. Biomol. Screening* 9: 141–146.

126 Lussenburg, B.M.A., Babel, L.C., Vermeulen, N.P.E., and Commandeur, J.N.M. (2005). Evaluation of alkoxyresorufins as fluorescent substrates for cytochrome P450BM3 and site-directed mutants. *Anal. Biochem.* 341: 148–155.

127 Burke, M.D., Thompson, S., Weaver, R.J. et al. (1994). Cytochrome-P450 specificities of alkoxyresorufin O-dealkylation in human and rat-liver. *Biochem. Pharmacol.* 48: 923–936.

128 Cheng, Q., Sohl, C.D., and Guengerich, F.P. (2009). High-throughput fluorescence assay of cytochrome P450 3A4. *Nat. Protoc.* 4: 1258–1261.

129 Morlock, L.K., Bottcher, D., and Bornscheuer, U.T. (2018). Simultaneous detection of NADPH consumption and H_2O_2 production using the Ampliflu (TM) Red assay for screening of P450 activities and uncoupling. *Appl. Microbiol. Biotechnol.* 102: 985–994.

130 Ruff, A.J., Dennig, A., Wirtz, G. et al. (2012). Flow cytometer-based high-throughput screening system for accelerated directed evolution of P450 monooxygenases. *ACS Catal.* 2: 2724–2728.

131 Li, R.J., Xu, J.H., Yin, Y.C. et al. (2016). Rapid probing of the reactivity of P450 monooxygenases from the CYP116B subfamily using a substrate-based method. *New J. Chem.* 40: 8928–8934.

132 Kubota, M., Nodate, M., Yasumoto-Hirose, M. et al. (2005). Isolation and functional analysis of cytochrome P450 CYP153A genes from various environments. *Biosci. Biotechnol., Biochem.* 69: 2421–2430.

133 Belsare, K.D., Ruff, A.J., Martinez, R. et al. (2014). P-Link: a method for generating multicomponent cytochrome P450 fusions with variable linker length. *Biotechniques* 57: 13.

134 Hoffmann, S.M., Weissenborn, M.J., Gricman, L. et al. (2016). The impact of linker length on P450 fusion constructs: activity, stability and coupling. *ChemCatChem* 8: 1591–1597.

135 Ba, L., Li, P., Zhang, H. et al. (2013). Semi-rational engineering of cytochrome P450sca-2 in a hybrid system for enhanced catalytic activity: insights into the important role of electron transfer. *Biotechnol. Bioeng.* 110: 2815–2825.

136 Harris, K.L., Thomson, R.E.S., Strohmaier, S.J. et al. (2018). Determinants of thermostability in the cytochrome P450 fold. *Biochim. Biophys. Acta, Proteins Proteomics* 1866: 97–115.

137 Salazar, O., Cirino, P.C., and Arnold, F.H. (2003). Thermostabilization of a cytochrome p450 peroxygenase. *ChemBioChem* 4: 891–893.

138 Voigt, C.A., Martinez, C., Wang, Z.G. et al. (2002). Protein building blocks preserved by recombination. *Nat. Struct. Biol.* 9: 553–558.

139 Li, Y., Drummond, D.A., Sawayama, A.M. et al. (2007). A diverse family of thermostable cytochrome P450s created by recombination of stabilizing fragments. *Nat. Biotechnol.* 25: 1051–1056.

140 Cirino, P.C. and Arnold, F.H. (2003). A self-sufficient peroxide-driven hydroxylation biocatalyst. *Angew. Chem. Int. Ed.* 42: 3299–3301.

141 Shapiro, M.G., Westmeyer, G.G., Romero, P.A. et al. (2010). Directed evolution of a magnetic resonance imaging contrast agent for noninvasive imaging of dopamine. *Nat. Biotechnol.* 28: 264–270.

142 Eiben, S., Bartelmas, H., and Urlacher, V.B. (2007). Construction of a thermostable cytochrome P450 chimera derived from self-sufficient mesophilic parents. *Appl. Microbiol. Biotechnol.* 75: 1055–1061.

143 Saab-Rincon, G., Alwaseem, H., Guzman-Luna, V. et al. (2018). Stabilization of the reductase domain in the catalytically self-sufficient cytochrome P450(BM3) by consensus-guided mutagenesis. *ChemBioChem* 19: 622–632.

144 Kumar, S., Zhao, Y.H., Sun, L. et al. (2007). Rational engineering of human cytochrome p450 2B6 for enhanced expression and stability: importance of a Leu(264)-> Phe substitution. *Mol. Pharmacol.* 72: 1191–1199.

145 Talakad, J.C., Wilderman, P.R., Davydov, D.R. et al. (2010). Rational engineering of cytochromes P450 2B6 and 2B11 for enhanced stability: insights into structural importance of residue 334. *Arch. Biochem. Biophys.* 494: 151–158.

146 Kumar, S., Sun, L., Muralidhara, B.K. et al. (2006). Engineering mammalian cytochrome P4502B1 by directed evolution for enhanced catalytic tolerance to temperature and dimethyl sulfoxide. *Protein Eng. Des. Sel.* 19: 547–554.

147 Behrendorff, J., Moore, C.D., Kim, K.H. et al. (2012). Directed evolution reveals requisite sequence elements in the functional expression of P450 2F1 in *Escherichia coli*. *Chem. Res. Toxicol.* 25: 1964–1974.

148 Vazquez-Albacete, D., Cavaleiro, A.M., Christensen, U. et al. (2017). An expression tag toolbox for microbial production of membrane bound plant cytochromes P450. *Biotechnol. Bioeng.* 114: 751–760.

Part III

Applications in Industrial Biotechnology

10

Protein Engineering Using Unnatural Amino Acids

Yang Yu[1], Xiaohong Liu[2], and Jiangyun Wang[2]

[1] *Institute of Biochemical Engineering, School of Chemistry and Chemical Engineering, Beijing Institute of Technology, Department of Chemical Engineering, 5 Zhongguancun South Street, Beijing, Haidian District, 100081, China*
[2] *Laboratory of RNA Biology Institute of Biophysics, Chinese Academy of Sciences, 15 Datun Road, Beijing, Chaoyang District, 100101, China*

10.1 Introduction

Proteins are the executors of biological functions in the cell, performing roles such as catalyzing reactions, transporting molecules, and providing structural support. Proteins are central to biotechnology research such as making potent biocatalysts and developing new materials. A wide variety of methods, extensively reviewed by major players in this field [1–14] and mentioned in this book, have been developed for protein engineering.

Protein engineering often involves substituting one amino acid residue in the protein with one of the 19 other proteinogenic amino acids. This approach is restricted by the limited functional groups available in the proteinogenic or natural amino acids. For example, of 20 proteinogenic amino acids, only a handful are capable of coordinating with metal ions. Many of these amino acids have unique structures and properties such as the imidazole sidechain of histidine and thiolate group of cysteine. This makes the engineering of ligands of metalloproteins very hard. Such restrictions in the diversity of structures, functional groups, and other properties in amino acids affect protein engineering, and an expansion of the amino acid alphabet is needed.

Incorporation of unnatural amino acids (UAAs) into proteins effectively complements current protein engineering methods and greatly expands the chemical structures available to protein chemists [15–23]. It allows protein engineering with an expanded alphabet to achieve improved or novel functions, such as enhancing protein stabilities, probing catalytic mechanisms, tuning enzymatic activities, and designing novel functions.

Protein Engineering: Tools and Applications, First Edition. Edited by Huimin Zhao.
© 2021 WILEY-VCH GmbH. Published 2021 by WILEY-VCH GmbH.

10.2 Methods for Unnatural Amino Acid Incorporation

UAA incorporation can be realized through different methods. Solid-phase peptide synthesis allows the introduction of UAAs, such as L-amino acids with different sidechains, D-amino acids, and β amino acids into the peptide chain [24]. Due to the accumulation of errors from each round of synthesis, peptides over 50 amino acid residues are challenging for routine synthesis. To overcome the size limit for peptide synthesis, short peptide fragments can be chemically synthesized and ligated through Staudinger ligation or native chemical ligation methods. Native chemical ligation takes place between a peptide thioester and a peptide with an N-terminal thiol group. The two peptides undergo a chemoselective transthioesterification reaction that can re-arrange the bond between the thiol and thioester to a native peptide bond through a spontaneous acyl transfer [24]. Expressed protein ligation (EPL) uses an engineered intein, a self-splicing protein domain to produce peptide thioester from a recombinantly expressed protein. The peptide thioester subsequently reacts with the thiol-containing peptide in a native-chemical-ligation-style reaction to form the full-length protein [25–27].

Besides chemical synthesis, certain chemical reactions could target specific natural amino acids and convert them into UAAs. In analogy to site-directed mutagenesis, this method is termed chemical mutagenesis. For example, serine could be converted to selenocysteine through phenylmethanesulfonyl fluoride activation followed by hydrogen selenide treatment to displace the sulfonate group with –SeH. After chemical mutagenesis of Ser in the active site to selenocysteine, subtilisin, a serine protease was converted to an acyltransferase [28]. Tyrosine could be converted to 3-nitrotyrosine (3-NO_2Y) through nitration reactions. The surface-exposed tyrosine in azurin was chemically mutated to 3-NO_2Y. The increase of the reduction potential of nitrotyrosine versus tyrosine facilitates electron hopping in the protein [29]. Chemical mutagenesis provides a facile approach for UAA incorporation, but it generally requires that the host protein is highly stable to endure the organic solvent in which the chemical reaction takes place and the chemical reagents added in the process.

While chemical synthesis and mutagenesis provide structure diversity of the protein building blocks, they often require sophisticated equipment and extra purification steps. The native protein synthesis machinery, including tRNA, amino-acyl tRNA synthetase (aaRS), mRNA, ribosome, and various protein factors (initiation, elongation, release factors, etc.), synthesizes protein in an astonishingly high speed (10–20 aa/s) and fidelity (10^{-4}–10^{-5} aa) [30, 31]. By altering some components of the system, the native protein synthesis machinery can be engineered to produce proteins with UAAs. The natural promiscuity of aminoacyl tRNA synthetases enables structural analogues of natural amino acids to be incorporated into proteins both *in vitro* and *in vivo*. By removing a natural amino acid from the growth media, supplementing it with an unnatural one, and using auxotrophic strains that cannot produce the natural amino acid being replaced, cells can be tricked to incorporate UAAs structurally similar to natural amino acids. This method of amino acid replacement is relatively easy to perform and does not require any reengineering of the translational machinery.

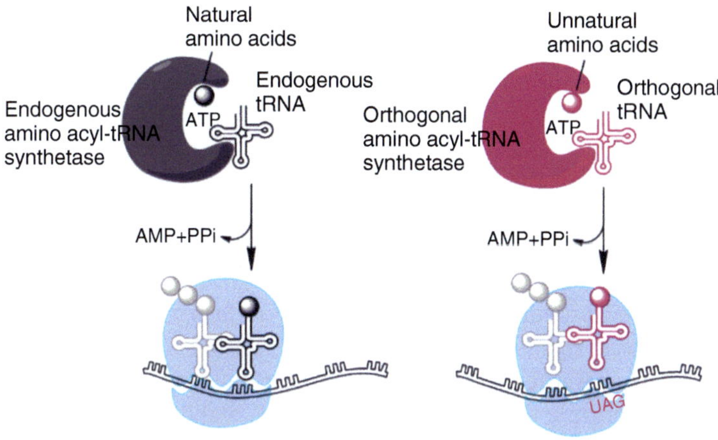

Figure 10.1 In the native translation process (left), tRNA is charged with the corresponding natural amino acid by aaRS, the anticodon loop of tRNA recognizes the corresponding codon on the mRNA, and as a result, the amino acid is added to the peptide chain during translation. Genetic codon expansion (right) introduces orthogonal tRNA and aaRS, which charges the tRNA with a specific UAA. tRNA recognizes a reassigned codon (UAG here), and the UAA it carries is added to the peptide. Source: Yu et al. [18]. © 2017, Springer Nature.

To further provide flexibility in UAA incorporation, genetic code reprogramming was realized in *in vitro* translation system through a ribozyme that can recognize tRNA and charge it with certain amino acids [32, 33]. UAA-charged tRNAs are synthesized using the ribozyme. Adding them into a carefully designed *in vitro* translation system enables incorporation of multiple UAAs into protein selectively [32]. This method can incorporate various UAAs that other biological methods are incapable of doing, like α-N-methyl, D- and β- amino acids.

To achieve site-specific incorporation and expand the diversity of UAAs, further engineering of the protein synthesis machinery is needed. An orthogonal tRNA/aaRS pair is introduced into an organism. "Being orthogonal" means the tRNA and the aaRS are mutually specific, do not crosstalk with the endogenous counterparts, and are compatible with the translation apparatus. The aaRS recognizes a specific UAA and then acylates the tRNA with the UAA. The tRNA recognizes a blank codon (codons that do not encode proteinogenic amino acids), which could be either amber stop codon (UAG), quadruplet codon, or from genomes with reassigned codons. With the help of ribosome and protein factors, the tRNA could deliver the UAA to the elongating peptide. Through the genetic code expansion process, a specific UAA is incorporated at the position corresponding to the blank codon (Figure 10.1) [15].

10.3 Applications of Unnatural Amino Acids in Protein Engineering

Protein engineering serves different purposes, such as increasing protein stability, tuning enzyme activity or selectivity, and even designing novel protein functions.

Protein engineering is based on the understanding of protein function, including its three-dimensional structure, dynamics, and reaction mechanism. Complementary to natural amino acid mutagenesis, UAA incorporation can be used in these areas.

10.3.1 Enhancing Stability

Increasing protein stability is a major target for engineering for several reasons. First, temperature directly affects the reaction rate. A 10-degree increase in reaction temperature will typically lead to a twofold increase in the reaction rate. Proteins with higher stability could work at higher temperatures with higher activity. In addition, increased stability makes the target protein tolerant to additional mutations. Many enzymes need to work in extreme conditions for real-world applications, such as organic solvents and extreme pHs. Stability enhancement makes the enzyme amenable for such conditions.

Fluorination of residues is considered to be beneficial to protein stability as it increases hydrophobicity while causing minimal structural perturbation. Lipase catalyzes ester hydrolysis and transacylation reactions in industrial biocatalysis, thus being an important target for protein engineering. Global fluorination of aromatic residues by introducing 4-fluorophenylalanine, 5-fluorotryptophan, and 3-fluorotyrosine into lipase B from *Candida antarctica* increased the shelf-life of the enzyme [34]. In another study, introducing 3-fluorophenylalanine and 4-fluorophenylalanine into *Thermoanaerobacter thermohydrosulfuricus* lipase increased lipase activity by 25% [35]. Similar strategies are also proven effective in S5 phosphotriesterase [36], organophosphate hydrolase [37], chloramphenicol acetyltransferase [38], and PvuII endonuclease [39].

Cytochrome P450s are a group of oxygen-activating enzymes catalyzing a wide range of reactions, including many key steps in natural product biosynthesis. It is hypothesized that reactive oxygen species (ROS) generated during P450 catalysis damages the protein, lowering its activity [40]. One of the vulnerable residues, Met in P450BM3 was replaced by its redox-inactive structural analogue, norleucine in all 13 positions of the protein. Although less thermostable than the wild type enzyme, the mutant showed a twofold increase in activity in the hydroxylation reaction of *para*-nitrophenoxy carboxylate using H_2O_2 as the oxidant [41]. Similarly, replacement of methionine by norleucine in adenylate kinase resulted in similar reactivity but much higher resistance to hydrogen peroxide [42].

10.3.2 Mechanistic Study Using Spectroscopic Methods

While improving the stability of a protein generally improves its activity, precise engineering to increase protein activity requires detailed knowledge of the reaction mechanism of the protein. Many spectroscopic methods for mechanistic study require probe molecules. Various UAAs can be used as spectroscopic probes, such as UAAs with NMR-active nuclei, or those with an unpaired electron for electronic paramagnetic resonance (EPR) spectroscopy, or fluorescent UAAs. By integrating these UAAs into a protein, researchers could label a protein with probes *in vivo* with minimal structural perturbation, and perform mechanistic studies using various spectroscopic methods [43].

Nuclear magnetic resonance (NMR) spectroscopy of protein relies on the use of stable isotopes with non-zero nuclear spin, such as ^{2}H, ^{13}C, and ^{15}N. Global incorporation of these isotopes into a protein can be realized by the addition of isotope-enriched chemicals in a minimal medium. A global incorporation method requires assigning all peaks, even when only a certain residue is of interest. Under the scenario where dynamics of a certain residue need to be studied by NMR spectroscopy, site-specific labeling is a preferred method, and genetic codon expansion can be used to introduce the isotope-labeled UAA into a protein.

UAAs containing new NMR-active isotopes can be introduced into a protein through genetic codon expansion. ^{19}F, the NMR-active isotope of fluorine, has 100% natural abundance. It does not naturally appear in a protein, and when presented in a protein, its chemical shift is very sensitive to the local environment, making it an ideal NMR probe for biomacromolecules. Different UAAs, including 4-fluorophenylalanine, 2,6-difluorotyrosine (F2Y), and 4-trifluoromethylphenylalanine, have been incorporated into a protein through genetic code expansion. Due to its structural similarity to the Tyr, F2Y can replace native Tyr residues in a protein without causing much structure perturbation [44]. In a proof of principle experiment, ^{19}F NMR spectra showed distinct chemical shifts for F2Y with and without phosphorylation [45]. Due to its sensitivity and minimal structural perturbation to the protein, F2Y can be used to study protein phosphorylation and other modification in the signal transduction process. For example, F2Y was introduced into arrestin, a protein downstream of G-protein coupled receptor (GPCR) protein for signal transduction [46]. ^{19}F NMR revealed the structure dynamics of arrestin upon binding of GPCR with different phosphorylation pattern, suggesting a receptor-phospho-selective mechanism for arrestin to recognize different phosphorylation pattern of GPCR and passing the signal to downstream protein (Figure 10.2). There are nearly 300 SH3-domain-containing proteins (SH3-CPs) as downstream effector proteins of 800 GPCRs. ^{19}F NMR of F2Y-incorporated arrestin revealed that interaction of GPCR and arrestin allosterically regulates the proline regions of β-arrestin 1, which in turn modulates recruitment of SH3 domains [47]. These studies demonstrate that UAAs as NMR probes could be used to probe protein conformational changes as well as protein-protein interactions.

EPR spectroscopy is a spectroscopic method for probing the coordination environment of metalloproteins, measuring the distances between labels, and studying enzyme reaction mechanisms. Tyr is a redox-active amino acid, which often acts as an electron/proton donor during electron transfer or catalysis. Ribonucleotide reductase (RNR) catalyzes the oxidation of ribonucleotide to deoxyribonucleotide, a process crucial for survival. The hallmark of type Ia RNR is its long-range (35Å) electron transfer between the di-iron center and the thiyl radical, involving several pathway Tyr residues as the electron/proton relay [48]. To study the thermodynamics of this process, a series of EPR-active tyrosine analogues with different redox potentials were used to replace the native Tyr residues in RNR [49]. The studies mapped out Tyr residues involving the electron transfer process and revealed a kinetic gating

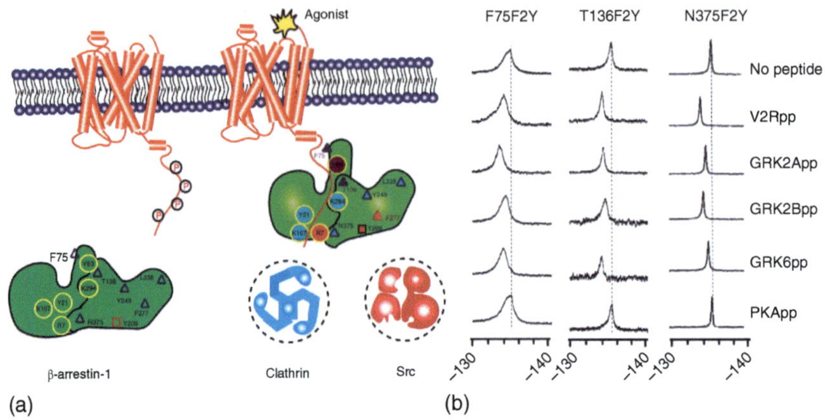

Figure 10.2 F2Y NMR probe revealed arrestin mediated signaling directed by phosphor-coding of the receptor [46]. (a) Arrestin mediates signal transduction from GPCR to its downstream effectors, including Clathrin and Src. (b) Different phosphorylation pattern in the tail region of GPCR affects arrestin structure, as revealed by ^{19}F NMR, using F2Y as the probe. Source: Yang et al. [46]. CC BY 4.0.

mechanism that regulates electron/proton transfer between α/β subunit of the protein [50].

While NMR and EPR spectroscopy provide sensitive means to probe protein structural change or reaction mechanisms, they often require extraction and purification of protein before data collection and sophisticated data interpretation process. Fluorescent spectroscopy is more labile for study protein function inside a cell. L-(7-hydroxycoumarin-4-yl) ethylglycine, which has a hydroxycoumarin sidechain, is the first fluorescent amino acid incorporated into a protein through genetic codon expansion [51]. The dansyl and prodan fluorophores are also genetically encoded in yeast and mammalian cells [52, 53]. These UAAs provide a handle to probe protein localization and dynamics, as well as to study the local environment of the amino acid residue using fluorescence microscopy. For example, the hydroxycoumarin-containing UAA was incorporated into the phosphotyrosine binding pocket of STAT3. Phosphorylation of STAT3 results in fluorescence increase of the UAA, making it a biosensor for STAT3 phosphorylation status [54]. In addition to the UAAs with fluorescence, those with bioorthogonal reaction groups could attach a fluorophore after the reaction. Cyclopropene or other alkene-bearing UAAs could perform photoclick reaction with tetrazole, forming fluorescent adducts [55, 56]. Such reactions provide a facile approach to study protein dynamics using fluorescence spectroscopy.

10.3.3 Tuning Catalytic Activity

Structural and mechanistic studies lead to the identification of key amino acid residues, which is crucial for catalysis and other protein functions. Altering these key residues is an efficient way for tuning protein activity, but it needs fine balancing: the new amino acids should be different from the native ones to cause activity change, and yet they should resemble the structures of the native ones to ensure minimal perturbation to the protein. With the diversity in structure and chemical properties, UAAs provide a unique advantage for such a task.

Being a redox-active residue, the Tyr residue in the heme-copper oxidase (HCO) is essential for its activity, as mutating it to Phe resulted in a non-functional enzyme. The function of a conserved Tyr residue in the HCO was studied in an engineered myoglobin (Mb)-based oxidase [57, 58]. The Tyr residue in the Mb-based oxidase is replaced with a series Tyr analogues, which have similar structure but different pK_a values and redox potentials. Activities of the mutants are inversely correlated with the pK_a values of the Tyr analogs. The active role of Tyr in catalysis is further confirmed by trapping of a reaction intermediate containing a tyrosyl radical or the radical of UAAs replacing the Tyr [57].

A Tyr residue in the active site of ketosteroid isomerase forms hydrogen bonds with the substrate and the reaction intermediate, altering the electrostatic field and contributing to the catalysis. Replacing the Tyr with 3-chlorotyrosine and other UAAs resulted in the change of electrostatic field, as measured by vibrational Stark spectroscopy [59, 60].

Due to the neutral pK_a value and metal coordination ability of imidazole sidechain, histidine often appears in the active site of enzymes for acid–base catalysis, forms hydrogen bonds with other residues, relays proton, or works as ligands in metalloproteins. These unique properties of His make it hard to replace with other natural amino acids. Various His analogues have been incorporated into protein through the genetic codon expansion method [61]. One of them, 3-methylhistidine (3-MeHis or Nδ-methylhistidine, NMH) has an imidazole sidechain with a methylated δN. As a result, it cannot form hydrogen bond with other residues, and 3-MeHis is forced to coordinate with metal through εN. In peroxidases, there is a conserved Asp residue, forming hydrogen bond with the heme-coordinating His. Such interaction “pushes” electrons to the distal site, facilitating the heterolytic cleavage of the O-O bond. Probing the effect of the hydrogen bond has not been successful since replacing either His or Asp with other natural amino acids resulted in an inactive protein. 3-MeHis, being structurally similar to His, can replace the native proximal His in ascorbate peroxidase (APX) [62]. The APX mutant with 3-MeHis showed up to fivefold increased total turnover number compared with its native counterpart.

Cys and Met are sulfur-containing amino acids. They often function in metalloproteins as metal ligands and cannot be replaced by other natural amino acids. As the element below sulfur in the periodic table, selenium has similar chemical properties. Selenocysteine (Sec) and selenomethionine (SeM) can replace their sulfur counterparts in protein without disturbing metal coordination. Sec naturally occurs in protein so it is regarded as the twenty-first natural amino acid, but genetically encoding Sec requires extra genetic elements and other enzymes. By adding selenocysteine insertion sequence and the UGA codon in the target gene, proximal Cys in cytochrome P450 (P450cam) was replaced by Sec [63]. Due to the stronger electron-donating ability of selenolate moiety, compared with the thiolate, the catalytic activity was decreased twofold and the electron transfer and catalysis were partially uncoupled [63]. The perturbation of electronic structure by Sec substitution is further utilized to characterized compound I of P450 [64].

In a copper-containing electron transfer protein, azurin, copper coordination environment directly influences its redox potential and other properties. By replacing the axial methionine with various isostructural UAAs using EPL method, the reduction potential of azurin was systematically tuned over a range of 200 mV [65, 66]. Furthermore, by introducing a strong ligand with similar size as methionine, the thiol-containing homocysteine UAA, at the same axial position, the blue copper protein azurin was converted to a red copper protein [67].

10.3.4 Tuning Selectivity

The hallmark of protein-based catalysts is their great selectivity in reaction, including enantio-, regio-, and stereoselectivity. Tuning these selectivities is an important topic for both academic and industrial researches. Mutagenesis using UAAs can greatly expand structural variability of protein, making them feasible for selectivity tuning.

Tyrosine analogs 4-amino-phenylalanine, 4-acetyl-phenylalanine, 2-benzyl-tyrosine, and 3-(2-napthyl)alanine were introduced into Cytochrome P450 (CYP102A1), which converts (S)-ibuprofen methylester into benzylic alcohol (62%) and allylic alcohol (38%) derivatives. One of the variants with 2-benzyl-tyrosine replacing Leu181 showed inverted selectivity with products of 15% benzylic alcohol and 85% allylic alcohol derivatives. Another variant with 3-(2-napthyl)alanine replacing Ala32 showed even higher selectivity towards allylic alcohol derivatives (95%) [68].

Diketoreductase reduces a variety of ketones to chiral alcohols. In its catalysis, Trp222 is the key residue for enantioselectivity control [69]. The residue is replaced with natural amino acids (Val, Leu, Met, Phe, and Tyr) or UAAs (4-cyano-L-phenylalanine, 4-methoxy-L-phenylalanine, 4-phenyl-L-phenylalanine, O-tert-butyl-L-tyrosine). A correlation of the size of the residue at the position and the enantio-preference is established. Variant with O-tert-butyl-L-tyrosine, which is bulkier than Trp, showed higher ee value, while the variant with 4-cyano-L-phenylalanine showed inverted enantioselectivity for R-isomer.

10.3.5 Enzyme Design

Designing an enzyme from scratch or repurposing an existing protein to one with new catalytic activity test our knowledge of the protein. UAAs can be used in the process to create certain structure features or to anchor cofactors for catalysis.

Native enzymes often use post-translationally modified residues for catalysis. These residues are hard to install in an artificial enzyme. UAAs can be used to mimic the structure of the post-translationally modified residues in an artificial enzyme. HCOs harbor a pair of covalently attached His and Tyr residues in the active site. The structure feature is considered key to the catalysis as it tunes the pK_a of the phenol group and positions the motif for copper binding. After studying the His-Tyr cross-link motif and mimicking it in a small molecule model complex [70], installing the feature into a model protein is crucial for mimicking and investigating the oxidase activity. A UAA, 2-amino-3-(4-hydroxy-3-(1H-imidazol-1-yl)phenyl)propanoic acid mimics the Tyr-His cross-link. Such UAA was incorporated into a redesigned myoglobin through genetic code expansion [71]. Together with other mutations, the myoglobin-based model (imiTyrCuBMb) recapitulated the structural features of the HCOs, and exhibited an oxygen consumption rate of 2.2 min^{-1}, with less than 6% converted to ROS.

Tyr also forms cross-link with Cys between the C3 ring carbon of Tyr and Sγ of Cys. Such cross-link appears in galactose oxidase, glyoxal oxidase, cysteine dioxygenase, sulfite reductase, and cytochrome *c* nitrite reductase. The study of model compounds indicated that the cross-link lowers the pK_a value and the reduction potential of Tyr, facilitating enzymatic reactions [72]. 2-amino-3-(4-hydroxy-3-(methylthio)phenyl)propanoic acid (MtTyr) mimics the Tyr-S bond in the cross-link [73]. The UAA MtTyr was incorporated into myoglobin through genetic code expansion, to mimic the active site of cytochrome *c* nitrite reductase, which has a heme and Tyr-Cys cross-link. The designed protein could reduce hydroxylamine, an intermediate in nitrite reduction at a rate of 800 min^{-1}, a reaction nitrite reductase performs.

As mentioned in Section 10.3.3, UAAs can be used to fine-tune certain properties of the residue without changing its structure. In addition to altering an existing enzyme's activity or selectivity, such UAAs can be used to create new activities in a protein scaffold. As the oxygen carrier, myoglobin exhibits some peroxidase activity. Replacing its proximal His with a His analogue, 3-methylhistidine (3-MeHis, NMH) causes minimum structure perturbation, but the methylation blocks hydrogen bonding between His and Ser92, causing the heme reduction potential to rise from 65 to 139 mV. The substitution leads to a 3.7-fold increase in the peroxidase activity of Mb. Based on the mutation, directed evolution further increased the peroxidase activity by 10-fold, leading to a k_{cat}/K_M value of 1.5×10^6 M^{-1} s^{-1} toward Amplex Red oxidation. Combining UAA incorporation and directed evolution, the engineered Mb surpasses the peroxidase activity of the native ascorbate oxidase (Figure 10.3) [74]. Replacing the proximal His with 3-MeHis also increases electrophilicity of the heme Fe. The heme Fe could then form a reactive carbenoid adduct for cyclopropanation reaction [75]. Heme enzymes catalyzing the same type of carbene transfer reactions have been designed through directed evolution and metal cofactor replacement [76, 77], showing that we can use different protein engineering strategies for the same target reaction.

There are many cases where protein engineering is achieved through modifying residues close to a cofactor, such as metal ion, metal complex, or organic molecules. Due to the difficulty in designing a high-affinity site for a certain cofactor, *de novo* design of a cofactor binding site in protein poses a great challenge for protein engineering. Such a task can be achieved through the incorporation of UAAs, either as a high-affinity metal ligand, or as a carrier for bio-orthogonal reaction groups.

Metal coordinating amino acids, such as Cys, Met, His, usually interacts with metal in a mono-dentate fashion. Designing a new metal site needs to precisely

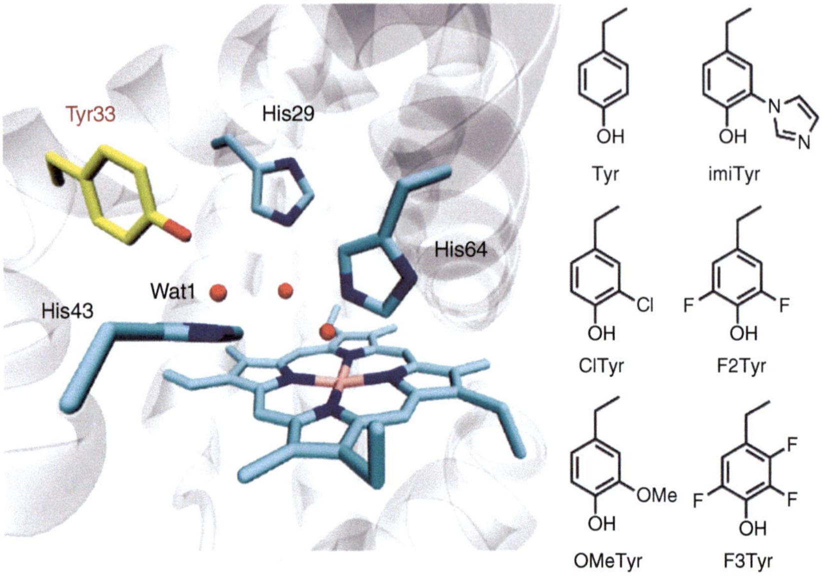

Figure 10.3 Unnatural amino acids used to probe the role of Tyr in an artificial oxidase. Source: Yu et al. [23]. © 2018, American Chemical Society.

position 3 or more ligands in the protein scaffold, which is a challenging task. Some UAAs have bipyridine, hydroxyquinoline, or pyrazolylphenol groups as side chains and could coordinate metal ions in a bidentate fashion with orders of magnitude higher affinity than the natural ones. Such UAAs alone can anchor metal ions in protein with reasonable affinity, without the need to design additional ligands at the site. For example, 2-amino-3-(8-hydroxyquinolin-5-yl)propanoic acid binds Cu(II) with a K_D of 0.1 fM, which is tighter than many native copper sites [78].

Bipyridylalanine (BpyAla) has bipyridine as the sidechain, a bidentate ligand often used in inorganic complexes. BpyAla was incorporated into *E. coli* catabolite activator proteins (CAP), a double-strand DNA binding protein. In the presence of a reductant, Cu(II) in the protein is able to cleave the bound dsDNA [79]. BpyAla was also introduced at the dimer interface of Lactococcus multidrug resistance regulator (LmrR) [80]. Upon binding of a copper ion, the engineered LmrR protein catalyzes Friedel–Crafts alkylation reaction with up to 83% ee and 94% conversion of 1 mM substrate over three days. In another study, computational design combining cluster model calculations (quantum mechanics), protein–ligand docking and molecular dynamics simulations enable precise placement of BpyAla in LmrR [81]. The designed enzyme is able to catalyze the hydration of α,β-unsaturated 2-acyl pyridine with up to 64% ee.

In addition to metal ions, larger cofactors can also be anchored in a protein through bioconjugation. To date, a series of reactions that can happen in living systems yet not interfere with native biochemical reactions have been developed, including Staudinger ligation, copper(I)-catalyzed or copper-free alkyne–azide cycloaddition, inverse-electron-demand Diels–Alder (IEDDA) reactions between tetrazines and strained alkenes, "photo-click" chemistry between tetrazoles and alkenes, etc. UAAs with functional groups for these reactions, such as 4-azido-phenylalanine (AzF), *p*-propargyloxyphenylalanine, and *N*-ε-acryllysine, can be incorporated into a specific site of a protein, enabling site-specific bioconjugation of cofactors. AzF was introduced into tHisF and phytase for anchoring of Rh_2-tetraacetate and Mn- and Cu-terpyridine complexes through strain-promoted azide–alkyne cycloaddition [82]. The same UAA was used to anchor a dirhodium tetracarboxylate complex into a thermostable prolyl oligopeptidase, after mutations of the protein scaffold to make room for the cofactor [83]. The designed enzyme with the dirhodium cofactor can catalyze enantioselective cyclopropanation of a broad range of olefins.

UAAs can be used to design proteins with novel photochemical properties. Nature develops photosystems to harvest and store photon energy. Through a set of orchestrated photochemical reactions, photon energy is converted to electron flow through photo-induced electron transfer and then stored as chemical energy through CO_2 fixation. The photo-induced electron transfer can be realized in an artificial enzyme by placing a redox center close to the fluorophore in a fluorescent protein. The redox center can be installed either by using UAAs (HqAla or PyTyr) as metal ligands to anchor redox-active metal ions [78, 84] or directly using a redox-active amino acid (4-fluoro-3-nitrophenylalanine, FNO_2Phe) [85]. Femtosecond transient absorption spectroscopy showed that the GFP149FNO_2Phe mutant has a photo-induced electron transfer rate of $(9.09 \pm 0.45) \times 10^{10}\ s^{-1}$, which is faster than the electron transfer (ET) rate between P700* and A0 in photosystem I [85]. To harvest light energy and convert it to chemical energy, a photosystem-mimicking enzyme needs long-lived excited state and charge-separated state, and a redox center with low reduction potential. Benzophenone has near 100% quantum efficiency in intersystem crossing from singlet excited state to triplet excited state. A UAA, benzophenone–alanine was introduced into the Tyr66 position of superfolder yellow fluorescent protein (sfYFP) to yield photosensitizer protein (PSP). The lifetime of the excited state of the sfYFP mutant increases 10^5-fold in comparison with that of the wild type protein. Moreover, the benzophenone anion radical has low redox potential, making the redox potential of the PSP lower than -1.14 V, greatly exceeding the lower limit for natural redox centers. Covalent attaching a Ni-terpyridine complex to PSP enables photo-catalyzed CO_2 reduction to CO with a 0.28% quantum yield [86]. The Ni-terpyridine-attached PSP protein successfully recapitulates the light-harvest and carbon fixation feature of the photosynthesis apparatus, suggesting the great potential of UAAs in enzyme design.

10.3.6 Protein Engineering Toward a Synthetic Life

UAA incorporation provides novel methods to alter protein functions, or even create new functions. Due to the central role of proteins in many cellular activities, encoding UAAs in a protein allows us to decode biological processes, or even create new forms of life.

One of the key features of biological processes is chirality. Except for glycine, which does not have a chiral center, amino acids in a protein are all L-enantiomer. Ribose in DNA and RNA is D-enantiomer, while L-form is not found in nature. There are questions about how chirality is evolved on earth and whether a mirror-image system exists. A D-DNA polymerase (African swine fever virus polymerase X), DNA polymerase composed of D-amino acids and glycine is assembled through peptide synthesis and native chemical ligation [87]. The D-DNA polymerase uses L-DNA as a template to replicate and transcribe the corresponding L-DNA and L-RNA strand, in an orthogonal fashion [88]. Moreover, an L-DNAzyme produced in the polymerase-catalyzed reaction showed the expected activity. The discovery points out a possibility to create mirror-image bio-molecules to perform basic biological functions or even forming a self-sustaining cell.

Biocontainment aims to confine harmful organisms in controlled laboratory conditions. In addition to current strategies including auxotrophy, inducible expression systems, and safety circuits, it is possible to construct synthetic auxotroph dependent on UAAs for biocontainment. In a recent study, six key enzymes in *E. coli* were redesigned to accommodate a bulky UAA, biphenylalaine [89]. The engineered *E. coli* could only survive in the presence of the UAA. Escape frequency is below the detection limit of 10^{-10}. Similarly, a synthetic virus was designed that its replication in the cell is dependent on the UAA. Stop codons were introduced into the genes encoding four structural proteins. The addition of UAA, N^{ε}-2-azidoethyloxycarbonyl-L-lysine allows the ribosome to read through the mRNA to generate a full-length protein. Otherwise, the premature termination happens and truncated proteins are synthesized, which prevents viral replication. The engineered virus can only replicate in the controlled environment supplemented with the UAA to generate virus with similar immune response, but cannot replicate in the host animal, which makes vaccination safer [90].

UAA incorporation through genetic code expansion relies on codon reassignment, since all 64 possible combinations of triplet codon are occupied. A pair of unnatural bases, d5SICS and dNaM, can be recognized by DNA and RNA polymerases. The introduction of a nucleoside triphosphate transporter into *E. coli* enables the cell to utilize the unnatural bases in DNA replication and transcription process [91]. Further engineering of the cellular translational machinery enables encoding UAAs with unnatural bases [92]. In theory, the introduction of the new base pair adds 61 unassigned codons, greatly enhancing the ability to expand genetic codon. Other techniques, such as chromosome synthesis and genetic editing, also enable reassignment of the genetic codon in a whole genome scale, freeing several codons for UAA incorporation [93, 94].

10.4 Outlook

So far, more than 200 UAAs can be incorporated into a protein through genetic codon expansion, and the number is still growing [95]. These UAAs cover a large chemical space and a variety of functionalities. It is possible to make a library of standardized parts for all UAAs and to realize different functions in a given protein in a plug-and-play fashion.

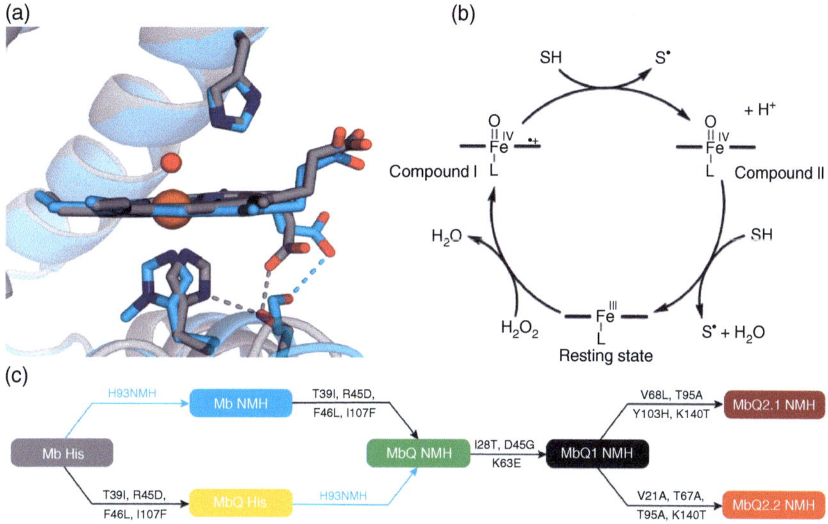

Figure 10.4 (a) Overlay of the active sites of Mb His (gray, PDB ID: 1A6K) and Mb NMH (PDB ID: 5OJ9). (b) Catalytic cycle of the peroxidase-catalyzed substrate (S) oxidation. (c) Directed evolution toward an engineered enzyme surpassing native enzyme's activity. Source: Pott et al. [74]. CC BY 4.0.

As reviewed in Chapters 1–7 of this book, there are many other methods for protein engineering, including directed evolution, high-throughput screening, and different computational methods. These methods can be used in combination with UAAs to create the desired protein.

Directed evolution mimics the natural selection process to evolve the target protein for certain purposes. Although it is technically challenging to build a mutant library with different UAAs, the directed evolution of a protein with a certain UAA at a fixed position is effective. As shown in Section 2.5 (Figure 10.4), UAA (3-MeHis) incorporation gave the target protein initial peroxidase activity. After that, several rounds of directed evolution further boosted that activity to surpass that of the native enzyme [74].

Surface display methods display protein variants that are subjected to screening and manipulation. They can be adapted for high throughput screening. Screening proteins with UAAs requires mutating the corresponding gene with a TAG codon and supplementing the growth medium with the corresponding UAA. BpyAla was introduced into a disulfide-linked peptide library and the library was subjected to phage display. A BpyAla-containing peptide with 0.34 μM affinity to Fe(II) was obtained after three rounds of selection [96] (Figure 10.5).

Rosetta is a computational software suite used for protein engineering, including enzyme *de novo* design, antibody design, protein and peptide design, etc. Most work in Rosetta focuses on using natural amino acids, the parameterization of UAAs is lacking. BpyAla, a metal coordinating UAA, is first parameterized in Rosetta. A metal site with BpyAla was constructed and installed in a host protein

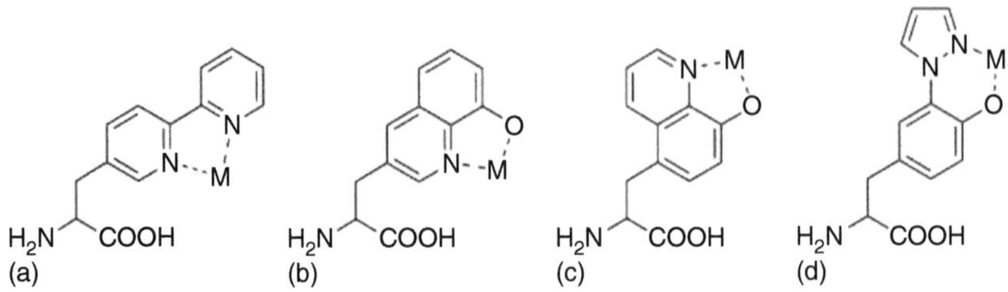

Figure 10.5 Some metal-coordinating UAAs and their potential coordinating mode. (a) (2,2-bipyridin-5yl)-alanine (Bpy-Ala); (b) (8-hydroxyquinolin-3-yl)alanine; (c) 2-amino-3-(8-hydroxyquinolin-5-yl)propionic acid (HqAla); (d) 2-amino-3-[4-hydroxy-3-(1*H*-pyrazol-1-yl)phenyl]propionic acid (PyTyr).

using Rosetta, and similar to proteins with natural amino acids, the experimentally determined structures matched the designed ones [97]. This work demonstrates the ability of Rosetta to design proteins with UAAs. Using a similar strategy, a Cu(II) coordinating site with BpyAla in LmrR was designed and constructed. The artificial enzyme can catalyze the hydration of α,β-unsaturated 2-acyl pyridine, as discussed in Section 2.5 [81].

10.5 Conclusions

As an expanded toolbox for protein engineering, UAA incorporation has been used in enhancing protein stability, probing mechanism, tuning catalytic activity, tuning selectivity, designing enzymes, and even engineering synthetic life. UAA incorporation is becoming a more and more standard practice for protein engineers, with modulated components for incorporation and an ever-growing selection of UAAs with different functional groups and utilities. For example, tRNA and aaRS of a particular UAA are usually integrated on one plasmid, UAA incorporation through genetic codon expansion usually only requires introducing a TAG codon at a specific site on target gene and supplement protein expression medium with the corresponding UAA. Many functionalities, including bio-orthogonal reaction, photoreactivity, spectroscopic probe, metal-chelating, and post-translational modification mimicking, have been realized in UAAs. It is possible to modularize these functionalities: plug-in a certain UAA and realize the function in the protein. In the future, UAA incorporation will play an important role in protein engineering, and greatly expand what the already very powerful molecules are capable of.

References

1 Baltzer, L., Nilsson, H., and Nilsson, J. (2001). De Novo design of proteins – what are the rules? *Chem. Rev.* 101: 3153–3164.

2 Ueno, T., Abe, S., Yokoi, N., and Watanabe, Y. (2007). Coordination design of artificial metalloproteins utilizing protein vacant space. *Coord. Chem. Rev.* 251: 2717–2731.

3 Lu, Y., Yeung, N., Sieracki, N., and Marshall, N.M. (2009). Design of functional metalloproteins. *Nature* 460: 855.

4 Bornscheuer, U.T., Huisman, G.W., Kazlauskas, R.J. et al. (2012). Engineering the third wave of biocatalysis. *Nature* 485: 185.

5 Reetz, M.T. (2013). The importance of additive and non-additive mutational effects in protein engineering. *Angew. Chem. Int. Ed.* 52: 2658–2666.

6 Kiss, G., Çelebi-Ölçüm, N., Moretti, R. et al. (2013). Computational enzyme design. *Angew. Chem. Int. Ed.* 52: 5700.

7 Reetz, M.T. (2013). Biocatalysis in organic chemistry and biotechnology: past, present, and future. *J. Am. Chem. Soc.* 135: 12480.

8 Yu, F., Cangelosi, V.M., Zastrow, M.L. et al. (2014). Protein design: toward functional metalloenzymes. *Chem. Rev.* 114: 3495–3578.

9 Korendovych, I.V. and DeGrado, W.F. (2014). Catalytic efficiency of designed catalytic proteins. *Curr. Opin. Struct. Biol.* 27: 113.

10 Hayashi, T., Sano, Y., and Onoda, A. (2015). Generation of new artificial metalloproteins by cofactor modification of native hemoproteins. *Isr. J. Chem.* 55: 76–84.

11 Lin, Y.-W. (2017). Rational design of metalloenzymes: from single to multiple active sites. *Coord. Chem. Rev.* 336: 1–27.

12 Fried, S.D. and Boxer, S.G. (2017). Electric fields and enzyme catalysis. *Annu. Rev. Biochem.* 86: 387–415.

13 Brandenberg, O.F., Fasan, R., and Arnold, F.H. (2017). Exploiting and engineering hemoproteins for abiological carbene and nitrene transfer reactions. *Curr. Opin. Biotechnol.* 47: 102–111.

14 Schwizer, F., Okamoto, Y., Heinisch, T. et al. (2018). Artificial metalloenzymes: reaction scope and optimization strategies. *Chem. Rev.* 118 (1): 142–231.

15 Liu, C.C. and Schultz, P.G. (2010). Adding new chemistries to the genetic code. *Annu. Rev. Biochem.* 79: 413–444.

16 Hu, C., Chan, S.I., Sawyer, E.B. et al. (2014). Metalloprotein design using genetic code expansion. *Chem. Soc. Rev.* 43: 6498–6510.

17 Chin, J.W. (2014). Expanding and reprogramming the genetic code of cells and animals. *Annu. Rev. Biochem.* 83: 379–408.

18 Yu, Y., Cui, C., Wang, J., and Lu, Y. (2017). Biosynthetic approach to modeling and understanding metalloproteins using unnatural amino acids. *Sci. China Chem.* 60: 188–200.

19 Hu, C., Yu, Y., and Wang, J. (2017). Improving artificial metalloenzymes' activity by optimizing electron transfer. *Chem. Commun.* 53: 4173–4186.

20 Mukai, T., Lajoie, M.J., Englert, M., and Söll, D. (2017). Rewriting the genetic code. *Annu. Rev. Microbiol.* 71: 557–577.

21 Wang, L. (2017). Engineering the genetic code in cells and animals: biological considerations and impacts. *Acc. Chem. Res.* 50: 2767–2775.

22 Agostini, F., Völler, J.-S., Koksch, B. et al. (2017). Biocatalysis with unnatural amino acids: enzymology meets xenobiology. *Angew. Chem. Int. Ed.* 56: 9680–9703.

23 Yu, Y., Hu, C., Xia, L., and Wang, J. (2018). Artificial metalloenzyme design with unnatural amino acids and non-native cofactors. *ACS Catal.* 8: 1851–1863.

24 Kent, S.B.H. (2009). Total chemical synthesis of proteins. *Chem. Soc. Rev.* 38: 338–351.

25 Muir, T.W., Sondhi, D., and Cole, P.A. (1998). Expressed protein ligation: a general method for protein engineering. *Proc. Natl. Acad. Sci. USA* 95: 6705.

26 Evans, T.C. Jr., Benner, J., and Xu, M.-Q. (1998). Semisynthesis of cytotoxic proteins using a modified protein splicing element. *Protein Sci.* 7: 2256.

27 Ayers, B., Blaschke, U.K., Camarero, J.A. et al. (1999). Introduction of unnatural amino acids into proteins using expressed protein ligation. *Biopolymers* 51: 343.

28 Wu, Z.P. and Hilvert, D. (1989). Conversion of a protease into an acyl transferase: selenolsubtilisin. *J. Am. Chem. Soc.* 111: 4513–4514.

29 Warren, J.J., Herrera, N., Hill, M.G. et al. (2013). Electron flow through nitrotyrosinate in *Pseudomonas aeruginosa* Azurin. *J. Am. Chem. Soc.* 135: 11151–11158.

30 Young, R. and Bremer, H. (1976). Polypeptide-chain-elongation rate in *Escherichia coli* B/r as a function of growth rate. *Biochem. J.* 160: 185–194.

31 Rosenberger, R.F. and Hilton, J. (1983). The frequency of transcriptional and translational errors at nonsense codons in the *lacZ* gene of *Escherichia coli*. *Mol. Gen. Genet.* 191: 207–212.

32 Goto, Y., Katoh, T., and Suga, H. (2011). Flexizymes for genetic code reprogramming. *Nat. Protoc.* 6: 779.

33 Murakami, H., Saito, H., and Suga, H. (2003). A versatile tRNA aminoacylation catalyst based on RNA. *Chem. Biol.* 10: 655.

34 Budisa, N., Wenger, W., and Wiltschi, B. (2010). Residue-specific global fluorination of Candida antarctica lipase B in Pichia pastoris. *Mol. Biosyst.* 6: 1630–1639.

35 Hoesl, M.G., Acevedo-Rocha, C.G., Nehring, S. et al. (2011). Lipase congeners designed by genetic code engineering. *ChemCatChem* 3: 213–221.

36 Baker, P.J. and Montclare, J.K. (2011). Enhanced refoldability and thermoactivity of fluorinated phosphotriesterase. *ChemBioChem* 12: 1845–1848.

37 Votchitseva, Y.A., Efremenko, E.N., and Varfolomeyev, S.D. (2006). Insertion of an unnatural amino acid into the protein structure: preparation and properties of 3-fluorotyrosine-containing organophosphate hydrolase. *Russ. Chem. Bull.* 55: 369–374.

38 Tatyana, P., Wen, Z.W., and Kim, M.J. (2006). Influence of global fluorination on chloramphenicol acetyltransferase activity and stability. *Biotechnol. Bioeng.* 94: 921–930.

39 Dominguez, M.A., Thornton, K.C., Melendez, M.G., and Dupureur, C.M. (2001). Differential effects of isomeric incorporation of fluorophenylalanines into *Pvu*II endonuclease. *Proteins J.* 45: 55–61.

40 Gray, H.B. and Winkler, J.R. (2015). Hole hopping through tyrosine/tryptophan chains protects proteins from oxidative damage. *Proc. Natl. Acad. Sci. USA* 112: 10920–10925.

41 Cirino, P.C., Tang, Y., Takahashi, K. et al. (2003). Global incorporation of norleucine in place of methionine in cytochrome P450 BM-3 heme domain increases peroxygenase activity. *Biotechnol. Bioeng.* 83: 729–734.

42 Gilles, A.M., Marlière, P., Rose, T. et al. (1988). Conservative replacement of methionine by norleucine in *Escherichia coli* adenylate kinase. *J. Biol. Chem.* 263: 8204–8209.

43 Ai, H.-w. (2012). Biochemical analysis with the expanded genetic lexicon. *Anal. Bioanal.Chem.* 403: 2089–2102.

44 Oyala, P.H., Ravichandran, K.R., Funk, M.A. et al. (2016). Biophysical characterization of fluorotyrosine probes site-specifically incorporated into enzymes: *E. coli* ribonucleotide reductase as an example. *J. Am. Chem. Soc.* 138: 7951–7964.

45 Li, F., Shi, P., Li, J. et al. (2013). A genetically encoded 19F NMR probe for tyrosine phosphorylation. *Angew. Chem. Int. Ed.* 52: 3958–3962.

46 Yang, F., Yu, X., Liu, C. et al. (2015). Phospho-selective mechanisms of arrestin conformations and functions revealed by unnatural amino acid incorporation and 19F-NMR. *Nat. Commun.* 6: 8202.

47 Yang, F., Xiao, P., Qu, C.-x. et al. (2018). Allosteric mechanisms underlie GPCR signaling to SH_3-domain proteins through arrestin. *Nat. Chem. Biol.* 14: 876–886.

48 Cotruvo, J.A. and Stubbe, J. (2011). Class I ribonucleotide reductases: metallocofactor assembly and repair in vitro and in vivo. *Annu. Rev. Biochem.* 80: 733–767.

49 Minnihan, E.C., Nocera, D.G., and Stubbe, J. (2013). Reversible, long-range radical transfer in *E. coli* class Ia ribonucleotide reductase. *Acc. Chem. Res.* 46: 2524–2535.

50 Ravichandran, K.R., Taguchi, A.T., Wei, Y. et al. (2016). A >200 meV uphill thermodynamic landscape for radical transport in *Escherichia Coli* ribonucleotide reductase determined using fluorotyrosine-substituted enzymes. *J. Am. Chem. Soc.* 138: 13706–13716.

51 Wang, J., Xie, J., and Schultz, P.G. (2006). A genetically encoded fluorescent amino acid. *J. Am. Chem. Soc.* 128: 8738–8739.

52 Summerer, D., Chen, S., Wu, N. et al. (2006). A genetically encoded fluorescent amino acid. *Proc. Natl. Acad. Sci. USA* 103: 9785–9789.

53 Lee, H.S., Guo, J., Lemke, E.A. et al. (2009). Genetic incorporation of a small, environmentally sensitive, fluorescent probe into proteins in *Saccharomyces cerevisiae*. *J. Am. Chem. Soc.* 131: 12921–12923.

54 Lacey, V.K., Parrish, A.R., Han, S. et al. (2011). A fluorescent reporter of the phosphorylation status of the substrate protein STAT3. *Angew. Chem. Int. Ed.* 50: 8692–8696.

55 Yu, Z., Pan, Y., Wang, Z. et al. (2012). Genetically encoded cyclopropene directs rapid, photoclick-chemistry-mediated protein labeling in mammalian cells. *Angew. Chem. Int. Ed.* 51: 10600–10604.

56 Li, F., Zhang, H., Sun, Y. et al. (2013). Expanding the genetic code for photoclick chemistry in *E. coli*, mammalian cells, and *A. thaliana*. *Angew. Chem. Int. Ed.* 52: 9700–9704.

57 Yu, Y., Lv, X., Li, J. et al. (2015). Defining the role of tyrosine and rational tuning of oxidase activity by genetic incorporation of unnatural tyrosine analogs. *J. Am. Chem. Soc.* 137: 4594–4597.

58 Yu, Y., Zhou, Q., Wang, L. et al. (2015). Significant improvement of oxidase activity through the genetic incorporation of a redox-active unnatural amino acid. *Chem. Sci.* 6: 3881.

59 Natarajan, A., Schwans, J.P., and Herschlag, D. (2014). Using unnatural amino acids to probe the energetics of oxyanion hole hydrogen bonds in the ketosteroid isomerase active site. *J. Am. Chem. Soc.* 136: 7643–7654.

60 Wu, Y. and Boxer, S.G. (2016). A critical test of the electrostatic contribution to catalysis with noncanonical amino acids in ketosteroid isomerase. *J. Am. Chem. Soc.* 138: 11890–11895.

61 Xiao, H., Peters, F.B., Yang, P.Y. et al. (2014). Genetic incorporation of histidine derivatives using an engineered pyrrolysyl-tRNA synthetase. *ACS Chem. Biol.* 9: 1092–1096.

62 Green, A.P., Hayashi, T., Mittl, P.R.E., and Hilvert, D. (2016). A chemically programmed proximal ligand enhances the catalytic properties of a heme enzyme. *J. Am. Chem. Soc.* 138: 11344–11352.

63 Aldag, C., Gromov, I.A., García-Rubio, I. et al. (2009). Probing the role of the proximal heme ligand in cytochrome P450cam by recombinant incorporation of selenocysteine. *Proc. Natl. Acad. Sci. USA* 106: 5481–5486.

64 Onderko, E.L., Silakov, A., Yosca, T.H., and Green, M.T. (2017). Characterization of a selenocysteine-ligated P450 compound I reveals direct link between electron donation and reactivity. *Nat. Chem.* 9: 623.

65 Berry, S.M., Ralle, M., Low, D.W. et al. (2003). Probing the role of axial methionine in the blue copper center of azurin with unnatural amino acids. *J. Am. Chem. Soc.* 125: 8760.

66 Garner, D.K., Vaughan, M.D., Hwang, H.J. et al. (2006). Reduction potential tuning of the blue copper center in *Pseudomonas aeruginosa* azurin by the axial methionine as probed by unnatural amino acids. *J. Am. Chem. Soc.* 128: 15608.

67 Clark, K.M., Yu, Y., Marshall, N.M. et al. (2010). Transforming a blue copper into a red copper protein: engineering cysteine and homocysteine into the axial position of azurin using site-directed mutagenesis and expressed protein ligation. *J. Am. Chem. Soc.* 132: 10093.

68 Kolev Joshua, N., Zaengle Jacqueline, M., Ravikumar, R., and Fasan, R. (2014). Enhancing the efficiency and regioselectivity of P450 oxidation catalysts by unnatural amino acid mutagenesis. *ChemBioChem* 15: 1001–1010.

69 Ma, H., Yang, X., Lu, Z. et al. (2014). The "Gate Keeper" role of Trp222 determines the enantiopreference of diketoreductase toward 2-chloro-1-phenylethanone. *PLoS One* 9: 2403–2404.

70 McCauley, K.M., Vrtis, J.M., Dupont, J., and van der Donk, W.A. (2000). Insights into the functional role of the tyrosine–histidine linkage in cytochrome C oxidase. *J. Am. Chem. Soc.* 122: 2403–2404.

71 Liu, X., Yu, Y., Hu, C. et al. (2012). Significant increase of oxidase activity through the genetic incorporation of a tyrosine-histidine cross-link in a myoglobin model of heme-copper oxidase. *Angew. Chem. Int. Ed.* 51: 4312–4316.

72 Verma, P., Pratt, R.C., Storr, T. et al. (2011). Sulfanyl stabilization of copper-bonded phenoxyls in model complexes and galactose oxidase. *Proc. Natl. Acad. Sci. USA* 108: 18600–18605.

73 Zhou, Q., Hu, M., Zhang, W. et al. (2013). Probing the function of the Tyr-Cys cross-link in metalloenzymes by the genetic incorporation of 3-methylthiotyrosine. *Angew. Chem. Int. Ed.* 52: 1203–1207.

74 Pott, M., Hayashi, T., Mori, T. et al. (2018). A noncanonical proximal heme ligand affords an efficient peroxidase in a globin fold. *J. Am. Chem. Soc.* 140: 1535–1543.

75 Hayashi, T., Tinzl, M., Mori, T. et al. (2018). Capture and characterization of a reactive haem–carbenoid complex in an artificial metalloenzyme. *Nat. Catal.* 1: 578–584.

76 Coelho, P.S., Brustad, E.M., Kannan, A., and Arnold, F.H. (2013). Olefin cyclopropanation via carbene transfer catalyzed by engineered cytochrome P450 enzymes. *Science* 339: 307.

77 Key, H.M., Dydio, P., Clark, D.S., and Hartwig, J.F. (2016). A biological catalysis by artificial haem proteins containing noble metals in place of iron. *Nature* 534: 534.

78 Liu, X., Li, J., Hu, C. et al. (2013). Significant expansion of the fluorescent protein chromophore through the genetic incorporation of a metal-chelating unnatural amino acid. *Angew. Chem. Int. Ed.* 52: 4805–4809.

79 Lee, H.S. and Schultz, P.G. (2008). Biosynthesis of a site-specific DNA cleaving protein. *J. Am. Chem. Soc.* 130: 13194–13195.

80 Drienovska, I., Rioz-Martinez, A., Draksharapu, A., and Roelfes, G. (2015). Novel artificial metalloenzymes by in vivo incorporation of metal-binding unnatural amino acids. *Chem. Sci.* 6: 770–776.

81 Drienovska, I., Alonso-Cotchico, L., Vidossich, P. et al. (2017). Design of an enantioselective artificial metallo-hydratase enzyme containing an unnatural metal-binding amino acid. *Chem. Sci.* 8: 7228–7235.

82 Yang, H., Srivastava, P., Zhang, C., and Lewis, J.C. (2014). A general method for artificial metalloenzyme formation through strain-promoted azide–alkyne cycloaddition. *ChemBioChem* 15: 223–227.

83 Srivastava, P., Yang, H., Ellis-Guardiola, K., and Lewis, J.C. (2015). Engineering a dirhodium artificial metalloenzyme for selective olefin cyclopropanation. *Nat. Commun.* 6: 7789.

84 Liu, X., Jiang, L., Li, J. et al. (2014). Significant expansion of fluorescent protein sensing ability through the genetic incorporation of superior photo-induced electron-transfer quenchers. *J. Am. Chem. Soc.* 136: 13094–13097.

85 Lv, X., Yu, Y., Zhou, M. et al. (2015). Ultrafast photoinduced electron transfer in green fluorescent protein bearing a genetically encoded electron acceptor. *J. Am. Chem. Soc.* 137: 7270–7273.

86 Liu, X., Kang, F., Hu, C. et al. (2018). A genetically encoded photosensitizer protein facilitates the rational design of a miniature photocatalytic CO_2-reducing enzyme. *Nat. Chem.* 10: 1201–1206.

87 Xu, W., Jiang, W., Wang, J. et al. (2017). Total chemical synthesis of a thermostable enzyme capable of polymerase chain reaction. *Cell Discovery* 3: 17008.

88 Wang, Z., Xu, W., Liu, L., and Zhu, T.F. (2016). A synthetic molecular system capable of mirror-image genetic replication and transcription. *Nat. Chem.* 8: 698.

89 Mandell, D.J., Lajoie, M.J., Mee, M.T. et al. (2015). Biocontainment of genetically modified organisms by synthetic protein design. *Nature* 518: 55.

90 Si, L., Xu, H., Zhou, X. et al. (2016). Generation of influenza A viruses as live but replication–incompetent virus vaccines. *Science* 354: 1170–1173.

91 Malyshev, D.A., Dhami, K., Lavergne, T. et al. (2014). A semi-synthetic organism with an expanded genetic alphabet. *Nature* 509: 385.

92 Zhang, Y., Ptacin, J.L., Fischer, E.C. et al. (2017). A semi-synthetic organism that stores and retrieves increased genetic information. *Nature* 551: 644.

93 Amiram, M., Haimovich, A.D., Fan, C. et al. (2015). Evolution of translation machinery in recoded bacteria enables multi-site incorporation of nonstandard amino acids. *Nat. Biotechnol.* 33: 1272.

94 Ostrov, N., Landon, M., Guell, M. et al. (2016). Design, synthesis, and testing toward a 57-codon genome. *Science* 353: 819–822.

95 Dumas, A., Lercher, L., Spicer, C.D., and Davis, B.G. (2015). Designing logical codon reassignment–expanding the chemistry in biology. *Chem. Sci.* 6: 50–69.

96 Day, J.W., Kim, C.H., Smider, V.V., and Schultz, P.G. (2013). Identification of metal ion binding peptides containing unnatural amino acids by phage display. *Bioorg. Med. Chem. Lett.* 23: 2598–2600.

97 Mills, J.H., Khare, S.D., Bolduc, J.M. et al. (2013). Computational design of an unnatural amino acid dependent metalloprotein with atomic level accuracy. *J. Am. Chem. Soc.* 135: 13393–13399.

11

Application of Engineered Biocatalysts for the Synthesis of Active Pharmaceutical Ingredients (APIs)

Juan Mangas-Sanchez[1,2,3], *Sebastian C. Cosgrove*[1,4], *and Nicholas J. Turner*[1]

[1] *Department of Chemistry, University of Manchester, Manchester Institute of Biotechnology, 131 Princess Street, Manchester M1 7DN, United Kingdom*
[2] *Institute of Chemical Synthesis and Homogeneous Catalysis, CSIC, Pedro Cerbuna 12, 50009, Zaragoza, Spain*
[3] *ARAID Foundation, Zaragoza, Spain*
[4] *Lennard-Jones Laboratory, School of Chemical and Physical Sciences, Keele University, Keele, Staffordshire ST5 5BG, United Kingdom*

11.1 Introduction

The use of enzymes in organic synthesis features many advantages. For example, enzymes are highly selective and operate under mild reaction conditions. Furthermore, advances in molecular biology, metagenomics, and bioinformatics as well as increased speed of gene synthesis and sequencing have resulted in a great expansion of the field to cover a much broader range of transformations that are of increased value to synthetic chemists. These developments have encouraged large chemical and pharmaceutical companies to consider biocatalysis as a core technology rather than a last resort when all other options have failed. Nevertheless, the use of enzymes at industrial scale still possesses limitations. Reaction conditions in Nature are often mild and, therefore, enzymes have not been challenged to work under the conditions normally required in industry, i.e. organic solvents, high substrate loadings or non-physiological pH, or temperature. Moreover, they can present low catalytic efficiency toward substrates that are structurally distinct from the compound(s) they evolved to transform. They also often present exquisite chemo and stereoselectivity in highly functionalized molecules. Nevertheless, the enzyme may not have the desired selectivity, or the stereoselectivity on non-natural substrates may sometimes not be as high. Directed evolution (DE) approaches introduce genetic diversity in proteins, followed by screening for improved characteristics, and have been proven to be a powerful tool to tackle all these challenges. Consequently, over the last 20 years, DE and other rational approaches for protein engineering have been implemented by academic and industrial groups and nowadays are routinely applied to improve enzyme performance [1, 2]. The impact that protein engineering has had on the chemistry community, and society as a whole, was underlined by the award of one-half of the 2018 Nobel Prize in Chemistry to

Protein Engineering: Tools and Applications, First Edition. Edited by Huimin Zhao.
© 2021 WILEY-VCH GmbH. Published 2021 by WILEY-VCH GmbH.

Frances H. Arnold, for pioneering the use of DE to generate improved biocatalysts. In this chapter, we review recent examples of the use of engineered enzymes for the preparation of active pharmaceutical ingredients (APIs).

11.1.1 Transferases

11.1.1.1 Transaminases

Transaminase (TA) enzymes mediate the transfer of primary amine groups between an amine and a carbonyl compound. TAs are pyridoxal phosphate (PLP) dependent and require a sacrificial nitrogen source for the reaction to proceed. As the reactions are in equilibrium, an important part of TA reactions is ensuring the equilibrium is far enough on the side of the product, so several methods have been developed to help drive this reaction forward. A landmark in the application of engineered enzymes in pharmaceutical production was the TA synthesis of a key intermediate for sitagliptin [3]. Sitagliptin is an antidiabetic drug, where the chemical route involved a Rh-catalyzed hydrogenation step to synthesize the key enantioenriched amine intermediate **2**. A comprehensive evolution campaign led to a TA that could not only produce the final compound **2** in 99.95% *ee*, but also operate at $200\,g\,l^{-1}$ substrate concentration, and with a 53% increase in productivity compared with the precious-metal route (Figure 11.1).

This pioneering example demonstrated how structure-guided engineering could transform the activity of an enzyme for industrial application.

Bornscheuer and coworkers demonstrated the engineering of an (*S*)-selective TA enzyme, which would accept bulky ketone precursors [4]. One challenge that TAs face is bigger substrates, so the group screened a host of TA enzymes for activity against some bulky amine substrates, some of which are intermediates in pharmaceutical drugs. The authors selected an (*S*)-selective TA from *Ruegeria* sp. TM1040 (termed 3FCR), as opposed to two more commonly used TA from *Vibrio fluvialis* and *Chromobacterium violaceum*, respectively. Although the wild type enzyme displays low activity toward the benchmark substrate 1-phenylethylamine [5], a variant bearing two substitutions (Y59W/T231A) demonstrated much increased activity over two substrates of interest. Using this variant as a starting point, substrate docking and energy minimization was used to identify three more amino acid residues that could be key for substrate or cofactor binding, i.e. Y87 (important for aromatic interactions with aromatic groups), Y152 (involved in coordination of the PLP), and P423 (hypothesized the rigidity imparted by proline would block the entrance to the active site). Gratifyingly, the introduction of mutations Y87F, Y152F, and P423H gave a vastly improved mutant, therefore justifying the rational approach adopted. Interestingly, the authors wished to demonstrate the identified mutations in this study could

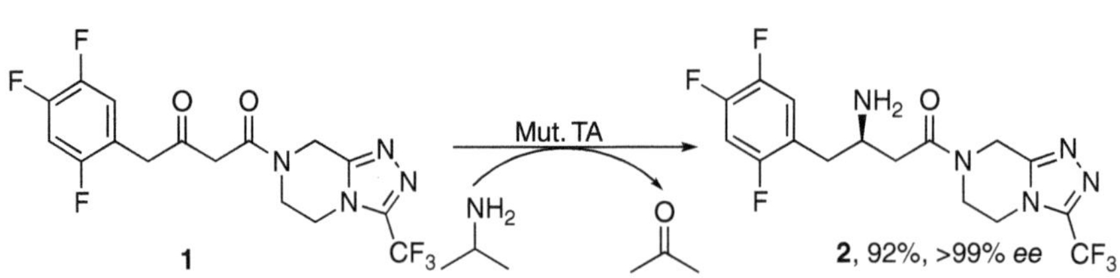

Figure 11.1 Transaminase catalyzed reductive amination to form sitagliptin.

be applied as a general tool to improve homologous TA enzymes. Remarkably, applying the same mutations to *Mesorhizobium loti maff 303099* (71.5% sequence ID with 3FCR) also delivered a much-improved mutant with activity towards several bulky amine targets. This was further applied to six novel proteins that had never been expressed before: all demonstrated some activity toward the substrates of interest, underlining the benefits of this engineering approach. The result was the production of six substrates that are challenging for TAs usually (no activity versus the wild-type), all in concentrations of 200–275 mM, >95% conversion for every substrate, and *ee* generally >94%.

11.1.2 Oxidoreductases

11.1.2.1 Ketoreductases

Ketoreductases (KREDs) or carbonyl reductases (CREDs) catalyze the NAD(P)H-dependent asymmetric reduction of carbonyl compounds to the corresponding alcohols. Even though large panels of wild-type enzymes are currently available from different suppliers, protein engineering is often needed. Atorvastatin, marketed under the name Lipitor **3**, a cholesterol-lowering drug in the statin family, has been subjected to different enzymatic routes for the preparation of key intermediates (Figure 11.2). In a collaboration between Delft University and Codexis, KRED and glucose dehydrogenase (GDH) were engineered to improve the activity and stability via DNA shuffling [7]. In the initial process employing the wild-type enzymes, 9 gL^{-1} of biocatalyst were required to convert 80 gL^{-1} of substrate **4** in 24 hours, far from suitable for a large-scale process. Through several rounds of DNA shuffling, a 7- and 13-fold increase in activity was obtained for the KRED and the GDH respectively, while maintaining the enantioselectivity (>99.5%), improving the space time yield by a factor of six and the catalyst yield by a factor of 20. For the next step, a

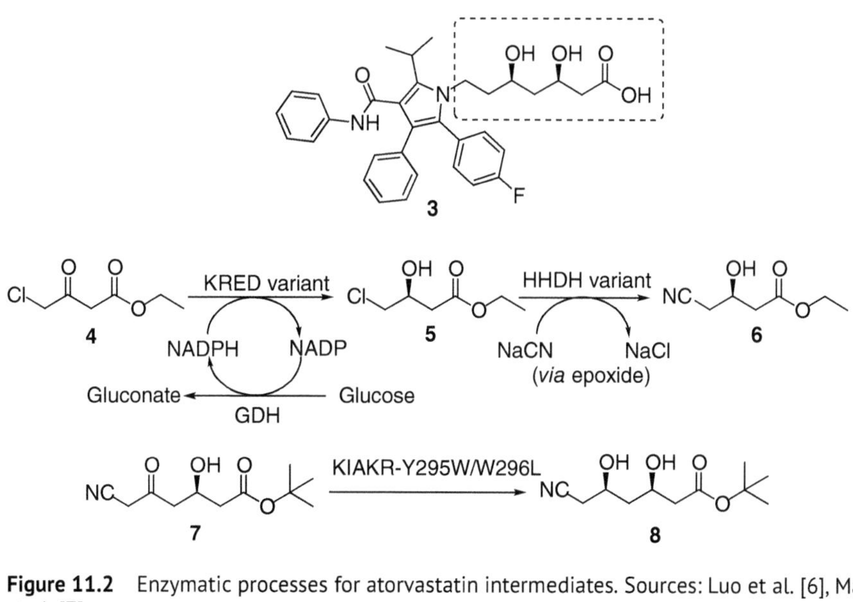

Figure 11.2 Enzymatic processes for atorvastatin intermediates. Sources: Luo et al. [6], Ma et al. [7].

halohydrin dehydrogenase (HHDH) that was previously engineered by the same authors [8] was employed. These enzymes catalyze the conversion of halohydrins to epoxides and the reverse reaction [9], but they can accept other nucleophiles [10]. In this way, the engineered HHDH was used to convert the enantiopure halohydrin **5** into the corresponding epoxide, followed by nucleophilic attack of cyanide to afford the final product **6**. Similarly, Luo *et al.* improved the KRED from *Kluyveromyces lactis* for the asymmetric reduction of another precursor **7** for atorvastatin through a rational approach [6]. In this case, via homology modeling, they identified two residues likely to have an effect in the catalytic activity due to their proximity to the carbonyl moiety and nicotinamide adenine dinucleotide (NADH). After two rounds of site-saturation mutagenesis, they identified a double variant, which presented a 13-fold increase in the specific activity. This variant was used in a process at 50 g l^{-1} substrate loading obtaining a total conversion to the final product **8** in 80 minutes.

(1*S*)-2-Chloro-1-(3,4-diflurorophenyl) ethanol **10** – (*S*)-CFPL – is a key intermediate in the synthesis of ticagrelor **11**, a platelet aggregation inhibitor currently produced by AstraZeneca. In a recent study, Zhao *et al.* performed an initial screening of 27 KREDs from *Chryseobacterium* sp. and identified *Ch*KRED20 as the most promising catalyst for the asymmetric reduction of 2-chloro-1-(3,4-difluorophenyl) ethenone **9** [11]. Nevertheless, the productivities obtained were insufficient for its implementation in an industrial process. By using a random approach, they constructed a library of 12 300 mutants, which resulted in the identification of two single-mutant variants with improved activity greater than 50%. They subsequently generated two saturation libraries at these two positions obtaining a variant, *Ch*KRED20-L205A, with a 10-fold increase in activity compared to that of the wild-type. Unfortunately, efforts aimed at combining the beneficial mutations did not provide any further improvement in the catalytic efficiency. This variant was then used at substrate concentrations up to 200 gL^{-1} obtaining total conversion after six hours and affording (*S*)-CFPL in 95% isolated yield and >99% *ee* (Figure 11.3).

A similar strategy was used by Codexis for the preparation of (*S*)-licarbazine [12]. They initially screened a panel of enzymes containing variants of the KRED from *Lactobacillus kefir* finding several of them capable of producing the desired product. The best performing mutants were subsequently tested under process conditions where only one converted the substrate albeit at very low productivity. This variant was selected and subsequently improved via DE using the in-house Code Evolver platform. This technology enables the quick development of engineered enzymes by combining artificial intelligence and high-throughput screening. After four rounds of evolution using the desired process conditions in the assay, a variant

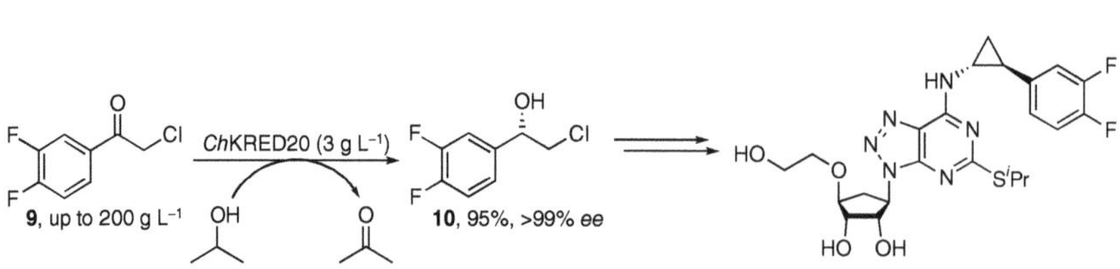

Figure 11.3 KRED catalyzed reduction of prochiral ketone for ticagrelor intermediate.

bearing 30 mutations was found to convert quantitatively the substrate obtaining (*S*)-licarbazine in >95% isolated yield, >99% *ee* at 100 gL^{-1} substrate, and 55 °C and 60% isopropanol on a 500 ml scale. Codexis also employed this platform to engineer KREDs for more challenging processes such as dynamic reductive kinetic resolution of substituted ketoesters to access β-lactam antibiotic intermediates using the KRED from *L. kefir* as a template for evolution [13]. After several rounds, they obtained a five-point variant (H40R/A94T/F147L/L199H/A202L) and they scaled up the process to 25 g of starting material, affording near quantitative yield using isopropanol as a sacrificial substrate (50% v/v), 1 gL^{-1} KRED and 0.14 $g\,L^{-1}$ cofactor loading.

Sulopenem **14** is a β-lactam broad spectrum antibiotic that was originally developed by Pfizer in the 1980s, but has since been licensed to Iterum Therapeutics and is currently awaiting initiation of phase three clinical trials in the United States. A key intermediate in the synthesis of the drug is tetrahydrothiophene-3-ol **13**, and was formerly synthesized chemically through a five-step route that used hazardous chemicals including hydrobromic acid and borane. It also only produced the desired product reliably in 96–98% enantiomeric excess [14]. Codexis and Pfizer developed a one-step biocatalytic reduction of the corresponding ketone **12** by evolving a KRED to have much improved stereoselectivity for the desired enantiomer [15]. After an initial screen of in-house and commercially available KREDs, the best variant offered 63% *ee* of the desired enantiomer, which was a naturally occurring (*R*)-selective KRED from *L. kefir*. Initially the substrate concentration of 40 $g\,L^{-1}$ and enzyme concentration of 4 $g\,L^{-1}$ gave full conversion in two hours, so improving the enantioselectivity was the primary goal. A combination of random mutagenesis, DNA shuffling, and ProSARs (protein sequence activity relationships) [8] analysis were applied to generate libraries that were first evaluated in an "first-tier" high-throughput screening that monitored nicotinamide adenine dinucleotide phosphate (NADPH) depletion. Low to no activity variants were eliminated. A "second-tier" screen produced enough material for chiral high performance liquid chromatography (HPLC) analysis to analyze enantioselectivity. The second-tier screens also tested for thermal stability. After this process the best variants were analyzed with ProSAR, which assesses large amounts of sequence-activity data to identify amino acid hot spots for evolution. The first round of evolution gave a variant, which delivered 80% *ee*. The process was repeated eight times, with an additional thermal challenge between rounds four and five (the protein was heated at 60 °C for 24 hours) giving a highly active, stereoselective, and thermo-stable variant. The final process was run on 130 kg Scales to deliver the product **13** in 99.4% *ee*, 88% yield, and >99% purity (Figure 11.4).

More recently, Reetz and coworkers used triple-code saturation mutagenesis (TCSM) to generate two alcohol dehydrogenases capable not only of reducing the tetrahydrothiophene substrate discussed earlier but also the tetrahydrofuran equivalent, an intermediate in the HIV drug amprenavir [16]. TCSM uses a rational approach to reduce the bottleneck of screening for mutagenesis. By using NNK codon degeneracy, 94 colonies have to be screened for a 95% library coverage, whereas by using reduced degeneracies, the amount of colonies for a high coverage is vastly reduced, especially when several sites are targeted at the same time. In TCSM three amino acids are carefully selected based on the need

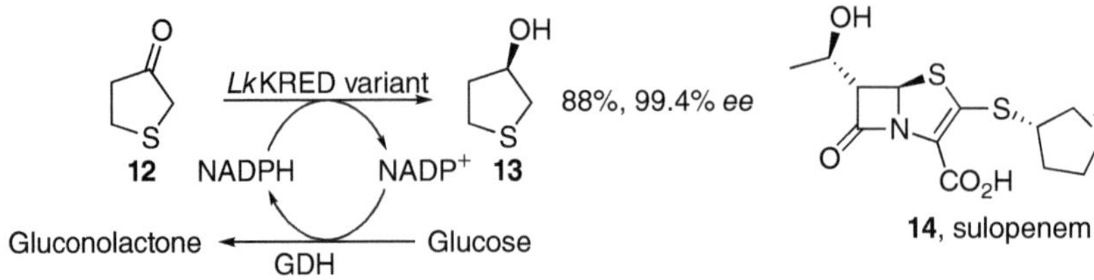

Figure 11.4 KRED catalyzed reduction of tetrahydrothiophene-3-one.

of the enzyme and the reaction (homology, mechanism, etc.). In this study, using the alcohol dehydrogenase (ADH) from *Thermoethanolicus brockii* (*Tb*SADH) as template, the (*R*)-selective ADH was generated by screening with the triple code valine–asparagine–leucine, and the (*S*)-ADH with valine–glutamine–leucine. The efficiency of this method meant only one round of evolution was necessary to generate two variants with opposing stereoselectivities. The authors noted that TCSM is particularly useful for engineering stereoselectivity, which was underlined by the evolution of two variants from the same wild-type parents with opposite stereoselectivity.

Vibegron **17** is a phase III clinical trial candidate for the treatment of overactive bladder syndrome (OAB). It is a β_3-adrenergic receptor agonist, recently discovered as an effective treatment of OAB and developed by Merck and coworkers [17]. The chemical route, which had been used to prepare >100 kg, proceeded in a 32% yield and was described as being atom inefficient. Researchers at Merck envisioned a biocatalytic step that could improve the efficiency of the synthetic route, with a dynamic kinetic reduction of an α-amino ketone **15** using a KRED viewed as more direct route to the key intermediate [18]. The precursor **15** was to be synthesized racemically. It was anticipated that the KRED could be used to resolve the stereocenter bearing the amino group in a dynamic resolution to afford a single diastereoisomer; however, the resolution only worked at elevated temperatures and pH, meaning a thermo- and pH-stable enzyme was required (Figure 11.5).

The researchers screened the commercially available KRED panel from Codexis to identify an initial target for evolution. A KRED termed KRED-P1B2 gave high

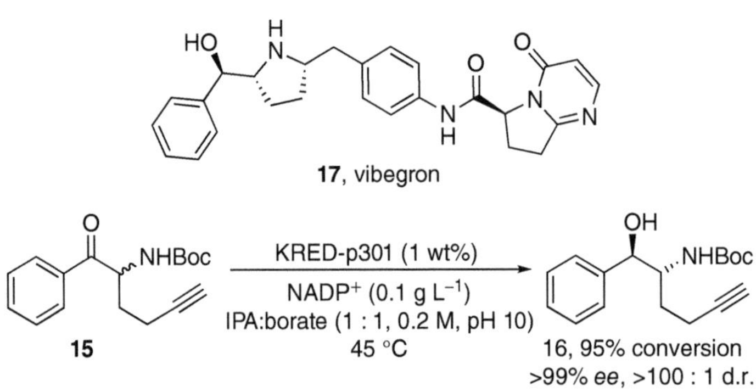

Figure 11.5 Biocatalytic synthesis of key fragment for vibegron.

stereoselectivity (>99% ee) and diastereoselectvity (16 : 1) at ambient conditions; however conversion and diastereoselectivity were reduced dramatically at pH 10 and 45 °C. Using rational design, seven amino acid positions in the active site of P1B2, specifically in the substrate-binding pocket, were identified and subjected to saturation mutagenesis. Seven unique mutations at five amino acids gave significant increases in activity over the wild type enzyme, with substitution of methionine 206 for a phenylalanine moiety identified as being particularly important for diastereoselectivity. Interestingly, the authors noted that $NADP^+$ degradation at elevated temperatures and pH may have been a major contributor to the lower conversion, so in parallel substitution of an alternative site of the enzyme was also explored. Histidine 40 was identified as being important to $NADP^+$ binding, based on *in silico* structural analysis. Swapping for an arginine validated this theory, with the H40R variant retaining activity at lower concentrations of cofactor. Second and third rounds of evolution in the second shell position (between 5 and 10 Å of the substrate) and other positions highlighted as hotspots using other evolution programs gave an additional nine mutations to obtain >50-fold improvement to the wild type enzyme. The final operating conditions were 50 gL^{-1} substrate concentration, 1 wt% enzyme loading with the reaction running at pH 10 and 45 °C. The desired amino alcohol **16** was obtained in a single diastereoisomer and has been scaled to a manufacturing scale.

11.1.2.2 Amino Acid Dehydrogenases

Amino acids constitute important building blocks for pharmaceuticals and other fine chemicals [19, 20] and hence the development of new biocatalysts for their preparation has attracted major attention [21–26]. In this context, amino acid dehydrogenases (AADHs) catalyze the NAD(P)H-dependent reductive amination of α-ketoacids to the corresponding α-amino acids. Bristol Myers Squibb designed a synthetic route for the preparation of (*R*)-5,5,5-trifluorovaline, an intermediate of a γ-secretase inhibitor for the treatment of Alzheimer's disease [27]. The authors envisaged two different routes from 5,5,5-trifluoro-2-oxopentanoic acid **18** to access the corresponding amino acid **19**. For the first one, a TA was employed using alanine as the amine donor. The need for a lactate dehydrogenase to drive the equilibrium as well as a second enzymatic cascade for NAD^+ recycling made the authors consider an AADH process for further development. The D-aminopimelic acid dehydrogenase from *Bacillus sphaericus* was selected for evolution, and its activity improved via site-directed mutagenesis. Using the AADH from *Corynebacterium glutamicum* as template, five key mutations previously described [28] in that enzyme to expand the substrate scope were introduced. A GDH from *Gluconobacter oxidans* was employed for NADH regeneration and genes encoding the AADH and the GDH were cloned into the same expression plasmid and co-expressed. Further optimization included the inactivation of a glutamate dehydrogenase from the expression host strain to eliminate the background reaction that was converting the starting material into the (*S*)-enantiomer. Under the final optimized conditions using a lysate containing both enzymes, at pH 9, 30 °C, 1.7 equiv of NH_4Cl and

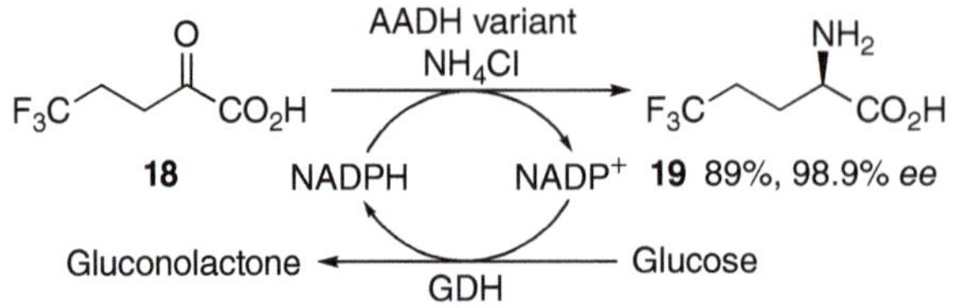

Figure 11.6 AADH catalyzed production of unnatural amino acids.

300 mM substrate loading, they prepared the desired (*R*)-amino acid **19** on a 50 g scale in 88.5% yield and 98.9% *ee* (Figure 11.6).

11.1.2.3 Cytochrome P450 Monoxygenases

Another limitation enzymes present is that, for many useful reactions, either the natural biocatalyst has not yet been discovered or simply the transformation is not found in Nature. Cytochrome P450 monoxygenases catalyze a vast array of oxidations of organic molecules such as hydroxylation, epoxidation, and heteroatom oxygenation among others [29, 30]. Mechanistically, a high-valent iron-oxene intermediate is responsible for the oxygen atom insertion into C—H and C=C bonds. With this in mind, the group of Arnold envisaged that by modifying this intermediate using diazoester reagents, carbene transfer reactions could be performed as well [31]. They found that variants of the cytochrome P450 from *Bacillus megaterium* ($P450_{BM3}$) could catalyze this transformation in moderate to high diastereoselectivity with total turnover numbers (TTNs) up to 364. This process was subsequently improved by modifying the reduction potential via axial ligand mutations, finding a C400S variant with much greater catalytic power [32] (Figure 11.7).

Further investigations led to another variant, BM3-star, that is capable of performing the cyclopropanation of *N*,*N*-diethyl-2-phenylacrylamide and ethyl diazoacetate to afford a precursor of the antidepressant levomilnacipran in 93% conversion (86% isolated yield) a diastereomeric ratio of 2 : 98 and 92% *ee* [33].

Hydroxylated tetralone compounds are important building blocks for many pharmaceuticals and natural products. For example, 4-hydroxy tetralone derivatives have been shown to demonstrate high activity against multi-drug resistant *Escherichia coli* strains [34], and are also present in several anti-inflammatory drugs. In an attempt

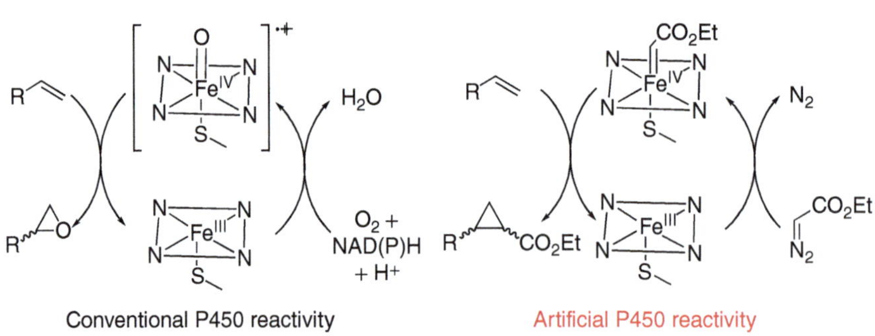

Figure 11.7 Rationale in re-tasking of P450 to mediate carbene transfer.

to generate a biocatalytic synthesis of these targets, Reetz and coworkers engineered a P450 monooxygenase for the (*S*)-selective hydroxylation of the tetralone precursor [35]. Starting with the wild type $P450_{BM3}$ from *B. megaterium*, and 25 previously reported variants, they studied the hydroxylation of 1-tetralone. After initial screens, several variants proved to have perfect regioselectivity as well as higher stereoselectivity for the correct (*S*)-enantiomer than the wild type enzyme. It was noted all the variants had a single substitution at A328, singling it out as a "hot-spot" for mutagenesis. In the first round of evolution, saturation mutagenesis of residue A328 with NNK degeneracy found A328Y had high regioselectivity and also afforded high levels of enantioselectivity (94% *ee*). Various other A328 variants were also shown to be highly selective for similar substrates with more electron-rich aromatic groups, smaller aliphatic rings and even indane and tetralin. Using this rational design approach, the authors were able to select from one of four A328 variants to be able to produce six hydroxylated building blocks in both high regio- and stereoselectivity (Figure 11.8).

11.1.2.4 Baeyer–Villiger Monoxygenases

Baeyer–Villiger oxidations are commonly used in synthetic chemistry. The lack of selectivity as well as the harsh conditions often needed emphasizes the benefits that enzymatic reactions at ambient temperatures can impart. Baeyer–Villiger monoxygenases (BVMOs) are flavin-dependent enzymes that not only catalyze the conversion of ketones into esters, but also the oxidation of heteroatoms [36]. In a series of different processes, Codexis engineered the cyclohexanone monooxygenase (CHMO) from *Acinetobacter* sp. to access Armodafinil, a drug used to treat sleep disorders, via thioether sulfoxidation [37]. In this system, NADPH needs to be converted back to $NADP^+$ so the process was coupled with an alcohol dehydrogenase and 2-propanol as a sacrificial substrate. Similarly, they engineered the same naturally occurring enzyme to prepare (*S*)-esomeprazole **27** – used to treat different gastric diseases – from pyrmetazole **26** as an alternative to the Kagan–Sharpless–Pitchen sulfoxidation [38] (Figure 11.9). They followed a DE

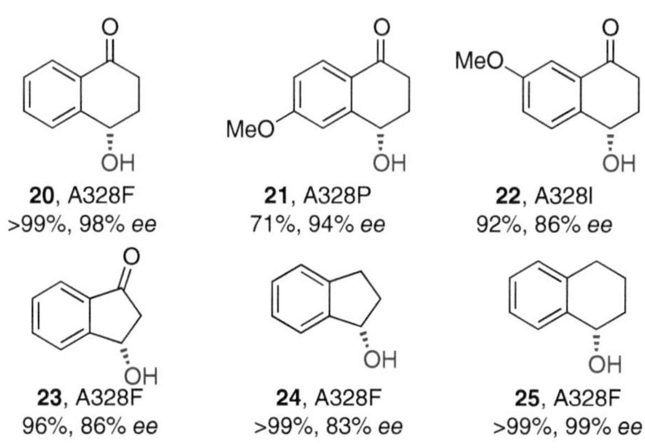

Figure 11.8 Panel of hydroxylated substrates synthesized using A328 variants of P450-BM3.

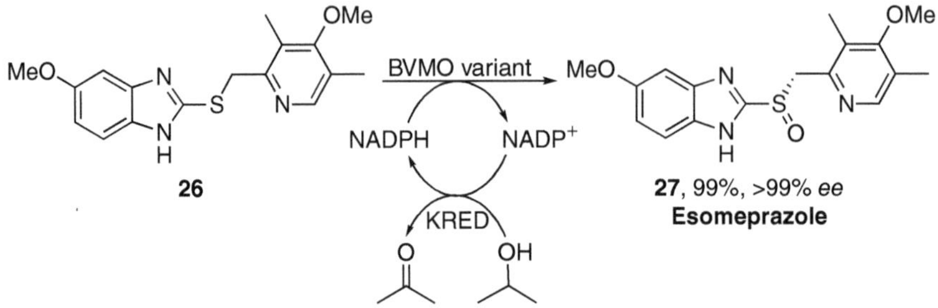

Figure 11.9 Application of an engineered BVMO for the synthesis of esomeprazole.

strategy employing their in-house Code Evolver technology. They focused on four different aspects, e.g. enzyme productivity, enantioselectivity, chemoselectivity, and cofactor loading. Different strategies were employed to introduce diversity during the evolution process and the conditions in the screening were modified in each round, introducing different evolutionary pressures and therefore targeting different aspects. They obtained a variant after 19 rounds of evolution with a 140 000-fold improvement in productivity, perfect stereoselectivity, and chemoselectivity and using only 0.1 $g\,l^{-1}$ cofactor.

11.1.2.5 Amine Oxidases

Amine oxidases (AOs) catalyze the oxidation of amines into imines and can be classified in two different families. Type I AOs are Cu-dependent enzymes that covalently bind the imine product, therefore rendering them not useful for synthetic purposes. On the other hand, type II AOs are flavin-dependent oxidoreductases and several of them have been used in organic synthesis [39]. A deracemization strategy is normally employed with these catalysts where one enantiomer is selectively oxidized to the corresponding imine in the presence of a non-selective reducing agent. The imine is then reduced leading to the accumulation of the unreactive enantiomer after several rounds of selective oxidation and non-selective reduction. By combining rounds of random mutagenesis and site-saturation libraries, the monoamine oxidase from *Aspergillus niger* (MAO-N) has been the object of extensive engineering to improve its substrate scope and synthetic applicability (Figure 11.10). For instance, the MAO-N D5 variant was found to be active towards substituted pyrrolidines. In particular, this variant was used to perform the oxidation of 3,4-substituted *meso*-pyrrolidines **28** that after cyanide addition to the alkene followed by hydrolysis afforded an intermediate for the preparation of telaprevir **30**, a drug used for the treatment of hepatitis C [42]. The determination of MAO-N D5 structure provided insights into more rational approaches toward evolution [43]. Saturation libraries were generated at specific positions in order to open the active site and therefore accept larger substrates. Using MAO-N D5 as a template, a variant bearing four additional mutations (MAO-N D9) was employed to

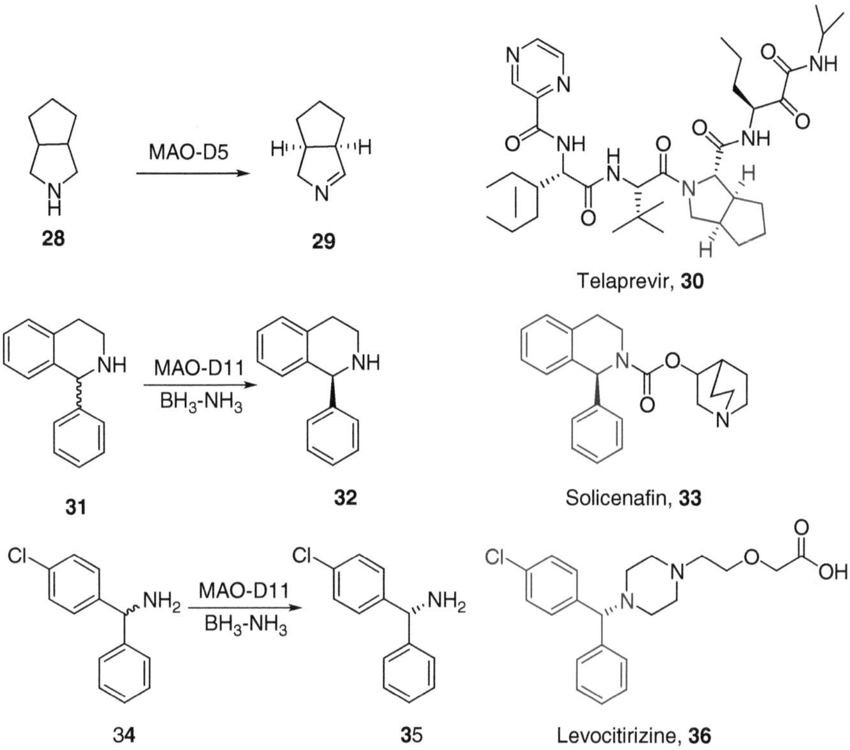

Figure 11.10 MAO-N variants for the preparation of different APIs. Sources: Ghislieri et al. [40], Rowles et al. [41].

deracemize the anti-tumor alkaloid crispine A [41]. Further engineering generated MAO-N D11, which was capable of effectively deracemizing intermediates used in the synthesis of the antihistamine solifenacin **33**, which is used in the treatment of overactive bladder, and levocetirizine **34** [40]. The cyclohexylamine oxidase (CHAO) from *Brevibacterium oxidans* was also evolved to increase the substrate scope towards tetrahydroquinolines through a semi rational approach [44]. Using the information from the crystal structure in complex with cyclohexanone and flavin adenine dinucleotide (FAD) [45], 11 residues around the active site were targeted and site-saturation libraries designed. Four variants were found to have a higher activity toward 2-methyltetrahydroquinoline (2-MTHQ) and combinatorial libraries on those positions were constructed via iterative saturation mutagenesis (ISM). After four rounds of evolution, a variant containing three substitutions (T198F/L199S/M226F) was selected and used for the deracemization of 2-MTHQ in combination with a non-selective reductant to yield the enantiopure product in three hours.

In a collaboration, Merck and Codexis engineered a variant of MAO-N for the stereoselective desymmetrization of a pyrrolidine intermediate in the drug boceprevir [46]. The chemical route to the key intermediate, a proline methyl ester derivative, required nine steps via the shortest direct route. The authors stated that four steps could be removed through a stereoselective desymmetrization with MAO, trapping the imine intermediate with a sulfonate salt and converting this to the nitrile, which could be hydrolyzed to the acid. MAO-N was attractive due to the dependency on FAD as a co-factor, reducing the complication of the process compared with metal-dependent AOs. Initial screens with MAOs from *Aspergillus niger* and *Aspergillus oryzae* showed the MAO-N had higher stability but lower activity than monoamine oxidase from *Aspergillus oryzae* (MAO-O), so MAO-N was chosen as the enzyme for optimization. A combination of random mutagenesis through epPCR and homology-inspired design was used to generate initial hits, revealing the variant K348Q/A289V as the best in this initial round of engineering, which was termed MAON156 by the authors. The second round took a parallel approach, combining MAON156 with other beneficial mutations from round one, and also through shuffling MAON156 with the *A. oryzae* homolog. This gave rise to two further variants, MAON274 and MAON291, obtained from the hybrid variation and hit shuffling, respectively. Two more rounds of evolution, using a combination of the previous techniques (and structure-guided evolution due to the crystal structure becoming available) delivered the final variant, MAON401, which had the beneficial mutations from each round giving a thermostable and highly active variant. The final process was run at a final concentration of 65 g l^{-1} substrate, 6 wt% enzyme (MAO-N and catalase), with pure oxygen used in the headspace of the reactor as this was found to increase the rate of reaction.

11.1.2.6 Hydroxylases

Proline hydroxylases are 2-oxoglutarate-dependent dioxygenases that catalyze the hydroxylation of proline to hydroxyproline and require 2-oxoglutarate and oxygen as co-substrates as well as Fe(II) as cofactor [47]. Using the wild-type *cis*-4-proline hydroxylase from *Sinorhizobium meliloti* as scaffold, Codexis carried out several rounds of evolution to obtain active variants toward the hydroxylation of (*S*)-pipecolic acid to afford (2*S*,5*S*)-5-hydroxypipecolic acid [48]. One of the variants was used to perform the hydroxylation at a 15 g scale obtaining a 94% conversion after 52 hours.

11.1.2.7 Imine Reductases

Imine reductases (IREDs) are NADPH-dependent enzymes that catalyze the asymmetric reduction of imines to chiral amines [49]. Pioneering work by Mitsukura et al. [50–52] followed by different academic and industrial groups have greatly expanded the number of enzymes available for the synthesis of secondary and tertiary amines through either cyclic imine reduction [53–59] or asymmetric reductive amination [60–66]. GlaxoSmithKline (GSK) have recently demonstrated the outstanding synthetic potential of these enzymes by evolving an IRED from their in-house collection [64] to prepare a lysine-specific demethylase-1 (LSD1) inhibitor GSK2879552 **37**, which is currently under investigation for the treatment of small cell lung cancer and acute leukemia [67]. They initially performed a screening of their in-house

IRED panel and identified IR-46 as the most promising biocatalyst for the reductive amination of **38** using **40** as the amine partner. However, due to the special operating conditions (low enzyme loading, high substrate concentration, and acidic pH), protein engineering was required to meet the commercial requirements. They initially focused their protein engineering effort on finding variants able to operate at decreased pH and increased substrate loading. Based on a structural homology model of IR-46 they targeted 256 out of 296 amino acid positions for single-site saturation mutagenesis, finding a Y142S variant named M1 that showed a 40-fold increased activity under these reaction conditions. Other beneficial mutations were also found in the first round that were combined in a second round creating eight libraries, which yielded a variant M2 (Y142S, L37Q, A187V, L201F, V215I, Q231F, S258N) containing seven mutations distributed across the structure. Further computational analysis identified several additional positions, which were targeted in a last round under the desired conditions of the final process. Bearing 13 mutations (Y142S; L37Q, A187V, L201F, V215I, Q231F, S258N; G44R, V92K, F97V, L198M, T260C, A303D), the M3 variant which was obtained showed a >38 000-fold improvement in activity over the wild-type under the process conditions. The process was carried out on a 20 l scale to afford 1.4 kg of the final product in 84.4% isolated yield and 99.7% *ee*. They finally envisioned a hydrogen-borrowing cascade to replace the Cu-catalyzed oxidation of the primary alcohol **39** by using a KRED that would work in a self-sufficient redox neutral cascade with the IRED without the need of an external cofactor regeneration system. Using the KRED from *Lactobacillus coleohominis*, the final product was isolated in 48.3% yield and in 99.5% *ee*. Despite this elegant approach, the authors envisaged further process development to make this alternative and greener route economically viable due to the thermodynamics of the fully reversible process not favoring the final product (Figure 11.11).

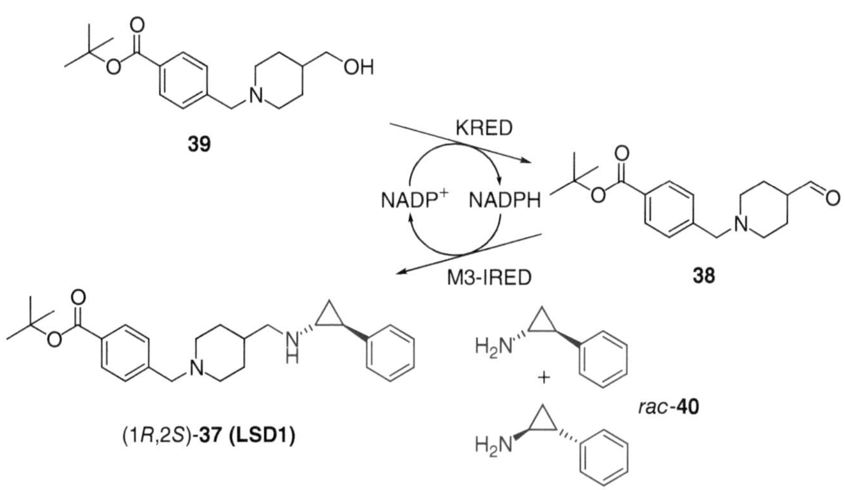

Figure 11.11 Hydrogen borrowing cascade employing an engineered imine reductase for the synthesis of LSD1.

11.1.3 Lyases

11.1.3.1 Ammonia Lyases

Another enzyme class that has been extensively studied for the preparation of amino acids is ammonia lyases. These enzymes catalyze the addition of ammonia onto double bonds to generate the corresponding L-amino acid in a reaction with high atom efficiency [68]. One member of this class, the 3-methylaspartate ammonia lyase (MAL), was engineered by Poelarends and coworkers to expand its substrate scope and access unnatural amino acids [69]. The crystal structure in complex with the natural substrate [70] was used to inform about interactions in the active site. It was found that the side chains of Q73 and Q172 were involved in the amino group binding site architecture, therefore saturation mutagenesis libraries were created in those positions finding a variant Q73A with much broader substrate scope. It was also found that three other residues were involved in the alkyl binding pocket so the same strategy was applied in those positions, finding a variant, L384A, that was able to accept non-native fumarate derivatives to access a family of aspartic acids. The Turner group has carried out extensive research in the modification of ammonia lyases for the preparation of amino acids. Using a homology model based on the phenylalanine ammonia lyase (PAL) from *Rhodotorula glutinis*, five CASTing (Combinatorial Active site Saturation Test) libraries were constructed for the PAL from *Rhodotorula graminis* (*Rgr*PAL) to increase the production of different L-phenylalanine derivatives [71]. Further engineering of two residues in the aryl binding pocket of the active site of the PAL from *Planctomyces brasiliensis* (*Pb*PAL) expanded the substrate scope for the production of challenging electron-rich L-phenylalanines [72]. Through site-saturation mutagenesis targeting residues around the binding site of *Pb*PAL, libraries were constructed and screened to increase the production of D-phenylalanine derivatives. Using *p*-nitrocinnamic acid as substrate, a H359Y variant showed a 3.5-fold increase in the formation of the D-enantiomer and was subsequently used in combination with a L-amino acid deaminase from *Proteus mirabilis* to access enantiopure D-phenylalanine derivatives in good to excellent conversions [73].

11.1.4 Isomerases

Isomerases catalyze the conversion of a molecule into its isomeric form. These enzymes are normally used industrially to transform cheap and abundant sugars into rare and expensive sugars with high potential in the pharmaceutical industry [74]. For instance, lactulose is a non-digestible disaccharide that is used for the treatment of chronic constipation and hepatic encephalopathy and cellobiose 2-epimerases are used for its production. In this context, Shen et al. engineered the cellobiose 2-epimerase from *Caldicellulosiruptor saccharolyticus* via random mutagenesis to improve lactulose production [75]. Through different rounds of error-prone PCR and colony screening, they obtained a variant containing five substitutions with a threefold increase in lactulose production. Similarly, a previously engineered variant of the D-tagatose epimerase from *Pseudomonas cichorii* [76] with increased thermostability was subjected to further evolution to increase its activity towards D-fructose and L-sorbose for the production of D-psicose and D-tagatose, respectively [77]. An initial round of error-prone PCR was performed, followed by site-saturation mutagenesis on the positions that showed increased K_{cat}. They obtained two variants, IDF10-3 and ILS6, with a 3.7- and 2-fold increase in TTN toward D-fructose and L-sorbose, respectively.

11.1.5 Hydrolases

11.1.5.1 Esterases

Esterases catalyze the stereoselective hydrolysis of esters to yield carboxylic acids. In particular, α-hydroxyacids are key intermediates in various drugs such as (*S*)-clopidogrel **43** than can be synthesized from (*R*)-2-hydroxy-2-(2′-chlorophenyl) acetic acid (*R*)-**42**. Ju et al. identified an esterase from *Pseudomonas putida* from a metagenomics library that was able to perform the kinetic resolution of (*rac*)-*α*-acetoxyphenylacetic **41** acid in high enantioselectivity (Figure 11.12) [78]. Nevertheless, the activity and stability of the wild-type enzyme do not meet the requirements for industrial application. Using three different crystal structures as templates, the authors constructed a model to provide insights into possible mutations to improve the catalytic efficiency [79]. Two residues, W187 and D287, were identified as hotspots for substitution and two single variants, D287A and W187H, were designed and tested to enlarge the binding pocket and increase the local polarity, respectively. Although an increase in activity was observed with the D287A variant, the enantioselectivity dropped significantly. On the other hand, the W187H variant showed a 14-fold increase in K_{cat} without any loss in enantioselectivity and was subsequently used for the kinetic resolution of a series of α-hydroxyacids at substrate concentrations up to 100 and $0.5\,g\,l^{-1}$ catalyst loading.

11.1.5.2 Haloalkane Dehalogenase

Haloalkane dehalogenase enzymes catalyze the conversion of alky halides to alcohols. The active site of these enzymes has been found to be water-free, lowering the energy for the conversion of the haloalkanes, and have found use in bioremediation of halogenated organic pollutants. This interesting transformation

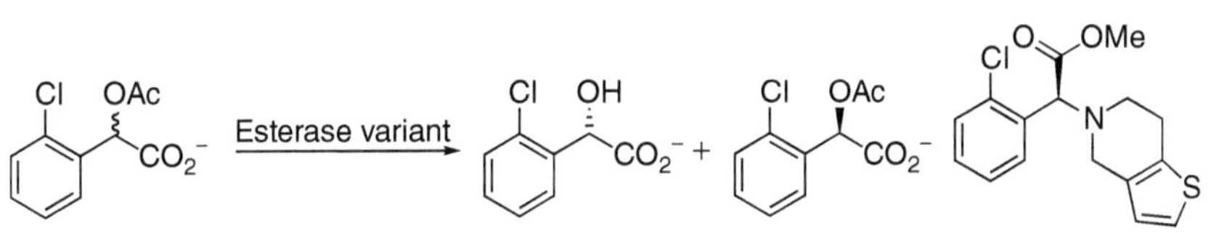

Figure 11.12 Esterase catalyzed resolution of α-hydroxy acids for the synthesis of (*S*)-clopidogrel intermediate.

has also found application in organic synthesis: the production of enantiopure alcohols from low-cost readily available chemical waste is an attractive option to the chemical industry [80]. Janssen and coworkers described a biocatalytic hydrolysis of 1,2,3-trichloropropane (TCP) **44**, a common industrial waste product that is highly toxic and hard to dispose of through traditional chemical methods [81]. The product they envisaged producing was enantiopure versions of 2,3-dichloropropan-1-ol (DCP) **45**, which can be converted to (*R*)- and (*S*)-epichlorohydrin **46**. Both epichlorohydrin enantiomers are high-value intermediates in numerous pharmaceutical syntheses, so conversion of waste products would represent a significant product (Figure 11.13).

The study began with a haloalkane dehalogenase from *Rhodococcus rhodochrous* (DhaA). In a previous study, a five-point mutant had been generated with improved activity toward **44**, but producing the (*R*)-enantiomer in only 13% *ee*. From the outset, it was stated there was no high throughput screen for this enzyme so limiting the number of variants was essential. To begin, a structural model of the five-point mutant with TCP docked was generated based on the published crystal structure of the wild-type enzyme. This initial study identified 16 residues that surrounded TCP when it was docked in the active site. To avoid repeated screening of the same variants, and to take advantage of synergistic effects of multiple mutations, saturation mutagenesis of the 16 active positions was coupled with 14 overlapping pair-wise libraries as well. In total, 6400 possible variants could be produced, and of the 98 that were picked and tested by chiral gas chromatography (GC), the results revealed hot spots that increased the enantioselectivity of the enzyme for both (*R*)- and (*S*)-enantiomers. For example, F168 was found to be beneficial for both enantiomers as when mutated to a tryptophan the (*R*)-enantiomer of DCP was obtained in 47% *ee*, and when it became a cysteine the (*S*)-DCP was produced in 78% *ee*. With this knowledge, the authors targeted once again through pair-wise saturation libraries with beneficial mutations combined again, and produced two more mutants with high selectivity for both enantiomers. Finally, a round of site-restricted mutagenesis was applied to the second-round mutants. This method proceeds with using partly undefined codons that only cover certain types of amino acids, hydrophobic, for example, giving site-specific amino acids. Coupled with some positions that were identified at distance from the active site, two variants with 13 and 17 amino acid substitutions for the production of (*R*)- and (*S*)-DCP were

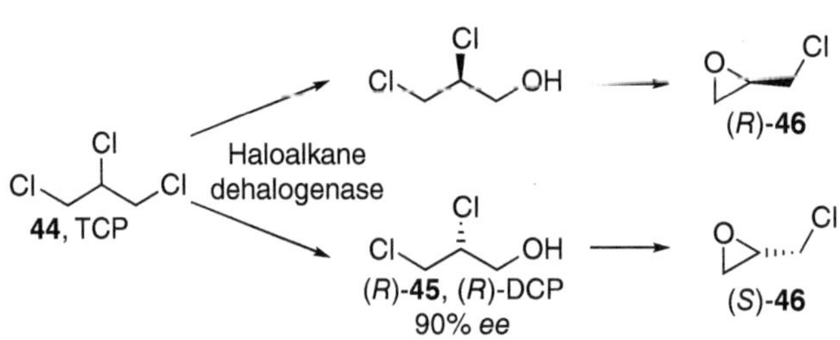

Figure 11.13 Bioremediation of TCP with haloalkane dehalogenase to afford enantiopure high-value intermediates.

obtained. This demonstrated the use of methods when high-throughput screening was not available.

11.1.6 Multi-enzyme Cascade

The similar conditions under which enzymes operate enable the combination of several enzymatic steps in cascade processes. This can facilitate the construction of synthetic complexity whilst avoiding tedious and laborious purification steps, as well as overcoming potential equilibrium issues that enzymes sometimes present. Very recently, Merck have developed an impressive nine-enzyme cascade for the preparation of the nucleoside analog islatravir **47**, currently under investigation for HIV treatment, where five of these enzymes were subjected to DE (Figure 11.14) [82].

Inspired by the bacterial nucleoside salvage pathway, they envisaged a five-step cascade process combining an alcohol oxidase (AOx), a pantothenate kinase (PanK), an aldolase, a phosphopentomutase (PPM), and a purine nucleoside phosphorylase (PNP) (Figure 11.14) to synthesize the final API **47** from ethynyl glycerol **51**. For the alcohol oxidation step, they selected the galactose oxidase (GOase) from *Fusarium graminearum* and focused on both activity and stereoselectivity for its evolution. After 12 rounds of evolution, they obtained a variant bearing 34 amino acid substitutions with 11-fold increased activity, reversed selectivity and improvements in protein expression, stability, and product inhibition. For the phosphorylation step, they selected the PanK from *E. coli* as the starting point. After three rounds of evolution, a variant with 10 amino acid changes was obtained, with a remarkable increase in activity (from <1% to >95% conversion) and increased selectivity that allowed for the selective phosphorylation of the primary alcohol **51** to

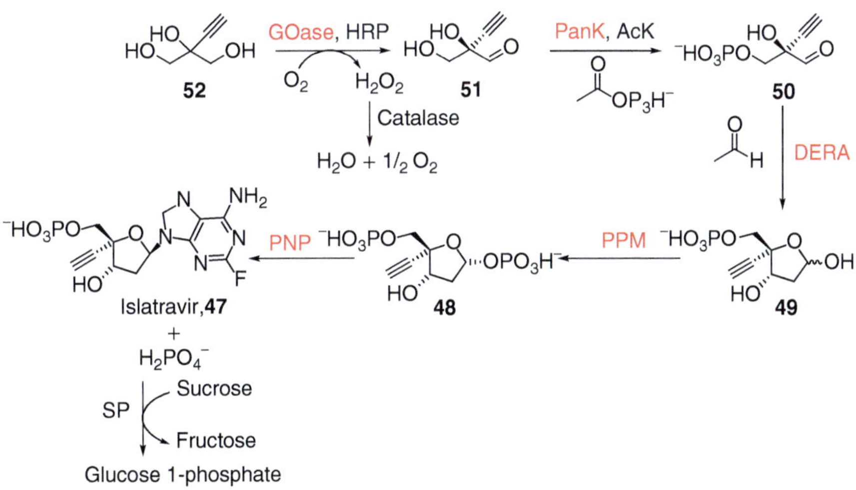

Figure 11.14 Five-step cascade employing five engineered enzymes (in red) and a total of nine enzymes for the synthesis of the HIV reverse transcriptase translocation inhibitor islatravir **47**.

the aldehyde **50**. For the C—C bond formation step to form the sugar backbone **49** via an aldol reaction between the acetaldehyde and **50**, they selected the deoxyribose 5-phosphate aldolase (DERA) from *Shewanella halifaxensis* as a template for evolution. The focus of the evolution process was to improve the tolerance toward acetaldehyde and after two rounds (11 amino acid substitutions) they obtained a variant able to tolerate >400 mM acetaldehyde, while retaining high activity. They subsequently needed a suitable enzyme to covert the 5-phosphate isomer **49** into the 1-phosphate **48** product. The PPM from *E. coli* was subjected to evolution by targeting residues in the active site using a homology model based on the *Bacillus cereus* PPM. A first round via single point libraries and a second round combining beneficial mutations from the previous round yielded a final variant with a 70-fold increase in activity over the wild-type bearing five substitutions. For the final glycosylation reaction to yield islatravir **47**, a PNP from *E. coli* was engineered. A variant with seven substitutions after four rounds of evolution was obtained with 350-fold improved activity. For thermodynamic reasons, the PNP and the PMP catalyzed reactions had to be performed simultaneously. Additional enzymes were added into the cascade. In the oxidation step, catalase was added in order to consume H_2O_2 generated in the reaction as a by-product. At the same time, horseradish peroxidase was also added in order to keep the appropriate oxidation state of the copper within the GOase active site. In the phosphorylation step, as phosphate source they employed acetyl phosphate combined with a thermostable acetate kinase (AcK) from *Thermotoga maritima*. Finally, to drive the equilibrium toward the formation of islatravir **47**, the sucrose phosphorylase (SP) from *Alloscardovia omnicolens* was used to deplete inorganic phosphate generated in the last step and shift the entire equilibrium forward. The whole cascade process containing five backbone enzymes and four assisting biocatalysts yielded islatravir **47** in 51% overall yield from ethynyl glycerol **52**.

11.2 Conclusions

The need for environmentally friendly chemical synthetic routes and the fact that naturally occurring enzymes are rarely suitable for industrial processes have made DE an essential tool to develop more efficient, selective, and robust biocatalysts. Even though the technology is well established, current methods still rely on a high degree of randomization due to the limited knowledge we have about enzyme catalysis. The development of new bioinformatics tools that can predict beneficial mutations, alongside faster screening methodologies, should reduce optimization times and therefore enable fast evolution of enzymes for different purposes. Processes such as those recently developed by GSK and Merck combining DE and enzymatic cascades would inspire alternative and new synthetic routes for the manufacture of chemicals in different industries in a more sustainable and efficient manner (Table 11.1).

Table 11.1 Selected examples of enzyme engineering approaches for the preparation of active pharmaceutical ingredients discussed in this chapter.

Product	Enzyme	Purpose	Engineering technique	References
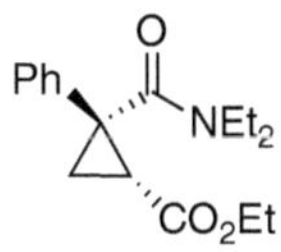 Levomilnacipran intermediate	P450	Increase cyclopropanation activity	Site-directed mutagenesis	[33]
 Telaprevir intermediate	Monoamino oxidase (MAO-N D5)	Expand substrate scope	Random mutagenesis	[42]
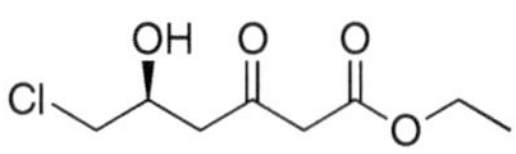 Atorvastatin intermediate	KRED and GDH	Increase activity and stability	DNA shuffling	[7]
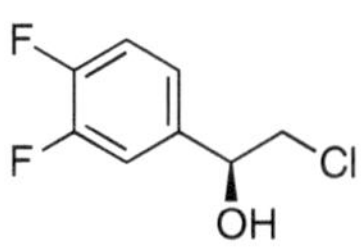 Ticagrelor intermediate	KRED	Increase activity	Site-directed mutagenesis	[11])
Unnatural aminoacids	MAL (methylaspartate ammonia lyase)	Expand substrate scope	Site-directed and saturation mutagenesis	[26, 69, 72, 73]
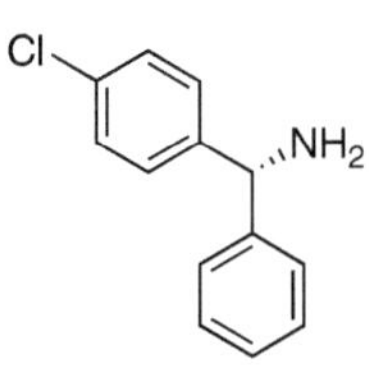 Levotirizine intermediate	Monoamino oxidase (MAO-N D9)	Expand substrate scope	Saturation mutagenesis	[40]
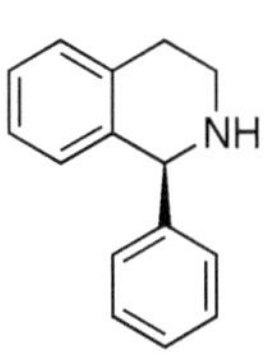 Solifenacin intermediate	Monoamino oxidase (MAO-N D11)	Expand substrate scope	Saturation mutagenesis	[40]

(Continued)

Table 11.1 (Continued)

Product	Enzyme	Purpose	Engineering technique	References
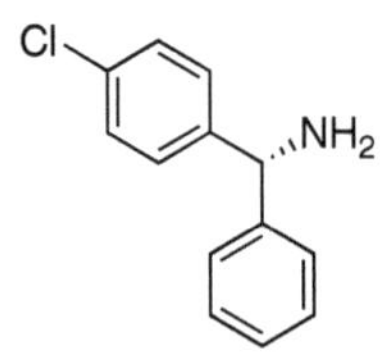 Levotirizine intermediate	Monoamino oxidase (MAO-N D11)	Expand substrate scope	Saturation mutagenesis	[40]
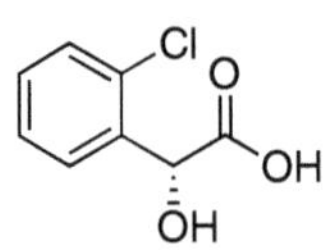 (*S*)-clopidogel intermediate	Esterase	Increase activity	Site-directed mutagenesis	[83]
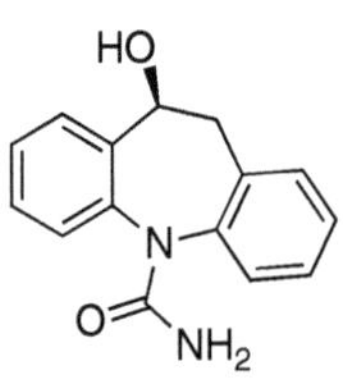 (*S*)-licarbazepine	KRED	Increase activity and operational stability	Directed evolution	[84]
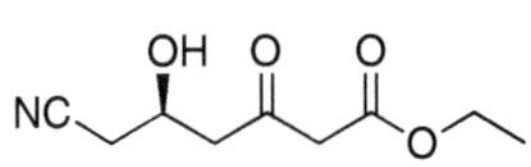 Atorvastatin intermediate	Halohydrin dehalogenase	Increase activity and operational stability	Different techniques to generate diversity	[8]
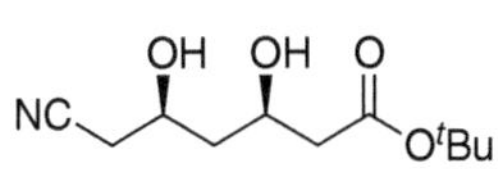 Atorvastatin intermediate	KRED	Increase activity	Site-saturation mutagenesis	[6]
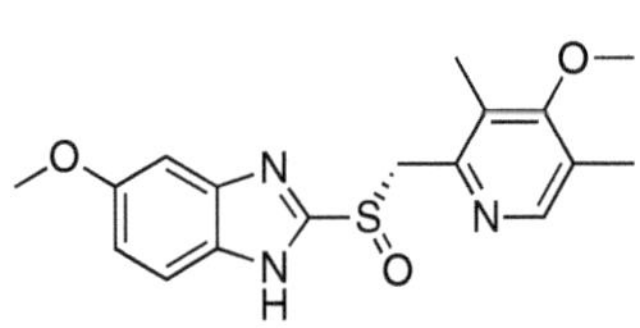 Esomeprazole Esomeprazole	BVMO	Increase activity	Code-evolver (directed evolution Codexis)	[38]

Table 11.1 (Continued)

Product	Enzyme	Purpose	Engineering technique	References
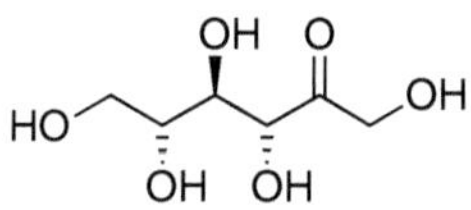 D-Psicose 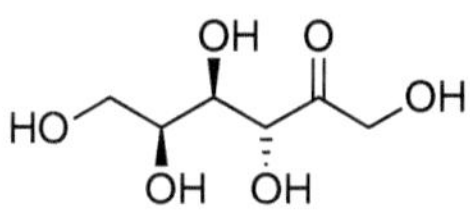 L-Tagatose	D-Tagatose epimerase	Increase activity	epPCR and site-saturation mutagenesis	[77]
Lactulose	Cellobiose 2-epimerase	Increase activity, thermostability	epPCR	[75]
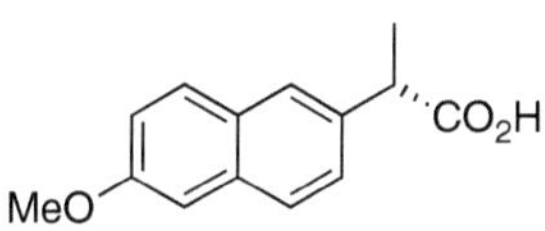 Naproxen	Arylmalonate decarboxylase	Inversion of stereochemistry and increased activity	Site-directed mutagenesis (MG)	[85]
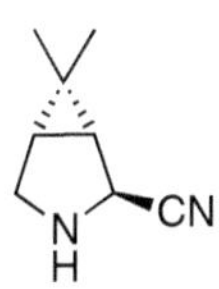 Boceprevir intermediate	Monoamine oxidase (MAO-N D5)	Increased activity, increased solubility and increased thermostability	ep-PCR and homology-inspired guided evolution	[46]
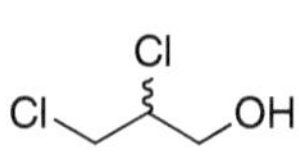 Epichlotohydrin precursors	Haloalkane dehalogenase	Improved and inverted stereoselectivity	Site-saturation and site-restricted mutagenesis	[81]
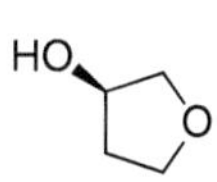 Amprenavir intermediate	Alcohol dehydrogenase	Improved and inverted stereoselectivity, and increased activity	Triple-code saturation mutagenesis	[16]
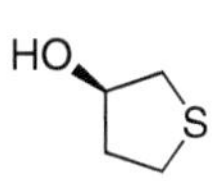 Sulopenem intermediate	Ketoreductase	Improved stereoselectivity	Random mutagenesis, gene shuffling, semi-synthetic shuffling, ProSAR	[15]

(Continued)

Table 11.1 (Continued)

Product	Enzyme	Purpose	Engineering technique	References
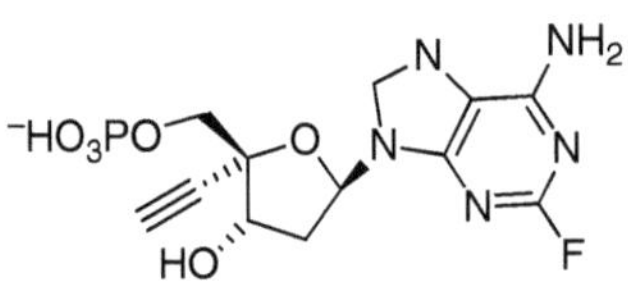 Islatravir	Multi-enzyme cascade	Improved activity, stability, expression, stereoselectivity, and decreased product inhibition	Site saturation mutagenesis, structure-guided engineering and recombination	[82]
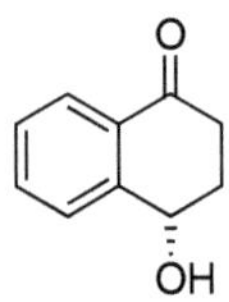 Common intermediate for anti-inflammatories	P450	Improved stereoselectivity	Structure-guided saturation mutagenesis	[35]
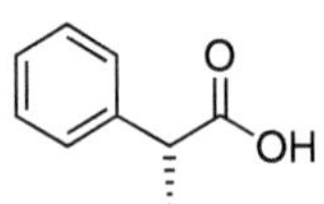 Profen derivatives	Lipase	Increased substrate scope and improved stereoselectivity	CASTing	[86]
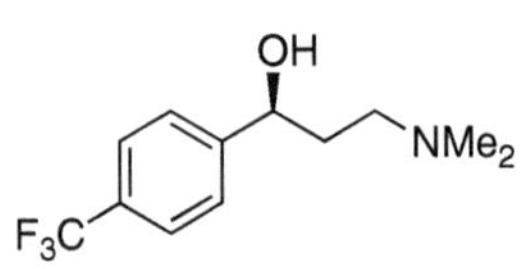 Fluoxetine intermediate	Carbonyl reductase	Improved and inverted stereoselectivity	CASTing	[87]
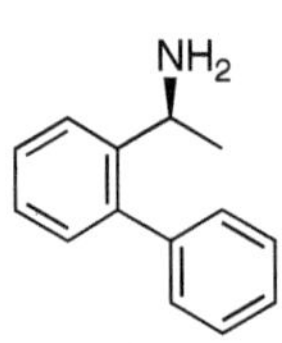 Bulky (*S*)-amine	Transaminase	Improved and inverted stereoselectivity	Structure-guided engineering	[88]
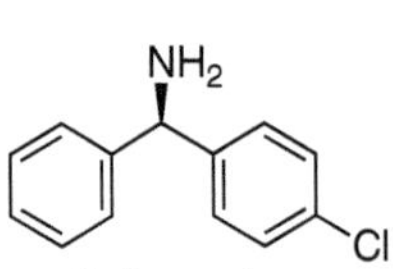 Bulky amines	Transaminase	Improved activity and stereoselectivity	Rational design	[4]

Table 11.1 (Continued)

Product	Enzyme	Purpose	Engineering technique	References
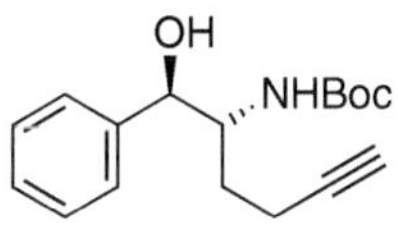 Vibegron intermediate	Ketoreductase	Increased pH stability, improved activity, and improved stereoselectivity	Rational design	[18]
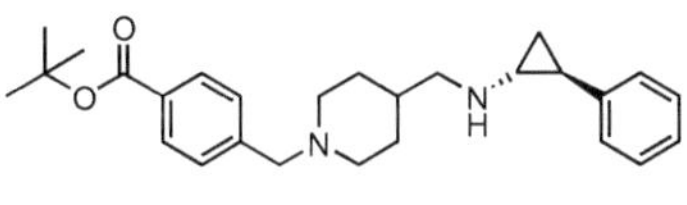 LSD1	Imine reductase	Increased pH stability, thermostability, and activity	Site-saturation mutagenesis, structure-guided engineering and recombination	[67]
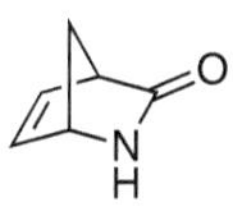 HIV-treatment intermediate	(+)-γ-Lactamase	Increased selectivity and increased thermostability	CAST and B-FITTER	[89]

Sources: Gao et al. [89], Ghislieri et al. [40].

References

1 Badenhorst, C.P.S. and Bornscheuer, U.T. (2018). Getting momentum: from biocatalysis to advanced synthetic biology. *Trends Biochem. Sci.* Elsevier Ltd. 43 (3): 180–198. https://doi.org/10.1016/j.tibs.2018.01.003.

2 Sheldon, R.A. and Pereira, P.C. (2017). Biocatalysis engineering: the big picture. *Chem. Soc. Rev.* Royal Society of Chemistry 46 (10): 2678–2691. https://doi.org/10.1039/c6cs00854b.

3 Savile, C.K., Janey, J.M., Mundorff, E.C. et al. (2010). Biocatalytic asymmetric synthesis of sitagliptin manufacture. *Science* 329 (July): 305–310. https://doi.org/10.1126/science.1188934.

4 Pavlidis, I.V., Weiß, M. S. Genz, M. et al. (2016). Identification of (*S*)-selective transaminases for the asymmetric synthesis of bulky chiral amines. *Nat. Chem.* Nature Publishing Group 8 (July): 1076–1082. https://doi.org/10.1038/NCHEM.2578.

5 Steffen-Munsberg, F., Vickers, C., Thontowi, A. et al. (2013). Connecting unexplored protein crystal structures to enzymatic function. *ChemCatChem* 5 (1): 150–153. https://doi.org/10.1002/cctc.201200544.

6 Luo, X., Wang, Y.J., Shen, W., and Zheng, Y.G., (2016). Activity improvement of a *Kluyveromyces lactis* aldo-keto reductase KlAKR via rational design. *J. Biotechnol.* Elsevier B.V. 224: 20–26. https://doi.org/10.1016/j.jbiotec.2016.03.008.

7 Ma, S.K., Gruber, J., Davis, C. et al. (2010). A green-by-design biocatalytic process for atorvastatin intermediate. *Green Chem.* 12 (1): 81–86. https://doi.org/10.1039/b919115c.

8 Fox, R.J., Davis, S.C., Mundorff, E.C. et al. (2007). Improving catalytic function by ProSAR-driven enzyme evolution. *Nat. Biotechnol.* 25 (3): 338–344. https://doi.org/10.1038/nbt1286.

9 van den Wijngaard, A.J., Reuvekamp, P.T.W., and Janssen, D.B. (1991). Purification and characterization of haloalcohol dehalogenase from *Arthrobacter* sp. strain AD2. *J. Bacteriol.* 173 (1): 124–129.

10 Nakamura, T., Nagasawa, T., Fujio, Y. et al. (1991). A new catalytic function of halohydrin hydrogen-halide-lyase, synthesis of β-hydroxynitriles from epoxides and cyanide. *Biochem. Biophys. Res. Commun.* 180 (1): 124–130. https://doi.org/10.1016/S0006-291X(05)81264-1.

11 Zhao, F.J., Liu, Y., Pei, X.Q. et al. (2017). Single mutations of ketoreductase *Ch*KRED20 enhance the bioreductive production of (1*S*)-2-chloro-1-(3,4-difluorophenyl) ethanol. *Appl. Microbiol. Biotechnol.* 101 (5): 1945–1952. https://doi.org/10.1007/s00253-016-7947-0.

12 Modukuru, N.K., Sukumaran, J., Collier, S.J. et al. (2014). Development of a practical, biocatalytic reduction for the manufacture of (*S*)-licarbazepine using an evolved ketoreductase. *Org. Process Res. Dev.* 18 (6): 810–815. https://doi.org/10.1021/op4003483.

13 Campopiano, O. et al. (2017). Ketoreductase polypeptides for the production of azetidinone. US Patent No. US7883879B2.

14 Ghosh, A.K., Thompson, W.J., Lee, H.Y. et al. (1993). Cyclic sulfolanes as novel and high affinity P2 ligands for HIV-1 protease inhibitors. *J. Med. Chem.* American Chemical Society 36 (7): 924–927. https://doi.org/10.1021/jm00059a019.

15 Liang, J., Mundorff, E., Voladri, R. et al. (2010). Highly enantioselective reduction of a small heterocyclic ketone: biocatalytic reduction of tetrahydrothiophene-3-one to the corresponding (*R*)-alcohol. *Org. Process Res. Dev.* 14 (1): 188–192. https://doi.org/10.1021/op9002714.

16 Sun, Z., Lonsdale, R., Ilie, A. et al. (2016). Catalytic asymmetric reduction of difficult-to-reduce ketones: triple-code saturation mutagenesis of an alcohol dehydrogenase. *ACS Catal.* 6 (3): 1598–1605. https://doi.org/10.1021/acscatal.5b02752.

17 Edmondson, S.D., Zhu, C., Kar, N.F. et al. (2016). Discovery of vibegron: a potent and selective β3 adrenergic receptor agonist for the treatment of overactive bladder. *J. Med. Chem.* 59 (2): 609–623. https://doi.org/10.1021/acs.jmedchem.5b01372.

18 Xu, F., Kosjek, B., Cabirol, F.L. et al. (2018). Synthesis of vibegron enabled by a ketoreductase rationally designed for high pH dynamic kinetic reduction. *Angew. Chem. Int. Ed.* 57 (23): 6863–6867. https://doi.org/10.1002/anie.201802791.

19 Metz, A.E. and Kozlowski, M.C. (2015). Recent advances in asymmetric catalytic methods for the formation of acyclic α,α-disubstituted α-amino acids. *J. Org. Chem.* 80 (1): 1–7. https://doi.org/10.1021/jo502408z.

20 Sato, T., Izawa, K., Aceña, J.L. et al. (2016). Tailor-made α-amino acids in the pharmaceutical industry: synthetic approaches to (1*R*,2*S*)-1-amino-2-vinylcyclopropane-1-carboxylic acid (Vinyl-ACCA). *Eur. J. Org. Chem.* 2016 (16): 2757–2774. https://doi.org/10.1002/ejoc.201600112.

21 Alexandre, F.R., Pantaleone, D.P., Taylor, P.P. et al. (2002). Amine-boranes: effective reducing agents for the deracemisation of DL-amino acids using L-amino acid oxidase from *Proteus myxofaciens*. *Tetrahedron Lett.* 43 (4): 707–710. https://doi.org/10.1016/S0040-4039(01)02233-X.

22 Drauz, K. (1997). Chiral amino acids. *Chimia* 6 (6): 310–314.

23 Komeda, H. and Asano, Y. (2000). Gene cloning, nucleotide sequencing, and purification anti characterization of the D-stereospecific amino-acid amidase from *Ochrobactrum anthropi* SV3. *Eur. J. Biochem.* 267 (7): 2028–2035. https://doi.org/10.1046/j.1432-1327.2000.01208.x.

24 Mathew, S., Bea, H., Nadarajan, S.P. et al. (2015). Production of chiral β-amino acids using ω-transaminase from *Burkholderia graminis*. *J. Biotechnol.* Elsevier B.V. 196–197: 1–8. https://doi.org/10.1016/j.jbiotec.2015.01.011.

25 Mathew, S., Nadarajan, S.P., Chung, T., Park, H.H., and Yun, H. (2016). Biochemical characterization of thermostable ω-transaminase from *Sphaerobacter thermophilus* and its application for producing aromatic β- and γ-amino acids. *Enzyme Microb. Technol.* Elsevier Inc. 87–88: 52–60. https://doi.org/10.1016/j.enzmictec.2016.02.013.

26 Weise, N.J., Parmeggiani, F., Ahmed, S.T., and Turner, N.J. (2015). The bacterial ammonia lyase EncP: a tunable biocatalyst for the synthesis of unnatural amino acids. *J. Am. Chem. Soc.* 137 (40): 12977–12983. https://doi.org/10.1021/jacs.5b07326.

27 Hanson, R.L., Johnston, R.M., Goldberg, S.L. et al. (2013). Enzymatic preparation of an *R*-amino acid intermediate for a γ-secretase inhibitor. *Org. Process Res. Dev.* 17 (4): 693–700. https://doi.org/10.1021/op400013e.

28 Vedha-Peters, K., Gunawardana, M., Rozzell, J.D., and Novick, S.J. (2006). Creation of a broad-range and highly stereoselective D-amino acid dehydrogenase for the one-step synthesis of D-amino acids. *J. Am. Chem. Soc.* 128 (33): 10923–10929. https://doi.org/10.1021/ja0603960.

29 Bernhardt, R. and Urlacher, V.B. (2014). Cytochromes P450 as promising catalysts for biotechnological application: chances and limitations. *Appl. Microbiol. Biotechnol.* 98 (14): 6185–6203. https://doi.org/10.1007/s00253-014-5767-7.

30 Girvan, H.M., Munro, A.W., Author, C., and Andrew Munro, A.W. (2016). Applications of microbial cytochrome P450 enzymes in biotechnology and synthetic biology. *Curr. Opin. Chem. Biol.* Elsevier Ltd 31: 136–145. https://doi.org/10.1016/j.cbpa.2016.02.018.

31 Coelho, P.S., Brustad, E.M., Kannan, A., and Arnold, F.H. (2013). Olefin cyclopropanation via carbene transfer catalyzed by engineered cytochrome P450 enzymes. *Science* 339 (6117): 307–310. https://doi.org/10.1126/science.1231434.

32 Coelho, P.S., Wang, Z.J., Ener, M.E. et al. (2013). A serine-substituted P450 catalyzes highly efficient carbene transfer to olefins in vivo. *Nat. Chem. Biol.* 9 (8): 485–487. https://doi.org/10.1038/nchembio.1278.

33 Wang, Z.J., Renata, H., Peck, N.E. et al. (2014). Improved cyclopropanation activity of histidine-ligated cytochrome P450 enables the enantioselective formal synthesis of levomilnacipran. *Angew. Chem. Int. Ed.* 53 (26): 6810–6813. https://doi.org/10.1002/anie.201402809.

34 Dwivedi, G.R., Upadhyay, H.C., Yadav, D.K. et al. (2014). 4-Hydroxy-α-tetralone and its derivative as drug resistance reversal agents in multi drug resistant *Escherichia coli*. *Chem. Biol. Drug Des.* 83 (4): 482–492. https://doi.org/10.1111/cbdd.12263.

35 Roiban, G.D., Agudo, R., Ilie, A. et al. (2014). CH-activating oxidative hydroxylation of 1-tetralones and related compounds with high regio- and stereoselectivity. *Chem. Commun.* 50 (92): 14310–14313. https://doi.org/10.1039/c4cc04925j.

36 Dong, J.J., Fernández-Fueyo, E., Hollmann, F. et al. (2018). Biocatalytic oxidation reactions: a chemist's perspective. *Angew. Chem. Int. Ed.* 57 (30): 9238–9261. https://doi.org/10.1002/anie.201800343.

37 Bong, Y.K. et al. (2011). Synthesis of Prazole Compounds. US Patent No. WO 2011/071982 A2.

38 Bong, Y.K., Song, S., Nazor, J. et al. (2018). Baeyer-Villiger monooxygenase-mediated synthesis of esomeprazole as an alternative for Kagan sulfoxidation. *J. Org. Chem.* 83 (14): 7453–7458. https://doi.org/10.1021/acs.joc.8b00468.

39 Batista, V.F., Galman, J.L., Pinto, D.C. et al. (2018). Monoamine oxidase: tunable activity for amine resolution and functionalization. *ACS Catal.* 8: 11889–11907. https://doi.org/10.1021/acscatal.8b03525.

40 Ghislieri, D., Green, A.P., Pontini, M. et al. (2013). Engineering an enantioselective amine oxidase for the synthesis of pharmaceutical building blocks and alkaloid natural products. *J. Am. Chem. Soc.* 135 (29): 10863–10869. https://doi.org/10.1021/ja4051235.

41 Rowles, I., Malone, K.J., Etchells, L.L. et al. (2012). Directed evolution of the enzyme monoamine oxidase (mao-n): highly efficient chemo-enzymatic deracemisation of the alkaloid (±)-crispinea. *ChemCatChem* 4 (9): 1259–1261. https://doi.org/10.1002/cctc.201200202.

42 Köhler, V., Bailey, K.R., Znabet, A. et al. (2010). Enantioselective biocatalytic oxidative desymmetrization of substituted pyrrolidines. *Angew. Chem. Int. Ed.* 49 (12): 2182–2184. https://doi.org/10.1002/anie.200906655.

43 Atkin, K.E., Reiss, R., Koehler, V. et al. (2008). The structure of monoamine oxidase from *Aspergillus niger* provides a molecular context for improvements in activity obtained by directed evolution. *J. Mol. Biol.* Elsevier Ltd 384 (5): 1218–1231. https://doi.org/10.1016/j.jmb.2008.09.090.

44 Li, G., Ren, J., Yao, P. et al. (2014). Deracemization of 2-methyl-1,2,3,4-tetrahydroquinoline using mutant cyclohexylamine oxidase obtained by iterative saturation mutagenesis. *ACS Catal.* 4 (3): 903–908. https://doi.org/10.1021/cs401065n.

45 Mirza, I.A., Burk, D.L., Xiong, B. et al. (2013). Structural analysis of a novel cyclohexylamine oxidase from *Brevibacterium oxydans* IH-35A. *PLoS One* 8 (3): 1–8. https://doi.org/10.1371/journal.pone.0060072.

46 Li, T., Liang, J., Ambrogelly, A. et al. (2012). Efficient, chemoenzymatic process for manufacture of the boceprevir bicyclic [3.1.0]proline intermediate based on amine oxidase-catalyzed desymmetrization. *J. Am. Chem. Soc.* 134 (14): 6467–6472. https://doi.org/10.1021/ja3010495.

47 Shibasaki, T., Mori, H., Chiba, S., and Ozaki, A. (1999). Microbial proline 4-hydroxylase screening and gene cloning. *Appl. Environ. Microbiol.* 65 (9): 4028–4031.

48 Chen, H. et al. (2013). Biocatalysts and methods for hydroxylation of chemical compounds. US Patent No. US20150118719A1.

49 Mangas-Sanchez, J., France, S.P., Montgomery, S.L. et al. (2017). Imine reductases (IREDs). *Curr. Opin. Chem. Biol.* Elsevier Ltd. 37: 19–25. https://doi.org/10.1016/j.cbpa.2016.11.022.

50 Mitsukura, K., Suzuki, M., Tada, K., Yoshida, T., and Nagasawa, T. (2010). Asymmetric synthesis of chiral cyclic amine from cyclic imine by bacterial whole-cell catalyst of enantioselective imine reductase. *Org. Biomol. Chem.* 8 (20): 4533–4535. https://doi.org/10.1039/c0ob00353k.

51 Mitsukura, K., Suzuki, M., Shinoda, S. et al. (2011). Purification and characterization of a novel (*R*)-imine reductase from *Streptomyces* sp. GF3587. *Biosci. Biotechnol. Biochem.* 75 (9): 1778–1782. https://doi.org/10.1271/bbb.110303.

52 Mitsukura, K., Kuramoto, T., Yoshida, T. et al. (2013). A NADPH-dependent (*S*)-imine reductase (SIR) from *Streptomyces* sp. GF3546 for asymmetric synthesis of optically active amines: purification, characterization, gene cloning, and expression. *Appl. Microbiol. Biotechnol.* 97 (18): 8079–8086. https://doi.org/10.1007/s00253-012-4629-4.

53 Aleku, G.A., Man, H., France, S.P. et al. (2016). Stereoselectivity and structural characterization of an imine reductase (IRED) from *Amycolatopsis orientalis*. *ACS Catal.* 6: 3380–3889. https://doi.org/10.1021/acscatal.6b00782.

54 France, S.P., Aleku, G.A., Sharma, M. et al. (2017). Biocatalytic routes to enantiomerically enriched dibenz[*c*,*e*]azepines. *Angew. Chem. Int. Ed.*: 15589–15593. https://doi.org/10.1002/anie.201708453.

55 Gand, M., Müller, H., Wardenga, R., and Höhne, M. (2014). Characterization of three novel enzymes with imine reductase activity. *J. Mol. Catal. B: Enzym.* Elsevier B.V. 110: 126–132. https://doi.org/10.1016/j.molcatb.2014.09.017.

56 Gröger, H. et al. (2017). Asymmetric biocatalytic reduction of cyclic imines: design and application of a Tailor-made whole-cell catalyst. *Heterocycles* 95 (2): 1261. https://doi.org/10.3987/COM-16-S(S)89.

57 Hussain, S., Leipold, F., Man, H. et al. (2015). An (*R*)-imine reductase biocatalyst for the asymmetric reduction of cyclic imines. *ChemCatChem* 7 (4): 579–583. https://doi.org/10.1002/cctc.201402797.

58 Lenz, M., Scheller, P.N., Richter, S.M. et al. (2016). Cultivation and purification of two stereoselective imine reductases from *Streptosporangium roseum* and

Paenibacillus elgii. *Protein Exp. Purif.* Elsevier Ltd. doi: https://doi.org/10.1016/j.pep.2016.05.003.

59 Scheller, P.N., Fademrecht, S., Hofelzer, S. et al. (2014). Enzyme toolbox: novel enantiocomplementary imine reductases. *ChemBioChem* 15 (15): 2201–2204. https://doi.org/10.1002/cbic.201402213.

60 Aleku, G.A., France, S.P., Man, H. et al. (2017). A reductive aminase from *Aspergillus oryzae*. *Nat. Chem.* Nature Publishing Group 9: 961–969.

61 France, S.P., Howard, R.M., Steflik, J. et al. (2017). Identification of novel bacterial members of the imine reductase enzyme family that perform reductive amination. *ChemCatChem* John Wiley & Sons, Ltd 10 (3): 510–514. https://doi.org/10.1002/cctc.201701408.

62 Huber, T., Schneider, L., Präg, A. et al. (2014). Direct reductive amination of ketones: structure and activity of S-selective imine reductases from streptomyces. *ChemCatChem* 6 (8): 2248–2252. https://doi.org/10.1002/cctc.201402218.

63 Lenz, M., Meisner, J., Quertinmont, L. et al. (2017). Asymmetric ketone reduction by imine reductases. *ChemBioChem* 18 (3): 253–256. https://doi.org/10.1002/cbic.201600647.

64 Roiban, G.D., Kern, M., Liu, Z. et al. (2017a). Efficient biocatalytic reductive aminations by extending the imine reductase toolbox. *ChemCatChem* 9 (24): 4475–4479. https://doi.org/10.1002/cctc.201701379.

65 Sharma, M., Mangas-Sanchez, J., France, S.P. et al. (2018). A mechanism for reductive amination catalyzed by fungal reductive aminases. *ACS Catal.*: 11534–11541. https://doi.org/10.1021/acscatal.8b03491.

66 Wetzl, D., Gand, M., Ross, A. et al. (2016). Asymmetric reductive amination of ketones catalyzed by imine reductases. *ChemCatChem*: 1–5. https://doi.org/10.1002/cctc.201600384.

67 Schober, M., MacDermaid, C., Ollis, A.A. et al. (2019). Chiral synthesis of LSD1 inhibitor GSK2879552 enabled by directed evolution of an imine reductase. *Nat. Catal.* Nature Publishing Group 2 (10): 909–915. https://doi.org/10.1038/s41929-019-0341-4.

68 Parmeggiani, F., Weise, N.J., Ahmed, S.T., and Turner, N.J. (2018). Synthetic and therapeutic applications of ammonia-lyases and aminomutases. *Chem. Rev.* 118 (1): 73–118. https://doi.org/10.1021/acs.chemrev.6b00824.

69 Raj, H., Szymański, W., De Villiers, J. et al. (2012). Engineering methylaspartate ammonia lyase for the asymmetric synthesis of unnatural amino acids. *Nat. Chem.* 4 (6): 478–484. https://doi.org/10.1038/nchem.1338.

70 Levy, C.W., Buckley, P.A., Sedelnikova, S. et al. (2002). Insights into enzyme evolution revealed by the structure of methylaspartate ammonia lyase. *Structure* 10 (1): 105–113. https://doi.org/10.1016/S0969-2126(01)00696-7.

71 Rowles, I., Groenendaal, B., Binay, B. et al. (2016). Engineering of phenylalanine ammonia lyase from *Rhodotorula graminis* for the enhanced synthesis of unnatural L-amino acids. *Tetrahedron* Elsevier Ltd. 72 (46): 7343–7347. https://doi.org/10.1016/j.tet.2016.06.026.

72 Ahmed, S.T., Parmeggiani, F., Weise, N.J. et al. (2018). Engineered ammonia lyases for the production of challenging electron-rich L-phenylalanines. *ACS Catal.* 8 (4): 3129–3132. https://doi.org/10.1021/acscatal.8b00496.

73 Parmeggiani, F., Lovelock, S.L., Weise, N.J. et al. (2015). Synthesis of D- and L-phenylalanine derivatives by phenylalanine ammonia lyases: a multienzymatic cascade process. *Angew. Chem. Int. Ed.* 54 (15): 4608–4611. https://doi.org/10.1002/anie.201410670.

74 Beerens, K., Desmet, T., and Soetaert, W. (2012). Enzymes for the biocatalytic production of rare sugars. *J. Ind. Microbiol. Biotechnol.* 39 (6): 823–834. https://doi.org/10.1007/s10295-012-1089-x.

75 Shen, Q., Zhang, Y., Yang, R. et al. (2016). Enhancement of isomerization activity and lactulose production of cellobiose 2-epimerase from *Caldicellulosiruptor saccharolyticus*. *Food Chem.* Elsevier Ltd 207: 60–67. https://doi.org/10.1016/j.foodchem.2016.02.067.

76 Bosshart, A., Hee, C.S., Bechtold, M. et al. (2015). Directed divergent evolution of a thermostable D-tagatose epimerase towards improved activity for two hexose substrates. *ChemBioChem* 16 (4): 592–601. https://doi.org/10.1002/cbic.201402620.

77 Bosshart, A., Wagner, N., Lei, L. et al. (2016). Highly efficient production of rare sugars D-psicose and L-tagatose by two engineered D-tagatose epimerases. *Biotechnol. Bioeng.* 113 (2): 349–358. https://doi.org/10.1002/bit.25547.

78 Ju, X., Yu, H.L., Pan, J. et al. (2010). Bioproduction of chiral mandelate by enantioselective deacylation of α-acetoxyphenylacetic acid using whole cells of newly isolated *Pseudomonas* sp. ECU1011. *Appl. Microbiol. Biotechnol.* 86 (1): 83–91. https://doi.org/10.1007/s00253-009-2286-z.

79 Park, S., Zheng, L., Kumakiri, S. et al. (2014). Development of DNA-based hybrid catalysts through direct ligand incorporation: toward understanding of DNA-based asymmetric catalysis. *ACS Catal.* 4 (11): 4070–4073. https://doi.org/10.1021/cs501086f.

80 Janssen, D.B. (2004). Evolving haloalkane dehalogenases. *Curr. Opin. Chem. Biol.* 8 (2): 150–159. https://doi.org/10.1016/j.cbpa.2004.02.012.

81 van Leeuwen, J.G.E., Wijma, H.J., Floor, R.J. et al. (2012). Directed evolution strategies for enantiocomplementary haloalkane dehalogenases: from chemical waste to enantiopure building blocks. *ChemBioChem* 13 (1): 137–148. https://doi.org/10.1002/cbic.201100579.

82 Huffman, M.A., Fryszkowska, A., Alvizo, O. et al. (2019). Design of an in vitro biocatalytic cascade for the manufacture of islatravir. *Science* 366 (6470): 1255–1259. https://doi.org/10.1126/science.aay8484.

83 Di Ma, B., Kong, X. D., Yu, H. L. et al. (2014). Increased catalyst productivity in α-hydroxy acids resolution by esterase mutation and substrate modification. *ACS Catal.* 4 (3): 1026–1031. https://doi.org/10.1021/cs401183e.

84 Ju, X., Tang, Y., Liang, X. et al. (2014). Development of a biocatalytic process to prepare (*S*)-*N*-boc-3-hydroxypiperidine. *Org. Process Res. Dev.* 18 (6): 827–830. https://doi.org/10.1021/op500022y.

85 Miyauchi, Y., Kourist, R., Uemura, D., and Miyamoto, K. (2011). Dramatically improved catalytic activity of an artificial (*S*)-selective arylmalonate decarboxylase by structure-guided directed evolution. *Chem. Commun.* 47 (26): 7503–7505. https://doi.org/10.1039/c1cc11953b.

86 Engström, K., Nyhlén, J., Sandström, A. G., and Bäckvall, J. E. (2010). Directed evolution of an enantioselective lipase with broad substrate scope for hydrolysis of α-substituted esters. *J. Am. Chem. Soc.* 132 (20): 7038–7042. https://doi.org/10.1021/ja100593j.

87 Zhang, D., Chen X., Chi, J. et al. (2015). Semi-rational engineering a carbonyl reductase for the enantioselective reduction of β-amino ketones. *ACS Catal.* 5 (4): 2452–2457. https://doi.org/10.1021/acscatal.5b00226.

88 Dourado, D.F.A.R., Pohle, S., Carvalho, A. T. P. et al. (2016). Rational design of a (*S*)-selective-transaminase for asymmetric synthesis of (1*S*)-1-(1,1′-biphenyl-2-yl) ethanamine. *ACS Catal.* 6 (11): 7749–7759. https://doi.org/10.1021/acscatal.6b02380.

89 Gao, S., Zhu, S., Huang, R. et al. (2018). Engineering the enantioselectivity and thermostability of a (+)-lactamase from microbacterium hydrocarbonoxydans for kinetic resolution of vince lactam (2-azabicyclo[2.2.1]hept-5-en-3-one). *Appl. Environ. Microbiol.* 84 (1): 1–13.

12

Directing Evolution of the Fungal Ligninolytic Secretome

Javier Viña-Gonzalez[1] and Miguel Alcalde[2]

[1] *EvoEnzyme S.L., Parque Científico de Madrid, Faraday 7, Cantoblanco Campus, 28049, Madrid, Spain*
[2] *Institute of Catalysis, CSIC, Department of Biocatalysis, Marie Curie 228049 Madrid, Spain*

12.1 The Fungal Ligninolytic Secretome

Present in wood, grass, and agricultural residues, lignocellulose is the most abundant feedstock in the biosphere. As such, within the new sustainable paradigm of the twenty-first century, lignocellulose biomass is considered the most ecological alternative to crude-oil for the production of fuel and chemicals [1]. Lignocellulose is made up of cellulose and hemicellulose (up to 90%), with lignin representing the third most abundant component at roughly 10–25%. While cellulose and hemicellulose are polysaccharides, lignin is a three-dimensional aromatic and hydrophobic biopolymer of phenylpropane units that strengthens the cell wall while protecting the plant against microbial attack [1, 2]. Due to its antimicrobial and antioxidant properties, lignin has many applications that range from the synthesis of value-added bioproducts to its use in nanotechnology [3–5]. Despite its widespread natural distribution, only a few organisms are capable of overcoming the recalcitrance of lignin to degrade it.

Lignin deconstruction by bacterial attack is a rare phenomenon, occurring slowly and requiring the active assistance of certain fungi [6, 7]. Indeed, basidiomycete white-rot fungi are essentially the only organisms capable of breaking down lignin, provoking two different patterns of decay: selective and simultaneous lignin degradation. The fungal mineralization of lignin is achieved by an extracellular enzymatic consortium, the ligninolytic secretome, assisted by fungal aromatic and phenolic metabolites, organic acids, and reactive oxygen species [6, 8, 9]. The taskforce of ligninolytic oxidoreductases – also known as ligninases – is mostly comprised of high-redox potential peroxidases (LiP, MnP, VP), peroxygenases (unspecific peroxygenase [UPO]), laccases, and enzymes that supply H_2O_2, Figure 12.1.

Lignin peroxidases (LiPs) catalyze the H_2O_2-mediated oxidation of non-phenolic lignin units via an external catalytic tryptophan connected to the heme domain through a long-range electron transfer pathway [10, 11]. Manganese peroxidases

Protein Engineering: Tools and Applications, First Edition. Edited by Huimin Zhao.
© 2021 WILEY-VCH GmbH. Published 2021 by WILEY-VCH GmbH.

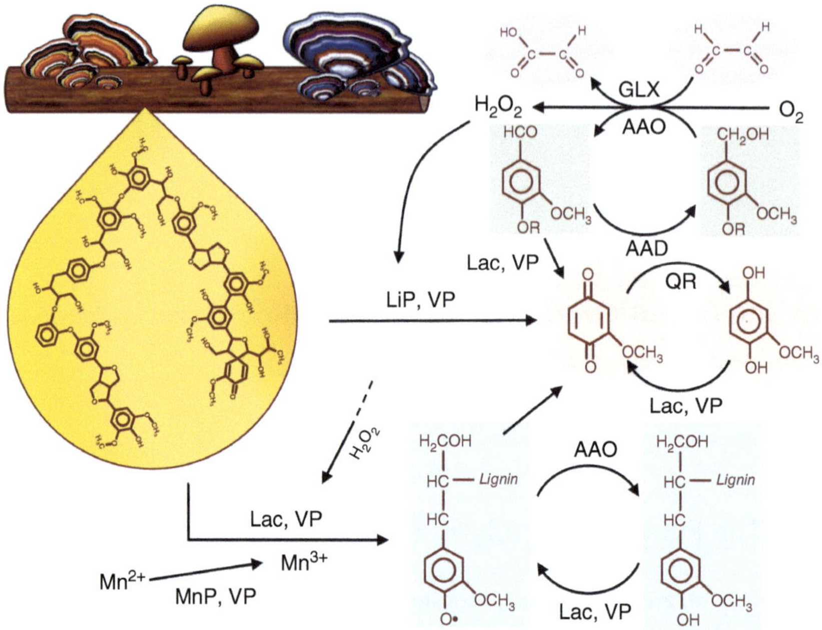

Figure 12.1 Interrelationships for the ligninolytic consortium of enzymes. Lac, laccase; LiP, lignin peroxidase; MnP, manganese peroxidase; VP, versatile peroxidase; GLX, glyoxal oxidase; AAO, aryl-alcohol oxidase; AAD, aryl-alcohol deshidrogenase; QR, quinone reductase.

(MnPs) transform Mn^{2+} into Mn^{3+}, which, once stabilized by chelation with different organic acids (e.g. oxalate, malate, or lactate), can act as a strong oxidant of phenolic structures in the lignin polymer [12]. Versatile peroxidases (VPs) integrate the structural and mechanistic features of LiP, MnP, and generic peroxidases, bringing together within the same protein scaffold the catalytic tryptophan, the Mn^{2+} binding site, and the heme domain for the oxidation of high-, medium-, and low-redox potential compounds. Consequently, VPs can oxidize both the phenolic and non-phenolic lignin moieties [13, 14]. UPOs are unique heme-thiolate mono(per)oxygenases that were recently added to this enzymatic pool of H_2O_2 users. UPOs may be involved in the oxidation of smaller lignin fragments by *O*-demethylation and cleavage of non-phenolic lignin units, suggesting a possible role in lignin combustion after the action of peroxidase counterparts [15, 16].

To instigate such complex oxidative process, white-rot fungi need to secrete several other oxidases that feed H_2O_2 to the peroxidases and peroxygenases. In the case of glyoxal oxidase (GLX), a copper radical oxidase, the production of H_2O_2 through the reduction of molecular oxygen is coupled to the oxidation by LiP of glycolaldehyde to oxalate, which in turn chelates the Mn^{3+} produced by MnP. By contrast, the aryl-alcohol oxidase (AAO) produces H_2O_2 by oxidizing both lignin compounds and fungal metabolites. In addition, AAO collaborates with mycelium dehydrogenases in an aromatic alcohol/aldehyde redox cycle to generate a continuous supply of peroxide to the enzymatic consortium [17, 18].

Last but not least, laccases are broad specificity blue copper oxidases that catalyze the oxidation of the phenolic elements of lignin. Moreover, the redox potential of such attack can be escalated toward non-phenolic units when low molecular weight compounds are introduced to act as diffusible redox mediators [19].

Uniquely situated in the carbon cycle and with one of the most complex enzymatic tasks in nature, the ligninolytic consortium has truly awakened the interest of biotechnologists. High-redox potential, substrate promiscuity, and high selectivity are certainly desired characteristics for industrial catalysts, and the potential to achieve such features opens a prospective landscape of applications that range from organic chemistry and decontamination to biomedical applications. Ligninases can be used in dozens of organic synthesis processes, from alkane functionalization, aromatic hydroxylation, and selective epoxidation to chiral resolution. Significantly, these enzymes offer solutions for bioremediation, degrading recalcitrant pollutants such as chlorinated benzenes, polycyclic aromatic hydrocarbons, phenolic pollutants, and many other xenobiotic agents. Lignocellulose biorefineries are also generating much interest, in which ligninases can be employed either as individual catalysts or as a coordinated synthetic ligninolytic secretome for the production of biofuels and chemicals [20, 21]. Moreover, ligninases can be used in biosensors and as chemical reporters in food, pulp, and paper industries and in cleaning and textile processes [20, 22–25].

With the aim of converting ligninases into true biotechnological catalysts, protein engineering can be used to sculpt their properties and adapt them to industrial standards. Currently, the best methodology to tackle such a challenge is without doubt directed evolution (from the pioneering work of Prof. Frances H. Arnold at the California Institute of Technology, recently awarded the Nobel Prize in Chemistry for this invention). In this conceptual twist to the well-known natural evolution algorithm, genetic diversity is first created by random mutagenesis and DNA recombination, and the mutant libraries generated are then screened in high-throughput assays (HTSs), repeating this process until the desired enzyme trait is attained [26, 27].

In the past 15 years, we have witnessed important advances in the laboratory evolution of different elements of the ligninolytic consortium, addressing different needs from bioremediation to green synthetic chemistry. In these enterprises, reliable heterologous functional expression platforms, *ad hoc* library creation methods, and specific screening assays and computational protocols have been applied to fully exploit the power of evolution, Figure 12.2 [20, 28–30]. Here, we summarize and discuss each of the strategies followed during this challenging journey to transfer ligninases from the forest to the laboratory bench by directing their evolution.

12.2 Functional Expression in Yeast

12.2.1 The Evolution of Signal Peptides

To be able to drive their evolution toward new properties, it is essential for ligninases to be expressed functionally in an adequate heterologous host and the

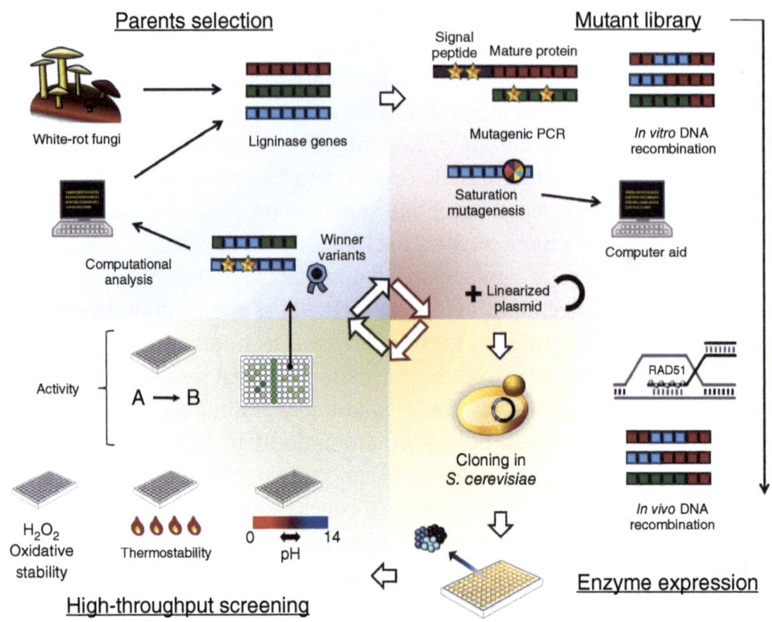

Figure 12.2 General scheme for a directed evolution round of ligninases. From one or several parental types, mutant libraries are created, cloned and functionally expressed in *S. cerevisiae*. Supernatants are assessed by HTS assays revealing the most promising candidates which are then characterized and submitted to a new evolution round. Computational analysis can be performed to rationalize mutations as well as to identify new hot-spot residues.

budding yeast *Saccharomyces cerevisiae* is the go-to organism for such effects. This ascomycete can express foreign eukaryotic proteins, performing correct folding and post-translational modifications that include glycosylation, disulfide bond formation, and N- and C-terminal processing. In addition, episomal vectors are available for this yeast, with high-transformation efficiencies, and there is a substantial tool-box for its genetic manipulation [31–33]. Moreover, *S. cerevisiae* can be easily handled within a high-throughput context in synthetic biology, making it one of the favored model organisms for directed evolution of eukaryotic enzymes. However, the lack of more specific chaperones, the scarcity of prosthetic groups, and serious bottlenecks in the Golgi compartment can limit heterologous expression in this yeast [34], a problem in the framework of such experiments. Yeast growth is strongly limited by oxygen availability and stirring, and given that library construction and screening is typically performed in microtiter plates, expression may also be limited in such conditions.

The most straightforward strategy to overcome these shortcomings is the directed evolution of the native signal sequence, such that the transit and processing of the exogenous polypeptide adapts to the subtleties of the secretory pathway. One of the first successful examples of this was the directed evolution of the medium-redox potential laccase from *Myceliophthora thermophila* (MtL) both in the signal peptide and the mature protein, enhancing secretion from scratch up to 18 mg l^{-1} in *S. cerevisiae*. In addition to 11 substitutions in the mature protein, several mutations were identified in the signal peptide that were introduced over 10 generations of evolution. Interestingly, not only the N-terminal but also the C-terminal of the MtL laccase was processed during secretion, specifically due to the H(c2)R mutation that created a new cleavage site for the *KEX*2 protease in the Golgi apparatus, facilitating the C-terminal maturation of the laccase and increasing secretion 10-fold [35]. The more recent directed evolution of an UPO from *Agrocybe aegerita* involved the inclusion of four mutations in a relatively short backbone sequence within the native signal peptide [36–38]. These mutations sat in the hydrophobic core of the leader peptide (F12Y-A14F-R15G-A21D), and they affected the interaction between the signal recognition particle and the signal peptide prior to cleavage. As such, these changes may benefit the correct insertion of the UPO chain into the membrane bilayer of the endoplasmic reticulum (ER). As a result, secretion was enhanced 27-fold, with total expression by *S. cerevisiae* of 8 mg l^{-1} and of over 200 mg l^{-1} by *Pichia pastoris* in a bioreactor [38].

Another useful approach to improve functional heterologous expression is to replace the native signal sequence with common yeast prepro-leaders, such as those from the mating α-factor, the invertase, the acid phosphatase, and the killer K1 toxin [39–42], or even by designing synthetic prepro-leaders through sequence-based semi rational approaches [43]. These prepro-leaders can then be subjected to directed evolution to further improve secretion, as in the cases of two high-redox potential laccases (HRPLs) from *Pycnoporus cinnabarinus* (PcL) and PM1 basidiomycete (PM1L), where the α-factor prepro-leader from *S. cerevisiae* replaced the native signal peptides. During the directed evolution of PcL for secretion, six cycles of random mutation, DNA recombination, and screening were employed to introduce five beneficial substitutions into the mature protein. Besides, six mutations were included in the α-factor prepro-leader (A9D in the pre-leader and F48S-S58G-G62R-E86G in the pro-leader), which led to a 40-fold improvement in secretion [44]. We carried out eight generations of directed evolution on PM1L to obtain a secretion mutant expressed at ~8 mg l^{-1}. Three of the 11 substitutions in this variant were located in the α-factor prepro-leader: V10D in the pre-leader, and N23K and A87T in the pro-leader. Interestingly, the V10D mutation, like the A9D substitution in PcL, produced a 2.2-fold improvement in secretion due to a reduced hydrophobicity of this domain, which may favor the translocation of the nascent polypeptide into the ER. Cleavage optimization also helped obtain a more soluble enzyme, as seen with the A87T mutation that increased hydrophilicity near the *KEX*2 protease cleavage site, potentially reinforcing the final processing of the α-factor pro-leader [45].

A more recent twist in prepro-leader design involved constructing chimeric leaders for the expression of AAO from *Pleurotus eryngii*. In this study, several AAO fusions harboring combinations of the pre- and pro-leaders from the α-factor and from the K1 killer toxin were explored: preproα-AAO, preαproK-AAO, preKproα-AAO, and preproK-AAO [46]. The most suitable synthetic preαproK chimeric peptide suppressed a deficient cleavage site for the *STE13* protease in the Golgi compartment to promote the correct N-terminal processing and to foster secretion. Recently, the preαproK-AAO was evolved by adding four new mutations to the peptide, one in the preα leader and three in the proK leader, producing final secretion levels of 4.5 $mg\,l^{-1}$ in *S. cerevisiae* and up to 25 $mg\,l^{-1}$ in *P. pastoris* in a fed-batch bioreactor [47].

All the examples mentioned earlier pursue the dream of generating a universal signal peptide to standardize the efficient secretion of foreign proteins in *S. cerevisiae.* Indeed, a pioneering study already addressed whether similarly evolved α-factor prepro-leaders improve the expression of very different proteins in yeast [48]. Encouraged by the results, we recently analyzed the effect of several evolved α-factor prepro-leaders on different laccases: PM1L, PcL, MtL, and C30L from *Trametes* sp. After evaluating the expression of 12 constructs, we found a correlation between the signal prepro-leaders and the laccase sequences from which they were evolved, highlighting a mild tendency among laccases but unfortunately, far from defining a true universal peptide leader [49].

12.2.2 Secretion Mutations in Mature Protein

While the evolution of signal peptides has proven to be successful to enhance the expression of ligninases, favoring the ER to Golgi transit and exocytosis, the accumulation of secretion mutations in the mature protein is also an invaluable strategy to circumvent folding and post-translational related hurdles. Secretion mutations foster solubility, stability, and also the kinetics and thermodynamics of intermediate protein states [50]. For instance, when the VP from *P. eryngii* was fused to the α-factor prepro-leader and subjected to joint directed evolution for expression in yeast, all the potential beneficial mutations that appeared at the α-factor prepro-leader were eventually ruled out by *in vivo* shuffling. As such, with only the four secretion mutations accumulated in the mature protein, the ultimate secretion variant was expressed at about 22 $mg\,l^{-1}$ [51]. These mutations improved folding and maturation by enhancing structural flexibility and thermostability. Likewise, only five substitutions in the mature UPO mutant were responsible for an enhancement above 40-fold in total activity following its directed evolution for secretion. These mutations were in hydrophobic regions and they were conservative in terms of polarity and charge, highlighting a relationship between secretion and stability through hydrophobic interactions, and the spatial readjustment of certain protein regions [37] Figure 12.3. In both the UPO and VP studies, the accumulation of beneficial mutations and their synergistic effect in the mature protein was achieved by *in vivo* shuffling of the best performers (with four mutations after four rounds of evolution for VP and five mutations after five rounds of evolution for UPO). Thus, this strategy helps to preclude the introduction of neutral or deleterious mutations within the scheme of evolution.

However, synergy between mutations is not always needed to produce a quantitative jump in secretion. In a laboratory evolution campaign of AAO for secretion, only one substitution at the catalytic pocket increased secretion almost 100-fold. The H91N ancestral/consensus mutation improved the anchoring of

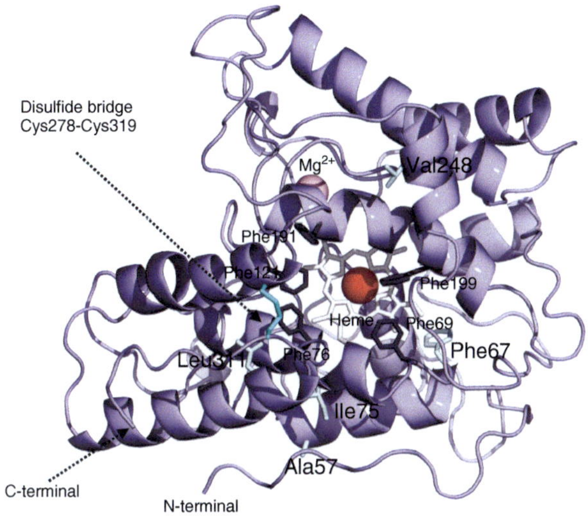

Figure 12.3 Mapping of the mutations (highlighted in blue) for functional expression and activity on the crystal structure of the evolved UPO from *Agrocybe aegerita*. The five aromatic residues involved in delimiting the entrance to the heme access channel and orienting the substrate to the heme are depicted in dark blue, the heme in white with the Fe^{3+} in red, the structural Mg^{2+} in pink, and the disulfide bridge in magenta (PDB entry 5OXU). Molina-Espeja et al. [30].

the FAD co-factor, again establishing a dependence on stability for functional expression [46].

12.2.3 The Importance of Codon Usage

At the genomic level, gene expression is also regulated by codon usage so that strongly expressed proteins present a bias toward frequently used codons, thereby avoiding restrictions due to tRNA availability [52]. Currently, ordering synthetic genes with optimized codon usage seems to be a reasonable option to favor expression, yet this approach is not always successful and there are many fruitless cases in different heterologous hosts. By contrast, the random introduction of silent mutations to favor codon usage can enhance the expression of different ligninases. This is particularly relevant for laccases like the PcL secretion variant that accumulated four silent mutations, all of them favored codon usage [44]. Similarly, the PM1L secretion mutant accumulated five silent mutations, three favored codon usage [45]. A particular interesting case study is the evolution of the MtL, in which seven of the nine silent mutations that appeared in selected variants during the first 10 generations of evolution toward functional expression were substitutions to a more commonly used codon [35]. Moreover, in the whole history of synthetic MtL evolution, the final mutant selected after 23 rounds of evolution was also the best-expressed laccase (evolved for secretion, organic solvent and pH tolerance, and for the synthesis of heteropolymeric dyes). Indeed, in the last three evolution campaigns, the acquisition of four new substitutions doubled secretion to

$37 \, mg \, l^{-1}$, half of which were silent mutations that favored protein translation and therefore, promoted its expression [53–55].

12.3 Yeast as a Tool-Box in the Generation of DNA Diversity

Among the most important advantages of *S. cerevisiae* for directed evolution is its high frequency of homologous DNA recombination that can be used as a molecular tool-box for development library creation methods. Thanks to the Rad51 recombinase (an ortholog of the bacterial recA) and other ancillary factors, the recombination machinery of *S. cerevisiae* allows DNA fragments with 40 homologous nucleotides to be assembled with roughly 60% efficiency [29, 32, 56]. Not only does this feature facilitate the *in vivo* cloning of libraries into linearized expression vectors in a single transformation step but also, it opens the door to boost genetic diversity. Thus, an ever-growing family of DNA recombination methods based on the *in vivo* gap repair mechanism in yeast is available in the laboratory to aid the evolution of ligninases and many other enzymes Figure 12.4.

Based on the engineering of specific overlapping homologous areas, the IVOE (In Vivo Overlap Extension) protocol has been developed as a fast and reliable way to

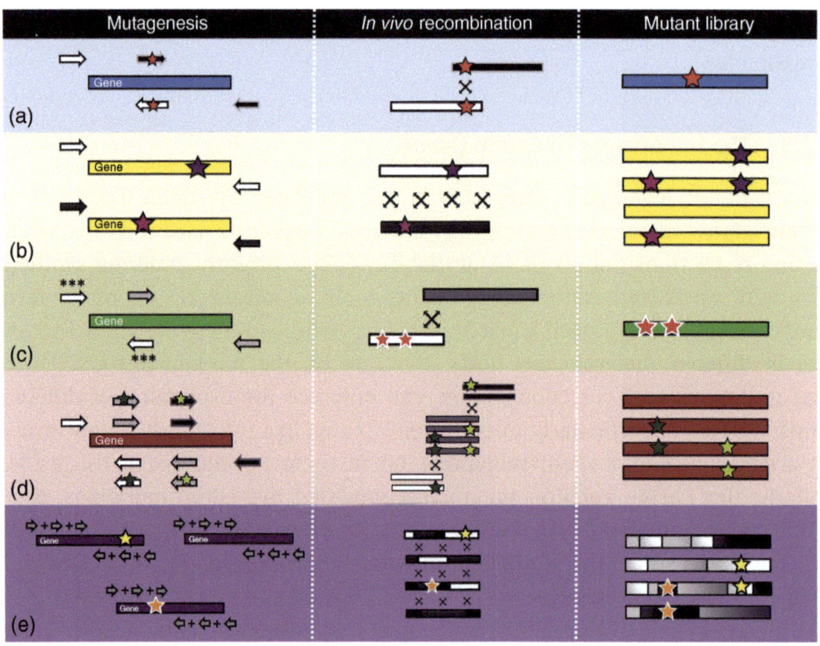

Figure 12.4 Library creation methodologies based on *S. cerevisiae*: (a) IVOE, (b) DNA shuffling, (c) MORPHING, (d) site directed recombination, and (e) mutagenic StEP + *in vivo* DNA shuffling. Nucleotide substitutions are depicted as stars. Possible recombination events are depicted as crosses.

achieve site directed mutagenesis (including deletion and insertion mutations), to construct site-directed combinatorial libraries, and to perform gene assembly, while the proof-reading apparatus of *S. cerevisiae* precludes the insertion of unwanted mutations [57] Figure 12.4a. Due to its simplicity and versatility, IVOE has become indispensable for our ligninase directed evolution campaigns [20, 29]. Likewise, the *in vivo* shuffling of beneficial mutations is systematically applied to the evolution of laccases, VPs, UPOs, and AAOs. As long as DNA sequence identities are above 50% and mutations are situated at a distance of at least 20 residues from each other, *in vivo* shuffling can be used to search for beneficial combinatorial effects among the mutants' offspring [58] Figure 12.4b. One of our most recent applications of this method was in the evolution of AAO for expression, where up to three beneficial mutations were brought together from three independent parental types in just one cycle of DNA shuffling [47].

With the aim of creating smarter libraries that place less of a demand on screening, MORPHING (Mutagenic Organized Recombination Process by Homologous IN vivo Grouping) has proved to be a useful mutagenesis/recombination technique in the evolution of ligninases [36]. Through MORPHING we can introduce random mutations and recombination events in specific segments while protecting the remaining protein structure from mutagenesis Figure 12.4c. Mutant segments as short as 30 amino acids are subjected to mutagenic PCR, whereas the remaining regions are amplified with high-fidelity polymerases. The resulting DNA fragments are flanked with homologous overhangs and they are spliced *in vivo* along with the linearized plasmid in a one-pot transformation step. While MORPHING has been successfully used to evolve signal peptides [36, 37, 47], it has also been applied in the focused evolution of many relevant biochemical properties of ligninases. One of the first examples of MORPHING was the improvement of the oxidative stability of VP (tolerance in the presence of H_2O_2) [59]. After the multiple structural alignment of peroxidases to locate H_2O_2-sensitive regions, a set of three independent segments was selected to introduce mutations using loads from one to five mutations per segment in order to identify important structural determinants for the oxidative stabilization of VP. MORPHING has also been used on the remaining members of the ligninolytic armory to improve a range of properties, from thermostability to substrate specificity and product selectivity [33, 46, 60, 61]. Another case of MORPHING involved adapting UPO for the selective synthesis of human drug metabolites (HDMs) [61]. After molecular docking simulations with the β blocker propranolol, a protein segment that contains important recognition motifs was subjected to MORPHING. As a result, a single substitution (F191S) at the entrance of the heme access channel was seen to produce a dramatic enhancement of 2 orders of magnitude in the catalytic efficiency for propranolol, with 99% regioselectivity during the synthesis of the HDM. One of the latest applications of MORPHING involved the enhancement of the thermostability of the PM1L. Based on the analysis of flexible surface loops by *B*-factor and molecular dynamic (MD) simulations, three MORPHING blocks were designed to collect mutations on such labile regions and the libraries were screened for thermostability. A double mutant with substitutions S264K and S356N increased the half-life at 70 °C by 31 minutes maintaining activity levels for the laccase [62].

DNA-puzzle is another helpful technique to recover lost mutations from independent protein regions while assessing potential epistatic effects. This method was first used during the final stage of the evolution of VP for oxidative stability. Accordingly, up to nine sequence blocks were amplified from three VP fragments (flanked with 50 bp homology overhangs), and they were recombined *in vivo* to produce variants with a good balance between activity and oxidative stability. Indeed, an increase in the half-life of 35 minutes was achieved in the presence of 3000 equiv of H_2O_2, with a 6 °C enhancement in thermostability [59].

The recombination machinery of yeast can also be harnessed to perform *in vivo* site-directed recombination, a useful method to evaluate beneficial mutations from different parental types and their reversions in a combinatorial mode Figure 12.4d. This approach involves designing oligonucleotides that contain a 50% mixture of parental and substituted amino acids, and it was recently applied to AAO after identifying 10 mutations to enhance the oxidation of secondary alcohols over three rounds of evolution [63]. From the pool of combinations, the three best variants presented a common mutational backbone. More interestingly, the best performer purged any other substitutions and left those three mutations as the best epistatic event, producing a final variant that showed a high-enantioselectivity (ee > 99%) and had a catalytic efficiency improved by 3 orders of magnitude versus the native enzyme.

The combination of *in vivo* and *in vitro* DNA recombination also provides an interesting approach to enhance the number of crossover events and thus, sequence diversity. CLERY (Combinational Libraries Enhanced by Recombination in Yeast) is a method that combines *in vivo* and *in vitro* DNA shuffling [64], and it was applied to design chimeric laccases from PM1L and PcL secretion mutants. Notably, several laccases in the pool of chimeras showed differences in terms of their pH activity profile and substrate affinity, as well as increased thermostability [65]. Similarly, we made used of mutagenic *in vitro* StEP (Staggered Extension Process [66] and *in vivo* shuffling when evolving VP for thermostability [51, 67], as well as to evolve PM1L for blood tolerance [68] Figure 12.4e.

Although traditionally adaptive evolution (i.e. by selection of the fittest) is the main focus in laboratory evolution experiments, the creation of genetic diversity by neutral genetic drift is also an appealing technique to introduce neutral mutations that can open up new evolutionary trajectories [69, 70]. By maintaining similar phenotypes in successive rounds of purifying selection (i.e. to keep the natural activity of the enzyme), the accumulation of neutral mutations allows different genotypes to be merged with new properties in terms of promiscuity and stability. However, neutral genetic drift is a time-consuming method that, unless ultrahigh-throughput screening protocols are available, requires an average of roughly 20 rounds of evolution to unmask new traits. Very recently, we combined genetic drift with *in vivo* shuffling to speed up the neutral evolution of UPO. After only eight rounds of purifying selection and shuffling, a highly functional family of neutral UPO variants was created, with 4.6 substitutions per clone and neutral mutations covering ~10% of the protein sequence. Up to 80% of the neutral mutations were located at the protein surface, with 25% of enriched mutations (i.e. those that appear in more than one neutral clone) addressing their possible role as consensus–ancestral mutations.

Interestingly, many of the neutrally evolved UPOs had a different activity profile and improved stability, both at high temperature and in the presence of organic co-solvents [71].

12.4 Bringing Together Evolutionary Strategies and Computational Tools

Recent developments have favored the confluence of *in vitro* evolution and computational studies. The *in silico* toolbox mainly, but not exclusively, contains Machine learning, QM/MM (Quantum Mechanics/Molecular Mechanics), MD, PELE (Protein Energy Landscape Exploration – a method based on the Monte Carlo algorithm to model protein dynamics and protein-ligand interactions), SCHEMA-RASSP (SCHEMA-Recombination as Shortest Path Problem) [72], consensus design, and ancestral resurrection. These tools pave the way for more efficient engineering and the application of a better rationale to laboratory evolution [29]. For example, PcL was subjected to saturation mutagenesis to improve turnover rates for several redox mediators, while molecular simulations showed that the F392N substitution widened the T1 cavity of the laccase and allowed substrate binding to be modulated [73]. Recent UPO evolutionary campaigns have also been aided by computational analysis. When this peroxygenase was evolved for the selective synthesis of 1-naphthol, QM/MM, and PELE revealed that the G241D substitution at the entrance of the heme channel produced conformational changes that affected the catalytic acid-base pair and substrate binding [74]. In our latest evolution experiment for the synthesis of HDMs from propranolol, the evolved UPO included only a single substitution (F191S). By combining PELE and MD, we concluded that the improvement in catalytic activity and total turnover number (over 200 000) was due to a reduction in the residence times at the binding site for the major reaction product, enhancing its production and purification [61].

However, computational support for enzyme evolution not only helps elucidate the phenotypic effect of an acquired substitution but may also be a useful tool to predict structural hot-spots for mutagenesis. Such a co-operative strategy has been followed to evolve PM1L toward a higher redox potential at the T1Cu site [75], running computer-aided evolution experiments based on ordering sequence positions according to the frequency of beneficial mutations. After screening 40 positions by PELE, two mutations in the first coordination sphere of the T1Cu were selected and confirmed experimentally. The 50 mV increase in the redox potential at the T1Cu site in the final PM1L variant was coupled to higher activity relative to a panel of high-redox potential mediators, and improved stability against temperature and pH. Indeed, *a posteriori*, computational analysis indicated that both the redox potential and substrate binding are fundamental to control the catalytic performance of laccases.

Another recent example of computer-guided evolution based on SCHEMA-RASSP has been reported, involving the generation of a family of thermostable laccase chimeras [76]. SCHEMA-RASSP is a computational algorithm for protein recombination to generate libraries with a maximal amount of folded chimeras and greater sequence diversity [77, 78]. In this case, SCHEMA-RASSP was combined with *in vivo* shuffling to generate a set of laccase chimeras with high sequence and functional diversity from three different orthologs. The closest homolog in the resulting family of chimeras varied by 46 amino acids, harboring multiple crossover events between and within the SCHEMA blocks, with the most thermostable variant displaying a fivefold enhancement of its half-life of thermal inactivation at 70 °C (up to 108 minutes).

Another handy computational strategy that can be applied to ligninase evolution is consensus design. The distribution of amino acids in multiple sequence alignments (MSAs) of homologous proteins reveals the most conserved positions in the different sequences. The identification and introduction of such consensus mutations can help stabilize enzymes, while enhancing secretion levels without jeopardizing catalysis [79]. We employed a consensus computational method to identify and insert consensus mutations in PM1L. After carrying out a MSA of 500 basidiomycete laccases, the most promising consensus mutations were selected, assessed in different screening assays and further explored by *in vivo* site-directed recombination. The most promising mutant harbored an ancestral consensus mutation that improved secretion, thermostability, and catalytic efficiency [80]. Following this exploration of similarities within abundant sequences, it is also possible to go way back in time and reconstruct ancestral enzymes based on phylogenetic inference. The resurrection of ancestral ligninases on the bench could provide directed evolution platforms with an ideal canvas for engineering, based on primitive enzyme characteristics like robustness and promiscuity. From the recent studies of ancestral ligninolytic peroxidases [81, 82] and the first example of laboratory evolution of a Precambrian enzyme [83], such exciting starting points give a wider meaning to *in vitro* evolution, acting as a "take two" for natural protein specialization under the rules laid down by researchers [84]. Our laboratory is currently deep into such an enterprise by evolving resurrected fungal laccases after successfully reconstructing the phylogenetic tree of PM1L, "*bringing back to life*" some of its ancient members [85].

12.5 High-Throughput Screening (HTS) Assays for Ligninase Evolution

As mentioned earlier, a good host organism for directed evolution must be easily adapted to a HTS format, classically using 96-well plates. Dozens of HTS assays have been developed for such purposes, establishing reliable protocols in terms of sensitivity, reproducibility, coefficients of variation, linearity, etc. Accordingly, robust HTS assays are available to select ligninolytic enzyme mutants with improved secretion, activity, stability, redox potential, shifted pH profile, and much more.

Inherent to the exploration of mutant libraries for the desired enzymatic traits, adequate secretion must first be achieved to obtain a steady signal before improving other relevant biochemical attributes. As already discussed, this is the first bottleneck that researchers must overcome when starting a directed evolution project with a new ligninase. As such, in most cases the HTS protocol employed must detect variants with Total Activity Improvements (TAI: the product of specific activity and secretion) in the first few rounds of evolution. In the case of VPs and laccases, the TAI has routinely been directly measured on the yeast supernatant with the colorimetric substrate ABTS (2,2′-azino-bis(3-ethylbenzothiazoline-6-sulphonic acid) [35, 44, 45, 51]. Conversely, the TAI for AAO was estimated indirectly by quantifying the H_2O_2 released after oxidation of the substrate using a coupled peroxidase activity assay (a cascade reaction that can be easily followed colorimetrically). Alternatively, H_2O_2 production can be assessed directly with the chemical FOX method based on the Fenton reaction [33, 46]. For more complex systems like UPO, with both peroxygenase (*P*) and peroxidase (*p*) activities, a dual screening assay had to be set up to control the *P*/*p* ratio (using ABTS and NBD [5-nitro-1,3-benzodioxole] as substrates, respectively) [37, 60].

In terms of evolving stability (against temperature, pH, non-conventional media, etc.), the selection of a suitable incubation threshold under the demanding conditions and the gradual enhancement of the selective pressure are two golden rules to take into account. Thermostability is a particularly relevant issue when developing industrial biocatalysts. If a heat stress step is added to the screening protocol after microfermentation, it is possible to detect variants with enhanced thermostability based on the residual activity (i.e. the activity of supernatants before and after thermal shock). Hence, it is necessary to determine an effective ratio of residual activity:initial activity so that a suitable temperature for heat shock is assigned to ensure that the residual activity stays at approximately one-third of the initial activity. To accelerate evolution, the selective pressure can be gradually increased, as for the evolution of VP thermostability where the screening temperature rose from 60 to 90 °C [51]. Finally, greater precision is achieved by further testing selected variants for their thermostability, plotting the T_{50} value (temperature at which the 50% of the initial activity is maintained after a 10 minutes incubation) [67].

Organic solvents are present in many chemical processes and the performance of enzymes in such artificial environments is a key feature that is susceptible to evolution, employing an adequate HTS assay to detect the most resistant and active variants. In the case of MtL, the best stability variant after five generations withstood higher concentrations of organic solvents thanks to increasing the selective pressure in the course of evolution from 20% to 60% of co-solvent. Indeed, co-solvent promiscuity was also evident in other water–solvent mixtures due to the combination of two co-solvents in the same dual HTS assay (acetonitrile and ethanol), each with a different chemical nature and polarity [35, 53]. The use of ionic liquids is more environmentally friendly and accordingly, laccases have also been evolved to be functionally active in the presence of 15% v/v 1-ethyl-3-methylimidazolium ethyl sulfate (enhancing the selective pressure up to 30%) [86].

VP, LiP, MnP, and UPO are inactivated by H_2O_2, a mechanism of inhibition common to all heme-containing peroxidases. Like thermostability, we evolved VP for oxidative stability by carefully designing a HTS assay based on the H_2O_2:enzyme molar ratio. In this assay, the parental types in each round of evolution were capable of retaining one-third of their initial activity after a given period in the presence of

H_2O_2 and the selective pressure was gradually enhanced to 0.6 mM H_2O_2 in the final generations [59].

Fungal ligninolytic peroxidases and laccase do not work in the neutral/alkaline pH range, yet they are strongly inactivated by modest concentrations of OH^-. As ligninolysis takes place at an acid pH due to the release of organic acids by white-rot fungi, nature has never applied natural selection for this property. However, there are a plethora of applications for alkaline ligninases, ranging from bioremediation and organic synthesis, to their use for biomedical purposes. To create alkaline VPs and laccases (MtL), the activity at alkaline pH relative to that at acid pH was used as the main discriminatory factor. Conceptually, we screened for clones with improvements at alkaline pH but that maintained their activity at acid pH, thereby shifting the pH activity profile of these enzymes. Through this approach, we found variants that worked both under neutral/basic conditions, as well as at an acidic pH, a versatility that is useful for many industrial and environmental applications [54, 87]. It proved more complicated to design the first blood-tolerant HRPL, where the pH of human blood (7.4) and high concentrations of NaCl (140–150 mM) precludes the use of the laccase in biosensors and biofuel cells. As plasma or blood are not suitable to screen mutant libraries in the framework of a directed evolution experiment, we prepared an *ad hoc* HTS assay based on these features in a buffer that simulated blood, albeit in the absence of coagulating agents and red blood cells [68]. With this system, we obtained a blood-tolerant laccase variant that was later applied to the first prototype of a wireless 3D-nanobiodevice for use on human blood [88].

In synthetic chemistry, enzymes can provide a repertory of solutions unavailable to traditional methods of chemical synthesis. These may include augmenting selectivity and efficiency, making such processes more environmentally friendly and/or less demanding energetically. We are applying evolution to adapt UPO to industrial needs. For example, this enzyme was recently evolved for 1-naphthol synthesis (an important agrochemical) using an *ad hoc* dual screening assay to quench the peroxidase activity on 1-naphthol and to promote the peroxygenase activity on naphthalene (the latter assessed by coupling the formation of 1-naphthol to a red azo dye) [74]. Following a similar HTS scheme, we also evolved UPO for the selective synthesis of the HDM 5′-OH propranolol from the β blocker propranolol. The HTS assay was based on the 4-aminoantypirine assay that coupled 5′-OH propranolol generation to a colorimetric response. To achieve the breakdown of peroxygenase and peroxidase activities, the reaction with propranolol was carried out in the presence and absence of ascorbic acid, a phenoxyl radical scavenger commonly used to return the effect of the peroxidase activity [61, 89].

We recently evolved laccases for the synthesis of C–N heteropolymeric dyes at alkaline pHs, a long sought after process in the textile sector. The HTS assay was based on oxidative cross-coupling between 2,5-diaminobenzenosulfonic acid (2,5-DABSA) and catechol at pH 8.0. After laccase oxidation, the colorimetric response of the C-heteropolymeric dye was dependent on pH interfering with the background of culture supernatants. To circumvent this, we prepared a selective expression medium (SEM) for enzyme production so that the standard two step fermentation in *S. cerevisiae* was reduced to a single step using galactose as both the single carbon source and the main inducer of expression [55]. SEM is convenient when screening must be done at the limits of the UV/Vis wavelengths, as we saw when evolving PM1L toward a higher redox potential. For this task we prepared an HTS assay based on the oxidation of high-redox potential mediators of different chemical natures and oxidation routes. In particular, the main assay was based on an inorganic transition metal complex acting as a high-redox potential mediator (octacyanomolybdate IV). To avoid substrate dependent bias during screening, the HTS was complemented with two colorimetric assays based on the oxidation of other synthetic mediators that acted as negative and positive controls [75].

12.6 Conclusions and Outlook

Not surprisingly, the potential of the fungal ligninolytic secretome opens new challenges associated with the herculean task of transforming natural oxidoreductases into industrial biocatalysts. Through the efforts to develop novel ligninases, the toolkit for directed and computational evolution has expanded, introducing strategies for functional expression, library creation methods, and HTS assays. These modern evolution platforms have also taken advantage of recent developments in ancestral enzyme resurrection, opening an exciting prospect in the field of synthetic biology with the possible assembly of different evolved ligninases into a single "white-rot yeast" (WRY). Such a synthetic microbe will be fundamental to produce the first evolved secretome aimed at processing lignocellulose biomass [20, 21]. Indeed, with the aim of achieving a more efficient delignification and inhibitor detoxification during the pretreatment of lignocellulose biomass, the WRY may constitute a more efficient and eco-friendly biological solution. Certainly, after the abundant individual efforts taken to evolve each one of the constituents of the consortium, the time has come to harmonize the action of the set of enzymes in an ideal synthetic secretome comprising enzymatic activities able to remove lignin as well as inhibitors (especially phenolic compounds). Crafting such machinery still presents several challenges, like tuning the different enzymatic mechanisms and adjusting protein dosage from secretion. With a unique composition of evolved ligninases not found together in nature and the capacity to be further adapted by top-notch directed evolution techniques to different scenarios, this synthetic secretome produced by WRY may become in the near future into a self-sufficient and highly versatile alternative to the existing biological and physico-chemical pretreatments reported to date (unpublished results).

Acknowledgments

This work was supported by the Spanish Government Grants BIO2016-79106-R-Lignolution, PID2019-106166RB-100-OXYWAVE, the Comunidad de Madrid Synergy CAM project Y2018/BIO-4738-EVOCHIMERA-CM, the CSIC Grant

PIE-201580E042, and the Bio Based Inustries Joint Undertakng under the European Union's Horizon 2020 Research and Innovation programme (grant agreement n° 886567-BIZENTE project).

References

1 Isikgor, F. and Becer, R. (2015). Lignocellulosic biomass: a sustainable platform for production of bio-based chemicals and polymers. *Polym. Chem.* 6: 4497–4559.

2 Zhou, C.H., Xia, X., Lin, C.X. et al. (2011). Catalytic conversion of lignocellulosic biomass to fine chemicals and fuels. *Chem. Soc. Rev.* 40 (11): 5588–5617.

3 Sims, R.E., Mabee, W., Saddler, J.N., and Taylor, M. (2010). An overview of second generation biofuel technologies. *Bioresour. Technol.* 101 (6): 1570–1580.

4 Xie, S., Ragauskas, A.J., and Yuan, J.S. (2016). Lignin conversion: opportunities and challenges for the integrated biorefinery. *Ind. Biotechnol.* 12 (3): 161–167.

5 Roopan, S.M. (2017). An overview of natural renewable bio-polymer lignin towards nano and biotechnological applications. *Int. J. Biol. Macromol.* 103: 508–514.

6 Sigoillot, J.C., Berrin, J.G., Bey, M. et al. (2012). Fungal strategies for lignin degradation. In: *Lignins: Biosynthesis, Biodegradation and Bioengineering* (eds. L. Jouanin and C. Lapierre), 263–308. Academic Press.

7 Brown, M.E. and Chang, M.C.Y. (2014). Exploring bacterial lignin degradation. *Curr. Opin. Chem. Biol.* 19: 1–7.

8 Martinez, A.T., Speranza, M., Ruiz-Dueñas, F.J. et al. (2005). Biodegradation of lignocellulosics: microbial, chemical, and enzymatic aspects of the fungal attack of lignin. *Int. Microbiol.* 8 (3): 195–204.

9 Martinez, A.T., Ruiz-Dueñas, F.J., Martinez, M.J. et al. (2009). Enzymatic delignification of plant cell wall: from nature to mill. *Curr. Opin. Biotechnol.* 20 (3): 348–357.

10 Harvey, P.J., Floris, R., Lundell, T. et al. (1992). Catalytic mechanisms and regulation of lignin peroxidase. *Biochem. Soc. Trans.* 20 (2): 345–349.

11 Falade, A.O., Nwodo, U.U., Iweriebor, B.C. et al. (2017). Lignin peroxidase functionalities and prospective applications. *MicrobiologyOpen* 6 (1): e00394.

12 Hatakka, A., Lundell, T., Hofrichter, M., and Maijala, P. (2003). Manganese peroxidase and its role in the degradation of wood lignin. In: *Applications of Enzymes to Lignocellulosics* (eds. S.D. Mansfield and J.N. Saddler), 230–243. American Chemical Society.

13 Camarero, S., Sarkar, S., Ruiz-Dueñas, F.J. et al. (1999). Description of a versatile peroxidase involved in the natural degradation of lignin that has both manganese peroxidase and lignin peroxidase substrate interaction sites. *J. Biol. Chem.* 274 (15): 10324–10330.

14 Ruiz-Dueñas, F.J., Morales, M., Garcia, E. et al. (2009). Substrate oxidation sites in versatile peroxidase and other basidiomycete peroxidases. *J. Exp. Bot.* 60 (2): 441–452.

15 Hofrichter, M., Ullrich, R., Pecyna, M.J. et al. (2010). New and classic families of secreted fungal heme peroxidases. *Appl. Microbiol. Biotechnol.* 87 (3): 871–897.

16 Hofrichter, M., Kellner, H., Pecyna, M.J., and Ullrich, R. (2015). Fungal unspecific peroxygenases: heme-thiolate proteins that combine peroxidase and cytochrome P450 properties. In: *Monooxygenase, Peroxidase and Peroxygenase Properties and Mechanisms of Cytochrome P450* (eds. E.G. Hrycay and S.M. Bandiera), 341–368. Springer.

17 Kersten, P. and Cullen, D. (2007). Extracellular oxidative systems of the lignin-degrading basidiomycete *Phanerochaete chrysosporium*. *Fungal Genet. Biol.* 44 (2): 77–87.

18 Hernandez-Ortega, A., Ferreira, P., and Martinez, A.T. (2012). Fungal aryl-alcohol oxidase: a peroxide-producing flavoenzyme involved in lignin degradation. *Appl. Microbiol. Biotechnol.* 93 (4): 1395–1410.

19 Mate, D.M. and Alcalde, M. (2015). Laccase engineering: from rational design to directed evolution. *Biotechnol. Adv.* 33 (1): 25–40.

20 Alcalde, M. (2015). Engineering the ligninolytic enzyme consortium. *Trends Biotechnol.* 33 (3): 155–162.

21 Martinez, A.T., Ruiz-Dueñas, F.J., Camarero, S. et al. (2017). Oxidoreductases on their way to industrial biotransformations. *Biotechnol. Adv.* 35 (6): 815–831.

22 Xu, F. (2005). Applications of oxidoreductases: recent progress. *Ind. Biotechnol.* 1 (1): 38–50.

23 Kunamneni, A., Plou, F.J., Ballesteros, A., and Alcalde, M. (2008). Laccases and their applications: a patent review. *Recent Pat. Biotechnol.* 2 (1): 10–24.

24 Karich, A., Ullrich, R., Scheibner, K., and Hofrichter, M. (2017). Fungal unspecific peroxygenases oxidize the majority of organic EPA priority pollutants. *Front. Microbiol.* 8: 1463.

25 Wang, Y., Lan, D., Durrani, R., and Hollmann, F. (2017). Peroxygenases en route to becoming dream catalysts. What are the opportunities and challenges? *Curr. Opin. Chem. Biol.* 37: 1–9.

26 Bornscheuer, U., Huisman, G., Kazlauskas, R. et al. (2012). Engineering the third wave of biocatalysis. *Nature* 485 (7397): 185–194.

27 Molina-Espeja, P., Viña-Gonzalez, J., Gomez-Fernandez, B.J. et al. (2016). Beyond the outer limits of nature by directed evolution. *Biotechnol. Adv.* 34 (5): 754–767.

28 Garcia-Ruiz, E., Mate, D.M., Gonzalez-Perez, D. et al. (2014). Directed evolution of ligninolytic oxidoreductases: from functional expression to stabilization and beyond. In: *Cascade Biocatalysis* (eds. S. Riva and W.D. Fessner), 1–22. Wiley-VCH.

29 Mate, D.M., Gonzalez-Perez, D., Mateljak, I. et al. (2017). The pocket manual of directed evolution: tips and tricks. In: *Biotechnology of Microbial Enzymes* (ed. B. Goutam), 185–213. Elsevier.

30 Molina-Espeja, P., Gomez de Santos, P., and Alcalde, M. (2017). Directed evolution of unspecific peroxygenase. In: *Directed Enzyme Evolution Advances and Applications* (ed. M. Alcalde), 127–143. Springer.

31 Pourmir, A. and Johannes, T.W. (2012). Directed evolution: selection of the host organism. *Comput. Struct. Biotechnol. J.* 2: e201209012.

32 Gonzalez-Perez, D., Garcia-Ruiz, E., and Alcalde, M. (2012). *Saccharomyces cerevisiae* in directed evolution: an efficient tool to improve enzymes. *Bioeng. Bugs* 3 (3): 172–177.

33 Viña-Gonzalez, J., Gonzalez-Perez, D., and Alcalde, M. (2016). Directed evolution method in *Saccharomyces cerevisiae*: mutant library creation and screening. *J. Visualized Exp.* 110: e53761.

34 Delic, M., Valli, M., Graf, A.B. et al. (2013). The secretory pathway: exploring yeast diversity. *FEMS Microbiol. Rev.* 37 (6): 872–914.

35 Bulter, T., Alcalde, M., Sieber, V. et al. (2003). Functional expression of a fungal laccase in *Saccharomyces cerevisiae* by directed evolution. *Appl. Environ. Microbiol.* 69 (8): 5037–5037.

36 Gonzalez-Perez, D., Molina-Espeja, P., Garcia-Ruiz, E., and Alcalde, M. (2014). Mutagenic organized recombination process by homologous in vivo grouping (MORPHING) for directed enzyme evolution. *PLoS One* 9 (3): e90919.

37 Molina-Espeja, P., Garcia-Ruiz, E., Gonzalez-Perez, D. et al. (2014). Directed evolution of unspecific peroxygenase from *Agrocybe aegerita*. *Appl. Environ. Microbiol.* 80 (11): 3496–3507.

38 Molina-Espeja, P., Ma, S., Mate, D.M. et al. (2015). Tandem-yeast expression system for engineering and producing unspecific peroxygenase. *Enzyme Microb. Technol.* 73–74: 29–33.

39 Bitter, G.A., Chen, K.K., Banks, A.R., and Lai, P.H. (1984). Secretion of foreign proteins from *Saccharomyces cerevisiae* directed by alpha-factor gene fusions. *Proc. Natl. Acad. Sci. U.S.A.* 81 (17): 5330–5334.

40 Chang, C.N., Matteucci, M., Perry, L.J. et al. (1986). *Saccharomyces cerevisiae* secretes and correctly processes human interferon hybrid proteins containing yeast invertase signal peptides. *Mol. Cell. Biol.* 6 (5): 1812–1819.

41 Cartwright, C.P., Zhu, Y.S., and Tipper, D.J. (1992). Efficient secretion in yeast based on fragments from K1 killer preprotoxin. *Yeast* 8 (4): 261–272.

42 Hashimoto, Y., Koyabu, N., and Imoto, T. (1998). Effects of signal sequences on the secretion of hen lysozyme by yeast: construction of four secretion cassette vectors. *Protein Eng.* 11 (2): 75–77.

43 Kjeldsen, T., Pettersson, A.F., Hach, M. et al. (1997). Synthetic leaders with potential BiP binding mediate high-yield secretion of correctly folded insulin precursors from *Saccharomyces cerevisiae. Protein Expression Purif.* 9 (3): 331–336.

44 Camarero, S., Pardo, I., Cañas, A.I. et al. (2012). Engineering platforms for directed evolution of laccase from *Pycnoporus cinnabarinus*. *Appl. Environ. Microbiol.* 78 (5): 1370–1384.

45 Mate, D., Garcia-Burgos, C., Garcia-Ruiz, E. et al. (2010). Laboratory evolution of high-redox potential laccases. *Chem. Biol.* 17 (9): 1030–1041.

46 Viña-Gonzalez, J., Gonzalez-Perez, D., Ferreira, P. et al. (2015). Focused directed evolution of aryl-alcohol oxidase in *Saccharomyces cerevisiae* by using chimeric signal peptides. *Appl. Environ. Microbiol.* 81 (18): 6451–6462.

47 Viña-Gonzalez, J., Elbl, K., Ponte, X. et al. (2018). Functional expression of aryl-alcohol oxidase in *Saccharomyces cerevisiae* and *Pichia pastoris* by directed evolution. *Biotechnol. Bioeng.* 115 (7): 1666–1674.

48 Rakestraw, J.A., Sazinsky, S.L., Piatesi, A. et al. (2009). Directed evolution of a secretory leader for the improved expression of heterologous proteins and full-length antibodies in *Saccharomyces cerevisiae. Biotechnol. Bioeng.* 103 (6): 1192–1201.

49 Mateljak, I., Tron, T., and Alcalde, M. (2017). Evolved α-factor prepro-leaders for directed laccase evolution in *Saccharomyces cerevisiae. Microb. Biotechnol.* 10 (6): 1830–1836.

50 Roodveldt, C., Aharoni, A., and Tawfik, D.S. (2005). Directed evolution of proteins for heterologous expression and stability. *Curr. Opin. Struct. Biol.* 15 (1): 50–56.

51 Garcia-Ruiz, E., Gonzalez-Perez, D., Ruiz-Dueñas, F.J. et al. (2012). Directed evolution of a temperature-, peroxide- and alkaline pH-tolerant versatile peroxidase. *Biochem. J.* 441 (1): 487–498.

52 Angov, E. (2011). Codon usage: nature's roadmap to expression and folding of proteins. *Biotechnol. J.* 6 (6): 650–659.

53 Zumarraga, M., Bulter, T., Shleev, S. et al. (2007). In vitro evolution of a fungal laccase in high concentrations of organic cosolvents. *Chem. Biol.* 14 (9): 1052–1064.

54 Torres-Salas, P., Mate, D.M., Ghazi, I. et al. (2013). Widening the pH activity profile of a fungal laccase by directed evolution. *ChemBioChem* 14 (8): 934–937.

55 Vicente, A.I., Viña-Gonzalez, J., Santos-Moriano, P. et al. (2016). Evolved alkaline fungal laccase secreted by *Saccharomyces cerevisiae* as useful tool for the synthesis of C–N heteropolymeric dye. *J. Mol. Catal. B: Enzym.* 134: 323–330.

56 Van Komen, S., Macris, M., Sehorn, M.G., and Sung, P. (2006). Purification and assays of *Saccharomyces cerevisiae* homologous recombination proteins. *Methods Enzymol.* 408: 445–463.

57 Alcalde, M. (2010). Mutagenesis protocols in *Saccharomyces cerevisiae* by in vivo overlap extension. In: *In vitro Mutagenesis Protocols* (ed. J. Braman), 3–14. Springer.

58 Oldenburg, K.R., Vo, K.T., Michaelis, S., and Paddon, C. (1997). Recombination-mediated PCR-directed plasmid construction in vivo in yeast. *Nucleic Acids Res.* 25 (2): 451–452.

59 Gonzalez-Perez, D., Garcia-Ruiz, E., Ruiz-Dueñas, F.J. et al. (2014). Structural determinants of oxidative stabilization in an evolved versatile peroxidase. *ACS Catal.* 4 (11): 3891–3901.

60 Mate, D.M., Palomino, M.A., Molina-Espeja, P. et al. (2017). Modification of the peroxygenative: peroxidative activity ratio in the unspecific peroxygenase from *Agrocybe aegerita* by structure-guided evolution. *Protein Eng. Des. Sel.* 30 (3): 191–198.

61 Gomez de Santos, P., Cañellas, M., Tieves, F. et al. (2018). Selective synthesis of the human drug metabolite 5′-hydroxypropranolol by an evolved self-sufficient peroxygenase. *ACS Catal.* 8 (6): 4789–4799.

62 Vicente, A.I., Viña-Gonzalez, J., Mateljak, I. et al. (2019). Enhancing thermostability by modifying flexible surface loops in an evolved high-redox potential laccase. *AlChE J.* 66: e16747.

63 Viña-Gonzalez, J., Jimenez-Lalana, D., Sancho, F. et al. (2019). Structure-guided evolution of aryl alcohol oxidase from *Pleurotus eryngii* for the selective oxidation of secondary benzyl alcohols. *Adv. Synth. Catal.* 361: 2514–2525.

64 Abecassis, V., Pompon, D., and Truan, G. (2000). High efficiency family shuffling based on multi-step PCR and in vivo DNA recombination in yeast: statistical and functional analysis of a combinatorial library between human cytochrome P450 1A1 and 1A2. *Nucleic Acids Res.* 28 (20): e88–e88.

65 Pardo, I., Vicente, A.I., Mate, D.M. et al. (2012). Development of chimeric laccases by directed evolution. *Biotechnol. Bioeng.* 109 (12): 2978–2986.

66 Zhao, H., Giver, L., Shao, Z. et al. (1998). Molecular evolution by staggered extension process (StEP) in vitro recombination. *Nat. Biotechnol.* 16 (3): 258.

67 Garcia-Ruiz, E., Mate, D., Ballesteros, A. et al. (2010). Evolving thermostability in mutant libraries of ligninolytic oxidoreductases expressed in yeast. *Microb. Cell Fact.* 9 (1): 17.

68 Mate, D.M., Gonzalez-Perez, D., Falk, M. et al. (2013). Blood tolerant laccase by directed evolution. *Chem. Biol.* 20 (2): 223–231.

69 Bloom, J.D., Romero, P.A., Lu, Z., and Arnold, F.H. (2007). Neutral genetic drift can alter promiscuous protein functions, potentially aiding functional evolution. *Biol. Direct* 2 (1): 17.

70 Gupta, R.D. and Tawfik, D.S. (2008). Directed enzyme evolution via small and effective neutral drift libraries. *Nat. Methods* 5 (11): 939.

71 Martin-Diaz, J., Paret, C., García-Ruiz, E. et al. (2018). Shuffling the neutral drift of unspecific peroxygenase in *Saccharomyces cerevisiae*. *Appl. Environ. Microbiol.* 84 (15): 00808–00818.

72 Monza, E., Acebes, S., Lucas, M.F., and Guallar, V. (2017). Molecular modeling in enzyme design, toward in silico guided directed evolution. In: *Directed Enzyme Evolution: Advances and Applications* (ed. M. Alcalde), 257–284. Springer.

73 Pardo, I., Santiago, G., Gentili, P. et al. (2016). Re-designing the substrate binding pocket of laccase for enhanced oxidation of sinapic acid. *Catal. Sci. Technol.* 6 (11): 3900–3910.

74 Molina-Espeja, P., Cañellas, M., Plou, F.J. et al. (2016). Synthesis of 1-naphthol by a natural peroxygenase engineered by directed evolution. *ChemBioChem* 17 (4): 341–349.

75 Mateljak, I., Monza, E., Lucas, M.F. et al. (2019). Increasing redox potential, redox mediator activity and stability in a fungal laccase by computer-guided mutagenesis and directed evolution. *ACS Catal.* 9 (5): 4561–4572.

76 Mateljak, I., Rice, A., Yang, K. et al. (2019). The generation of thermostable fungal laccases chimeras by SCHEMA-RASPP structure-guided recombination in vivo. *ACS Synth. Biol.* 8 (4): 833–843.

77 Voigt, C.A., Martinez, C., Wang, Z.-G. et al. (2002). Protein building blocks preserved by recombination. *Nat. Struct. Mol. Biol.* 9 (7): 553.

78 Endelman, J.B., Silberg, J.J., Wang, Z.G., and Arnold, F.H. (2004). Site-directed protein recombination as a shortest-path problem. *Protein Eng. Des. Sel.* 17 (7): 589–594.

79 Porebski, B.T. and Buckle, A.M. (2016). Consensus protein design. *Protein Eng. Des. Sel.* 29 (7): 245–251.

80 Gomez-Fernandez, B.J., Risso, V.A., Sanchez-Ruiz, J.M., and Alcalde, M. (2020). Consensus design of an evolved high-redox potential laccase. *Front. Bioeng. Biotechnol.* 8: 1-11.

81 Ayuso-Fernandez, I., Martínez, A.T., and Ruiz-Dueñas, F.J. (2017). Experimental recreation of the evolution of lignin-degrading enzymes from the Jurassic to date. *Biotechnol. Biofuels* 10 (1): 67.

82 Ayuso-Fernández, I., Ruiz-Dueñas, F.J., and Martinez, A.T. (2018). Evolutionary convergence in lignin-degrading enzymes. *Proc. Natl. Acad. Sci. U.S.A.* 115 (25): 6428–6433.

83 Gomez-Fernandez, B.J., Garcia-Ruiz, E., Martin-Diaz, J. et al. (2018). Directed-in vitro-evolution of Precambrian and extant Rubiscos. *Sci. Rep.* 8 (1): 5532.

84 Alcalde, M. (2017). When directed evolution met ancestral enzyme resurrection. *Microb. Biotechnol.* 10 (1): 22–24.

85 Gomez-Fernandez, B., Risso, V.A., Rueda, A. et al. (2020). Ancestral resurrection and directed evolution of fungal laccases. *Appl. Environ. Microbiol.* 86, e00778-20.

86 Liu, H., Zhu, L., Bocola, M. et al. (2013). Directed laccase evolution for improved ionic liquid resistance. *Green Chem.* 15 (5): 1348–1355.

87 Gonzalez-Perez, D., Mateljak, I., Garcia-Ruiz, E. et al. (2016). Alkaline versatile peroxidase by directed evolution. *Catal. Sci. Technol.* 6 (17): 6625–6636.

88 Falk, M., Alcalde, M., Bartlett, P.N. et al. (2014). Self-powered wireless carbohydrate/oxygen sensitive biodevice based on radio signal transmission. *PLoS One* 9 (10): e109104.

89 Otey, C.R. and Joern, J.M. (2003). High-throughput screen for aromatic hydroxylation. In: *Directed Enzyme Evolution: Screening and Selection Methods* (eds. F.H. Arnold and G. Georgiou), 141–148. Springer.

13

Engineering Antibody-Based Therapeutics: Progress and Opportunities

Annalee W. Nguyen and Jennifer A. Maynard

The University of Texas at Austin, Department of Chemical Engineering, 200 East Dean Keeton, Austin, TX, USA

13.1 Introduction

Therapeutic antibodies continue to dominate the biopharmaceutical market with sales topping US$130 billion in 2018 [1]. Monoclonal antibodies received 53% of biopharmaceutical Federal Drug Administration (FDA) approvals from 2015 to 2018 [2] and continue to enjoy robust growth. In addition to market measures of the importance of antibodies to medicine, seven Nobel Prizes have been awarded in recognition of antibody-therapy related research, starting with the very first award in 1901 to von Behring and Kitasato for polyclonal antisera therapy and culminating with the recent 2018 awards to Winter for antibody phage display and to Allison and Honjo for antibody immune checkpoint blockade therapy. Although most approved therapies consist of unmodified monoclonal antibodies, other forms such as antibody-drug conjugates (ADCs), bispecific antibodies, and chimeric antigen receptors (CARs) are becoming established derivative markets that benefit from antibody engineering and development efforts.

Discovery and development of antibody therapeutics have changed greatly since the first monoclonal antibody therapeutic, OKT3, or muromonab-CD3 was approved in 1985. OKT3 is a fully murine IgG2a anti-CD3 immune suppressor used to prevent transplant rejection by blocking cytotoxic T-cell function. It had an unusual production method: hybridoma cell lines were injected into the mouse peritoneum to produce the protein in ascites. This first approved monoclonal antibody therapeutic had an unknown sequence at the time, and as a result of its non-human source was highly immunogenic with a short serum half-life. For these reasons, it was discontinued for human therapeutic use in 2010.

In 1998, a new era of blockbuster engineered antibody therapeutics emerged with the approval of two humanized antibodies (palivizumab and trastuzumab) causing significantly fewer side effects and setting a new standard for this class of pharmaceuticals. Human or humanized molecules efficiently and reproducibly manufactured using highly defined cell culture conditions are now expected. When introduced into patients, antibodies must exhibit low immunogenicity, high

Protein Engineering: Tools and Applications, First Edition. Edited by Huimin Zhao.
© 2021 WILEY-VCH GmbH. Published 2021 by WILEY-VCH GmbH.

stability, and long serum half-life. While binding specificity remains a key focus for antibody engineering efforts, the breadth of the antibody-based therapeutic market requires engineering to fine tune the therapeutic effects and manufacturability of any potential therapeutic. To provide an overview of current antibody development strategies, this chapter will include a description of antibody-based therapeutic formats (Section 13.2), then focus on current antibody engineering approaches for antibody discovery (Section 13.3), therapeutic optimization (Section 13.4), and product manufacturability (Section 13.5).

13.2 Antibody Formats

By far the most common format for monoclonal antibody therapeutics is human IgG1 (Section 13.2.1). However, in addition to the standard full-length monoclonal antibody structure, several antibody-based therapeutics with alternate formats have been approved. Nine ADCs (Section 13.2.2) and two bispecific antibodies (Section 13.2.3) are on the market. The first very low molecular weight single domain antibody (Section 13.2.4) was approved in 2019, and two CAR cellular therapies (Section 13.2.5) were approved in 2017, both representing completely new applications of the antibody binding site. While the vast majority of the ~80 antibody-based therapies are derived from IgG1s, several IgG2 and IgG4 subclass antibodies have also been approved [3].

13.2.1 Human IgG1 Structure

The prototypical human IgG1 (Figure 13.1a) consists of four polypeptide chains: two identical heavy chains (~50 kDa each) and two identical light chains (~25 kDa each), for a total 150 kDa molecular weight. Each chain consists of a series of immunoglobulin fold domains, each comprised of two anti-parallel β sheets stabilized by an internal disulfide bond. The heavy chain includes four immunoglobulin fold domains (the variable heavy or V_H domain and three constant heavy domains: C_H1, C_H2, C_H3) and the light chain includes two Ig fold domains (variable light, V_L, and constant light, C_L). The light and heavy chains form a hetero-dimer tethered by an inter-chain disulfide bond between C_L and C_H1. Two hetero-dimers then homo-dimerize via interactions between the C_H2 and C_H3 domains with additional disulfide bonds linking the heavy chain hinge regions to form an intact immunoglobulin.

The antigen binding site is formed by the V_L–V_H interface: each domain presents three variable loops called complementarity determining regions (CDRs), responsible for direct contact with antigen. Opposite the two antigen binding sites is the conserved crystallizable fragment (Fc), consisting of the heavy chain C_H2–C_H3 regions. The Fc is responsible for recruiting effector immune functions by interacting with receptors on immune cells and conferring a long-serum half-life. This modular structure allows substantial flexibility in design of antibody-based therapeutics, from minimal antigen binding constructs consisting of V_L and V_H domains joined by a flexible polypeptide linker to form a single chain variable

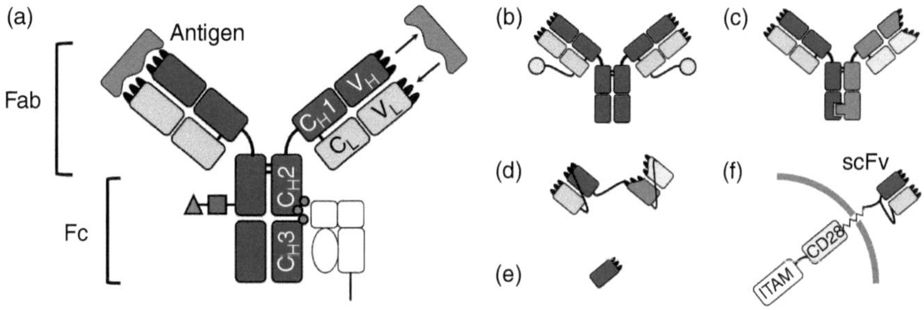

Figure 13.1 Therapeutic antibody formats. Several antibody-based therapeutics with designs tailored to their application are on the market or in development. The most common format is the (a) human IgG1 monoclonal antibody, shown with the heavy chains in dark gray, the light chains in light gray, and the six discrete immunoglobulin domains labeled. The Fab arms are formed by a heterodimer of the light chain V_L and C_L domains with the heavy domain V_H and C_H1, stabilized by an inter-chain disulfide bond. Within the Fab, the antigen binding site is formed by the V_H/V_L interface, with three CDRs per domain (black fill) directly mediating antigen binding. The Fc is comprised of homo-dimerized C_H2 and C_H3 domains, typically with an *N*-glycosylation site at residue N297 within each C_H2 domain (indicated by the triangle and square). The Fc region mediates effector functions through its interactions with FcRs (white fill) and complement proteins. (b) Antibody-drug conjugates (ADCs) use the antibody to localize a drug (circles) to diseased tissues and cells. Bispecific antibodies may be formed with (c) engineered heavy and light chains resulting in a full-length antibody with two specificity for two different antigens, (d) a dimer of scFvs as in the BiTE(R) (Amgen) technology, among many other methods. (e) Single domain antibodies comprised of a single V_H domain are an emerging format for therapeutics due to their very small size and high stability. Finally, (f) chimeric antigen receptors (CARs) comprise a tumor-targeting scFv antigen binding site linked to transmembrane and intracellular signaling domains that are recombinantly expressed on patient T cells.

fragment (scFv) to complex bispecific designs able to interact with multiple antigens simultaneously.

13.2.2 Antibody-Drug Conjugates

ADCs comprise an antibody tethered to a cytotoxic small molecule (Figure 13.1b); the antibody binds an antigen present on a specific cell type, increasing the local ADC concentration and essentially delivering the small molecule to target cells. Protein engineering efforts for ADCs focus on identification of permissible antibody sites and chemistries to support conjugation of the small molecule to the antibody. The ADCs on the market rely on thiol- and amine-coupling chemistry for attachment of cargo to solvent-exposed cysteine or lysine residues on the antibody. This typically requires introduction of a single, non-native cysteine residue, which is then prone to mispairing with cysteines normally involved in structurally important disulfide bonds. Because numerous solvent-exposed lysine residues are typically present on an antibody, lysine coupling produces a heterogeneous mixture in which the small molecule is coupled to one of several antibody sites. The use of non-natural amino acids to generate diverse reactive groups and increase linkage specificity is under active investigation and has considerable promise [4].

Surface exposed residues in the C_H2, C_H3, or C_L regions are considered reasonable coupling locations as they are unlikely to interfere with antigen binding by the V_H/V_L interface. However, coupling a small molecule to the Fc domain may impact Fc effector functions important for the therapeutic mechanism used by the ADC. To address this issue, the Herceptin IgG1 constant regions were screened by phage display to identify sites tolerant to the introduction of a free cysteine. This THIOMAB platform identified potential coupling sites that do not interfere with antigen binding or Fc receptor (FcR) binding [5]. To ameliorate the impact of cysteine mispairing, all disulfide bonds are reduced and re-oxidized after purification, improving the yield of properly folded protein and ensuring the introduced cysteine is available for conjugation.

13.2.3 Bispecific Antibodies

Bispecific antibodies are a variation of the typical IgG1 antibody in which each antigen binding site binds a different antigen (Figure 13.1c,d). This feature can be useful for targeting multiple epitopes within the same disease state or acting as a molecular "staple" to bring two antigens – or, in the case of surface-exposed receptors, cells, together. Bispecific antibodies have received the most attention for use in cancer therapies, but their abilities to target infectious agents and toxins may also provide enhanced efficacy as compared with a combination of two monoclonal antibodies while reducing the risk of pathogen escape variants [6].

The first bispecific antibody was approved by the FDA in 2014. Named blinatumomab, this protein consists of two scFvs joined into a single protein by flexible polypeptide linker (Figure 13.1d) that acts as a bispecific T-cell engager (BiTE®, Amgen). One scFv targets the CD3 surface receptor on T cells while the other targets the CD19 surface receptor on B cells. By bringing the T and B cells in close contact, blinatumomab triggers T-cell killing of the adjacent B cell, resulting in a potent therapy for B cell malignancies such as leukemia. A major benefit of the BiTE structure from a production standpoint is that the protein can be expressed as a single polypeptide with little opportunity for mispairing of the two V_H and V_L regions. The lack of an Fc in the BiTE format reduces the risk of cytokine storm due to T-cell over-stimulation, but it also results in a very short serum half-life (<three hours) [7] and thus requires use of a continuous infusion pump to sustain therapeutic serum concentrations.

Generation of full-length bispecific antibodies with a different binding specificity in each arm requires engineering to support heterodimeric instead of homodimeric pairing between the two distinct heavy chains. A widely used method employs complementary "knob-into-hole" or charge mutations to encourage proper pairing in each Fc. In this case, each heavy-light chain pair is produced in a separate cell culture process. The two heavy and light chain pairs are expressed separately. The purified proteins are then partially reduced to break the disulfide bonds stabilizing any undesirable mono-specific antibodies, and a mixture of both half-antibodies re-oxidized to form the bi-specific antibody [8, 9]. This procedure results in mainly full length antibody with different binding specificities in each Fab arm as well as an intact Fc region to mediate effector functions and a long serum-half life (Figure 13.1c).

A bigger challenge for production of full-length bispecific antibodies has been proper pairing of the light and heavy chains when all four chains are expressed in a single cell line. One strategy is to engineer biased binding into the Fab interface between the V_L–C_L and V_H–C_H1 regions, analogous to strategies used to support heavy chain heterodimerization [10, 11]. Combining the engineered "knob-into-hole" Fcs with these variants that promote heavy-light chain pairing has resulted in Roche's CrossMAb technology. Several of these are currently in clinical trials, including cibisatamab with anti-carcinoembryonic antigen and anti-CD3 binding for solid tumor indications in phase IB and faricimab with anti-Ang2 and anti-VEGF-A binding for macular degeneration in phase III.

Another popular approach is to use the same light chain for both binding sites. This strategy is successful for antibodies in which the light chain makes minor contributions to antigen binding [12]. Any heavy chain homodimers can then be removed during purification. This is the approach taken with emicizumab, a bispecific that mimics the role of factor VIII to treat hemophilia A, which received FDA approval in 2017 [13]. There are a number of different bispecific formats and strategies for their consistent production described in Brinkman and Kontermann [14] and in Sedykh et al. [15].

13.2.4 Single Domain Antibodies

Though still in relatively early development, there is considerable enthusiasm surrounding single domain antibodies (Figure 13.1e). Nanobody formats (~15 kDa) are based on the variable heavy chain domain of naturally occurring camelid antibodies that lack a light chain. Their high stability and small size are very attractive for tumor penetration and other applications. Moreover, nanobodies typically have long CDR H3 loops which are able to access epitopes unavailable to traditional antibodies. A major breakthrough in the field came with the development of a humanized nanobody scaffold in which the protein is resurfaced with human-like V_H framework regions, but is capable of accepting non-human nanobody CDRs [16]. Further work generated a library of >10^9 humanized nanobody variants, which is important to alleviate reliance on costly camel or llama immunization strategies [17]. The first FDA approval for a nanobody was granted in February of 2019 for Ablynx's caplacizumab, a nanobody dimer developed to treat a rare blood clotting disorder. Several other nanobody therapeutics are in development [18].

13.2.5 Chimeric Antigen Receptors

Finally, CARs are a new and exciting application of scFvs to generate synthetic T-cell receptors (Figure 13.1f). When introduced into a patient's T cells, these synthetic receptors may provide long-lasting control of diseased cells. The general design includes an extracellular tumor targeting scFv fused to a transmembrane domain and intracellular T-cell signaling domains capable of activating T cells. Several signaling designs have been evaluated and can include various combinations of CD28, 41BB, CD3ζ, and other domains. When the scFv regions bind antigens on a target cell surface, the CARs cluster and the T cell is activated, ultimately resulting in degranulation and killing of the target tumor cell. The two FDA-approved CARs (Yescarta™ and Kymriah™) both target B cell malignancies with reports of up to 80% relapse-free survival rates at 6 months and 59% at 12 months [19]. In this early development of CAR technology, scFvs have been isolated from full-length antibodies as "off the shelf" antigen binders, but the stability and monomericity of scFvs vary widely. This suggests opportunities to improve CAR performance by engineering antibody scFvs for this specific application.

Due to the impressive clinical outcomes, many strategies are being explored to generate next-generation CAR-T constructs, which have more tightly controlled activation, allow for modular design, or limit the emergence of tumor resistance. For instance, to improve control over T-cell function, patient T cells can be transformed with a universal CAR-T construct specific for a non-human sequence (e.g. a yeast transcription factor) [20]. After administration to the patient, injection of a bifunctional CAR ligand-Fab fusion protein, in which the Fab domain binds a ligand unique to the target cell, serves to redirect the CAR-T to the target cell. This approach allows control of CAR-T cell activation by infusion of a protein with predictable pharmacokinetics to avoid toxicity and tune CAR-T responses in real time. This general approach can be used with scFvs, BiTEs, peptides, and other molecules as switches in the UniCAR system [21]. In a similar path toward universal or adaptable CARs, FDA approved anti-CD19 CARs can be redirected to other cancers via CD19 ectodomains fused to an anti-cancer antigen scFvs [22]. Finally, to address relapse due to loss of CD19 expression after treatment with approved anti-CD19 CAR-T cells, bispecific CARs using two fused scFvs to generate "OR" gate functionality have been generated and may reduce the likelihood of escape. This has been demonstrated with CD19/CD20 [23] and BCMA/CS1 [24] bispecificity in developmental efforts optimizing B cell cancer CAR therapy.

13.3 Antibody Discovery

Regardless of the format, antibodies must bind a relevant therapeutic target. It is often unclear what that target should be for a specific disease state and the challenges of target identification are discussed in Section 13.3.1. Once a target is identified, relatively well-established methods have been developed for isolating high affinity antibodies (Section 13.3.2) using synthetic libraries (Section 13.3.2.1), immunization (Section 13.3.2.3), or hybrid approaches (Section 13.3.2.6).

13.3.1 Antibody Target Identification

Selection of an appropriate biomarker or antigen epitope for antibody targeting is by far the greatest challenge in the development of an antibody-based therapeutic for a specific disease. Disease targets for antibody therapeutics generally fall into three major areas: cancer, immune disorders (Section 13.3.1.1), and infectious diseases (Section 13.3.1.2). No reliable high throughput methods for identifying promising antibody targets exist, and development of antibody therapeutics always requires detailed knowledge of each disease system.

13.3.1.1 Cancer and Autoimmune Disease Targets

For cancer therapies, the necessity of targeting self-antigens that are overexpressed on tumor tissues but present at lower levels on healthy tissues can result in significant side effects. Approved therapeutics typically target over-expressed growth factor receptors (e.g. epidermal growth factor receptor [EGFR] and human epidermal growth factor receptor 2 [HER2]), cell-type specific receptors (e.g. CD19 and CD20), or immune regulatory receptors (e.g. CTLA-1, PD-1, PD-L1). Adjuvant trastuzumab (Herceptin®) therapy, for example, is quite effective with therapy responsible for a 23–35% improvement in overall survival, but its target, HER2, is also expressed on the surface of cardiomyocytes. This results in ~13% of treated patients experiencing cardiotoxic effects due to on-target-off-tumor binding to HER2 on cardiomyocytes with a typical combined therapy (trastuzumab with paclitaxel) [25]. The use of antibodies to suppress immune responses in autoimmune diseases can disrupt important normal immune processes. On the other hand, there are several examples of antibodies used for cancer therapy or to regulate immune responses that initiated dangerous uncontrolled immune activation ("cytokine storm") [26, 27] or new autoimmune disease [28]. Thus, the ideal target is a neoantigen that is expressed only in tumor or diseased tissues and is shared among multiple patients; however, identification of such molecules is challenging.

13.3.1.2 Infectious Disease Targets

The relatively few approved antibodies targeting infectious diseases are less likely to cause life-threatening side effects since the antigens are generally unique in the human body, but there are other challenges to identification of susceptible antibody targets. Infectious pathogens often secrete copious amounts of soluble proteins that can act as immune decoys, eliciting an impotent antibody response. This occurs with soluble antigen production of protein G in respiratory syncytial virus (RSV) [29] and a glycoprotein (sGP) in Ebola virus [30], and may also be true of the *Bordetella pertussis* filamentous hemagglutinin (Fha) [31]. More effective antibodies target surface-bound antigens that will initiate opsonophagocytosis and pathogen killing, but these epitopes are often concealed by surface carbohydrates.

For some bacterial infections, targeting secreted toxins can reduce the problematic symptoms of disease, thereby allowing normal immune functions to clear the bacteria. In this case, efficacy is dependent upon antibodies blocking the receptor binding site, enzyme active site, or other epitopes essential for intoxication. For example, infection with *B. pertussis*, the causative agent of whooping cough, results in high lung bacterial colonization levels but severe disease results from dysregulated white blood cell chemotaxis mediated by the pertussis toxin [32]. Both immunization and infection result in high titers of anti-pertussis toxin antibodies, of which only a fraction are protective [33]. Treatment of *B. pertussis* challenged baboons with a mixture of humanized 1B7 and 11E6 monoclonal antibodies, which target non-overlapping protective epitopes on pertussis toxin, halted clinical symptoms of disease [34], while prophylactic treatment with 1B7 alone protected neonatal baboons from clinical disease after subsequent infection [35]. In a similar way, viral pathogens are susceptible to antibodies that block viral/host cell fusion and or lock viral fusion proteins in nonfunctional conformations [36].

13.3.2 Screening for Target-Binding Antibodies

Once a target has been chosen, a variety of well-established methods exist to generate antibodies with high affinity for the target. Antibody binding specificity and affinity are determined by the CDRs within the variable regions of each chain. These have variable length and can contain up to 91 total amino acids: CDR-L1 contains 10–17 residues, CDR-L2 has 8–12 residues, CDR-L3 has 7–13 residues, CDR-H1 has 10–16 residues, CDR-H2 has 8–15 residues, and CDR-H3 spans 5–18 residues [37]. A library that randomizes each position within the CDR and allows for all possible CDR lengths would have astronomical size ($\gg 20^{91}$ or 2.5×10^{118}). Further complicating matters, each CDR loop except CDR-H3 forms predictable canonical structures based on the residue present at several positions [37], limiting the useful sequence space to those compatible with functional loop structures. Engineers therefore employ strategies that limit library size while maximizing the fraction of sequences forming functional antibody structures. These can include using existing sequence/structure information to build synthetic libraries (Section 13.3.2.1), taking advantage of the robust antibody generation mechanisms present in human or animal hosts (Section 13.3.2.2), or using in vitro methods to mimic the immune system in a hybrid antibody affinity maturation approach (Section 13.3.2.3).

13.3.2.1 Synthetic Library Derived Antibodies

Protein engineers can routinely make libraries containing between 10^8 and 10^{11} genetic variants in antibody phage display systems [38] (for a review on antibody isolation using phage display, see Frenzel et al. [39]). Because these libraries constitute a mere fraction of the entire sequence space possible by CDR randomization and CDR length variation, various methods to limit library sizes and bias libraries toward functional CDRs are employed. In many cases, the parent sequences used to generate the library are limited to a few germline sequences, residues are altered in only some CDRs, residues are randomized to allow sampling of a limited number of amino acids (such as small hydrophobic residues), and/or only the CDR-H3 is allowed to vary in length [38, 40].

Despite these restrictions, libraries of this type consistently identify binders to various antigens with affinities in the 0.1–1000 nM range. The Dyax (now Shire) Fab phage libraries have been particularly successful at generating fully human antibodies with at least four (avelumab, necitumumab, ramucirumab, and lanadelumab) that have received FDA approval. To create their libraries, Dyax combined human donor light chain sequences from peripheral blood mononuclear cells and a single germline heavy chain framework with donor CDR-H3 sequences and targeted mutagenesis in the CDR-H1 and CDR-H2 loops [41]. The HuCal Fab phage libraries (Bio-Rad/MorphoSys) include several human heavy and light chain germline sequences presenting CDRs representative of those possible through natural genetic recombination, with the most recent iteration comprising 4.5×10^{10} members [42]. The first HuCal-derived antibody, guselkumab, received FDA approval in 2017 (Janssen Biotech/Johnson & Johnson). Yeast display technology (Adimab) has enjoyed growing popularity with pharmaceutical companies as an alternative to phage display that allows for large libraries, eukaryotic protein folding machinery, and flow cytometry-based selection schemes. Several clinical stage antibodies were discovered using this platform with one currently in phase III trials [43]. Adimab's technology relies heavily on diverse pre-immune human immune repertoires with a focus on diversification of the CDR-H3 region [44].

13.3.2.2 Host-Derived Antibodies

The immune system has an impressive capacity to continuously evolve panels of well-behaved antibodies with high affinity and high specificity after exposure to an antigen of interest. Harnessing this power to generate therapeutic antibodies requires identification of hosts that have been naturally exposed to or immunized with the antigen (Section 13.3.2.3), pairing and screening of the responsive V_L and V_H regions (Section 13.3.2.4), followed by additional engineering to improve affinity, specificity, stability or other features, such as humanization, if desired (Section 13.3.2.5).

13.3.2.3 Immunization

Animal immunization is an older, but highly effective method for generating antibodies and is especially useful for generating responses against human antigens, which are poorly immunogenic in humans. Typically, mice (or other animals, such as camels to isolate nanobodies) are immunized and boosted with the antigen of interest, spleens, or circulating lymphocytes are collected for B-cell isolation, the V_L and V_H regions are amplified *en masse*, paired, and screened to identify those binding antigen. At a minimum, the variable regions are appended with human constant regions to create mouse-human chimeric antibodies. More commonly, the murine variable sequences are humanized by retaining the mouse CDRs but replacing as much of the mouse framework sequence as possible with the corresponding human germline sequence [34].

This approach has evolved to use humanized mice with grafted human antibody sequences that directly generate a human antibody repertoire after vaccination. The first successful humanized mouse platforms randomly inserted a small subset of human germline variable sequences into the mouse genome while knocking out the corresponding endogenous mouse genes [45]. Medarex (Bristol Myers Squibb) utilized this platform to develop no fewer than 10 FDA approved antibodies [46]. The XenoMouse (Abgenix) offered access to greater antibody diversity by introducing the majority of the human heavy and kappa light chain germline sequences [47]. These mice resulted in several FDA approved human antibodies, including panitumumab, durvalumab, brodalumab, denosumab, and evolocumab, but the human constant regions were incompatible with the rest of the mouse immune system and resulted in a generally muted immune response after immunization. Fully human antibody production yield was low in mice and somatic hypermutation in the resulting clones was minimal [48].

To address this, the Kymouse [49], the VelocImmune mouse [50], and the Trianni mouse [51] encode the complete human germline variable region repertoire and mouse constant regions. Even with these newer mice, some antigens have sufficient homology between humans and mice that it is difficult to break the tolerizing anti-self-antibody response. The introduction of human variable repertoires in chickens [52] and cows [53] are promising efforts to elicit immune responses to highly conserved targets in an animal phylogenetically distant from humans.

While immunization requires several weeks of waiting for antibody titers to rise, the extensive diversity of the endogenous germline sequences and *in vivo* affinity maturation has yet to be matched by synthetic methods. FDA approved antibodies discovered through immunization are still far more numerous than those found through screening synthetic libraries, and there is evidence to suggest that these antibodies have more favorable biophysical properties [54].

13.3.2.4 Pairing the Light and Heavy Variable Regions

Both variable regions contribute to antigen binding in most cases and have complementary interfaces, which stabilize the binding site. For these reasons, it is helpful to identify the light and heavy chains forming a binding site in a single B cell; however, most methods amplify these genes *en masse* from pooled B cells. Because the endogenous V_L and V_H sequences are located on different chromosomes, reconstituting the native B cell pairing after immunization is challenging.

In the early days, hybridomas were created by fusing splenic B cells with myeloma cells, retaining appropriate pairing and producing large quantities of antibodies in the absence of sequence information. As antibody technology has matured, it has become apparent that hybridomas do not always produce a single antibody clone and thus require screening even when cloning from the relatively homogeneous hybridoma [55].

Current demands for product reproducibility and purity have determined that any antibody currently moving toward clinical trials is expressed in a highly defined and controlled cell culture environment, in a cell line with no endogenous antibody production, and with stable integration of a known antibody sequence into its genome. In order to achieve this with current recombinant library-based discovery approaches, V_L and V_H sequences are randomly combined and then screened by either phage display or high-throughput sequencing to identify compatible pairs. Screening immune libraries with phage display is well described [56], and is a fast, inexpensive method to isolate antibody binders after immunization.

More recent developments in V_L–V_H pairing involve sequencing of the immune repertoire. The first approach involved immortalization of B cells, followed by flow cytometric sorting, single-cell reverse–transcriptase polymerase chain reaction (RT-PCR), and sequencing [57–60]. This has evolved into isolation of single B cells in emulsion droplets, recovering the V_L and V_H encoding messenger ribo-nucleic acid (mRNA) on the same poly-T coated bead, and amplifying the V_L and V_H with a linker to retain pairing [61]. Another common approach is to use barcoded primers for amplification of the V_L and V_H regions, which link sequences from a single cell via a common sequence [62]. Finally, high-throughput next-generation sequencing itself may identify pairs by matching the frequency of V_L and V_H genes observed in the immune repertoire of bone marrow plasma cells [63]. After high-throughput sequencing, highly abundant sequences are presumed to contain antibodies resulting from immunization. Individual antibodies are produced recombinantly using V_L and V_H pairs with similar abundance to screen for antigen binding. This method is extremely rapid and may be applied to human donor peripheral blood mononuclear cells, providing useful antibody repertoire data. It is also enables discovery of potential therapeutic antibodies in elite responders and specific disease states [64].

13.3.2.5 Humanization

While not required when antibodies are discovered from human repertoires or humanized mouse immune libraries, humanization to reduce the risk of anti-drug antibodies and provide human effector functions is necessary for antibodies derived from murine and other non-human sources. In the simplest form, appending human constant regions to non-human variable regions will partially accomplish this, but immunogenicity can still be significant and fully human antibodies are preferred. Humanization efforts involve computational comparison of the non-human sequence or structure with homologous human mature or germline antibody sequences to find more human-like substitutions, especially for solvent-exposed residues [65, 66]. A small set of potential humanized variants are created, and then screened for maintenance of antigen binding affinity and production yields [34].

Although the World Health Organization assigns International Nonproprietary Names (containing -o- for murine, -xi- for chimeric, -zu- for humanized, or -u- for human) based on sequence, the practical meaning of these designations is not clear [67]. "Human" designated antibodies must have ≥85% sequence identity with known human germline sequences, while "humanized" antibodies must have human constant regions and a "V region amino acid sequence which, analyzed as a whole, is closer to human than to other species." While pharmaceutical companies desire human designations during early development, the designation does not automatically imply the absence of immunogenic potential. Ultimately, immunogenicity begins to be assessed in phase I clinical trials, and is impacted by many factors [68]. While fully mouse antibodies are highly problematic and consistently generate human anti-mouse antibody responses, the exact sequence and structure requirements to make antibodies that are physiologically accepted as human are yet unknown. Even fully human antibodies can generate anti-drug antibodies: in one study nearly 30% of patients treated with the human antibody adalimumab had problematic anti-drug responses that affected therapy [69]. A discussion of current efforts to reduce the immunogenicity of antibody-based therapeutics is presented in Section 13.4.4.

13.3.2.6 Hybrid Approaches to Antibody Discovery

While these approaches have yet to yield approved antibodies, researchers are trying to combine the benefits of synthetic libraries (avoiding slow and cumbersome immunization protocols) and immunization (hands-off affinity maturation that results in highly developable antibodies). To achieve this, antibody display libraries have been introduced into mammalian cell lines and co-expressed with activation-induced cytidine deaminase, the protein responsible for somatic hyper-mutation of antibodies *in vivo*. As the library is screened by flow cytometry for antigen binding over several rounds, the antibody sequence is perpetually mutated with each cell division, allowing accumulation of new mutations and continued antibody affinity maturation. Unfortunately, the activation-induced cytidine deaminase (AID)-mediated mutation rate is quite low and the initial mammalian library sizes are restricted to about 10^6 variants, so discovery of antibodies from a completely naïve library can be laborious [70].

13.4 Therapeutic Optimization of Antibodies

After initial identification of antibodies capable of binding the target of interest described in Section 13.3, affinity maturation and specificity enhancement may be required. Traditional affinity maturation techniques are very well described in Tabasinezhad et al. along with new computational and next-generation sequencing techniques for maturation [71]. In this section, we will discuss several factors in addition to target binding specificity and affinity that determine the efficacy of an antibody-based therapeutic for the application of choice. Specifically, antibody serum half-life (Section 13.4.1), effector function activation (Section 13.4.2), tissue localization (Section 13.4.3), and immunogenicity (Section 13.4.4) all appear important to ensure high potency. Many of these factors result from Fc interactions with FcR on cellular surfaces and soluble complement components (Table 13.1 and Figure 13.2).

13.4.1 Serum Half-Life

Endogenous antibodies have an *in vivo* half-life of about 21 days [84], which is quite long for serum proteins. Antibodies are too large for renal clearance [85] and rely on recycling by the neonatal Fc receptor (FcRn) to avoid lysosomal degradation as well as clogging of the renal filtration apparatus [86]. FcRn, widely expressed on the surface of endothelial and hematopoietic cells, interacts with binding sites in the human IgG1 Fc (Figure 13.2d). The Fc–FcRn interaction is relatively low affinity at neutral serum pH, but increases dramatically as the pH of the endosome decreases. While the soluble contents of the endosome are transferred to lysosomes for degradation, IgG complexed with membrane-bound FcRn are transferred to recycling endosomes and then the cell surface where IgG is released back into the serum. In therapeutic applications, long serum half-life reduces the frequency of doses required and could reduce the absolute dose of antibody required. The IgG1 Fc has been engineered to

Table 13.1 Antibody-receptor interactions of interest.

Receptor	Affinity (K_D) to human IgG1 Fc	Cellular expression [72]	Function
FcγRIa (CD64)	High (~5 nM) [73]	• Eosinophils • Neutrophils • Monocytes • Dendritic cells • Macrophages	• Unclear function • Potent cytotoxic effects [74]
FcγRIIa (CD32a)	Moderate (~500 nM) [73]	• Platelets • Eosinophils • Neutrophils • Monocytes • Dendritic cells • Macrophages	• Clearing invading bacteria and viruses (phagocytosis) • Clearing immune complexes [75]
FcγRIIb (CD32b)	Low (~3 μM) [73]	• B-cells • Eosinophils • Neutrophils • Mast cells • Basophils • Dendritic cells • Macrophages	• Inhibitory • Possible role in memory response generation [76]
FcγRIIIa (CD16a)	Moderate (~200 nM) for 158F allele Low (~1.5 μM) for 158 V allele [73]	• Natural killer cells • Mast cells • Eosinophils • Monocytes • Dendritic cells • Macrophages	• Killing infected and cancerous self-cells (ADCC) [77]
FcγRIIIb (CD16b)	Low (~3 μM) [73]	• Neutrophils	• Activating neutrophils [78]
FcRn (with β_2m)	Low (~1 μM at pH 6, >10 μM at pH 7.4) [79]	Widely expressed in • Epithelial cells • Hematopoietic cells	• Recycling of antibodies and albumin resulting in long half-life
C1q	Moderate (~50 nM) [80]	• Soluble complement system component	• Activating complement for wide-ranging immune activation including CDC
C3b	N/A – covalent deposition	• Soluble complement system component	• Clearing antibody-antigen immune complexes
Protein A	High (~5 nM) [81]	• N/A	• Antibody purification

Sources: van der Poel et al. [74], Shashidharamurthy et al. [75], Nimmerjahn and Ravetch [76].

both extend (Section 13.4.1.1) and reduce (Section 13.4.1.2) its natural half-life, but these changes often impact effector functions (Section 13.4.1.3) as well.

13.4.1.1 Antibody Half-Life Extension

Several half-life extending Fc mutations have been described. Three mutations (M252Y/S254T/T256E, referred to as YTE or sometimes MST) to the Fc resulted

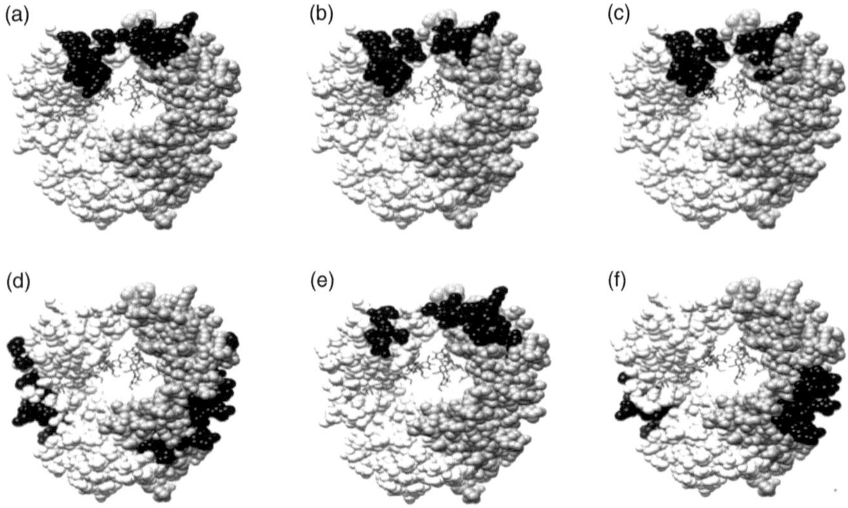

Figure 13.2 Details of the receptor binding sites on human IgG1 Fc. The human IgG1 Fc region comprised of a C_H2-C_H3 homo-dimer is shown in each panel (PDB ID 1HZH). One C_H2-C_H3 chain is shown in white space-fill and the other in gray space-fill. Residues highlighted in black space-fill indicate the Fc residues within 5 Å of bound (a) FcγRIa (according to PDB 4S4M, 4W4O, and 4ZNE), (b) FcγRIIa and FcγRIIb (according to PDB IDs 3RY6, 3WJJ, and 3WJL), (c) FcγRIIIa and FcγRIIIb (according to PDB IDs 1E4K and 5VU0), (d) the FcRn and β_2 microglobulin complex (according to PDB ID 4N0U and the rat 1I1A), (e) C1q (according to PDB ID 5FCZ), and (f) *Staphylococcus* protein A (according to PDB ID 5U4Y). The glycosylation on residue N297 (shown in stick model) is required for binding to FcγRIa, FcγRIIa and FcγRIIb, FcγRIIIa and FcγRIIIb. Only FcγRIIIa and FcγRIIIb binding are impacted by the specific glycoforms present, with binding increased when the antibody presents an afucosylated carbohydrate [73]. Antibody binding to C1q depends on not just the presence of glycosylation, but is impacted by the specific glycoforms present and is increased with galactosylation [73]. Binding to FcRn [82] and protein A [83] does not depend on Fc glycosylation. Sources: Dekkers et al. [73], Geuijen et al. [82], Gaza-Bulseco et al. [83].

in a half-life threefold extension in cynomolgus monkeys [87] and similar results in humans [88]. The longer circulation results from increased affinity between Fc and FcRn in the acidic endosome relative to neutral serum [89]. Medimmune is currently evaluating the YTE version of its anti-RSV antibody D25 in phase II clinical trials [36]. Prophylaxis with the unmodified antibody requires five monthly doses, which is expected to be reduced to a single dose per season for an extended half-life variant. Another set of Fc mutations extending IgG1 half-life by similar mechanisms include M428L/N434S (called LS), which was tested in an anti-vascular endothelial growth factor (VEGF) antibody and shown to have increased tumor killing over unmodified antibody in a mouse model [90]. The LS Fc integrated into an antibody targeting HIV also extended antibody half-life in non-human primates [91]. A newly developed DHS (L309D/Q311H/N434S) Fc, which outlasts YTE in humanized mice by ~50% may confer >100 days circulation half-life in humans [92].

13.4.1.2 Antibody Half-Life Reduction

In an interesting extension of this work, antibody Fc regions have been engineered to block endogenous antibodies from binding FcRn. These engineered Fcs have increased affinity for FcRn at all pH levels, successfully compete with endogenous antibodies for FcRn binding, and thereby confer rapid clearance of endogenous circulating antibodies [93]. These antibodies that block FcRn (called "AbDegs") have potential utility to treat antibody-driven autoimmune disorders and alleviated arthritis [94] and encephalomyelitis [95] in mouse models. A therapeutic based on this technology, efgartigimod, is in clinical development for the treatment of several autoimmune disorders [96].

13.4.1.3 Effect of Half-Life Modification on Effector Functions

Recently, alterations to the Fc to modulate FcRn binding were investigated for impact on other FcR interactions [97]. Despite the relatively distinct binding sites of FcRn relative to the other FcRs (Figure 13.2a–d), all of the mutations described earlier except DHS reduced binding to the other FcRs tested and impaired antibody activation of downstream effector functions such as cytotoxicity and phagocytosis. Depending on the application, these effects may be detrimental to clinical outcomes. Even in the case of anti-viral and anti-toxin antibodies, when high antibody affinity and titer seem most important for efficacy, the interaction of antibodies with FcRs can impact the quality of the endogenous antibody response to infection [98]. AbDegs are unlikely to be negatively impacted by reduced effector functions as their activity is driven by Fc–FcRn binding.

13.4.2 Effector Functions

As we learn more about antibody interactions with pathogens, tumor cells, and the rest of the immune system, it has become apparent that antibodies can protect from disease in many ways (Figure 13.3). These mechanisms include simply blocking ligand access to a binding or active site; killing invading pathogens through complement-dependent cytotoxicity (CDC); killing diseased cells through antibody-dependent cellular cytotoxicity (ADCC) and antibody-dependent cellular phagocytosis (ADCP); clearing immune complexes of antibodies bound to viruses or toxins; and interference with cellular signaling as either an agonist or antagonist. In applications where the target must be killed or cleared, Fc interactions with receptors or complement proteins are responsible for these activities.

Thoughtful understanding and consideration of the mechanism used by a therapeutic antibody is required to guide engineering for optimal performance. For example, if the general class of an antibody is anti-cancer, it may benefit from enhanced ADCC if the target is expressed on tumor cells, but may suffer if the target is a checkpoint blockade receptor on T cells. Engineering the antibody Fc for specific effector functions has become a major consideration for therapeutic antibodies in cancer (Section 13.4.2.1), infectious diseases (Section 13.4.2.2), and autoimmune diseases (Section 13.4.2.3). Several methods have been used to engineer the IgG1 Fc (Section 13.4.2.4) and are reviewed in greater detail by Wang et al. [99].

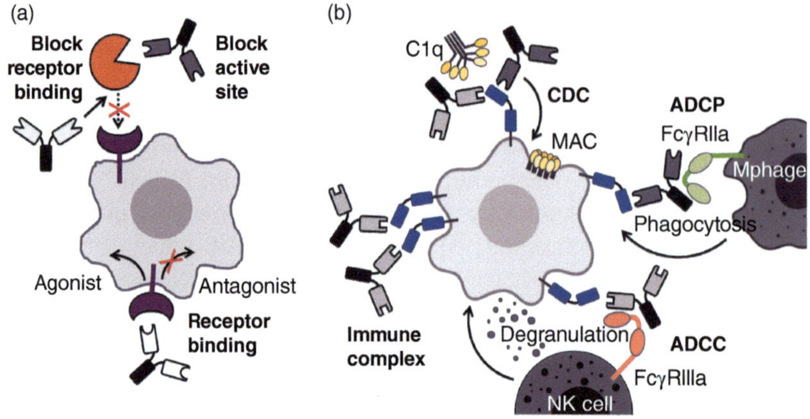

Figure 13.3 Mechanisms of antibody-mediated therapeutic effects. (a) Fc-independent antibody effects. Antibodies can bind soluble or membrane-bound antigens to inhibit their catalytic activity by binding directly to the active site or an adjacent allosteric site (*block active site*). They can also prevent an antigen's interactions with other proteins or receptors (*block receptor binding*). Finally, antibodies can bind receptors to block binding of the natural ligand or to trigger signaling and exert agonistic or antagonistic effects (*receptor binding*). (b) Fc-dependent antibody effects. Multiple antibodies coating a single antigen or particle (an "*immune complex*") increase the local concentration of Fc domains, which then exhibit increased avidity for complement and low affinity Fcγ receptors. Small immune complexes are mainly cleared when Fcs become attached to red blood cells via complement C3b, and the red blood cells hand the complexes off to the liver for destruction. Immune complexes may also be cleared through antibody-dependent phagocytosis (ADCP): binding to FcγRIIa on macrophage cells leads to endocytosis of small complexes or phagocytosis of large immune complexes. Opsonized particles can also trigger complement dependent cytotoxicity (CDC) through binding of C1q and subsequent formation of a membrane attached complex (MAC) to lyse cells/enveloped viruses. They can also trigger antibody-dependent cellular cytotoxicity (ADCC) clearance through multivalent binding to and clustering of FcγRIIIa receptors on natural killer (NK) cells.

13.4.2.1 Effector Function Considerations for Cancer Therapeutics

For antibodies targeting over-expressed tumor antigens, activation of ADCC, ADCP, and CDC following binding of the antibody Fc to activating FcRs (FcγRIa, FcγRIIa, FcγRIIIa, and FcγRIIIb) has been implicated in the therapeutic mechanism. This activity appears relevant for several FDA-approved antibodies (e.g. anti-HER2 [100], anti-EGFR [101], anti-PD-L1 [102]) and has been enhanced for other antibodies. For example, two anti-CD20 antibodies used for the treatment of leukemia and lymphoma, obinutuzumab (FDA approved) and ocaratuzumab (in clinical trials), were engineered for improved affinity to FcγRIIIa. For obinutuzumab [103], the type of Fc glycosylation was altered for higher affinity FcγRIIIa binding, while for ocaratuzumab [104–106], the Fc was altered with two amino acid substitutions for enhanced binding to a low affinity FcγRIIIa-158F allelic variant.

An alternate strategy for enhanced effector function is use of other antibody isotypes. One interesting approach generated an IgG1/immunoglobulin A (IgA) (IgGA) hybrid Fc to capitalize on the long half-life of the human IgG Fc and its binding to

FcγRs and the incorporate the strong cytotoxic activation of the IgA Fc upon binding to FcαRI [107]. The integrated IgA Fc retained high affinity for the complement C1q protein and was responsible for enhanced complement activation in the Herceptin IgGA tested. Strategic use of alternate subclasses of IgG such as IgG2, IgG3, or IgG4, which have variable binding affinity to FcRs may be another way to influence effector function. IgG2 and IgG4 both have generally reduced FcγR binding, whereas IgG3 has stronger binding to FcγRIIa relative to other subclasses but exhibits a reduced serum half-life of only about seven days [3].

Bispecific antibodies such as those used to bridge T cells and tumor cells via anti-CD3 and anti-tumor variable regions, respectively, would benefit from the long circulation afforded by the Fc/FcRn interaction, but not other inflammatory FcγR interactions. Amino acid substitutions to reduce FcγR binding may be important in this situation, and Fcs have been engineered with significantly reduced effector functions (SEFL variants A287C/N297G/L306C or R292C/N297G/V302C, and LALA-PG variant L234A/L235A/P329G), but unaltered FcRn binding and half-life and high stability [108–110]. An extended half-life "Fc-silenced" IgG1 has also been developed for applications such as prophylaxis against infection in which long-term epitope blocking, but not induction of effector functions is important [111].

13.4.2.2 Effector Function Considerations for Infectious Disease Prophylaxis and Therapy

In infectious disease applications, activation of effector functions may be helpful. Antibodies recognizing antigens on the bacterial surface exert bactericidal activities by evoking CDC. Virally infected cells may depend on many of the same anti-cancer mechanisms described above when the viral proteins are present on the surface of infected cells [112]. When antibodies are directed at the viral particle itself or secreted bacterial toxins, the long antibody half-life and immune complex removal mechanisms reliant on complement component C3b deposition and FcγRIIa binding are important for effective clearance, but the utility of other IgG effector functions is less clear.

Targeting bacteria and viruses directly presents additional complications and opportunities for engineering, since evolution has armed some pathogens with weapons to evade antibodies. For instance, *Pseudomonas aeruginosa* and *Staphylococcus aureus* strains produce proteases, which specifically cleave antibodies [113–116]. Several bacteria express outer membrane Fc-binding proteins, effectively cloaking the bacterium in antibody oriented such that it is unable to initiate downstream effector functions. Among many other examples, cytomegalovirus uses viral FcRs to deplete and repurpose virus-specific antibodies [117], and *Staphylococcus* expresses protein A and *Streptococcus* expresses protein G, which both bind antibody Fc domains [118]. Protein A and protein G are both important affinity chromatography reagents used in antibody purification, so manipulating the binding sites for these proteins would complicate downstream processing. The protein A binding site also overlaps significantly with FcRn binding (Figure 13.2f), and manipulation of this site could impact both of these important interactions.

13.4.2.3 Effector Function Considerations for Treating Autoimmune Disease

Not all FcRs are immune activating: Fc binding to FcγRIIb can result in reduction in inflammation and effector recruitment. By enhancing the affinity of Fc for FcγRIIb, antibodies targeting the CD19 surface marker on B cells reduced the symptoms of lupus erythematosus [119]. Biasing antibodies toward immune suppression by Fc engineering may be an effective way to inhibit autoimmunity in a tissue specific way. Some antibodies for autoimmune disease act as a sink for inflammatory cytokines (e.g. the anti-TNFα monoclonal antibody [mAb] adalimumab) or block inflammatory receptors (e.g. anti-IL-6R mAb tocilizumab) and the role of the Fc effector functions in this situation is again unclear. An excellent summary of the role of the Fc in autoimmune disease is presented by Li and Kimberly [120].

13.4.2.4 Approaches to Engineering the Effector Functions of the IgG1 Fc

Engineering of antibody effector functions has generally been approached with computational and rational design strategies based on published crystal structures of Fc–FcR complexes, followed by low-throughput screening of individual variants [121–125]. Fc glycosylation impacts binding to several effector FcRs by altering conformation [126, 127] as well as antibody stability [128]. However, *Escherichia coli* bacteria do not perform *N*-glycosylation and yeast hyper-glycosylate proteins, so screening large Fc libraries presents a significant challenge for traditional methods. Yeast display has been used to affinity mature the human IgG Fc for FcγRIIIa binding while avoiding FcγRIIb binding [129], but no comparison of the apparent affinity of the Fc variant displayed on yeast relative to full-length Chinese hamster ovary (CHO) cell produced Fc variants was provided. Ribosome display has also been used to engineer aglycosylated IgG1 Fcs with higher affinity binding to FcγRIIIa [130]. The identified mutations improved affinity in glycosylated Fcs as well and resulted in an altered oligosaccharide profile. Aglycosylated Fc was similarly engineered using *E. coli* display, but the impact on stability and half-life is unclear [131]. In contrast, mammalian display offers a more relevant system for Fc engineering. Glycosylation can play a major role binding interactions that can go unnoticed when antibodies screened in an aglycosylated format are then produced in CHO cells [132], so a logical approach is to screen variants in the CHO system used for most antibody production.

Substantial progress has been made toward engineering non-mammalian systems for human-like antibody production. Prokaryotic expression of functional full-length antibodies in the *E. coli* cytoplasm was achieved with Fc mutations that allow FcR engagement in the absence of glycosylation [133]. Introduction of mammalian glycosylation pathways into bacteria is also possible [134, 135], though the technology is not yet capable of replicating antibody glycosylation patterns. Yeast glycosylation patterns typically results in short serum half-life and high immunogenicity of proteins, but humanization of the yeast pathways can result in much more human-like molecules [136]. In fact, *Pichia pastoris* has been engineered for human glycosylation with a single glycoform of choice at the IgG1 N297 site [137]. This technology has allowed surface display and high-throughput engineering of full-length antibodies on yeast [138], but it has not been widely utilized.

Glycoengineering of the N297-linked carbohydrate has been motivated mainly by differences seen in the glycosylation and activity of antibodies produced *in vivo* versus in mammalian cell culture. Afucosylated Fc has been extensively shown to bind FcγRIIIa with greater affinity and activate more potent ADCC [139]. Again, these efforts are guided by rational design, and high-throughput methods of glycosylation library screening are not available. Glycosylation is typically manipulated by metabolic engineering approaches [140], cell culture conditions, and additives, not protein engineering [141].

13.4.3 Tissue Localization

Antigen targeting can provide exquisite molecular specificity; however in some disease states, especially cancer and autoimmune disease, the disease-associated self-antigens may be expressed in healthy tissues as well. This "on target, off tumor" lack of selectivity is a major problem for many of the most recent immunotherapeutic platforms such as checkpoint inhibitor blockade therapies [142], bispecific antibodies [143], and CARs [144]. One innovative approach to adding tissue selectivity is peptide-conjugated antibodies (Probodies™, CytomX) in which the peptide occludes the binding site until tumor-associated proteases activate the drug by releasing the peptide. Peptide bacterial display libraries have been used both to find peptides that block antibody binding sites [145] and to develop linkers susceptible to specific tumor protease cleavage [146]. Several Probody™ candidates are in preclinical and clinical trials [147].

Delivery to diseased tissues has the potential to enhance efficacy and reduce side effects. In infectious diseases, pathogens often invade by penetrating the mucus membranes and associated tissues. IgA and immunoglobulin M (IgM) both interact strongly with the polymeric Ig receptor, resulting in their active transport to the mucosal surface. By integrating the polymeric Ig receptor binding site into IgG1, an antibody targeting *P. aeruginosa* achieved higher lung localization than unmodified IgG1 [148]. Delivery to the brain is especially challenging as only 0.1–0.2% of circulating antibody is estimated to access the central nervous system [149]. Co-opting existing active transport systems to transport antibodies across the blood-brain barrier has been fairly successful: this has been achieved by generating a bispecific antibody in which one arm binds the existing transferrin receptor [150]. High-throughput screening of antibody libraries has also identified other targets that may be more specific than the transferrin receptor [151]. Engineering antibody-based therapeutics capable of reaching therapeutic levels in the brain is a high-priority area of ongoing research.

13.4.4 Immunogenicity

Mouse antibodies are known to generate a strong human anti-mouse antibody response, but the mouse variable regions in human-mouse chimeric antibodies can also be problematic. Severe immune reactions to the chimeric antibody cetuximab were caused by a mouse germline encoded glycosylation site in the V_H region

[152]. Humanization of antibodies is not sufficient to ensure that a therapeutic antibody will not be recognized as a foreign antigen for clearance by the immune system, though it certainly reduces the likelihood of anti-drug antibody generation [153]. Predictably reducing immunogenicity is challenging and the most developed approaches involve reducing the likelihood of T-cell recognition of the antibody sequence (Section 13.4.4.1) and minimizing aggregation (Section 13.4.4.2).

13.4.4.1 Reducing T-Cell Recognition

While antibody biophysical properties and formulation may contribute to immune activation, the antigenic region of otherwise well-behaved antibodies is thought to reside primarily in the variable regions. To generate a strong anti-drug response, the peptides resulting from intracellular antibody proteolysis must be amenable to loading on major histocompatibility complex (MHC) class II molecules for presentation and activation of helper T cells. Researchers can engineer antibody CDR sequences to avoid MHC class II presentation in an effort to reduce immunogenicity. Computational approaches seem well-suited to this problem and they can predict known MHC II peptide motifs and the fit of a peptide into the MHC by structural analysis [154]. Direct testing of peptides against donor T-cell panels can also be informative to determine sequences that may be altered for reduced immunogenicity [155]. Unfortunately, this analysis is typically performed post hoc and methods to incorporate low immunogenicity sequences into libraries or bias immune libraries against immunogenicity are currently limited.

13.4.4.2 Reducing Aggregation

Immunogenicity can be exacerbated by the presence of protein aggregates. By reducing the tendency of antibodies to aggregate through protein engineering or formulation, the immunogenicity of a therapeutic antibody may be minimized [156]. Aggregation is typically mediated by CDR interactions [157] and we have some understanding of the residues and motifs that either predispose or discourage antibody aggregation [158]. By engineering out or precluding aggregation prone sequences during library design and screening, immunogenicity risk may be minimized and manufacturability improved for selected antibodies. Unfortunately, the aggregation prone regions may not be sequentially contiguous, requiring crystal structures to predict and disrupt aggregation prone patches [157]. When possible, computational efforts to predict aggregation allow targeted mutagenesis to reduce aggregation and maintain affinity [159], and can be used to predict high viscosity in the therapeutic formulation [160]. Empirical testing of biophysical properties of preclinical antibodies may provide a means to screen large numbers of candidates for various developability "red flags," which would at least minimize the cost of advancing problematic antibodies to clinical trials [54].

13.5 Manufacturability of Antibodies

Cell culture technology has advanced to the point that antibodies can be produced with titers up to 13 $g\,l^{-1}$ [161] in mammalian cells. Most of these advances are due to technological improvements in cell culture engineering, but antibody sequence can also contribute greatly to variability in production yields and homogeneity of the final product. Although the regulatory pathway is fairly clear for standard antibody therapeutics and similar in cost to other drugs, the cost of antibody production remains high. Though yeast have been engineered to produce full-length antibodies with human-like glycosylation, reported yields from manufacturing scale bioreactors arc $<2\,\mathrm{g\,l^{-1}}$ [162], nearly an order of magnitude lower than the best mammalian systems. Ensuring antibodies developed for therapy will be well-behaved, high expressing candidates and continuing to optimize antibody production processes is expected to reduce the cost of therapy and expand the number of targets accessible to antibody therapy. In this section yield improvement strategies (Section 13.5.1) and alternative antibody production methods (Section 13.5.2) are discussed.

13.5.1 Increasing Antibody Yield

13.5.1.1 Codon Usage

Silent alterations in the antibody DNA sequence that do not cause amino acid changes can greatly impact the transcriptional and translational efficiency of the antibody sequence. Human codon optimization for expression in a mouse SP2 cell line substantially improved antibody yield [163] over the human germline usage, and species matched codon optimization is typically preferred (e.g. hamster optimized codons for CHO production). Even when human germline sequences are used to generate fully human therapeutic antibodies, the codon usage in the germline may be skewed to encourage CDR mutations during somatic hypermutation [164], which may negatively impact expression levels. Controlling codon usage can also modulate the translation rate, supporting proper folding of the protein product and reduced aggregation [165]. Other DNA sequence changes such as those that influence mRNA stability, GC content, or codon-pair usage can further improve transgene expression in mammalian cells [166] (GeneArt, Qiagen).

13.5.1.2 Signal Peptide Optimization

For antibody secretion during mammalian cell production, the heavy and light chain sequences must each carry an *N*-terminal signal peptide. The signal-recognition particle binds the signal peptide during translation. This particle then localizes the ribosome-mRNA complex to the endoplasmic reticulum membrane where the ribosome translocates the nascent polypeptide into the endoplasmic reticulum lumen. When five commercial antibodies were compared for transient expression levels from CHO cells with eight different heavy chain signal peptides and two different light chain signal peptides, the transfection-normalized expression yield varied up to eightfold for a given antibody [167]. Interestingly, the best signal peptide pair was not consistent for the five antibodies tested. When signal peptide and DNA sequence optimization (using GenScript's Multiparameter Gene Optimization algorithm) were combined for pertuzumab production, expression yields increased by eightfold overall, mostly due to sequence optimization [168].

13.5.1.3 Expression Optimization

While high-throughput screening capability for every factor that ensures robust manufacturability of an antibody is not available, we now have various methods to screen antibodies in the production host. By screening surface displayed antibodies in CHO cells, human embryonic kidney (HEK) cells, or other mammalian cell production lines for binding properties, expression issues are simultaneously screened *ipso facto*. Many mammalian display methods use HEK cells, which are less relevant for antibody production than CHO cells [169] and/or cumbersome viral infection techniques for screening libraries on the mammalian cell surface [170–174]. We recently reported a simple, plasmid transfection-based CHO cell surface display system for screening antibody Fab variants [132].

This work highlighted that non-mammalian screening systems may select for sequences that have decreased function after mammalian post-translational processing and eliminate functional sequences due to less sophisticated processing and folding machinery. By screening directly in CHO cells, the resulting antibodies had high stability (T_m), expression level, and binding affinity for the target antigen. For convenience, mammalian libraries are limited to approximately 10^6 unique variants. While this is several orders of magnitude smaller than phage libraries, this is also several orders of magnitude larger than the expression screening capabilities in microtiter plates. Mammalian antibody display libraries are likely too small for screening of completely naïve antibody libraries, but they can be useful for specificity redirection or screening preselected libraries to help avoid problems during scale up for manufacturing production. Mammalian display may also be well-suited to replace phage display screening of immune libraries as the immune repertoire may be quite small after immunization [63], and the mammalian platform ideal for excluding non-productive V_L–V_H combinations.

13.5.2 Alternative Production Methods

Although 84% of mAbs that received approval between 2015 and 2018 are produced in CHO cells [2], new hosts are emerging as potentially lower cost alternatives. In particular, plant-based antibody expression has proved to be capable of producing highly effective therapeutic antibodies targeting Ebola and rabies viruses [175]. They are notably able to produce antibody with relevant post-translational modifications including afucosylated Fc glycosylation [176]. The scalability of plant-produced therapeutics is promising for cost reduction, but development of the technology for large-scale production is still needed [177].

Utilizing the patient's own cells to produce antibody through mRNA delivery is a particularly cost effective strategy for antibody production. MedImmune (AstraZeneca) has recently begun development of mRNA delivery methodologies with Ethris GmbH [178], and has published some early work evaluating gene delivery for *in vivo* antibody production of antibodies to prevent *P. aeruginosa* infection in mice [179]. Moderna (also with AstraZeneca and other companies) is in development of many therapeutic mRNA applications as well. While this would bypass many of the traditional hurdles associated with antibody production, it is not clear that product homogeneity could be guaranteed and the regulatory pathway is uncertain.

13.6 Conclusions

In the two decades since our review article entitled "Antibody Engineering" [180] was published, the field has dramatically changed. Engineering antibodies for specific binding, a major focus of that publication, has become nearly rote, while the antibody Fc (mentioned only once in the 2000 review and over 50 times here) and its many functions has taken center stage. Challenges such as recognizing effective antibody targets remain important, while new challenges such as manipulating the downstream effects of that targeting have emerged. In all cases, a detailed and sophisticated understanding of the immune system and disease states is required to generate useful therapeutics. Additional consideration of the incredible cost of therapeutic protein development and production motivate attention to developability throughout the antibody engineering pipeline. There are many opportunities for protein engineers to improve discovery platforms, therapeutic efficacy, and manufacturability of antibodies. As we continue to learn about the interactions between antibodies and the immune system, we expect these opportunities to expand.

Acknowledgments

This work was supported by grants from the NIH (F32 GM111018 to AWN and AI122753 to JAM) and the Welch Foundation (F-1767 to JAM). The authors thank Dr. Ellen Wagner for assistance with Figure 13.1.

References

1 Monoclonal Antibodies (MAbS) Global Market Report 2020. (2019). Research and markets. https://www.marketresearch.com/Business-Research-Company-v4006/Monoclonal-Antibodies-MAbS-Global-12801247 (accessed 11 December 2020).

2 Walsh, G. (2018). Biopharmaceutical benchmarks 2018. *Nat. Biotechnol.* 36 (12): 1136–1145.

3 Irani, V., Guy, A.J., Andrew, D. et al. (2015). Molecular properties of human IgG subclasses and their implications for designing therapeutic monoclonal antibodies against infectious diseases. *Mol. Immunol.* 67 (2 Pt A): 171–182.

4 Jain, N., Smith, S.W., Ghone, S., and Tomczuk, B. (2015). Current ADC linker chemistry. *Pharm. Res.* 32 (11): 3526–3540.

5 Bhakta, S., Raab, H., and Junutula, J.R. (2013). Engineering THIOMABs for site-specific conjugation of thiol-reactive linkers. *Methods Mol. Biol.* 1045: 189–203.

6 Wagner, E.K., Wang, X., Bui, A., and Maynard, J.A. (2016). Synergistic neutralization of pertussis toxin by a bispecific antibody in vitro and in vivo. *Clin. Vaccine Immunol.* 23 (11): 851–862.

7 Zhu, M., Wu, B., Brandl, C. et al. (2016). Blinatumomab, a bispecific T-cell engager (BiTE®) for CD-19 targeted cancer immunotherapy: clinical pharmacology and its implications. *Clin. Pharmacokinet.* 55 (10): 1271–1288.

8 Labrijn, A.F., Meesters, J.I., de Goeij, B.E. et al. (2013). Efficient generation of stable bispecific IgG1 by controlled Fab-arm exchange. *Proc. Natl. Acad. Sci. U.S.A.* 110 (13): 5145–5150.

9 Ridgway, J.B., Presta, L.G., and Carter, P. (1996). "Knobs-into-holes" engineering of antibody CH3 domains for heavy chain heterodimerization. *Protein Eng.* 9 (7): 617–621.

10 Lewis, S.M., Wu, X., Pustilnik, A. et al. (2014). Generation of bispecific IgG antibodies by structure-based design of an orthogonal Fab interface. *Nat. Biotechnol.* 32 (2): 191–198.

11 Verdino, P., Atwell, S., and Demarest, S.J. (2018). Emerging trends in bispecific antibody and scaffold protein therapeutics. *Curr. Opin. Chem. Eng.* 19: 107–123.

12 Van Blarcom, T., Lindquist, K., Melton, Z. et al. (2018). Productive common light chain libraries yield diverse panels of high affinity bispecific antibodies. *MAbs* 10 (2): 256–268.

13 Oldenburg, J., Mahlangu, J.N., Kim, B. et al. (2017). Emicizumab prophylaxis in hemophilia A with inhibitors. *N. Engl. J. Med.* 377 (9): 809–818.

14 Brinkmann, U. and Kontermann, R.E. (2017). The making of bispecific antibodies. *MAbs* 9 (2): 182–212.

15 Sedykh, S.E., Prinz, V.V., Buneva, V.N., and Nevinsky, G.A. (2018). Bispecific antibodies: design, therapy, perspectives. *Drug Des. Dev. Ther.* 12: 195–208.

16 Vincke, C., Loris, R., Saerens, D. et al. (2009). General strategy to humanize a camelid single-domain antibody and identification of a universal humanized nanobody scaffold. *J. Biol. Chem.* 284 (5): 3273–3284.

17 Moutel, S., Bery, N., Bernard, V. et al. (2016). NaLi-H1: a universal synthetic library of humanized nanobodies providing highly functional antibodies and intrabodies. *Elife* 5.

18 Arbabi-Ghahroudi, M. (2017). Camelid single-domain antibodies: historical perspective and future outlook. *Front. Immunol.* 8: 1589.

19 Maude, S.L., Laetsch, T.W., Buechner, J. et al. (2018). Tisagenlecleucel in children and young adults with B-cell lymphoblastic leukemia. *N. Engl. J. Med.* 378 (5): 439–448.

20 Rodgers, D.T., Mazagova, M., Hampton, E.N. et al. (2016). Switch-mediated activation and retargeting of CAR-T cells for B-cell malignancies. *Proc. Natl. Acad. Sci. U.S.A.* 113 (4): E459–E468.

21 Bachmann, M. (2019). The UniCAR system: a modular CAR T cell approach to improve the safety of CAR T cells. *Immunol. Lett.* 211: 13–22.

22 Klesmith, J.R., Su, L., Wu, L. et al. (2019). Retargeting CD19 chimeric antigen receptor T cells via engineered CD19-fusion proteins. *Mol. Pharm.* 16 (8): 3544–3558.

23 Zah, E., Lin, M.Y., Silva-Benedict, A. et al. (2016). T cells expressing CD19/CD20 bispecific chimeric antigen receptors prevent antigen escape by malignant B cells. *Cancer Immunol. Res.* 4 (6): 498–508.

24 Zah, E., Nam, E., Bhuvan, V. et al. (2020). Systematically optimized BCMA/CS1 bispecific CAR-T cells robustly control heterogeneous multiple myeloma. *Nat. Commun.* 11 (1): 2283.

25 Onitilo, A.A., Engel, J.M., and Stankowski, R.V. (2014). Cardiovascular toxicity associated with adjuvant trastuzumab therapy: prevalence, patient characteristics, and risk factors. *Ther. Adv. Drug Saf.* 5 (4): 154–166.

26 Sgro, C. (1995). Side-effects of a monoclonal antibody, muromonab CD3/orthoclone OKT3: bibliographic review. *Toxicology* 105 (1): 23–29.

27 Morgan, R.A., Yang, J.C., Kitano, M. et al. (2010). Case report of a serious adverse event following the administration of T cells transduced with a chimeric antigen receptor recognizing ERBB2. *Mol. Ther.* 18 (4): 843–851.

28 Michot, J.M., Bigenwald, C., Champiat, S. et al. (2016). Immune-related adverse events with immune checkpoint blockade: a comprehensive review. *Eur. J. Cancer* 54: 139–148.

29 Bukreyev, A., Yang, L., and Collins, P.L. (2012). The secreted G protein of human respiratory syncytial virus antagonizes antibody-mediated restriction of replication involving macrophages and complement. *J. Virol.* 86 (19): 10880–10884.

30 Mohan, G.S., Li, W., Ye, L. et al. (2012). Antigenic subversion: a novel mechanism of host immune evasion by Ebola virus. *PLoS Pathog.* 8 (12): e1003065.

31 Cherry, J.D. and Paddock, C.D. (2014). Pathogenesis and histopathology of pertussis: implications for immunization. *Expert Rev. Vaccines* 13 (9): 1115–1123.

32 Carbonetti, N.H. (2015). Contribution of pertussis toxin to the pathogenesis of pertussis disease. *Pathog. Dis.* 73 (8): ftv073.

33 Sutherland, J.N. and Maynard, J.A. (2009). Characterization of a key neutralizing epitope on pertussis toxin recognized by monoclonal antibody 1B7. *Biochemistry* 48 (50): 11982–11993.

34 Nguyen, A.W., Wagner, E.K., Laber, J.R. et al. (2015). A cocktail of humanized anti-pertussis toxin antibodies limits disease in murine and baboon models of whooping cough. *Sci. Transl. Med.* 7 (316): 316ra195.

35 Nguyen, A.W., DiVenere, A.M., Papin, J.F. et al. (2020). Neutralization of pertussis toxin by a single antibody prevents clinical pertussis in neonatal baboons. *Sci. Adv.* 6 (6): eaay9258.

36 Salazar, G., Zhang, N., Fu, T.M., and An, Z. (2017). Antibody therapies for the prevention and treatment of viral infections. *NPJ Vaccines* 2: 19.

37 North, B., Lehmann, A., and Dunbrack, R.L. Jr. (2011). A new clustering of antibody CDR loop conformations. *J. Mol. Biol.* 406 (2): 228–256.

38 Sidhu, S.S., Li, B., Chen, Y. et al. (2004). Phage-displayed antibody libraries of synthetic heavy chain complementarity determining regions. *J. Mol. Biol.* 338 (2): 299–310.

39 Frenzel, A., Kugler, J., Helmsing, S. et al. (2017). Designing human antibodies by phage display. *Transfus. Med. Hemother.* 44 (5): 312–318.

40 Sall, A., Walle, M., Wingren, C. et al. (2016). Generation and analyses of human synthetic antibody libraries and their application for protein microarrays. *Protein Eng. Des. Sel.* 29 (10): 427–437.

41 Hoet, R.M., Cohen, E.H., Kent, R.B. et al. (2005). Generation of high-affinity human antibodies by combining donor-derived and synthetic complementarity-determining-region diversity. *Nat. Biotechnol.* 23 (3): 344–348.

42 Prassler, J., Thiel, S., Pracht, C. et al. (2011). HuCAL PLATINUM, a synthetic Fab library optimized for sequence diversity and superior performance in mammalian expression systems. *J. Mol. Biol.* 413 (1): 261–278.

43 Guy Van Meter (2018). Adimab initiates update on clinical progression of partner programs. https://www.adimab.com/sites/default/files/press/adimab_pr_01_11_2018.pdf (accessed 11 December 2020).

44 Vasquez M, Feldhaus M, Gerngross TU, Wittrup DK, Adimab, LLC (Lebanon, NH, USA). (2014). Rationally designed, synthetic antibody libraries and uses therefor. United States Patent 8 691 730, filed 12 September 2008 and issued 8 April 2014.

45 Lonberg, N., Taylor, L.D., Harding, F.A. et al. (1994). Antigen-specific human antibodies from mice comprising four distinct genetic modifications. *Nature* 368 (6474): 856–859.

46 Booth, B. (2017). Human antibody discovery: of mice and phage. *Forbes* 11: 2017.

47 Mendez, M.J., Green, L.L., Corvalan, J.R. et al. (1997). Functional transplant of megabase human immunoglobulin loci recapitulates human antibody response in mice. *Nat. Genet.* 15 (2): 146–156.

48 Bruggemann, M., Osborn, M.J., Ma, B. et al. (2015). Human antibody production in transgenic animals. *Arch. Immunol. Ther. Exp. (Warsz)* 63 (2): 101–108.

49 Lee, E.C., Liang, Q., Ali, H. et al. (2014). Complete humanization of the mouse immunoglobulin loci enables efficient therapeutic antibody discovery. *Nat. Biotechnol.* 32 (4): 356–363.

50 Murphy, A.J., Macdonald, L.E., Stevens, S. et al. (2014). Mice with megabase humanization of their immunoglobulin genes generate antibodies as efficiently as normal mice. *Proc. Natl. Acad. Sci. U.S.A.* 111 (14): 5153–5158.

51 Harriman WD, Etches R, Leighton P, inventors; Crystal Bioscience Inc. (Emeryville, CA, US), assignee. Transgenic animal for production of antibodies having minimal CDRs. United States patent 8592644. 2013.

52 Ching, K.H., Collarini, E.J., Abdiche, Y.N. et al. (2018). Chickens with humanized immunoglobulin genes generate antibodies with high affinity and broad epitope coverage to conserved targets. *MAbs* 10 (1): 71–80.

53 Schmautz, C., Hillreiner, M., Ballweg, I. et al. (2018). Stimulated enrichment of *Clostridium difficile* specific IgA in mature cow's milk. *PLoS One* 13 (4): e0195275.

54 Jain, T., Sun, T., Durand, S. et al. (2017). Biophysical properties of the clinical-stage antibody landscape. *Proc. Natl. Acad. Sci. U.S.A.* 114 (5): 944–949.

55 Bradbury, A.R.M., Trinklein, N.D., Thie, H. et al. (2018). When monoclonal antibodies are not monospecific: hybridomas frequently express additional functional variable regions. *MAbs* 10 (4): 539–546.

56 Krebber, A., Bornhauser, S., Burmester, J. et al. (1997). Reliable cloning of functional antibody variable domains from hybridomas and spleen cell repertoires employing a reengineered phage display system. *J. Immunol. Methods* 201 (1): 35–55.

57 Meijer, P.J., Andersen, P.S., Haahr Hansen, M. et al. (2006). Isolation of human antibody repertoires with preservation of the natural heavy and light chain pairing. *J. Mol. Biol.* 358 (3): 764–772.

58 Smith, K., Garman, L., Wrammert, J. et al. (2009). Rapid generation of fully human monoclonal antibodies specific to a vaccinating antigen. *Nat. Protoc.* 4 (3): 372–384.

59 Tiller, T., Meffre, E., Yurasov, S. et al. (2008). Efficient generation of monoclonal antibodies from single human B cells by single cell RT-PCR and expression vector cloning. *J. Immunol. Methods* 329 (1–2): 112–124.

60 Yu, X., Tsibane, T., McGraw, P.A. et al. (2008). Neutralizing antibodies derived from the B cells of 1918 influenza pandemic survivors. *Nature* 455 (7212): 532–536.

61 DeKosky, B.J., Kojima, T., Rodin, A. et al. (2015). In-depth determination and analysis of the human paired heavy- and light-chain antibody repertoire. *Nat. Med.* 21 (1): 86–91.

62 Busse, C.E., Czogiel, I., Braun, P. et al. (2014). Single-cell based high-throughput sequencing of full-length immunoglobulin heavy and light chain genes. *Eur. J. Immunol.* 44 (2): 597–603.

63 Reddy, S.T., Ge, X., Miklos, A.E. et al. (2010). Monoclonal antibodies isolated without screening by analyzing the variable-gene repertoire of plasma cells. *Nat. Biotechnol.* 28 (9): 965–969.

64 Lavinder, J.J., Horton, A.P., Georgiou, G., and Ippolito, G.C. (2015). Next-generation sequencing and protein mass spectrometry for the comprehensive analysis of human cellular and serum antibody repertoires. *Curr. Opin. Chem. Biol.* 24: 112–120.

65 Bujotzek, A., Lipsmeier, F., Harris, S.F. et al. (2016). VH-VL orientation prediction for antibody humanization candidate selection: a case study. *MAbs* 8 (2): 288–305.

66 Hanf, K.J., Arndt, J.W., Chen, L.L. et al. (2014). Antibody humanization by redesign of complementarity-determining region residues proximate to the acceptor framework. *Methods* 65 (1): 68–76.

67 Abhinandan, K.R. and Martin, A.C. (2007). Analyzing the "degree of humanness" of antibody sequences. *J. Mol. Biol.* 369 (3): 852–862.

68 Jones, T.D., Carter, P.J., Pluckthun, A. et al. (2016). The INNs and outs of antibody nonproprietary names. *MAbs* 8 (1): 1–9.

69 Bartelds, G.M., Krieckaert, C.L., Nurmohamed, M.T. et al. (2011). Development of antidrug antibodies against adalimumab and association with disease activity and treatment failure during long-term follow-up. *J. Am. Med. Assoc.* 305 (14): 1460–1468.

70 Bowers, P.M., Horlick, R.A., Neben, T.Y. et al. (2011). Coupling mammalian cell surface display with somatic hypermutation for the discovery and maturation of human antibodies. *Proc. Natl. Acad. Sci. U.S.A.* 108 (51): 20455–20460.

71 Tabasinezhad, M., Talebkhan, Y., Wenzel, W. et al. (2019). Trends in therapeutic antibody affinity maturation: from in-vitro towards next-generation sequencing approaches. *Immunol. Lett.* 212: 106–113.

72 Takai, T. (2002). Roles of Fc receptors in autoimmunity. *Nat. Rev. Immunol.* 2 (8): 580–592.

73 Dekkers, G., Treffers, L., Plomp, R. et al. (2017). Decoding the human immunoglobulin G-glycan repertoire reveals a spectrum of Fc-receptor- and complement-mediated-effector activities. *Front. Immunol.* 8: 877.

74 van der Poel, C.E., Spaapen, R.M., van de Winkel, J.G., and Leusen, J.H. (2011). Functional characteristics of the high affinity IgG receptor, FcγRI. *J. Immunol.* 186 (5): 2699–2704.

75 Shashidharamurthy, R., Zhang, F., Amano, A. et al. (2009). Dynamics of the interaction of human IgG subtype immune complexes with cells expressing R and H allelic forms of a low-affinity Fcγ receptor CD32A. *J. Immunol.* 183 (12): 8216–8224.

76 Nimmerjahn, F. and Ravetch, J.V. (2006). Fcγ receptors: old friends and new family members. *Immunity* 24 (1): 19–28.

77 Yeap, W.H., Wong, K.L., Shimasaki, N. et al. (2016). CD16 is indispensable for antibody-dependent cellular cytotoxicity by human monocytes. *Sci. Rep.* 6: 34310.

78 Garcia-Garcia, E., Nieto-Castaneda, G., Ruiz-Saldana, M. et al. (2009). FcγRIIA and FcγRIIIB mediate nuclear factor activation through separate signaling pathways in human neutrophils. *J. Immunol.* 182 (8): 4547–4556.

79 Borrok, M.J., Wu, Y., Beyaz, N. et al. (2015). pH-dependent binding engineering reveals an FcRn affinity threshold that governs IgG recycling. *J. Biol. Chem.* 290 (7): 4282–4290.

80 Moore, G.L., Chen, H., Karki, S., and Lazar, G.A. (2010). Engineered Fc variant antibodies with enhanced ability to recruit complement and mediate effector functions. *MAbs* 2 (2): 181–189.

81 Choe, W., Durgannavar, T.A., and Chung, S.J. (2016). Fc-binding ligands of immunoglobulin G: an overview of high affinity proteins and peptides. *Materials (Basel)* 9 (12): 994.

82 Geuijen, K.P.M., Oppers-Tiemissen, C., Egging, D.F. et al. (2017). Rapid screening of IgG quality attributes – effects on Fc receptor binding. *FEBS Open Bio* 7 (10): 1557–1574.

83 Gaza-Bulseco, G., Hickman, K., Sinicropi-Yao, S. et al. (2009). Effect of the conserved oligosaccharides of recombinant monoclonal antibodies on the separation by protein A and protein G chromatography. *J. Chromatogr. A* 1216 (12): 2382–2387.

84 Morell, A., Terry, W.D., and Waldmann, T.A. (1970). Metabolic properties of IgG subclasses in man. *J. Clin. Invest.* 49 (4): 673–680.

85 Scott, R.P. and Quaggin, S.E. (2015). Review series: the cell biology of renal filtration. *J. Cell Biol.* 209 (2): 199–210.

86 Akilesh, S., Huber, T.B., Wu, H. et al. (2008). Podocytes use FcRn to clear IgG from the glomerular basement membrane. *Proc. Natl. Acad. Sci. U.S.A.* 105 (3): 967–972.

87 Zhu, Q., McLellan, J.S., Kallewaard, N.L. et al. (2017). A highly potent extended half-life antibody as a potential RSV vaccine surrogate for all infants. *Sci. Transl. Med.* 9 (388): eaaj1928.

88 Griffin, M.P., Khan, A.A., Esser, M.T. et al. (2017). Safety, tolerability, and pharmacokinetics of MEDI8897, the respiratory syncytial virus prefusion F-targeting monoclonal antibody with an extended half-life, in healthy adults. *Antimicrob. Agents Chemother.* 61 (3): e01714–e01716.

89 Dall'Acqua, W.F., Kiener, P.A., and Wu, H. (2006). Properties of human IgG1s engineered for enhanced binding to the neonatal Fc receptor (FcRn). *J. Biol. Chem.* 281 (33): 23514–23524.

90 Zalevsky, J., Chamberlain, A.K., Horton, H.M. et al. (2010). Enhanced antibody half-life improves in vivo activity. *Nat. Biotechnol.* 28 (2): 157–159.

91 Ko, S.Y., Pegu, A., Rudicell, R.S. et al. (2014). Enhanced neonatal Fc receptor function improves protection against primate SHIV infection. *Nature* 514 (7524): 642–645.

92 Lee, C.H., Kang, T.H., Godon, O. et al. (2019). An engineered human Fc domain that behaves like a pH-toggle switch for ultra-long circulation persistence. *Nat. Commun.* 10 (1): 5031.

93 Vaccaro, C., Zhou, J., Ober, R.J., and Ward, E.S. (2005). Engineering the Fc region of immunoglobulin G to modulate in vivo antibody levels. *Nat. Biotechnol.* 23 (10): 1283–1288.

94 Patel, D.A., Puig-Canto, A., Challa, D.K. et al. (2011). Neonatal Fc receptor blockade by Fc engineering ameliorates arthritis in a murine model. *J. Immunol.* 187 (2): 1015–1022.

95 Challa, D.K., Bussmeyer, U., Khan, T. et al. (2013). Autoantibody depletion ameliorates disease in murine experimental autoimmune encephalomyelitis. *MAbs* 5 (5): 655–659.

96 Beth DelGiacco and Joke Comijn Argenx Reports Third Quarter 2020 Financial Results and Provides Business Update. https://www.argenx.com/news/argenx-reports-third-quarter-2020-financial-results-and-provides-business-update (accessed 11 December 2020).

97 Grevys, A., Bern, M., Foss, S. et al. (2015). Fc engineering of human IgG1 for altered binding to the neonatal Fc receptor affects Fc effector functions. *J. Immunol.* 194 (11): 5497–5508.

98 Lofano, G., Gorman, M.J., Yousif, A.S. et al. (2018). Antigen-specific antibody Fc glycosylation enhances humoral immunity via the recruitment of complement. *Sci. Immunol.* 3 (26): eaat7796.

99 Wang, X., Mathieu, M., and Brezski, R.J. (2018). IgG Fc engineering to modulate antibody effector functions. *Protein Cell* 9 (1): 63–73.

100 Barok, M., Isola, J., Palyi-Krekk, Z. et al. (2007). Trastuzumab causes antibody-dependent cellular cytotoxicity-mediated growth inhibition of submacroscopic JIMT-1 breast cancer xenografts despite intrinsic drug resistance. *Mol. Cancer Ther.* 6 (7): 2065–2072.

101 Monteverde, M., Milano, G., Strola, G. et al. (2015). The relevance of ADCC for EGFR targeting: a review of the literature and a clinically-applicable method of assessment in patients. *Crit. Rev. Oncol. Hematol.* 95 (2): 179–190.

102 Boyerinas, B., Jochems, C., Fantini, M. et al. (2015). Antibody-dependent cellular cytotoxicity activity of a novel anti-PD-L1 antibody avelumab (MSB0010718C) on human tumor cells. *Cancer Immunol. Res.* 3 (10): 1148–1157.

103 Mossner, E., Brunker, P., Moser, S. et al. (2010). Increasing the efficacy of CD20 antibody therapy through the engineering of a new type II anti-CD20 antibody with enhanced direct and immune effector cell-mediated B-cell cytotoxicity. *Blood* 115 (22): 4393–4402.

104 Bowles, J.A., Wang, S.Y., Link, B.K. et al. (2006). Anti-CD20 monoclonal antibody with enhanced affinity for CD16 activates NK cells at lower concentrations and more effectively than rituximab. *Blood* 108 (8): 2648–2654.

105 Forero-Torres, A., de Vos, S., Pohlman, B.L. et al. (2012). Results of a phase 1 study of AME-133v (LY2469298), an Fc-engineered humanized monoclonal anti-CD20 antibody, in FcγRIIIa-genotyped patients with previously treated follicular lymphoma. *Clin. Cancer Res.* 18 (5): 1395–1403.

106 Cheney, C.M., Stephens, D.M., Mo, X. et al. (2014). Ocaratuzumab, an Fc-engineered antibody demonstrates enhanced antibody-dependent cell-mediated cytotoxicity in chronic lymphocytic leukemia. *MAbs* 6 (3): 749–755.

107 Kelton, W., Mehta, N., Charab, W. et al. (2014). IgGA: a “cross-isotype” engineered human Fc antibody domain that displays both IgG-like and IgA-like effector functions. *Chem. Biol.* 21 (12): 1603–1609.

108 Jacobsen, F.W., Stevenson, R., Li, C. et al. (2017). Engineering an IgG scaffold lacking effector function with optimized developability. *J. Biol. Chem.* 292 (5): 1865–1875.

109 Liu, L., Jacobsen, F.W., Everds, N. et al. (2017). Biological characterization of a stable effector functionless (SEFL) monoclonal antibody scaffold in vitro. *J. Biol. Chem.* 292 (5): 1876–1883.

110 Schlothauer, T., Herter, S., Koller, C.F. et al. (2016). Novel human IgG1 and IgG4 Fc-engineered antibodies with completely abolished immune effector functions. *Protein Eng. Des. Sel.* 29 (10): 457–466.

111 Borrok, M.J., Mody, N., Lu, X. et al. (2017). An “Fc-silenced” IgG1 format with extended half-life designed for improved stability. *J. Pharm. Sci.* 106 (4): 1008–1017.

112 Lee, W.S. and Kent, S.J. (2018). Anti-HIV-1 antibody-dependent cellular cytotoxicity: is there more to antibodies than neutralization? *Curr. Opin. HIV AIDS* 13 (2): 160–166.

113 Fick, R.B. Jr., Baltimore, R.S., Squier, S.U., and Reynolds, H.Y. (1985). IgG proteolytic activity of *Pseudomonas aeruginosa* in cystic fibrosis. *J. Infect. Dis.* 151 (4): 589–598.

114 Fick, R.B. Jr., Naegel, G.P., Squier, S.U. et al. (1984). Proteins of the cystic fibrosis respiratory tract. Fragmented immunoglobulin G opsonic antibody causing defective opsonophagocytosis. *J. Clin. Invest.* 74 (1): 236–248.

115 Heck, L.W., Alarcon, P.G., Kulhavy, R.M. et al. (1990). Degradation of IgA proteins by *Pseudomonas aeruginosa* elastase. *J. Immunol.* 144 (6): 2253–2257.

116 Galdino, A.C.M., de Oliveira, M.P., Ramalho, T.C. et al. (2019). Anti-virulence strategy against the multidrug-resistant bacterial pathogen *Pseudomonas aeruginosa*: pseudolysin (elastase B) as a potential druggable target. *Curr. Protein Pept. Sci.* 20 (5): 471–487.

117 Corrales-Aguilar, E., Trilling, M., Hunold, K. et al. (2014). Human cytomegalovirus Fcγ binding proteins gp34 and gp68 antagonize Fcγ receptors I, II, and III. *PLoS Pathog.* 10 (5): e1004131.

118 Woof, J.M. (2016). Immunoglobulins and their receptors, and subversion of their protective roles by bacterial pathogens. *Biochem. Soc. Trans.* 44 (6): 1651–1658.

119 Horton, H.M., Chu, S.Y., Ortiz, E.C. et al. (2011). Antibody-mediated coengagement of FcγRIIb and B cell receptor complex suppresses humoral immunity in systemic lupus erythematosus. *J. Immunol.* 186 (7): 4223–4233.

120 Li, X. and Kimberly, R.P. (2014). Targeting the Fc receptor in autoimmune disease. *Expert Opin. Ther. Targets* 18 (3): 335–350.

121 Lazar, G.A., Dang, W., Karki, S. et al. (2006). Engineered antibody Fc variants with enhanced effector function. *Proc. Natl. Acad. Sci. U.S.A.* 103 (11): 4005–4010.

122 Mimoto, F., Igawa, T., Kuramochi, T. et al. (2013). Novel asymmetrically engineered antibody Fc variant with superior FcγR binding affinity and specificity compared with afucosylated Fc variant. *MAbs* 5 (2): 229–236.

123 Richards, J.O., Karki, S., Lazar, G.A. et al. (2008). Optimization of antibody binding to FcγRIIa enhances macrophage phagocytosis of tumor cells. *Mol. Cancer Ther.* 7 (8): 2517–2527.

124 Shields, R.L., Namenuk, A.K., Hong, K. et al. (2001). High resolution mapping of the binding site on human IgG1 for FcγRI, FcγRII, FcγRIII, and FcRn and design of IgG1 variants with improved binding to the FcγR. *J. Biol. Chem.* 276 (9): 6591–6604.

125 Vafa, O., Gilliland, G.L., Brezski, R.J. et al. (2014). An engineered Fc variant of an IgG eliminates all immune effector functions via structural perturbations. *Methods* 65 (1): 114–126.

126 Mimura, Y., Katoh, T., Saldova, R. et al. (2018). Glycosylation engineering of therapeutic IgG antibodies: challenges for the safety, functionality and efficacy. *Protein Cell* 9 (1): 47–62.

127 Woof, J.M. and Burton, D.R. (2004). Human antibody-Fc receptor interactions illuminated by crystal structures. *Nat. Rev. Immunol.* 4 (2): 89–99.

128 Buck, P.M., Kumar, S., and Singh, S.K. (2013). Consequences of glycan truncation on Fc structural integrity. *MAbs* 5 (6): 904–916.

129 Stavenhagen, J.B., Gorlatov, S., Tuaillon, N. et al. (2007). Fc optimization of therapeutic antibodies enhances their ability to kill tumor cells in vitro and controls tumor expansion in vivo via low-affinity activating Fcγ receptors. *Cancer Res.* 67 (18): 8882–8890.

130 Stewart, R., Thom, G., Levens, M. et al. (2011). A variant human IgG1-Fc mediates improved ADCC. *Protein Eng. Des. Sel.* 24 (9): 671–678.

131 Jo, M., Kwon, H.S., Lee, K.H. et al. (2018). Engineered aglycosylated full-length IgG Fc variants exhibiting improved FcγRIIIa binding and tumor cell clearance. *MAbs* 10 (2): 278–289.

132 Nguyen, A.W., Le, K.C., and Maynard, J.A. (2018). Identification of high affinity HER2 binding antibodies using CHO Fab surface display. *Protein Eng. Des. Sel.* 31 (3): 91–101.

133 Robinson, M.P., Ke, N., Lobstein, J. et al. (2015). Efficient expression of full-length antibodies in the cytoplasm of engineered bacteria. *Nat. Commun.* 6: 8072.

134 Natarajan, A., Jaroentomeechai, T., Cabrera-Sanchez, M. et al. (2020). Engineering orthogonal human *O*-linked glycoprotein biosynthesis in bacteria. *Nat. Chem. Biol.* 16: 1062–1070.

135 Valderrama-Rincon, J.D., Fisher, A.C., Merritt, J.H. et al. (2012). An engineered eukaryotic protein glycosylation pathway in *Escherichia coli*. *Nat. Chem. Biol.* 8 (5): 434–436.

136 Hamilton, S.R., Davidson, R.C., Sethuraman, N. et al. (2006). Humanization of yeast to produce complex terminally sialylated glycoproteins. *Science* 313 (5792): 1441–1443.

137 Li, H., Sethuraman, N., Stadheim, T.A. et al. (2006). Optimization of humanized IgGs in glycoengineered *Pichia pastoris*. *Nat. Biotechnol.* 24 (2): 210–215.

138 Shaheen, H.H., Prinz, B., Chen, M.T. et al. (2013). A dual-mode surface display system for the maturation and production of monoclonal antibodies in glyco-engineered Pichia pastoris. *PLoS One* 8 (7): e70190.

139 Nimmerjahn, F. and Ravetch, J.V. (2007). Antibodies, Fc receptors and cancer. *Curr. Opin. Immunol.* 19 (2): 239–245.

140 Weikert, S., Papac, D., Briggs, J. et al. (1999). Engineering Chinese hamster ovary cells to maximize sialic acid content of recombinant glycoproteins. *Nat. Biotechnol.* 17 (11): 1116–1121.

141 Sha, S., Agarabi, C., Brorson, K. et al. (2016). *N*-glycosylation design and control of therapeutic monoclonal antibodies. *Trends Biotechnol.* 34 (10): 835–846.

142 Villadolid, J. and Amin, A. (2015). Immune checkpoint inhibitors in clinical practice: update on management of immune-related toxicities. *Transl. Lung Cancer Res.* 4 (5): 560–575.

143 Kebenko, M., Goebeler, M., Wolf, M. et al. (2018). A multicenter phase 1 study of solitomab (MT110, AMG 110), a bispecific EpCAM/CD3 T-cell engager

(BiTE®) antibody construct, in patients with refractory solid tumors. *OncoImmunology* 7 (8): e1450710.

144 Bonifant, C.L., Jackson, H.J., Brentjens, R.J., and Curran, K.J. (2016). Toxicity and management in CAR T-cell therapy. *Mol. Ther. Oncolytics* 3: 16011.

145 Erster, O., Thomas, J.M., Hamzah, J. et al. (2012). Site-specific targeting of antibody activity in vivo mediated by disease-associated proteases. *J. Controlled Release* 161 (3): 804–812.

146 Boulware, K.T. and Daugherty, P.S. (2006). Protease specificity determination by using cellular libraries of peptide substrates (CLiPS). *Proc. Natl. Acad. Sci. U.S.A.* 103 (20): 7583–7588.

147 Desnoyers, L.R., Vasiljeva, O., Richardson, J.H. et al. (2013). Tumor-specific activation of an EGFR-targeting probody enhances therapeutic index. *Sci. Transl. Med.* 5 (207): 207ra144.

148 Borrok, M.J., DiGiandomenico, A., Beyaz, N. et al. (2018). Enhancing IgG distribution to lung mucosal tissue improves protective effect of anti-*Pseudomonas aeruginosa* antibodies. *JCI Insight* 3 (12): e97844.

149 Yu, Y.J. and Watts, R.J. (2013). Developing therapeutic antibodies for neurodegenerative disease. *Neurotherapeutics* 10 (3): 459–472.

150 Yu, Y.J., Atwal, J.K., Zhang, Y. et al. (2014). Therapeutic bispecific antibodies cross the blood-brain barrier in nonhuman primates. *Sci. Transl. Med.* 6 (261): 261ra154.

151 Stutz, C.C., Zhang, X., and Shusta, E.V. (2014). Combinatorial approaches for the identification of brain drug delivery targets. *Curr. Pharm. Des.* 20 (10): 1564–1576.

152 Chung, C.H., Mirakhur, B., Chan, E. et al. (2008). Cetuximab-induced anaphylaxis and IgE specific for galactose-α-1,3-galactose. *N. Engl. J. Med.* 358 (11): 1109–1117.

153 Hwang, W.Y. and Foote, J. (2005). Immunogenicity of engineered antibodies. *Methods* 36 (1): 3–10.

154 Desai, D.V. and Kulkarni-Kale, U. (2014). T-cell epitope prediction methods: an overview. *Methods Mol. Biol.* 1184: 333–364.

155 Harding, F.A., Stickler, M.M., Razo, J., and DuBridge, R.B. (2010). The immunogenicity of humanized and fully human antibodies: residual immunogenicity resides in the CDR regions. *MAbs* 2 (3): 256–265.

156 Ratanji, K.D., Derrick, J.P., Dearman, R.J., and Kimber, I. (2014). Immunogenicity of therapeutic proteins: influence of aggregation. *J. Immunotoxicol.* 11 (2): 99–109.

157 Lee, C.C., Perchiacca, J.M., and Tessier, P.M. (2013). Toward aggregation-resistant antibodies by design. *Trends Biotechnol.* 31 (11): 612–620.

158 Fernandez-Escamilla, A.M., Rousseau, F., Schymkowitz, J., and Serrano, L. (2004). Prediction of sequence-dependent and mutational effects on the aggregation of peptides and proteins. *Nat. Biotechnol.* 22 (10): 1302–1306.

159 Shan, L., Mody, N., Sormani, P. et al. (2018). Developability assessment of engineered monoclonal antibody variants with a complex self-association behavior using complementary analytical and in silico tools. *Mol. Pharm.* 15 (12): 5697–5710.

160 Sharma, V.K., Patapoff, T.W., Kabakoff, B. et al. (2014). In silico selection of therapeutic antibodies for development: viscosity, clearance, and chemical stability. *Proc. Natl. Acad. Sci. U.S.A.* 111 (52): 18601–18606.

161 Matthews, C.B., Wright, C., Kuo, A. et al. (2017). Reexamining opportunities for therapeutic protein production in eukaryotic microorganisms. *Biotechnol. Bioeng.* 114 (11): 2432–2444.

162 Ye, J., Ly, J., Watts, K. et al. (2011). Optimization of a glycoengineered *Pichia pastoris* cultivation process for commercial antibody production. *Biotechnol. Prog.* 27 (6): 1744–1750.

163 Carton, J.M., Sauerwald, T., Hawley-Nelson, P. et al. (2007). Codon engineering for improved antibody expression in mammalian cells. *Protein Expression Purif.* 55 (2): 279–286.

164 Saini, J. and Hershberg, U. (2015). B cell variable genes have evolved their codon usage to focus the targeted patterns of somatic mutation on the complementarity determining regions. *Mol. Immunol.* 65 (1): 157–167.

165 Quax, T.E., Claassens, N.J., Soll, D., and van der Oost, J. (2015). Codon bias as a means to fine-tune gene expression. *Mol. Cell* 59 (2): 149–161.

166 Fath, S., Bauer, A.P., Liss, M. et al. (2011). Multiparameter RNA and codon optimization: a standardized tool to assess and enhance autologous mammalian gene expression. *PLoS One* 6 (3): e17596.

167 Haryadi, R., Ho, S., Kok, Y.J. et al. (2015). Optimization of heavy chain and light chain signal peptides for high level expression of therapeutic antibodies in CHO cells. *PLoS One* 10 (2): e0116878.

168 Ramezani, A., Mahmoudi Maymand, E., Yazdanpanah-Samani, M. et al. (2017). Improving Pertuzumab production by gene optimization and proper signal peptide selection. *Protein Expr. Purif.* 135: 24–32.

169 Frenzel, A., Hust, M., and Schirrmann, T. (2013). Expression of recombinant antibodies. *Front. Immunol.* 4: 217.

170 Beerli, R.R., Bauer, M., Buser, R.B. et al. (2008). Isolation of human monoclonal antibodies by mammalian cell display. *Proc. Natl. Acad. Sci. U.S.A.* 105 (38): 14336–14341.

171 Bowers, P.M., Horlick, R.A., Kehry, M.R. et al. (2014). Mammalian cell display for the discovery and optimization of antibody therapeutics. *Methods* 65 (1): 44–56.

172 Ho, M., Nagata, S., and Pastan, I. (2006). Isolation of anti-CD22 Fv with high affinity by Fv display on human cells. *Proc. Natl. Acad. Sci. U.S.A.* 103 (25): 9637–9642.

173 Li, C.Z., Liang, Z.K., Chen, Z.R. et al. (2012). Identification of HBsAg-specific antibodies from a mammalian cell displayed full-length human antibody library of healthy immunized donor. *Cell. Mol. Immunol.* 9 (2): 184–190.

174 Mason, D.M., Weber, C.R., Parola, C. et al. (2018). High-throughput antibody engineering in mammalian cells by CRISPR/Cas9-mediated homology-directed mutagenesis. *Nucleic Acids Res.* 46 (14): 7436–7449.

175 Hiatt, A., Whaley, K.J., and Zeitlin, L. (2014). Plant-derived monoclonal antibodies for prevention and treatment of infectious disease. *Microbiol. Spectr.* 2 (1): AID-0004-2012.

176 Zeitlin, L., Pettitt, J., Scully, C. et al. (2011). Enhanced potency of a fucose-free monoclonal antibody being developed as an Ebola virus immunoprotectant. *Proc. Natl. Acad. Sci. U.S.A.* 108 (51): 20690–20694.

177 Buyel, J.F., Twyman, R.M., and Fischer, R. (2017). Very-large-scale production of antibodies in plants: the biologization of manufacturing. *Biotechnol. Adv.* 35 (4): 458–465.

178 AstraZeneca and MedImmune enter strategic collaboration with Ethris to develop mRNA therapies for respiratory diseases (press release).

179 Patel, A., DiGiandomenico, A., Keller, A.E. et al. (2017). An engineered bispecific DNA-encoded IgG antibody protects against *Pseudomonas aeruginosa* in a pneumonia challenge model. *Nat. Commun.* 8 (1): 637.

180 Maynard, J. and Georgiou, G. (2000). Antibody engineering. *Annu. Rev. Biomed. Eng.* 2: 339–376.

14

Programming Novel Cancer Therapeutics: Design Principles for Chimeric Antigen Receptors

Andrew J. Hou[1] *and Yvonne Y. Chen*[1,2,3]

[1] *University of California–Los Angeles, Department of Chemical and Biomolecular Engineering, 420 Westwood Plaza, Boelter Hall 5531, Los Angeles, CA 90095, USA*
[2] *University of California–Los Angeles, Department of Microbiology, Immunology, and Molecular Genetics, 420 Westwood Plaza, Boelter Hall 5531, Los Angeles, CA 90095, USA*
[3] *Parker Institute for Cancer Immunotherapy Center at UCLA, 420 Westwood Plaza, Boelter Hall 5531, Los Angeles, CA 90095, USA*

14.1 Introduction

Adoptive transfer of T cells engineered to express a chimeric antigen receptor (CAR) has achieved remarkable results in the treatment of B-cell malignancies, and in 2017, the United States Food and Drug Administration (FDA) approved Kymriah (tisagenlecleucel, Novartis) and Yescarta (axicabtagene ciloleucel, Kite Pharma) for the treatment of patients with relapsed or refractory B-cell precursor acute lymphoblastic lymphoma (ALL) and large B-cell lymphoma, respectively [1, 2]. However, the broad applicability of CAR-T cell therapy toward other cancer types has proved to be a more elusive task, and much research in the field has been dedicated to engineering T cells with enhanced properties, such as greater target specificity, increased persistence, and reduced toxicity. Many factors, such as T-cell differentiation status, *ex vivo* expansion conditions, and preconditioning regimens, can greatly impact treatment outcomes. Here, we will focus on how the composition of the CAR molecule itself may impact treatment efficacy. Complementary to the topics covered here, detailed discussions on clinical efficacy and toxicities of CAR-T cell therapies can be found in several excellent reviews [3–6].

CARs are, as their name suggests, fusion molecules whose basic structure can be broken down into four basic parts: (i) an antigen-recognition domain, (ii) an extracellular spacer, (iii) a transmembrane domain, and (iv) a signaling domain. In this review, we present design considerations for each component of a CAR, as well as provide an overview of novel receptor modalities that hold promise for expanding the toolbox of cell-based immunotherapies.

Protein Engineering: Tools and Applications, First Edition. Edited by Huimin Zhao.
© 2021 WILEY-VCH GmbH. Published 2021 by WILEY-VCH GmbH.

14.2 Metrics to Evaluate CAR-T Cell Function

Before delving into the details of building a CAR, it is helpful to first discuss the metrics that are typically used to distinguish a high-performing CAR from a low-performing CAR (Figure 14.1). A CAR is only as good as the immune cell that expresses it, and accordingly, a CAR is judged by how well it programs immune-cell function. Though CARs have been introduced in various types of immune cells such as natural killer cells and hematopoietic stem cells, we will limit our discussion here to T cells, the most prevalent "chassis" for CARs.

At its core, the true determinant of CAR-T cell function is their ability to eradicate target tumor cells *in vivo* with minimal toxicity. However, given the relatively low throughput of *in vivo* studies, several *in vitro* assays have been developed to predict how engineered CAR-T cells might function once they have been adoptively transferred. Typical *in vitro* assays measure CAR-T cell activation (as assessed through expression of surface markers such as CD25 or CD69), proliferation, and cytokine production upon antigen challenge. Target-cell lysis assays are also commonly performed to assess CAR-T cell cytotoxicity. The *in vitro* "stress test," whereby CAR-T cells are repeatedly challenged with antigen-presenting target cells over the course of several days, has gained popularity as a method to more accurately simulate *in vivo* conditions. In these stress tests, continued target-cell clearance and T-cell persistence even after multiple rounds of antigen challenge are considered indicators of high-performing CARs [7–9].

Results from the assays mentioned so far all provide positive indications of CAR-T cell function (e.g., the more target cell lysis, the better). However, there are also

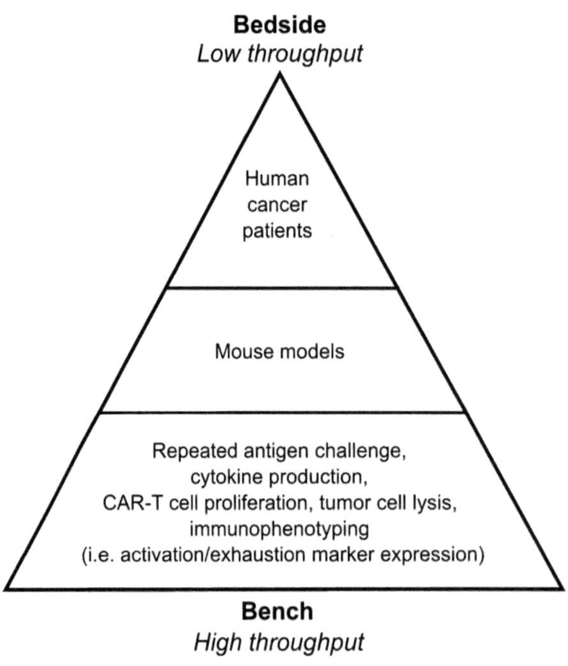

Figure 14.1 Metrics of CAR-T cell performance. CAR engineering typically begins at the bench, with high throughput assays to characterize CAR-T-cell function *in vitro*, prior to *in vivo* studies in tumor-bearing mice.

assays whose results are negative indicators of CAR-T cell function. For instance, prolonged expression of exhaustion markers such as PD-1, TIM-3, and LAG-3 is considered a negative predictor of CAR-T cell efficacy *in vivo*. Receptor signaling in the absence of antigen stimulation, known as tonic signaling, has also been associated with increased T-cell exhaustion and limited *in vivo* persistence [10, 11]. Therefore, in addition to assaying for antigen-specific CAR signaling, it is also important to ensure that CARs do not strongly signal in an antigen-independent manner.

In a recent study, Xiong et al. demonstrated that the quality of the immunological synapse (IS) formed upon antigen challenge may serve as an indicator of CAR-T cell performance *in vivo* [12]. Specifically, using both a glass-supported planar lipid bilayer system and a newly developed vertical cell pairing system, the authors measured various parameters of IS. High quality synapses, as defined in the study, were characterized by accumulation of the structural protein F-actin, presence of signaling molecules such as phosphorylated zeta chains, polarization of lytic granules, and clustering of tumor antigen. Interestingly, these properties directly correlated with *in vivo* CAR-T cell performance, whereas standard lysis and cytokine production assays did not. Though the equipment required to make IS measurements is not readily available, nor has a standardized scoring system of IS quality been developed, the study does provide an interesting biophysical framework for understanding CAR function.

With these metrics of CAR performance in mind, we now proceed to highlight each individual CAR component (Figure 14.2) and their role in tuning CAR function.

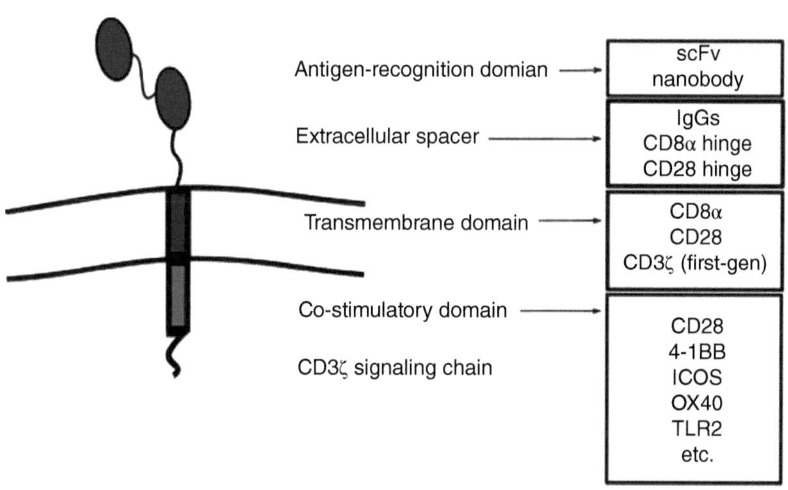

Figure 14.2 Basic structure of a chimeric antigen receptor. Starting from the N-terminus, chimeric antigen receptors consist of (1) an antigen-recognition domain, (2) an extracellular spacer, (3) a transmembrane domain, and (4) a signaling domain. The signaling domain consists of a CD3ζ signaling chain and, in second- and third-generation CARs, one or two co-stimulatory domains, respectively.

14.3 Antigen-Recognition Domain

The antigen-recognition domain classically consists of a single-chain variable fragment (scFv) comprising heavy- and light-chain variable domains. However, other ligand-binding motifs, such as nanobodies, have also been used. Nanobodies, which consist of a single heavy-chain variable domain, may be advantageous due to their smaller genetic footprint compared with scFv sequences, but nanobodies are less readily available. On the other hand, a wide array of antibodies have been developed to target various tumor antigens, and when building a CAR, it often suffices to simply attach the scFv to the downstream CAR components. If no antibodies exist for the target antigen, one might employ a variety of screening methods to obtain the desired result. Such screening methods have been extensively reviewed elsewhere [13]. The antigen-recognition domain is responsible for determining target-cell specificity, and careful tuning and engineering of the antigen-recognition domain can produce safer and stronger CARs.

14.3.1 Tuning the Antigen-Recognition Domain to Manage Toxicity

As more and more CARs begin to enter the clinic, the use of murine scFv-based CARs has sparked concerns of immunogenicity. This complicates direct "plug-and-play" of existing scFv sequences into CARs, since many existing scFv sequences are derived from murine antibodies. To address this challenge, some groups have employed methods such as DNA phage display [14] or yeast display [15] to screen through libraries of humanized scFv candidates. The high throughput of such methods should enable rapid identification of humanized scFv sequences that bind to cognate antigen as well as – if not better than – their murine counterparts.

Besides alleviating concerns of immunogenicity, careful engineering of the antigen-recognition domain can also address safety concerns arising from on-target, off-tumor toxicities. In settings where the targeted tumor antigen is ubiquitously expressed on normal, healthy tissue, it may be beneficial to consider tuning down target-binding affinity, as opposed to the conventional approach of maximizing binding affinity. As a case in point, one patient suffered lethal toxicity following treatment with anti-HER2 CAR-T cells, which reacted with healthy pulmonary tissue and triggered severe cytokine release syndrome [16]. To minimize risk of such on-target, off-tumor toxicities, some groups have performed affinity tuning of anti-HER2 and anti-epidermal growth factor receptor (EGFR) scFv sequences prior to incorporation within their respective CARs [17, 18]. In these studies, the authors demonstrated that CAR-T cells with lower-affinity scFv sequences were better able to discriminate between tumor cells, which overexpress the target antigen, and healthy cells, which express the target antigen at a lower level.

14.3.2 Incorporation of Multiple Antigen-Recognition Domains to Engineer "Smarter" CARs

In recent years, there has been increasing interest in engineering CARs capable of integrating multiple input signals and executing logical operations. Logical operations such as OR gates are of particular interest as a tool to prevent antigen escape, which is a mechanism of immune evasion whereby tumor cells downregulate expression of antigens being targeted by CAR-T cells. By programming CAR-T cells to behave as OR gates (i.e., functional output can be triggered by either input A or input B), anti-tumor function is retained even when expression of one antigen is lost. As an example, Zah et al. constructed a bispecific CD19/CD20 CAR whose antigen-recognition domain consisted of an anti-CD19 scFv tethered to an anti-CD20 scFv (Figure 14.3a) [9]. T cells expressing the CD19/CD20 bispecific CAR exhibited superior clearance of $CD19^+/CD20^+$ lymphoma xenografts compared with T cells expressing the CD19 CAR alone. Importantly, mice treated with CD19 CAR-T cells alone eventually succumbed to tumor relapse due to the outgrowth of $CD19^-$ tumor cells, whereas relapse was not observed in mice treated with CD19/CD20 bispecific CAR-T cells. In similar efforts to minimize antigen escape by malignant B cells, others have built OR-gate CARs targeting CD19 and CD22. Of note, the authors found that simply tethering an anti-CD19 scFv to an anti-CD22 scFv failed to produce CAR-T cells capable of controlling $CD19^+/CD22^+$ tumor growth *in vivo*. Instead, optimal *in vivo* performance was achieved when the CAR was engineered with a "loop" structure, where the CD22-targeting scFv was placed between the heavy and light chains of the CD19-targeting scFv (Figure 14.3a) [19]. In fact, multiple studies have underscored the fact that the structure of the CAR molecule needs to be optimized for its specific antigen(s) [9, 20–22].

Toward programming logic-gated CARs targeting other types of cancer, Hegde et al. have constructed a "tandem CAR" (TanCAR) capable of recognizing both HER2 and IL13Rα2, which are expressed in glioblastomas [23]. Notably, the extracellular component of the TanCAR was modeled *in silico*, which predicted not only that the two antigen-targeting moieties would have minimal interactions with each other, but also that they were spaced sufficiently far apart to accommodate simultaneous binding of both tumor antigens. In contrast to the previously discussed CD19/CD20 bispecific CAR-T cells, a synergistic effect was observed when TanCAR-T cells were challenged with both HER2 and IL13Rα2, with greater cytokine production compared to not only either antigen alone, but also the sum of cytokine levels produced when challenged with HER2 only and IL13Rα2 only. As such, the HER2/IL13Rα2 TanCAR cannot be precisely described based on Boolean logic, as it exhibits characteristics of both OR gates (i.e., the CAR responds to either HER2 or IL13Rα2) and AND gates (i.e., maximum output is triggered only when both input signals are present). Taken together, work by multiple groups have demonstrated the feasibility of engineering CARs that can integrate multiple input signals as a means to expand T-cell recognition capability [9, 19, 23].

One should note that trade-offs can exist between enhancing safety and enhancing target-cell recognition. For instance, tuning down scFv/target binding affinity in order to mitigate on-target, off-tumor toxicities also increase the risk that engineered CAR-T cells fail to recognize tumor cells, particularly if expression of tumor antigen is downregulated. On the other hand, bispecific, OR-gated CARs can prevent antigen escape by targeting multiple antigens, but doing so can also increase the risk of on-target, off-tumor toxicities. Therefore, careful biological considerations are

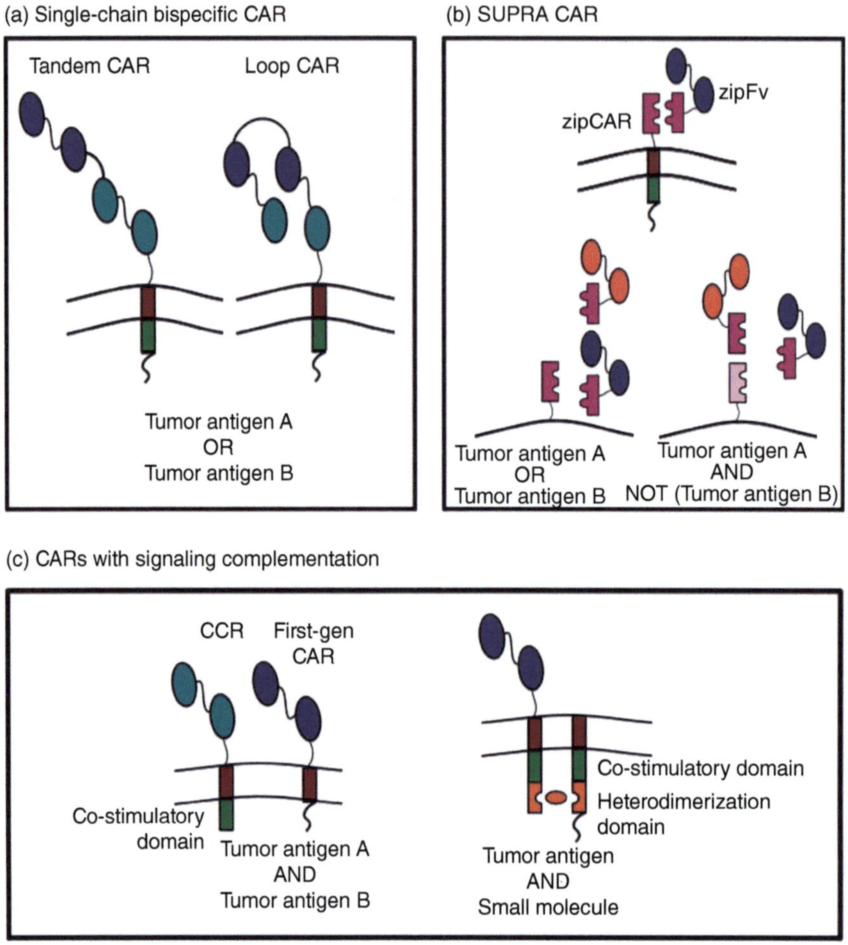

Figure 14.3 Programming CARs capable of executing molecular logic. (a) Single-chain bispecific CARs can be engineered by incorporating multiple scFv sequences into the CAR extracellular domain. This can be done in either a "tandem" or a "loop" format. Single-chain bispecific CARs that have been optimized to target each antigen with high efficiency can achieve OR-gate signaling, such that presence of one or the other antigen would be sufficient to trigger CAR activation. (b) SUPRA CARs, whose binding specificities are dictated by leucine zippers, can be programmed to execute OR-gated or "cell selector" logical functions. (c) AND-gated CARs can be engineered by leveraging the requirement for both ζ-chain and co-stimulatory signaling for full T-cell activation. Specifically, AND-gated logic can be achieved through pairing of a chimeric co-stimulatory receptor (CCR) with a first-generation (ζ-chain only) CAR. Alternatively, small molecule-gated CARs can be engineered through a split design, where ligand-induced dimerization is required to trigger T-cell activation.

warranted when designing T-cell engineering approaches. For instance, since CD20 is only expressed on mature B cells (which already express CD19) and B-cell aplasia is generally well-tolerated, a CD19/CD20 OR-gate CAR can provide safeguards against antigen escape without increasing the risk of off-tumor toxicity compared with the conventional CD19 CAR.

14.3.3 Novel Antigen-Recognition Domains to Enhance CAR Modularity

To further increase the modularity of the CAR architecture, other groups have engineered novel antigen-recognition domains that enable easy swapping of antigen targets without necessitating re-design and re-cloning of a new CAR [24–27]. The antigen-recognition domain of these CARs consists of an scFv that binds a peptide epitope rather than a tumor antigen. These CARs are cross-linked to tumor antigen by an intermediate switch molecule, which is a fusion between an antigen-binding protein – such as an antibody or scFv – and the peptide epitope that binds to the CAR. Though such strategies still rely on having an antibody or scFv sequence readily available, they remove the potential uncertainties associated with reformatting antigen-binding proteins into the context of a CAR. Furthermore, a major potential advantage of these "switchable" CARs is that they could provide a safeguard against toxicity, as CAR-T cell activity *in vivo* could be modulated by varying the dosage of the switch molecule. A major limitation is that continuous administration of the switch molecule is likely required for CAR-T cells to maintain anti-tumor function over a sustained period of time.

Cho et al. recently developed SUPRA CARs to incorporate affinity tuning, combinatorial antigen sensing, and universality all in one platform [28]. Similar to the modular switchable CARs described earlier, the SUPRA CAR is split into two components: a zipCAR that is expressed by T cells and a zipFv that comprises an scFv-leucine zipper fusion (Figure 14.3b). Target-cell recognition is programmed through the scFv, while the leucine zipper binds the zipFv to zipCAR-expressing T cells. As with the switchable CARs, the split architecture enables facile swapping of targeting domains. However, compared with the switchable CARs, the SUPRA CAR design encodes a broader range of parameters that can be tuned to modulate T-cell function, owing to the repertoire of different leucine zipper pairs with varying binding affinities. This enables one to not only tune the strength of T-cell response to a single antigen, but also program more complex logical functions. For instance, the authors demonstrated a "cell selector," which enables CAR-T cells to selectively target $HER2^+Axl^-$ cells while sparing $HER2^+Axl^+$ cells. Such logic is programmed by employing an anti-HER2 zipFv that binds more strongly to the anti-Axl zipFv than to the zipCAR. Therefore, when both HER2 and Axl are present on the target-cell surface, zipCAR conjugation to the cognate anti-HER2 zipFv is outcompeted by the higher affinity anti-Axl zipFv.

14.3.4 Engineering CARs that Target Soluble Factors

Though most CARs to-date target antigens that are immobilized on the cell surface, Chang et al. recently demonstrated a class of CARs capable of responding to soluble ligands [22]. Notably, such CARs require ligand-mediated receptor dimerization in order to signal properly. CARs that respond to soluble ligands may be useful as a strategy to resist and modify the tumor microenvironment, which contains factors such as transforming growth factor (TGF)-β that suppresses anti-tumor T-cell function. Indeed, T cells expressing a TGF-β-responsive CAR are potently activated, rather than suppressed, by TGF-β. Furthermore, TGF-β CAR-expressing $CD4^+$ T cells support the *in vitro* anti-tumor function of both NY-ESO T-cell receptor (TCR)- and CD20 CAR-expressing $CD8^+$ T cells [29]. CARs targeting soluble factors may therefore serve as a promising approach to enhance the efficacy of adoptive T-cell therapy.

14.4 Extracellular Spacer

The extracellular spacer links the antigen-recognition domain with the intracellular signaling domain of a CAR. Although the spacer serves a predominantly structural rather than ligand-binding or signaling function, the properties of the extracellular spacer can profoundly influence the performance of a CAR. The most commonly used sequences for the extracellular spacer come from either CD8α hinge or immunoglobulins such as IgG1 and IgG4. IgG-based spacers can also be lengthened by incorporating the CH2–CH3 domain from the Fc region. Multiple studies have reported that a change in spacer length can alter both *in vitro* and *in vivo* properties of CAR-T cells, though the optimal spacer length appears to vary for different CAR constructs [8, 21, 30]. For instance, long-spacer CD171 or GD2 CAR-T cells exhibit increased fratricidal activation-induced cell death (AICD), resulting in poor *in vivo* persistence compared with T cells expressing short-spacer CARs targeting the same antigens [30, 31]. In contrast, optimal function of an ROR1-specific CAR requires a long spacer [8]. Importantly, long spacers incorporating the unmodified CH2–CH3 domain have been shown to trigger an innate immune response *in vivo* via Fc receptor binding, but mutations to the CH2 domain that abolish Fc receptor binding could be incorporated to enhance T-cell persistence [8, 32, 33]. However, such modifications may not necessarily improve poor *in vivo* persistence caused by T-cell-intrinsic, fratricidal AICD, which has been attributed to Fas–FasL interactions [30].

The optimal spacer length can be rationally designed based on the observation that long spacers tend to be preferable for CARs whose scFv sequences target epitopes located close to the cell membrane. For example, Guest et al. systematically compared CARs targeting neural cell adhesion molecule (NCAM), 5T4, CD19, and carcinoembryonic antigen (CEA), each with either no spacer or a long IgG1 hinge-CH2–CH3 spacer [21]. The results indicated that T cells expressing long-spacer CARs exhibited enhanced cytotoxicity and interferon (IFN)-γ production against target cells expressing NCAM and 5T4, whose binding epitopes lie proximal to the target-cell membrane. On the other hand, T cells expressing CARs targeting CD19 and CEA, whose binding epitopes lie farther away from the cell membrane, showed superior *in vitro* functionality without an extracellular spacer. Interestingly, anti-CEA CAR-T cells challenged with target cells expressing a truncated form of CEA still did not require the extracellular spacer for optimal function. Here, the authors note that the requirement for the long spacer may have been bypassed by the fact that the target epitope lies on the N-terminus of the CEA antigen, and therefore has little steric hindrance for scFv binding. These results suggest that the effect of the extracellular spacer is impacted not only by the target epitope's proximity to the cell surface, but also its accessibility.

It is perhaps unsurprising that the extracellular spacer can have such an impact on CAR performance. Several studies have demonstrated that mechanical force plays a role in triggering T-cell activation upon TCR engagement with peptide-MHC (major histocompatibility complex) [34–38]. Specifically, the extension of bond lifetime between the TCR and peptide-MHC under force – known as a catch bond – has been shown to be critical for triggering downstream T-cell signaling events [35]. If one were to assume that the mechanics of CAR signaling is analogous to that of TCR signaling, the mechanical properties of the CAR extracellular spacer, such as length and rigidity, could presumably alter the magnitude of force transduced upon receptor engagement with target antigen. Weaker forces, in turn, may fail to adequately extend bond lifetime, resulting in sub-optimal T-cell activation. The importance of the extracellular spacer is also supported by the kinetic segregation model of T-cell signaling, which posits that exclusion of the bulky receptor phosphatase CD45 from the IS is required to enable downstream signal transduction upon TCR recognition of target peptide-MHC [39, 40]. Under this model, optimal CAR-T cell signaling could be achieved by modulating the length of the extracellular spacer such that the distance between a CAR and its cognate antigen is sufficiently small to exclude CD45 from the IS [41]. A detailed discussion on the mechanism of CAR signaling can be found in prior reviews [3].

The aforementioned CD19/CD20 OR-gate CAR [9] provides an example of how one could rationally design the extracellular spacer of a CAR. Since the bispecific CAR incorporates two different scFv sequences, the membrane-proximal scFv can be considered as a "spacer" for the membrane-distal scFv. Given that the target epitope for CD20 lies close to the tumor-cell surface, one might anticipate that an anti-CD20 CAR would function optimally with a long spacer. Combined with the fact that CD19-targeting CARs function best with a short extracellular spacer [8], one might predict that a bispecific CAR functions optimally with the CD20-targeting scFv in the membrane-distal position and the CD19-targeting scFv in the membrane-proximal position, and this prediction was indeed confirmed by experimental observations *in vitro*. To further tune $CD20^+$ target-cell recognition, the linker joining the CD19 and CD20 scFv sequences was also tuned in both length and flexibility. It was found that $CD20^+$ target cell recognition could be enhanced – without detriment to $CD19^+$ target cell recognition – by increasing either scFv linker length or rigidity. Increased linker length might serve to decrease steric hindrance imposed upon the CD20-targeting scFv by the neighboring

CD19-targeting scFv. Though one might intuitively expect that increased rigidity would limit the ability of the anti-CD20 scFv to access the target epitope, the rigid spacer may also potentially enhance CAR signaling by facilitating force transduction.

14.5 Transmembrane Domain

Like the extracellular spacer, the transmembrane domain is often overlooked as merely a structural component of a CAR. The CD28 and CD8α transmembrane regions have been the most commonly used moieties for this purpose. Recent work, however, suggests that the transmembrane domain may also shape CAR function. CD19 CARs incorporating a CD8α hinge and transmembrane domain were shown to induce less AICD *in vitro* compared with CARs with a CD28 hinge and transmembrane domain, though no significant differences were observed *in vivo* [42]. Furthermore, it is unclear whether the observed differences in AICD are a result of differences in the transmembrane domain or differences in the hinge region. Further studies are required to better elucidate the functional role, if any, of the transmembrane domain. Given that the transmembrane domain of membrane proteins are known to play critical roles in protein folding and dimerization/multimerization [43], it would not be surprising if future studies establish a clearer link between the transmembrane domain and CAR function.

14.6 Signaling Domain

14.6.1 First- and Second-Generation CARs

Among the CAR components discussed so far, the signaling domain has by far received the most attention, as extensive studies have investigated how different stimulatory/co-stimulatory domains might alter CAR function. In first-generation CARs, the signaling component consisted solely of the intracellular domain of the CD3ζ chain. Though first-generation CARs could program T cells to lyse target cells and, in some instances, produce cytokines in response to target antigen, first-generation CAR-T cells demonstrated limited anti-tumor efficacy and persistence in clinical studies [44–46].

In order to boost the signaling potency of CAR-T cells, co-stimulatory elements were eventually incorporated in addition to the CD3ζ signaling chain to generate what are now referred to as second- and third-generation CARs. To date, the most widely studied CARs contain either the CD28 or 4-1BB co-stimulatory domains in addition to the CD3ζ signaling domain (hereafter denoted as 28z and BBz, respectively). Not surprisingly, delivery of different co-stimulatory signals can have a profound impact on CAR-T cell phenotype and function. Comparison of the metabolic profiles of anti-CD19 28z and BBz CAR-T cells revealed that CD28 co-stimulation programs cells toward a glycolytic metabolism, resulting in a more effector phenotype, whereas 4-1BB co-stimulation induces greater oxidative metabolism, resulting in more memory phenotype [47]. These differences may form the basis for findings, from a separate study, that BBz CAR-T cells targeting GD2 exhibit less tonic signaling and consequently, a less exhausted phenotype compared with 28z CAR-T cells targeting the same antigen [11]. Consistent with these observations, clinical studies have demonstrated long-term persistence of anti-CD19 BBz CAR-T cells, which could be detected in circulation up to four years post-infusion [48–51]. In stark contrast, multiple studies have shown that anti-CD19 28z CAR-T cells rarely persist for more than 60 days [52, 53].

Though these results suggest that 4-1BB-mediated co-stimulation may confer superior anti-tumor properties to CAR-T cells, studies directly comparing BBz and 28z CAR-T cells in human patients have been much more limited, and variabilities in other confounding factors such as CAR-T cell manufacturing protocols, study design, and patient background may contribute to the differences observed in studies using BBz CAR-T cells versus 28z CAR-T cells. Indeed, in one study where patients were treated with a 1 : 1 mix of BBz and 28z CAR-T cells, the expansion and persistence of infused BBz CAR-T cells relative to the 28z CAR-T cells varied from patient to patient [54]. It is also worth noting that the majority of the studies discussed were conducted using CD19 CARs. Since the effect of the co-stimulatory domain may be context-dependent, it is advisable to test multiple co-stimulatory domains when designing CARs targeting novel antigens. Indeed, reports in the literature suggest that CAR-T cells with CD28 co-stimulation, but not 4-1BB co-stimulation, are able to resist TGF-β-mediated immune suppression [55, 56]. Others have reported that the addition of a CD28 co-stimulatory domain enhances CAR-T cell resistance to suppression mediated by regulatory T cells (Tregs) [57]. As both TGF-β and Tregs are known to play a role in the tumor microenvironment, these reports suggest that in the context of solid tumors, CD28 co-stimulation might in fact produce a more potent anti-tumor response than 4-1BB co-stimulation.

14.6.2 Combinatorial Co-stimulation

Instead of choosing between different co-stimulatory domains, one could also consider incorporating multiple co-stimulatory domains in what is referred to as a third-generation CAR. As demonstrated by multiple groups, the combination of CD28 and 4-1BB co-stimulatory domains in a single CAR can enhance T-cell persistence and anti-tumor function compared with both second-generation CD28- and 4-1BB-based CARs, respectively [58–60]. To systematically study the optimal configuration for combinatorial CD28 and 4-1BB co-stimulation, Zhao et al. engineered a panel of T cells expressing either a first-generation CAR alongside both 4-1BB ligand (4-1BBL) and CD28 ligand (CD80), a 28z CAR alongside 4-1BBL, a BBz CAR alongside CD80, or a third-generation 28BBz CAR [7]. Their results indicate that co-expression of 4-1BBL with a 28z CAR produces the most persistent T cells *in vivo*, accompanied by superior tumor clearance. The authors postulate that these differences in function can be explained by differential dynamics of CD28 and 4-1BB expression in T cells: whereas CD28 expression gradually diminishes over time, 4-1BB expression is sustained. Therefore, while CD80 may initially trigger CD28 signaling, it would eventually lose its co-stimulatory effect as CD28 is downregulated. On the other hand, 4-1BBL can continue to activate 4-1BB signaling over prolonged periods of time. More importantly, however, this work suggests that optimal T-cell activation is contingent on not only the type of co-stimulatory signal delivered, but also the geometry of the various signaling components relative to each other. Indeed, this is consistent with the fact that CAR function is significantly compromised when CD28 is moved from its typical membrane-proximal position to the membrane-distal position [61].

14.6.3 Other Co-stimulatory Domains: ICOS, OX40, TLR2

Beyond CD28 and 4-1BB, others have also tested CARs with co-stimulatory signals delivered by inducible T-cell co-stimulator (ICOS), OX40, and toll-like receptor 2 (TLR2). Guedan et al. demonstrated that a third-generation CAR incorporating both ICOS and 4-1BB co-stimulatory domains exhibits superior *in vivo* persistence and anti-tumor function compared with both second-generation BBz and ICOSz CARs [62]. In one of the earliest studies describing a third-generation CAR, Pulè et al. demonstrated that addition of an OX40 co-stimulatory domain to a second-generation anti-GD2 28z CAR enhances *in vitro* T-cell proliferation and cytokine production [63]. A separate study demonstrated that incorporation of OX40 co-stimulation in a CEA-targeting 28z CAR reduces production of the immunosuppressive cytokine interleukin (IL)-10, while maintaining high production of anti-tumor cytokines such as IFN-γ and IL-2 [64]. Decreased production of IL-10 may be particularly desirable for treatment of solid tumors, where elevated levels of IL-10 have been correlated with poor patient prognosis [65–67]. TLR2 co-stimulation enhances both *in vitro* and *in vivo* function of 28z CARs targeting CD19 and mesothelin [68]. This may be due to enhanced T-cell infiltration of tumors, as transcriptome analysis by RNA sequencing revealed that TLR2 co-stimulation upregulates genes associated with adhesion, migration, and synaptic transmission [68].

As with third-generation 28BBz CARs, optimal function of third-generation CARs incorporating ICOS, OX40, and TLR2 also depends on the configuration of the various signaling domains. The enhanced *in vivo* persistence of the third-generation ICOSBBz CAR described by Guedan et al. requires the ICOS signaling domain to be in the membrane-proximal position [62]. Interestingly, repression of CD28-mediated IL-10 production by a third-generation OX40-CD28z CAR seems to require OX40 to be downstream of the CD3ζ chain [63, 64]. Though different configurations of the third-generation TLR2-CD28z CAR were not tested, the construct evaluated by Lai et al. contains the TLR2 domain downstream of the CD3ζ chain.

14.6.4 Additional Considerations for CAR Signaling Domains

Although incorporation of co-stimulatory elements in second- and third-generation CARs has certainly proved an effective means to enhance the potency of CAR signaling in T cells, there is still more room for improvement as optimal T-cell activation requires a combination of not only TCR stimulation (delivered through the

CD3ζ chain in the context of a CAR) and co-stimulation, but also cytokine support [69]. To address this gap, Kagoya et al. have recently incorporated a truncated cytoplasmic domain from the IL-2 receptor β-chain (IL-2Rβ) in a second-generation anti-CD19 28z CAR [70]. Addition of a truncated IL-2Rβ domain and modification of the C-terminus of the CD3ζ chain to include the STAT3-binding YXXQ motif results in activation of the JAK-STAT signaling pathway in CAR-T cells upon antigen encounter. Consequently, this produces superior proliferation and *in vivo* anti-tumor function compared with both 28z and BBz CAR-T cells. In the configuration shown by Kagoya et al., the truncated IL-2Rβ domain lies between the CD28 and CD3ζ signaling domains, and it remains to be seen how changing the positioning of the IL-2Rβ domain might impact CAR function. It also remains to be seen whether incorporation of JAK-STAT signaling domains to a 4-1BB-based CAR would result in similar improvements in anti-tumor functions.

Interestingly, a recent phosphoproteomic study comparing 28z and BBz CARs suggests that differences in CAR performance are not, in fact, attributable to differences in the signaling pathways activated, but rather arise from differences in the kinetics and intensity of signaling [71]. This very recent finding will likely be examined by follow-up studies, but the implication that signaling strength and kinetics, rather than signaling type, is the main driver of differences in CAR performance could drastically alter the way we approach CAR engineering.

As is the case for the antigen-recognition domain, design of the signaling elements in a CAR requires a careful balance between potency and safety, which may shift depending on the cancer type being targeted and the relevant clinical concerns. For instance, augmentation of CAR signaling potency by incorporating multiple co-stimulatory domains may be desirable and perhaps even necessary for effective treatment of solid tumors, which inhabit an immunosuppressive niche. However, increased potency also increases the risk of toxicities associated with cytokine release syndrome. Indeed, co-stimulatory signals from both CD28 and 4-1BB likely contributed to the lethally high levels of inflammatory cytokines produced following treatment with an anti-HER2 CAR [16].

The requirement for multiple signals to trigger optimal T-cell activation has been leveraged in order to engineer T cells that act as AND gates, which can enhance safety by reducing ON-target, OFF-tumor effects (Figure 14.3c). As a proof-of-concept, T cells were engineered to co-express a first-generation anti-CD19 CAR with an anti-prostate-specific membrane antigen (PSMA) chimeric co-stimulatory receptor (CCR), where the CCR consists of PSMA-targeting scFv fused to a CD28 co-stimulatory domain [72]. Using this approach, signaling from neither the CAR nor the CCR alone was sufficient to potentiate a T-cell response. Instead, T cells were only fully activated when both the CAR and CCR engaged their cognate antigen. Using this same principle, others have built a small molecule-gated CAR by splitting the CAR into two separate chains, with one chain lacking the CD3ζ signaling domain and the other lacking the extracellular antigen-recognition domain. Each chain encodes a heterodimerization domain, such that upon binding to a cognate small molecule, the two chains assemble into a signaling competent structure with both ζ-chain and co-stimulatory domains required for full T-cell activation [73]. As with the bispecific CARs described earlier, these works demonstrate the potential clinical utility of T cells engineered to execute logical functions. However, very careful tuning of the ligand-binding domain's binding affinity was required to achieve the AND-gate response, thus presenting a potential challenge for the clinical translation of this design strategy.

14.7 High-Throughput CAR Engineering

Though the modularity of the CAR makes it amenable to rational design by "plug-and-play" of the different components discussed earlier, it does not preclude the possibility of engineering better CARs through high-throughput library screening and directed evolution. For instance, Alonso-Camino et al. utilized a screening platform whereby a library of CARs with varying scFv sequences was introduced into Jurkat T cells [74, 75]. Upon challenge with antigen-presenting target cells, cells with the highest surface expression of the activation marker CD69 were enriched by fluorescence-activated cell sorting (FACS)-sorted. Successive rounds of selection were performed before sequence analysis of sorted clones and identification of top-performing scFv sequences. Lipowska-Bhalla et al. used a similar platform based on Jurkat-cell activation to screen a library of anti-CEA scFv sequences formatted into a CAR. However, rather than directly using expanded, FACS-sorted cells in successive rounds of screening, scFv sequences producing the highest CD69 expression after each round of screening were recovered and subcloned into a retroviral vector backbone before being introduced into a fresh population of Jurkat cells for the next round of screening [76]. Importantly, incorporation of the subcloning step between rounds of screening yielded greater enrichment of functional scFv sequences compared to directly screening from round-to-round using expanded, FACS-sorted cells [76].

In another study, Duong et al. constructed a library of anti-HER2 CARs consisting of randomly varying cytoplasmic domains. In their library, 14 different signaling and adaptor molecules, including CD28, 4-1BB, and ICOS, were chosen and combined in random order and number. CAR libraries were retrovirally integrated into Jurkat cells and screened for CD69 upregulation upon antibody-mediated cross-linking of c-myc tags fused to the extracellular domain of each CAR. Jurkat clones with the highest CD69 expression were selected by FACS-sorting and each clone was subsequently validated for both CD69 upregulation and IL-2 secretion in response to $HER2^+$ 24JK target cells. Interestingly, results from the screen identified a novel CAR architecture composed of DAP10, CD27, and CD3ζ domains, and primary human T cells expressing these CARs exhibited *in vitro* effector function in response to tumor antigen [77].

Though these studies provided promising initial findings, a more thorough investigation of the function of the identified CARs, particularly *in vivo*, is necessary in order to fully demonstrate the clinical utility of these screening platforms. With respect to the screening method itself, some caution should be exercised as it remains unclear whether Jurkat-cell activation, as evaluated through CD69 upregulation, is fully predictive of productive primary human T-cell activation. One should also consider whether screening for only the most highly activated T cells would truly yield the most effective clinical product, particularly with regards to propensity for T-cell exhaustion as well as safety concerns. As with all high-throughput library screens, careful choice of an assay whose readout is strongly correlated with the desired properties is paramount. Future iterations of CAR library screening may benefit from sorting based on a *panel* of surface markers, which may better correlate with T-cell function compared with a single marker alone. Alternatively, as gene-editing efficiency continues to improve, one might be able to perform high-throughput library screens directly in primary human T cells, as was recently demonstrated [78].

14.8 Novel Receptor Modalities

Though the CAR has demonstrated remarkable efficacy and is the farthest along in clinical applications, other synthetic receptors are currently being developed that promise to further advance the field of T-cell engineering (Figure 14.4). In efforts to combat immunosuppressive mechanisms commonly encountered in the solid-tumor microenvironment, some groups have developed chimeras consisting of an extracellular receptor that binds an inhibitory molecule, coupled with an intracellular, co-stimulatory signaling domain (Figure 14.4a). Such "inverted receptors" include a PD-1–CD28 chimera [79], an IL-4–IL-7 chimera [80–82], and a TGF-βRII–4-1BB chimera [82]. In a recent study, T cells were engineered to co-express both the IL-4 and TGF-β inverted receptors alongside a first-generation, prostate stem-cell antigen (PSCA)-targeting CAR [82]. The combination of the two inverted receptors and the CAR acts as an AND-gate, where T-cell activation is only triggered when tumor antigen, IL-4, and TGF-β are all present, with the latter two cytokines serving as "signatures" of the immunosuppressive tumor milieu.

Another class of recently developed receptors, known as SynNotch, has also demonstrated potential to program anti-tumor T cell function [83–85]. Inspired by the Notch-Delta signaling pathway prominent in development, the SynNotch receptor consists of an extracellular ligand-binding domain tethered via the Notch core, which spans the cell membrane, to a synthetic transcription factor (Figure 14.4b). Upon ligand binding, the Notch core is cleaved, releasing the transcription factor and initiating downstream signaling events. Like CARs, SynNotch is a modular receptor, whose specificity can be programmed by swapping ligand-binding domains. Unlike CARs, however, SynNotch receptor's transcriptional and consequently functional outputs are synthetic rather than endogenously programmed. This enables researcher-specified system outputs by choice of transcription factor and engineering of the downstream genetic circuits that the transcription factor regulates. For instance, Roybal et al. have demonstrated that T cells expressing a CD19-targeting SynNotch could drive their own differentiation into a Th1 phenotype by over-expressing the transcription factor Tbet upon antigen encounter. T cells expressing the CD19 SynNotch receptor could also be programmed to secrete tumor necrosis factor-related apoptosis-inducing ligands (TRAILs) and a PD-1-targeting

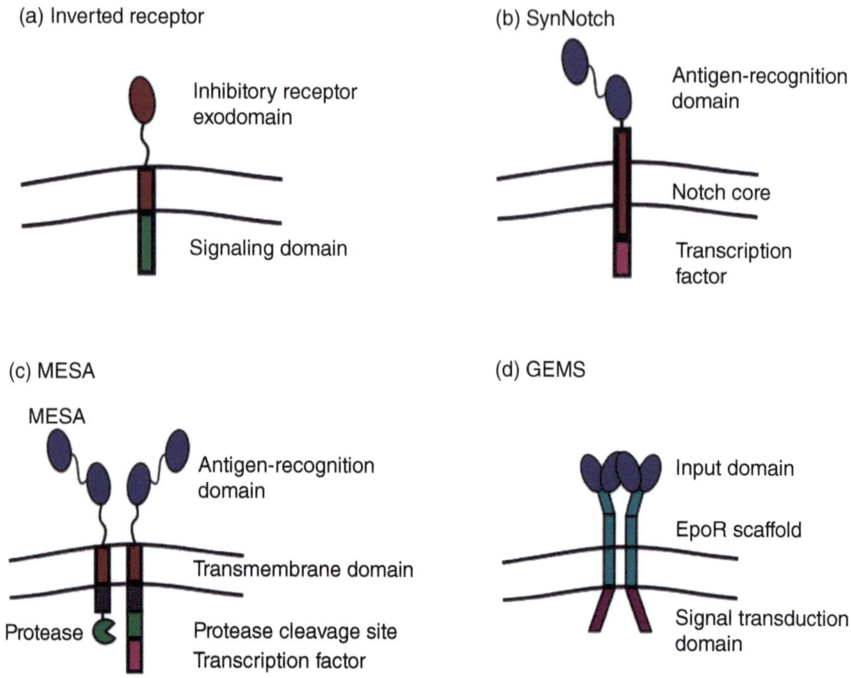

Figure 14.4 Novel receptor modalities for engineering anti-tumor T cells. (a) Inverted receptors are fusions between the ectodomain of inhibitory receptors such as PD-1 and the cytoplasmic domain of co-stimulatory domains such as CD28. In this fashion, normally immunosuppressive signals can be converted into stimulatory signals. (b) SynNotch receptors are chimeras between an antigen-recognition domain, the Notch core, and a transcription factor. Antigen binding triggers cleavage of the Notch core, releasing the transcription factor and initiating downstream signaling events. (c) MESA receptors are composed of a target chain (TC) and a protease chain (PC). Ligand binding induces dimerization of the two chains, resulting in protease-mediated cleavage of the TC and subsequent release of the transcription factor. (d) The GEMS receptor platform consists of an input domain, an EpoR scaffold, and a signaling domain. Dimerization of the input domains triggers activation of endogenous signaling pathways. Transcriptional output can be programmed by a synthetic transgene.

antibody, respectively, upon target-cell recognition. Though such customizable outputs are useful for programming T cells with novel therapeutic functions, they also necessitate the introduction of sizable transgenes into T cells. As such, this strategy can be limited in scope and complexity compared to CAR-mediated activation of endogenous T-cell signaling pathways, which trigger an array of different functional outputs such as cytokine production, proliferation, and cytotoxicity.

While SynNotch receptors respond to immobilized, surface-bound cellular ligands, others have engineered a cellular receptor platform, the Modular Extracellular Sensor Architecture (MESA), capable of responding to soluble factors [86]. The MESA receptor is composed of two separate chains: a target chain (TC) and a protease chain (PC). Both chains contain an extracellular ligand-binding domain, with the intracellular domain consisting either of a membrane-tethered transcription factor for the TC or a membrane-tethered protease for the PC (Figure 14.4c). Ligand-binding induces receptor dimerization, resulting in protease-mediated cleavage of the transcription factor associated with the TC, thereby activating transcription of downstream genetic circuits. T cells engineered to express a vascular endothelial growth factor (VEGF)-responsive MESA have been shown to secrete IL-2 in response to VEGF, which is normally immunosuppressive [87]. MESA receptors can furthermore be multiplexed to execute molecular logical functions [86, 88, 89]. As with SynNotch receptors, MESA receptor functionality is constrained in part by the amount of transgenic material that can be introduced into live cells to encode for synthetic genetic outputs.

Recent work to engineer receptor platforms with even greater modularity has been described by Scheller et al. [90]. Termed the Generalized Extracellular Molecule Sensor (GEMS), the receptors consist of modular input and signaling domains that are fused to an inactive erythropoietin receptor scaffold (Figure 14.4d). The GEMS platform was shown to respond to a wide variety of biomolecular inputs, ranging from small molecules such as nicotine to cancer markers such as prostate-specific antigen, with the only requirement being that the input molecule induces receptor dimerization. The intracellular domain is adopted from signaling chains from cytokine receptors such as IL-6RB and VEGFR2, which activates endogenous signal transduction pathways. However, as with SynNotch and MESA receptors, transcriptional output can further be programmed by introduction of synthetic transgenes.

In this review, we have outlined the basic structure of a CAR, with a part-by-part breakdown of different parameters to consider in order to optimize CAR performance. The modularity of a CAR makes it particularly conducive to rational design, thus presenting an attractive engineering-based approach to program T cells with potent anti-tumor function. Indeed, CAR-T cell engineering provides a compelling demonstration of the potential clinical impact of synthetic biology, and undoubtedly, future advances in gene editing, gene circuit design, and high-throughput screening methods promise to greatly accelerate CAR development and its translation from the benchtop to the clinic.

References

1 Approved Drugs. (2018). FDA approves tisagenlecleucel for B-cell ALL and tocilizumab for cytokine release syndrome. https://www.fda.gov/Drugs/InformationOnDrugs/ApprovedDrugs/ucm574154.htm (accessed 11 November 2020).

2 Press Announcements. (2018). FDA approves CAR-T cell therapy to treat adults with certain types of large B-cell lymphoma. https://www.fda.gov/NewsEvents/Newsroom/PressAnnouncements/ucm581216.htm (accessed 11 November 2020).

3 Chang, Z.L. and Chen, Y.Y. (2017). CARs: synthetic immunoreceptors for cancer therapy and beyond. *Trends Mol. Med.* 23 (5): 430–450.

4 Brudno, J.N. and Kochenderfer, J.N. (2018). Chimeric antigen receptor T-cell therapies for lymphoma. *Nat. Rev. Clin. Oncol.* 15 (1): 31–46.

5 Neelapu, S.S., Tummala, S., Kebriaei, P. et al. (2018). Chimeric antigen receptor T-cell therapy – assessment and management of toxicities. *Nat. Rev. Clin. Oncol.* 15 (1): 47–62.

6 June, C.H., O'Connor, R.S., Kawalekar, O.U. et al. (2018). CAR T cell immunotherapy for human cancer. *Science* 359 (6382): 1361–1365.

7 Zhao, Z., Condomines, M., van der Stegen, S.J.C. et al. (2015). Structural design of engineered costimulation determines tumor rejection kinetics and persistence of CAR T cells. *Cancer Cell* 28 (4): 415–428.

8 Hudecek, M., Sommermeyer, D., Kosasih, P.L. et al. (2015). The non-signaling extracellular spacer domain of chimeric antigen receptors is decisive for in vivo antitumor activity. *Cancer Immunol. Res.* 3 (2): 125–135.

9 Zah, E., Lin, M.-Y., Silva-Benedict, A. et al. (2016). T cells expressing CD19/CD20 bispecific chimeric antigen receptors prevent antigen escape by malignant B cells. *Cancer Immunol. Res.* 4 (6): 498–508.

10 Frigault, M.J., Lee, J., Basil, M.C. et al. (2015). Identification of chimeric antigen receptors that mediate constitutive or inducible proliferation of T cells. *Cancer Immunol. Res.* 3 (4): 356–367.

11 Long, A.H., Haso, W.M., Shern, J.F. et al. (2015). 4-1BB costimulation ameliorates T cell exhaustion induced by tonic signaling of chimeric antigen receptors. *Nat. Med.* 21 (6): 581–590.

12 Xiong, W., Chen, Y., Kang, X. et al. (2018). Immunological synapse predicts effectiveness of chimeric antigen receptor cells. *Mol. Ther.* 26 (4): 963–975.

13 Farajnia, S., Ahmadzadeh, V., Tanomand, A. et al. (2014). Development trends for generation of single-chain antibody fragments. *Immunopharmacol. Immunotoxicol.* 36 (5): 297–308.

14 Sommermeyer, D., Hill, T., Shamah, S.M. et al. (2017). Fully human CD19-specific chimeric antigen receptors for T-cell therapy. *Leukemia* 31 (10): 2191–2199.

15 Hua, C.K., Gacerez, A.T., Sentman, C.L., and Ackerman, M.E. (2017). Development of unique cytotoxic chimeric antigen receptors based on human scFv targeting B7H6. *Protein Eng. Des. Sel.* 30 (10): 713–721.

16 Morgan, R.A., Yang, J.C., Kitano, M. et al. (2010). Case report of a serious adverse event following the administration of T cells transduced with a chimeric antigen receptor recognizing ERBB2. *Mol. Ther.* 18 (4): 843–851.

17 Liu, X., Jiang, S., Fang, C. et al. (2015). Affinity-tuned ErbB2 or EGFR chimeric antigen receptor T cells exhibit an increased therapeutic index against tumors in mice. *Cancer Res.* 75 (17): 3596–3607.

18 Caruso, H.G., Hurton, L.V., Najjar, A. et al. (2015). Tuning sensitivity of CAR to EGFR density limits recognition of normal tissue while maintaining potent antitumor activity. *Cancer Res.* 75 (17): 3505–3518.

19 Qin, H., Ramakrishna, S., Nguyen, S. et al. (2018). Preclinical development of bivalent chimeric antigen receptors targeting both CD19 and CD22. *Mol. Ther. Oncolytics* 11: 127–137.

20 Hudecek, M., Lupo-Stanghellini, M.-T., Kosasih, P.L. et al. (2013). Receptor affinity and extracellular domain modifications affect tumor recognition by

ROR1-specific chimeric antigen receptor T cells. *Clin. Cancer Res.* 19 (12): 3153–3164.

21 Guest, R.D., Hawkins, R.E., Kirillova, N. et al. (2005). The role of extracellular spacer regions in the optimal design of chimeric immune receptors: evaluation of four different scFvs and antigens. *J. ImmunoTher.* 28 (3): 203–211.

22 Chang, Z.L., Lorenzini, M.H., Chen, X. et al. (2018). Rewiring T-cell responses to soluble factors with chimeric antigen receptors. *Nat. Chem. Biol.* 14 (3): 317–324.

23 Hegde, M., Mukherjee, M., Grada, Z. et al. Tandem CAR T cells targeting HER2 and IL13Rα2 mitigate tumor antigen escape. *J. Clin. Invest.* 126 (8): 3036–3052.

24 Rodgers, D.T., Mazagova, M., Hampton, E.N. et al. (2016). Switch-mediated activation and retargeting of CAR-T cells for B-cell malignancies. *Proc. Natl. Acad. Sci. U.S.A.* 113 (4): E459–E468.

25 Ma, J.S.Y., Kim, J.Y., Kazane, S.A. et al. (2016). Versatile strategy for controlling the specificity and activity of engineered T cells. *Proc. Natl. Acad. Sci. U.S.A.* 113 (4): E450–E458.

26 Raj, D., Yang, M.-H., Rodgers, D. et al. (2018). Switchable CAR-T cells mediate remission in metastatic pancreatic ductal adenocarcinoma. *Gut* 68 (6): 1052–1064.

27 Bachmann, D., Aliperta, R., Bergmann, R. et al. (2017). Retargeting of UniCAR T cells with an in vivo synthesized target module directed against CD19 positive tumor cells. *Oncotarget* 9 (7): 7487–7500.

28 Cho, J.H., Collins, J.J., and Wong, W.W. (2018). Universal chimeric antigen receptors for multiplexed and logical control of T cell responses. *Cell* 173 (6): 1426–1438.e11.

29 Hou, A.J., Chang, Z.L., Lorenzini, M.H. et al. (2018). TGF-β–responsive CAR-T cells promote anti-tumor immune function. *Bioeng. Transl. Med.* 3 (2): 75–86.

30 Künkele, A., Johnson, A.J., Rolczynski, L.S. et al. (2015). Functional tuning of CARs reveals signaling threshold above Which CD8+ CTL antitumor potency is attenuated due to cell Fas–FasL-dependent AICD. *Cancer Immunol. Res.* 3 (4): 368–379.

31 Gargett, T., Yu, W., Dotti, G. et al. (2016). GD2-specific CAR T cells undergo potent activation and deletion following antigen encounter but can be protected from activation-induced cell death by PD-1 blockade. *Mol. Ther.* 24 (6): 1135–1149.

32 Hombach, A., Hombach, A.A., and Abken, H. (2010). Adoptive immunotherapy with genetically engineered T cells: modification of the IgG1 Fc "spacer" domain in the extracellular moiety of chimeric antigen receptors avoids "off-target" activation and unintended initiation of an innate immune response. *Gene Ther.* 17 (10): 1206–1213.

33 Jonnalagadda, M., Mardiros, A., Urak, R. et al. (2015). Chimeric antigen receptors with mutated IgG4 Fc spacer avoid Fc receptor binding and improve T cell persistence and antitumor efficacy. *Mol. Ther.* 23 (4): 757–768.

34 Kim, S.T., Takeuchi, K., Sun, Z.-Y.J. et al. (2009). The αβ T cell receptor is an anisotropic mechanosensor. *J. Biol. Chem.* 284 (45): 31028–31037.

35 Liu, B., Chen, W., Evavold, B.D., and Zhu, C. (2014). Accumulation of dynamic catch bonds between TCR and agonist peptide-MHC triggers T cell signaling. *Cell* 157 (2): 357–368.

36 Ma, Z., Janmey, P.A., and Finkel, T.H. (2008). The receptor deformation model of TCR triggering. *FASEB J.* 22 (4): 1002–1008.

37 Feng, Y., Brazin, K.N., Kobayashi, E. et al. (2017). Mechanosensing drives acuity of αβ T-cell recognition. *Proc. Natl. Acad. Sci. U.S.A.* 114 (39): E8204–E8213.

38 Sibener, L.V., Fernandes, R.A., Kolawole, E.M. et al. (2018). Isolation of a structural mechanism for uncoupling T cell receptor signaling from peptide-MHC binding. *Cell* 174 (3): 672–687.e27.

39 Davis, S.J. and van der Merwe, P.A. (2006). The kinetic-segregation model: TCR triggering and beyond. *Nat. Immunol.* 7 (8): 803–809.

40 Chang, V.T., Fernandes, R.A., Ganzinger, K.A. et al. (2016). Initiation of T cell signaling by CD45 segregation at "close contacts". *Nat. Immunol.* 17 (5): 574–582.

41 James, J.R. and Vale, R.D. (2012). Biophysical mechanism of T-cell receptor triggering in a reconstituted system. *Nature* 487 (7405): 64–69.

42 Alabanza, L., Pegues, M., Geldres, C. et al. (2017). Function of novel anti-CD19 chimeric antigen receptors with human variable regions is affected by hinge and transmembrane domains. *Mol. Ther.* 25 (11): 2452–2465.

43 Volynsky, P.E., Polyansky, A.A., Fakhrutdinova, G.N. et al. (2013). Role of dimerization efficiency of transmembrane domains in activation of fibroblast growth factor receptor 3. *J. Am. Chem. Soc.* 135 (22): 8105–8108.

44 Gong, M.C., Latouche, J.B., Krause, A. et al. (1999). Cancer patient T cells genetically targeted to prostate-specific membrane antigen specifically lyse prostate cancer cells and release cytokines in response to prostate-specific membrane antigen. *Neoplasia* 1 (2): 123–127.

45 Till, B.G., Jensen, M.C., Wang, J. et al. (2008). Adoptive immunotherapy for indolent non-Hodgkin lymphoma and mantle cell lymphoma using genetically modified autologous CD20-specific T cells. *Blood* 112 (6): 2261–2271.

46 Kershaw, M.H., Westwood, J.A., Parker, L.L. et al. (2006). A phase i study on adoptive immunotherapy using gene-modified T cells for ovarian cancer. *Clin. Cancer Res.* 12 (20): 6106–6115.

47 Kawalekar, O.U., O'Connor, R.S., Fraietta, J.A. et al. (2016). Distinct signaling of coreceptors regulates specific metabolism pathways and impacts memory development in CAR T cells. *Immunity* 44 (2): 380–390.

48 Kalos, M., Levine, B.L., Porter, D.L. et al. (2011). T cells with chimeric antigen receptors have potent antitumor effects and can establish memory in patients with advanced leukemia. *Sci. Transl. Med.* 3 (95): 95ra73.

49 Maude, S.L., Frey, N., Shaw, P.A. et al. (2014). Chimeric antigen receptor T cells for sustained remissions in leukemia. *N. Engl. J. Med.* 371 (16): 1507–1517.

50 Porter, D.L., Hwang, W.-T., Frey, N.V. et al. (2015). Chimeric antigen receptor T cells persist and induce sustained remissions in relapsed refractory chronic lymphocytic leukemia. *Sci. Transl. Med.* 7 (303): 303ra139.

51 Maude, S.L., Laetsch, T.W., Buechner, J. et al. (2018). Tisagenlecleucel in children and young adults with B-cell lymphoblastic leukemia. *N. Engl. J. Med.* 378 (5): 439–448.

52 Brentjens, R.J., Rivière, I., Park, J.H. et al. (2011). Safety and persistence of adoptively transferred autologous CD19-targeted T cells in patients with relapsed or chemotherapy refractory B-cell leukemias. *Blood* 118 (18): 4817–4828.

53 Kochenderfer, J.N., Dudley, M.E., Kassim, S.H. et al. (2015). Chemotherapy-refractory diffuse large B-cell lymphoma and indolent B-cell malignancies can be effectively treated with autologous T cells expressing an anti-CD19 chimeric antigen receptor. *J. Clin. Oncol.* 33 (6): 540–549.

54 Cheng, Z., Wei, R., Ma, Q. et al. (2018). In vivo expansion and antitumor activity of coinfused CD28- and 4-1BB-engineered CAR-T cells in patients with B cell leukemia. *Mol. Ther.* 26 (4): 976–985.

55 Koehler, H., Kofler, D., Hombach, A., and Abken, H. (2007). CD28 costimulation overcomes transforming growth factor-β-mediated repression of proliferation of redirected human CD4+ and CD8+ T cells in an antitumor cell attack. *Cancer Res.* 67 (5): 2265–2273.

56 Golumba-Nagy, V., Kuehle, J., Hombach, A.A., and Abken, H. (2018). CD28-ζ CAR T cells resist TGF-β repression through IL-2 signaling, which can be mimicked by an engineered IL-7 autocrine loop. *Mol. Ther.* 26 (9): 2218–2230.

57 Loskog, A., Giandomenico, V., Rossig, C. et al. (2006). Addition of the CD28 signaling domain to chimeric T-cell receptors enhances chimeric T-cell resistance to T regulatory cells. *Leukemia* 20 (10): 1819–1828.

58 Carpenito, C., Milone, M.C., Hassan, R. et al. (2009). Control of large, established tumor xenografts with genetically retargeted human T cells containing CD28 and CD137 domains. *Proc. Natl. Acad. Sci. U.S.A.* 106 (9): 3360–3365.

59 Zhong, X.-S., Matsushita, M., Plotkin, J. et al. (2010). Chimeric antigen receptors combining 4-1BB and CD28 signaling domains augment PI3kinase/AKT/Bcl-XL activation and CD8+ T cell-mediated tumor eradication. *Mol. Ther.* 18 (2): 413–420.

60 Wang, J., Jensen, M., Lin, Y. et al. (2007). Optimizing adoptive polyclonal T cell immunotherapy of lymphomas, using a chimeric T cell receptor possessing CD28 and CD137 costimulatory domains. *Hum. Gene Ther.* 18 (8): 712–725.

61 Finney, H.M., Lawson, A.D., Bebbington, C.R., and Weir, A.N. (1998). Chimeric receptors providing both primary and costimulatory signaling in T cells from a single gene product. *J. Immunol.* 161 (6): 2791–2797.

62 Guedan, S., Posey, A.D., Shaw, C. et al. (2018). Enhancing CAR T cell persistence through ICOS and 4-1BB costimulation. *JCI Insight* 3 (1): 1–17. Available from: https://www.ncbi.nlm.nih.gov/pmc/articles/PMC5821198.

63 Pulè, M.A., Straathof, K.C., Dotti, G. et al. (2005). A chimeric T cell antigen receptor that augments cytokine release and supports clonal expansion of primary human T cells. *Mol. Ther.* 12 (5): 933–941.

64 Hombach, A.A., Heiders, J., Foppe, M. et al. (2012). OX40 costimulation by a chimeric antigen receptor abrogates CD28 and IL-2 induced IL-10 secretion by redirected CD4+ T cells. *Oncoimmunology* 1 (4): 458–466.

65 Lech-Maranda, E., Bienvenu, J., Michallet, A.-S. et al. (2006). Elevated IL-10 plasma levels correlate with poor prognosis in diffuse large B-cell lymphoma. *Eur. Cytokine Netw.* 17 (1): 60–66.

66 Mustea, A., Könsgen, D., Braicu, E.I. et al. (2006). Expression of IL-10 in patients with ovarian carcinoma. *Anticancer Res.* 26 (2C): 1715–1718.

67 Wang, H., Wang, L., Chi, P. et al. (2016). High level of interleukin-10 in serum predicts poor prognosis in multiple myeloma. *Br. J. Cancer* 114 (4): 463–468.

68 Lai, Y., Weng, J., Wei, X. et al. (2018). Toll-like receptor 2 costimulation potentiates the antitumor efficacy of CAR T cells. *Leukemia* 32 (3): 801–808.

69 Kershaw, M.H., Westwood, J.A., and Darcy, P.K. (2013). Gene-engineered T cells for cancer therapy. *Nat. Rev. Cancer* 13 (8): 525–541.

70 Kagoya, Y., Tanaka, S., Guo, T. et al. (2018). A novel chimeric antigen receptor containing a JAK-STAT signaling domain mediates superior antitumor effects. *Nat. Med.* 24 (3): 352–359.

71 Salter, A.I., Ivey, R.G., Kennedy, J.J. et al. (2018). Phosphoproteomic analysis of chimeric antigen receptor signaling reveals kinetic and quantitative differences that affect cell function. *Sci. Signal.* 11 (544): eaat6753.

72 Kloss, C.C., Condomines, M., Cartellieri, M. et al. (2013 Jan). Combinatorial antigen recognition with balanced signaling promotes selective tumor eradication by engineered T cells. *Nat. Biotechnol.* 31 (1): 71–75.

73 Wu, C.-Y., Roybal, K.T., Puchner, E.M. et al. (2015). Remote control of therapeutic T cells through a small molecule-gated chimeric receptor. *Science* 350 (6258): aab4077.

74 Alonso-Camino, V., Sánchez-Martín, D., Compte, M. et al. (2009). Lymphocyte display: a novel antibody selection platform based on T cell Activation. *PLoS One* 4 (9). Available from: https://www.ncbi.nlm.nih.gov/pmc/articles/PMC2747005.

75 Alonso-Camino, V., Sánchez-Martín, D., Compte, M. et al. (2013). CARbodies: human antibodies against cell surface tumor antigens selected from repertoires displayed on T cell chimeric antigen receptors. *Mol. Ther. Nucleic Acids* 2 (5): e93.

76 Lipowska-Bhalla, G., Gilham, D.E., Hawkins, R.E., and Rothwell, D.G. (2013). Isolation of tumor antigen-specific single-chain variable fragments using a chimeric antigen receptor bicistronic retroviral vector in a mammalian screening protocol. *Hum. Gene Ther. Methods* 24 (6): 381–391.

77 Duong, C.P.M., Westwood, J.A., Yong, C.S.M. et al. (2013). Engineering T cell function using chimeric antigen receptors identified using a DNA library approach. *PLoS One* 8 (5): e63037.

78 Shifrut, E., Carnevale, J., Tobin, V. et al. (2018). Genome-wide CRISPR screens in primary human t cells reveal key regulators of immune function. *Cell* 175 (7): 1958–1971. Available from: http://biorxiv.org/content/early/2018/08/03/384776.abstract.

79 Prosser, M.E., Brown, C.E., Shami, A.F. et al. (2012). Tumor PD-L1 co-stimulates primary human CD8(+) cytotoxic T cells modified to express a PD1:CD28 chimeric receptor. *Mol. Immunol.* 51 (3–4): 263–272.

80 Leen, A.M., Sukumaran, S., Watanabe, N. et al. (2014). Reversal of tumor immune inhibition using a chimeric cytokine receptor. *Mol. Ther.* 22 (6): 1211–1220.

81 Mohammed, S., Sukumaran, S., Bajgain, P. et al. (2017, 58). Improving chimeric antigen receptor-modified T cell function by reversing the immunosuppressive tumor microenvironment of pancreatic cancer. *Mol. Ther.* 25 (1): 249.

82 Sukumaran, S., Watanabe, N., Bajgain, P. et al. (2018). Enhancing the potency and specificity of engineered T cells for cancer treatment. *Cancer Discov.*: CD-17-1298.

83 Morsut, L., Roybal, K.T., Xiong, X. et al. (2016). Engineering customized cell sensing and response behaviors using synthetic notch receptors. *Cell* 164 (4): 780–791.

84 Roybal, K.T., Rupp, L.J., Morsut, L. et al. (2016). Precision tumor recognition by T cells with combinatorial antigen-sensing circuits. *Cell* 164 (4): 770–779.

85 Roybal, K.T., Williams, J.Z., Morsut, L. et al. (2016). Engineering T cells with customized therapeutic response programs using synthetic notch receptors. *Cell* 167 (2): 419–432.e16.

86 Daringer, N.M., Dudek, R.M., Schwarz, K.A., and Leonard, J.N. (2014). Modular extracellular sensor architecture for engineering mammalian cell-based devices. *ACS Synth. Biol.* 3 (12): 892–902.

87 Johnson, B.F., Clay, T.M., Hobeika, A.C. et al. (2007). Vascular endothelial growth factor and immunosuppression in cancer: current knowledge and potential for new therapy. *Expert Opin. Biol. Ther.* 7 (4): 449–460.

88 Schwarz, K.A., Daringer, N.M., Dolberg, T.B., and Leonard, J.N. (2017). Rewiring human cellular input-output using modular extracellular sensors. *Nat. Chem. Biol.* 13 (2): 202–209.

89 Hartfield, R.M., Schwarz, K.A., Muldoon, J.J. et al. (2017). Multiplexing engineered receptors for multiparametric evaluation of environmental ligands. *ACS Synth. Biol.* 6 (11): 2042–2055.

90 Scheller, L., Strittmatter, T., Fuchs, D. et al. (2018). Generalized extracellular molecule sensor platform for programming cellular behavior. *Nat. Chem. Biol.* 14 (7): 723.

Part IV

Applications in Medical Biotechnology

15

Development of Novel Cellular Imaging Tools Using Protein Engineering

Praopim Limsakul[1,2], Chi-Wei Man[1], Qin Peng[1,3], Shaoying Lu[1], and Yingxiao Wang[1]

[1] *University of California, San Diego, Institute of Engineering in Medicine, Department of Bioengineering, 9500 Gilman Dr, La Jolla, CA 92093, USA*
[2] *Prince of Songkla University, Faculty of Science, Songkhla 90112, Thailand*
[3] *Institute of Systems and Physical Biology, Shenzhen, Bay Laboratory, ShenZhen 518132, China*

15.1 Introduction

Protein engineering has been extensively used to improve and modify the structure and function of proteins for a wide range of applications, especially as cellular imaging tools that can be used to study cellular processes in living cells. Due to the complexity of cellular processes, multiple analytical methods, including genetically encoded fusion proteins, are used to elucidate their fundamental mechanisms. In particular, fluorescent proteins (FPs), which are derived from *Aequorea victoria* jellyfish green fluorescent protein (GFP) (avGFP) and other non-jellyfish proteins (e.g. *Discosoma* red fluorescent protein (DsRed)), have been utilized as fusion tags or markers for studying protein localization and interaction in living cells [1, 2]. In addition to FPs, other protein tags with attached chemical functionalities (e.g. SNAP-tag [3]) or other protein scaffolds (e.g. nanobody and monobody [4]) are also used to capture and interact with particular proteins or polypeptides, allowing for the incorporation of a variety of fluorescent labels.

The applications of these FPs and protein tags as noninvasive tags reveal new aspects of protein dynamics and cellular processes using optical microscopy. Fluorescence microscopy is extensively used to noninvasively observe the dynamics of biological molecules; however, its spatial resolution is limited by the diffraction of light which is a major weakness of fluorescence microscopy [5]. To overcome the diffraction-limited resolution of traditional optical microscopy methods such as confocal and wide-field microscopy, super-resolution techniques were recently utilized. Super-resolution microscopy makes it possible to image nanoscale subcellular structures by employing photoswitchable FPs, which can be stochastically turned ON by activation at a particular wavelength [6, 7]. Thus, engineering FPs and protein tags to improve their photophysical properties can improve both conventional fluorescence microscopy and super-resolution microscopy. Protein

Protein Engineering: Tools and Applications, First Edition. Edited by Huimin Zhao.
© 2021 WILEY-VCH GmbH. Published 2021 by WILEY-VCH GmbH.

engineering incorporating a combination of rational design, structure-based mutagenesis, and iterative cycles of directed evolution (random gene diversification followed by selection of clones with desired properties) is a key method to improve functionality, activity, stability, photophysical, and biochemical properties of FPs and protein tags for a wide range of applications [8]. In this chapter, we describe the development of engineered FPs and protein tags, followed by their imaging applications in cell biology and biomedicine.

15.2 Cellular Imaging Tools Developed by Protein Engineering

Cellular imaging probes have been extensively utilized for visualizing and quantifying biological processes at cellular levels in living cells. The probes generally consist of two functional parts: a detectable domain, and a specificity-determining domain which recognizes biologically relevant biomolecules. The detectable domain determines the imaging modality, for example, magnetic resonance imaging [9], fluorescence imaging [10], bioluminescence [11], ultrasound [12], nuclear imaging [13], or single photon emission computed tomography [14]. Here, we focus on FPs for detectable domains and their use in fluorescence microscopy. Utilizing directed evolution, many FPs with a variety of photophysical and biochemical properties have been developed for imaging purpose. To further extend a broader application of cellular imaging tools, antibody and protein scaffolds have also been optimized to be selective and specific to target biomolecules.

15.2.1 Fluorescent Proteins

FPs have been widely used to genetically label biomolecules to monitor cellular processes in living systems using fluorescence microscopy. Over the past decades, FPs have been modified using protein engineering to improve their photophysical and biochemical properties. Genetic manipulation has been performed on avGFP and other non-jellyfish proteins to create a vast array of FPs spanning the visible spectrum with improved brightness, stability, and chromophore maturation [1, 2]. Additionally, mutations have been used to improve gene expression by optimizing codon usage as well as to create monomeric FPs from their naturally oligomeric forms. Here, we review top-performing FPs that have been recently developed. For FPs developed prior to this, we recommend the reader to the following reference [15].

15.2.1.1 The FP Color Palette

Blue fluorescent proteins (BFPs) generally suffer from moderate brightness because of chemical instability of their chromophore [16]. Recently, mTagBFP2 has been developed through rational design [16] to reduce the conformational flexibility of the blue chromophore due to aliphatic amino acid residues proximal to Tyr64. The result showed that a single I174A substitution in mTagBFP2 resulted in improved brightness and stability compared with the parental mTagBFP [17] because it may provide a free space near Tyr64 that allows for hydrogen bonds to form and stabilize the Tyr64 side chain.

mTurquoise2 [18] is a brighter cyan fluorescent protein (CFP) that was recently engineered from Super Cyan Fluorescent Protein 3A (SCFP3A). Thr65 of SCFP3A [19] was mutated to enhance intrinsic molecular brightness, and fluorescence lifetime-based screening was used to optimize quantum yield, resulting in mTurquoise (SCFP3A(T65Y)). To further improve mTurquoise, Ile146 was mutated to Phe [20] to make it more rigid and stabilize the proximal chromophore. mTurquoise2 is the result of these two amino acid mutations.

Many yellow-green FPs have also been improved from avGFP. The mClover series is an example that was rationally designed to improve quantum yield by modifying the hydrogen bonding network surrounding the chromophore of wild-type GFP and residues involved in photostability, folding, and monomericity [21]. As a result of optimization, the most recent monomeric Clover 3 was developed, which is brighter and more photostable than the parental Clover [22]. LanYFP is a yellow-green FP that was optimized from the cephalochordate *Branchiostoma lanceolatum* (Allele Biotechnology). The LanYFP tetramer was engineered using structural modeling to substitute side chains with positive charges to monomerize the protein. To increase the brightness, random libraries were created, transformed into *Escherichia coli* and then selected for brightness. Using this method, fluorescence of the monomerized proteins was rescued, and mNeonGreen was created. With 21 amino acid substitutions, mNeonGreen is the brightest monomeric yellow-green FP that has been reported to date [23].

Red fluorescent proteins (RFPs) are attractive as reporters for *in vivo* imaging because of their low autofluorescence, light-scattering, and phototoxicity. mRuby is a series of RFPs that has been constantly improved. mRuby2 [21] was developed from mRuby [24], which was originally engineered from eqFP611 – from the sea anemone *Entacmaea quadricolor*, by mutating (i) amino acids involved in tetrameric structure, (ii) the chromophore of eqFP611, and (iii) interior residues to reduce fast photobleaching due to isomerization. After mRuby2, the brighter mRuby3 [22] was semi-rationally modified from its parental mRuby2 with 21 amino acid substitutions mainly located on the external barrel involved in either protein folding or chromophore brightness [21]. mKate2, an improved version of mKate, has been widely used in fusion protein constructs [25]; however, it tends to form a dimer at high concentrations. Thus, FusionRed, a completely monomeric version of mKate2, was developed by mutating amino acid residues that affect oligomerization and protein maturation [26]. Furthermore, mScarlet, a recently reported monomeric RFP, was optimized based on mCherry and multiple other naturally occurring RFPs to increase planarity of chromophores, modify numerous hydrogen bonding networks, and break dimerization interfaces at the outer barrel surface [27].

15.2.1.2 Photocontrollable Fluorescent Proteins

Another photophysical property of FPs is the ability to switch between different fluorescent states. After the first observation of photoswitching in wild-type

GFP and its mutants was reported [28], many subsequent FPs were reported whose fluorescent properties could be changed in response to irradiation with light of a specific wavelength, intensity, or duration [29]. Combining rational structure-guided and random mutagenesis to improve photophysical properties of FPs, photoactivatable, photoconvertible, and reversibly photoswitchable FPs have been developed for use in many imaging applications.

Photoactivatable fluorescent proteins (PAFPs) are irreversibly convertible from a non-fluorescent (dark) state to a fluorescent (bright) state in response to illumination at a specific wavelength. The conversion from dark-to-green or dark-to-red fluorescent states provides high contrast over background after photoactivation. Recently, many dark-to-red PAFPs have been developed, such as PAmCherry1 [30], PATagRFP [31], and PAmKate [6]. The photoactivation mechanism of PAmCherry involves decarboxylation of the conserved Glu222 through the photo-Kolbe reaction, while the mechanism of PATagRFP and PAmKate photoactivation remains unclear.

Photoconvertible fluorescent proteins (PCFPs) undergo irreversible color conversion from one fluorescent state to another in response to illumination at a specific wavelength. Although several color conversions have been developed, including cyan-to-green, green-to-red, or orange-to-far-red, most commonly used PCFPs are photoconvertible from green-to-red, for example, mEos2 [32], mEos3.2 [33], mClavGR2 [34], and mMaple [35]. Green-to-red PCFPs generally share a common His-Tyr-Gly derived chromophore structure and a common photoconversion mechanism. Irradiation with UV light causes irreversible cleavage near the chromophore and subsequent extension of the electron conjugation system, resulting in a new emission peak in the red spectrum [7].

In contrast to photoactivation and photoconversion where phototransformation involves covalent modification of FPs, reversibly photoswitching fluorescent proteins (rsFPs) require only conformational rearrangements of the chromophore and its surrounding environment to repeatedly switch between non-fluorescent and fluorescent states. The photoswitching mechanism in rsFPs is generally based on light-induced cis–trans isomerization, which is an intrinsic property of the isolated chromophore. Reversibly switchable enhanced green fluorescent protein (EGFP) mutant, rsEGFP2, exhibits several fold faster switching kinetics than the well-known Dronpa [36, 37]. Similarly, Dreiklang, evolved from Citrine, can switch between green and yellow fluorescence upon irradiation [38]. For RFPs, the first reported monomeric red rsFPs are rsCherry, rsCherryRev [39]. Both of them were derived from mCherry but have opposite switching modes. Despite low brightness and photoswitching contrast of rsCherries, rsTagRFP has high brightness and photoswitching contrast [40].

In summary, these PAFPs, PCFPs, and rsFPs are photocontrollable FPs whose photophysical properties can be controlled by light. PAFPs irreversibly undergo activation from a non-fluorescent state to a fluorescent state, while PCFPs irreversibly undergo color conversion from one fluorescent state to another. On the other hand, rsFPs undergo repeated cycles between fluorescent and non-fluorescent states. The high contrast between two fluorescent states upon light irradiation of specific wavelengths of these photocontrollable FPs provides specificity of labeling proteins of interest. Thus, photocontrollable FPs have become increasingly popular tools for protein tracking and super-resolution imaging.

15.2.1.3 Other Engineered Fluorescent Proteins

Circularly Permuted Fluorescent Proteins (cpFPs) are made by covalently connecting the N- and C-termini of FPs with a short linker peptide and are subsequently cleaved at a specific site on the protein to generate a new terminus [41]. With gaps in the β-can structure of FPs, the chromophore of cpFPs is more sensitive to its physicochemical environment, which enables this type of FP to act as an environment-sensitive biosensor. Connected to a target protein, cpFPs have been used as single-FP intensiometric biosensors [8]. For example, cpFPs can alter their pK_a, enabling them to act as a pH sensor [41]. Additionally, attaching calcium-binding domains (e.g. calmodulin-binding M13 fragment from myosin light-chain kinase) to cpFP fragments resulted in the development of Pericams and GCaMPs to sense calcium levels [42]. After circularly permuted GFP (cpGFP) was introduced, other colors of cpFPs have been developed [43, 44] by mainly using random mutagenesis on the loop regions of wild-type mCherry and mKate followed by fluorescence screening of bacterial colonies.

Split Fluorescent Proteins consist of two nonfluorescent fragments, which are non fluorescent in split form, but when these two fragments undergo self-complementation, a fluorophore is formed and becomes fluorescent. For example, split GFP contains GFP1-10 fragment (amino acid 1–214) of super-folder GFP, which contains three residues (Ser65, Tyr66, Gly67) that constitute the GFP chromophore and is nonfluorescent by itself because chromophore maturation requires the conserved Glu222 residue located on GFP11 fragment (amino acid 215–230) [45]. By fusing split GFP to biomolecules of interest, it can be used for visualization and quantification of protein subcellular location and activity [46]. To enable multicolor imaging, mNeongreen and sfCherry have also been developed into split FPs using directed evolution, resulting in a yellow-green-colored split mNeonGreen2 and a red-colored split super-folder Cherry2, respectively [46].

15.2.2 Antibodies and Protein Scaffolds

In addition to FPs, antibodies and protein scaffolds can also be applied for cellular imaging. Cellular imaging probes are typically comprised of two functional domains: a detectable domain, which produces signal, and a specificity-determining domain, which senses a target molecule with selectivity. Protein engineering of antibodies and other protein scaffolds can be applied to develop specificity-determining domains.

15.2.2.1 Antibodies

Antibodies are often used as probes for cellular imaging due to their specificity in recognizing antigens. Antibodies are heterodimers comprised of a heavy chain and a light chain. These protein chains combine to perform two general functions. The first is to provide specificity toward a target antigen, and the second is to elicit an immune response. The fragment antigen-binding (Fab) domain performs the former function and consists of a portion of the heavy chain and the entire light chain. The fragment crystallizable (Fc) domain performs the latter function and consists of the remaining heavy chain sections. It is the Fc domain that is responsible for determining antibody function and class (IgM, IgA, IgG, IgD, and IgE) [47], and while Fc domain engineering is a considerably important field of study [48], molecular imaging mainly focuses on engineering the Fab region and its derivatives for their binding specificity. When conjugated to imaging agents such as FPs or radioactive isotopes, antibodies can use their specificity to target ligands; however, due to their large size, full-length antibodies have several disadvantages when used as imaging agents *in vivo*. Weighing at ~150 kDa, their larger size also leads to a higher chance of immunogenicity [49]. Furthermore, inclusion of the Fc domain can trigger immune responses such as cytokine release and complement cascade activation. Finally, larger sized proteins are more difficult to produce and result in lower yields. Due to the aforementioned reasons, a whole range of antibody fragments have been engineered to retain the binding specificity of full-length antibodies but with lower molecular weight.

15.2.2.2 Antibody-Like Protein Scaffolds

There are various types of antibody-like fragments that have been engineered for imaging agents including Fabs [50], single-chain variable fragment (scFvs) [51], diabodies [52], minibodies [53], nanobodies [54], monobodies [55], and other single-domain antibodies. Fabs are the binding portion of full-length antibodies (~55 kDa), consisting of two single domains from the heavy chain and two single domains from the light chain. These retain the binding affinity of full-length antibodies but do not contain the signaling Fc domain. scFvs (~28 kDa) consist of V_H and V_L single domains from heavy and light chains connected via a peptide linker. scFvs are popular antibody derivatives because they maintain the six binding loops present on the Fab but are half the size of Fabs. Diabodies and minibodies are multivalent antibody derivatives, which mean that they have multiple antigen binding sites. This allows them to bind to a single antigen in multiple locations or to multiple antigens, a term referred to as avidity [56]. Nanobody is a single domain antibody, and monobody is an antibody single domain-like fragment derived from human fibronectin type III. Weighing at ~15 and ~10 kDa, respectively, nanobody and monobody are the smallest out of the entire suite of antibody-like fragments, which allow them to potentially target antigens buried within concave topologies [57]. They are easy to express, minimize steric hindrance, and retain the binding capabilities of their full-length parents.

15.2.2.3 Directed Evolution

To recognize target antigens, antibodies and antibody-like fragments need to be affinity-matured toward the target antigen, usually through the process of directed evolution. Directed evolution uses an iterative cycle of diversification and selection. During the first step of diversification, the antibody fragment DNA is randomly mutated with different methods. Error prone polymerase chain reaction (PCR) [58] uses low fidelity polymerases, increased magnesium/manganese salts, and/or mutagenic deoxynucleoside triphosphate (dNTP) analogs; however, this method of generating mutants is not ideal if mutational space is limited. Focused mutagenesis strategies make better use of mutational space and involve using oligonucleotides containing degenerate codons. One prerequisite to this method of mutagenesis is that it requires prior knowledge of the protein being mutated, for example, a crystal structure. Paired with computational methods such as protein docking and virtual alanine-scanning mutagenesis [59], focused mutagenesis is a powerful tool to generate mutant libraries.

After library generation, mutant libraries need to be screened for improved affinity using screening platforms such as ribosome display, phage surface display, and yeast surface display. Ribosome display uses ribosomes to translate a mutagenic library [60]. Antibody fragments are translated from mRNA and trapped in the ribosome due to the lack of a stop codon in the mRNA. The translated protein is then bound to the target protein, and the mRNA is recovered. Ribosome display has a large potential mutational space of up to 10^{15} library members. Phage surface display involves fusion of library member proteins to the bacteriophage minor coat protein and then "panning" for high affinity library members using immobilized target proteins [61]. Phage display possesses a mutational space of 10^{10} library members. Yeast surface display involves expressing library members on the yeast surface using the α-agglutinin adhesion receptor, and then target protein is bound to the surface of yeast cells usually with a FP to be used in fluorescence activated cell sorting (FACS) [62]. Yeast surface display possesses a mutational space of 10^8 library members, which is less than the previous two display methods; however, yeast surface display is advantageous because it is compatible with FACS, which is relatively fast and possesses a high probability of recovery for high-affinity mutants. Furthermore, yeast codon-optimization and post-translational modification (PTM) pathways (e.g. N-linked glycosylation and oxidative protein folding) are more similar to those of humans. This can lead to the generation of high-affinity antibody fragments (in the femtomolar range) [63], which can be powerful imaging tools used in conjunction with fluorescent or radioactive detection domains in imaging probes [64–67].

15.2.3 Genetically Encoded Non-fluorescent Protein Tags

In addition to fusing biomolecules with FPs, labeling them with synthetic probes in living cells is another powerful tool that is employed in live-cell imaging. Genetically encoded protein tags are encoded into cells to represent self-labeling enzyme tags, which recruit the attachment of an exogenously added synthetic ligand. A requirement for these tags is that they are bio-orthogonal, which means that they do not interact with the cell's endogenous biological molecules. Classical examples of these protein tags are the SNAP-tag, the CLIP-tag, and the HaloTag [3]. Upon addition of benzylguanine-, benzylcytosine-, and chloroalkane-fused substrates, these substrates are respectively bound to SNAP, CLIP, and Halo tags. Recently, Sun et al. were able to engineer the SNAP and CLIP tags ($SNAP_f$ and

$CLIP_f$) for faster labeling [68]. To reduce background from unbound substrate, encoded protein tag systems can be designed with quenchers. Jing et al. designed a trimethoprim quencher fluorophore substrate to be added to *E. coli* dihydrofolate reductase-tagged proteins [69]. Upon interaction of the enzyme and substrate molecule, the quencher is released, resulting in fluorescence. Protein tags have also been designed to be used for long-term imaging experiments. On the other end of the spectrum, protein tags can be engineered to smaller sizes to allow for easier expression and easier endogenous incorporation via clustered regularly interspaced short palindromic repeat (CRISPR)-mediated homology-directed repair. Kamiyama et al. designed GFP11 and sfCherry11, small split FPs (16 and 18 amino acids, respectively) that can recruit super-folder GFP and sfCherry [46]. Their small size makes them advantageous for use with CRISPR-mediated homology-directed repair. Enzymatic labeling is another option that can specifically label peptides with detectable probes. Some of the commonly used enzymes for specific labeling include biotin ligase [70], formylglycine-generating enzyme [71], and *O*-GlcNAc transferase [72].

15.3 Application in Cellular Imaging

With the advent of FPs and protein scaffolds for genetic labeling of biomolecules, genetically encoded biosensors have revolutionized research on cell biology, by allowing researchers to visualize and quantify a wide range of biological substances and events in living cells with high spatiotemporal resolutions [73–77]. In general, the main component of biosensors is a protein scaffold, which acts as a recognition module for the target analyte, and a fluorescent transducer, which can convert the protein–analyte interaction into a fluorescent signal. Different types of biosensors can be constructed because protein scaffolds in biosensors can be engineered to specifically recognize relevant biomolecules [78]. Thus, protein engineering has not only been applied to alter the properties of FPs [79] and protein scaffolds [55, 80, 81], but it has also been used to create novel fluorescent biosensors or probes to reveal cellular processes in cell biology as well as to diagnose diseases in medical applications.

15.3.1 Cell Biology Applications

This section summarizes the application of novel fluorescent biosensors to examine subcellular localization and cell signaling, which is essential for the mechanistic study of molecular functions and cell behavior in the context of cell biology.

15.3.1.1 Localization

FPs or protein-based biosensors have been used to target various subcellular compartments, enabling organelle visualization, studies of fusion–fission events, and local monitoring of cellular parameters [82]. To visualize small subcellular structures tens of nanometers in size, super-resolution fluorescence microscopy has been

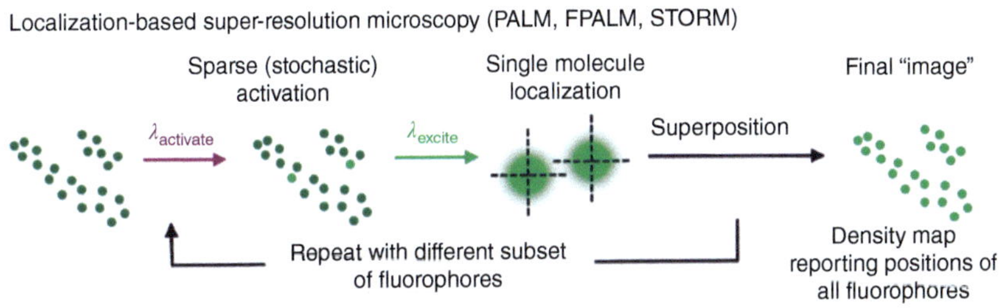

Figure 15.1 The basic principle of localization-based super-resolution microscopy. Only a few molecules are activated at a given time and recorded for high precision detection of local positions. The final "image" is a density map of all fluorophore positions. Source: Adapted from Nienhaus and Nienhaus [87].

used instead of traditional diffraction-limited light microscopy, which is unable to visualize objects with sizes smaller than 200 nm in the lateral direction and 500 nm in the axial direction [5]. PCFPs can switch their optical properties and are great markers for localization-based super-resolution microscopy (e.g. photo-activation localization microscopy (PALM), fluorescence photo-activation localization microscopy (FPALM), and stochastic optical reconstruction microscopy (STORM)) [83–86], because they can be activated irreversibly via a photo-induced chemical reaction that either turns on their emission or changes their emission wavelength to localize the FP-labeled molecules with high precision (Figure 15.1) [87]. After a photoactivated mCherry (PAmCherry1) was introduced for high-resolution imaging [88], Bo Huang's group merged the substitutions located on the 1–10 fragments with the sequence of sfCherry$2_{1\text{-}10}$ to generate PAsfCherry$2_{1\text{-}10}$. Combining PAsfCherry$2_{1\text{-}10}$ with FP_{11} tag knock-in method, they can perform STORM imaging of endogenous proteins tagged by $sfCherry_{11}$. Specifically, endogenous clathrin light chain A (CLTA) in human embryonic kidney (HEK) 293T cells and clathrin-coated pits were clearly observed in the STORM images [89]. PALM and STORM methods have also been applied to study signaling microclusters in T cells, e.g. the localization of linker for activation of T cells (LAT), and the SH2 domain-containing protein of 76 kDa (SLP-76) with either Dronpa or PAmCherry fluorophore tags [90].

15.3.1.2 Cell Signaling

In addition to localizing subcellular molecules with high spatial resolution, FPs and protein scaffolds can also be used to monitor biomolecular signaling and activities with high temporal resolution.

Calcium signaling is a crucial trigger for many biological processes in cells, and impacts nearly every aspect of cellular life [91]. Genetically encoded calcium indicators (GECI) can be precisely targeted to intraluminal locations and used to measure organellar Ca^{2+} content within the selected organelle [92]. A GECI consists of a Ca^{2+}-sensing protein (e.g. calmodulin (CaM) [93] or troponin C [TnC] [94]), and one or two FPs (e.g. single FP-based or Förster resonance energy transfer (FRET)-based GECIs). The representative single FP-based GECIs are GCaMP [95] (Figure 15.2a)

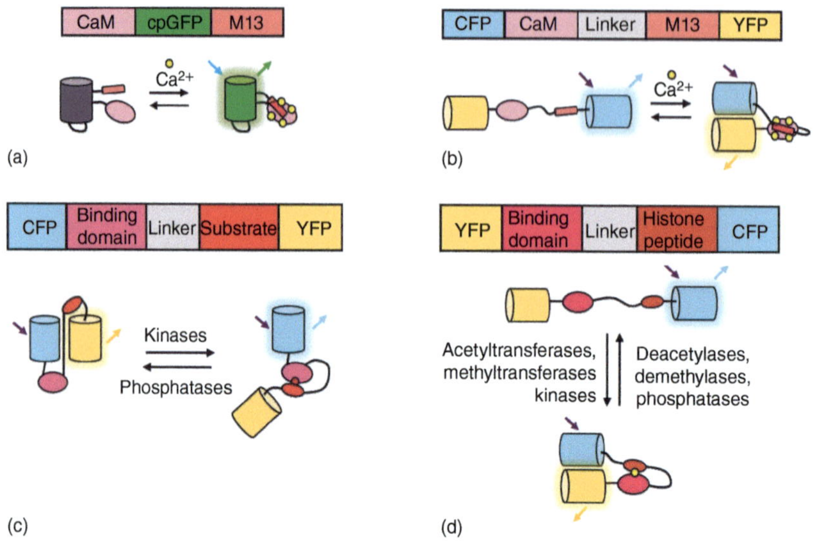

Figure 15.2 A schematic representation of different biosensors and their activation mechanism including (a) a single FP-based GECI (e.g. GCaMP), (b) a FRET-based GECI (e.g. Cameleon), (c) a FRET-based biosensor for protein kinases, and (d) a FRET-based biosensor for histone modifications.

and Pericam [42], which fuse the M13, a peptide sequence from myosin light chain kinase, and CaM to the new N and C termini of cpGFP and circularly permuted YFP (cpYFP), respectively [96]. Upon the binding of Ca^{2+} on CaM, M13 peptide binds to the Ca^{2+}-bound state of CaM, which induces intramolecular conformational rearrangement resulting in an increase in fluorescence intensity. Green fluorescent GCaMPs have proven extremely useful for imaging Ca^{2+} activities in various neural systems [97, 98]. Red fluorescent GECIs have also been developed using semi-rational design and directed evolution methods, e.g. K-GECO1 [26]. K-GECO1, which is the most recent red GECI, based on the eqFP578 variant FusionRed, performs with high sensitivity and fast kinetics for Ca^{2+} activity measurement *in vivo* [99]. In addition to the single-FP intensiometric Ca^{2+} biosensors, the Cameleon family of GECIs is based on ratiometric FRET biosensors [93] (Figure 15.2b). The general design consists of CaM and M13 flanked by a FRET pair, e.g. CFP and yellow fluorescent protein (YFP). Binding of M13 peptide to the Ca^{2+}-bound state of CaM promotes intramolecular conformational change that brings two FPs into close proximity, allowing FRET to occur. The two functional elements CaM/M13 switch and the FRET pair can be independently engineered and evolved for improved performance of the FRET-based Ca^{2+} biosensors. For example, the yellow Cameleon-Nano sensor, which contains the optimized Ca^{2+}-sensing domain with higher Ca^{2+} affinities (K_d = 15–140 nM) and large dynamic range (1450%), is very sensitive to detect the subtle changes of Ca^{2+} concentration [100]. Moreover, the Twitch sensor, with reduced number of calcium binding sites, high-affinity calcium binding, and optimized structure by functional screening, is highly sensitive that is suitable for *in vivo* imaging [101]. Several ratiometric GECIs have been developed to precisely quantify Ca^{2+} concentrations in different subcellular locations such as cytosol, mitochondria, endoplasmic reticulum (ER), or nucleus [102, 103].

Protein phosphorylation through activities of protein kinases is also an important cellular regulatory mechanism for cell signaling. To monitor the activity of signaling network regulators such as Src-family kinase (SFK) at the cell's leading edge, the SFK monobody biosensor has been developed. It consists of an engineered monobody specific for activated SFKs with a bright dye whose fluorescence increases upon target binding [104]. In this study, the SFK monobody biosensor showed the localization and transient activation of SFKs at the cell edge and in platelet-derived growth factor (PDGF)-induced dorsal ruffles. It also revealed that endogenous SFKs are in the active conformation within ruffles specifically during actin-based protrusion [104]. In addition to studying localization of SFK activity, proper subcellular localization of focal adhesion kinase (FAK), where the activity is regulated at subcellular compartments, is also crucial for many cellular processes. To visualize the FAK activity at different membrane microdomains, the novel FRET-based FAK biosensor was targeted into (with Lyn tag) or outside (with KRas tag) of detergent-resistant membrane (DRM) regions of the plasma membrane (Figure 15.2c). This study discovered that FAK is activated at membrane microdomains with distinct molecular mechanisms in response to different physiological stimuli [105]. Using the C-terminal FA-targeting (FAT) region of FAK, FAT biosensors were developed to visualize and quantify the local kinase activities at individual focal adhesion sites in live cells with FAT-Src and FAT-FAK biosensors [106–108].

PTMs of proteins are crucial events in signaling transduction and control of cellular functions. Specifically, modifications of histone proteins including acetylation, phosphorylation, and methylation play important roles in epigenetic regulation [109, 110]. Thus, FRET-based biosensors have been developed for visualizing dynamic activities of each histone modification. These biosensors share similar design; both consist of a CFP/YFP FRET pair, a full-length or partial sequence of histone peptide (e.g. H2B, H3, or H4) that can be modified by endogenous enzymes, binding domains that bind to modified histone peptide, and a flexible linker (Figure 15.2d). The order of each component can be varied to obtain large FRET dynamic range before and after the interaction between modified histone peptides and binding domains upon the activation of histone-modifying enzymes. For histone acetylation, the acetylation of lysines is highly dynamic and regulated by the opposing action of two families of enzymes, e.g. histone acetyltransferases (HATs) and histone deacetylases (HDACs) [111]. Currently, only a few genetically encoded Venus/CFP FRET-based probes have been developed to monitor the acetylation of histone H3 on K9/14 [112] and H4 on K5/K8 and K12 [113–115], controlled by HATs or HDACs in living cells. For histone phosphorylation, it takes place on residues including serine, threonine, and tyrosine, predominantly in the N-terminal histone tails, and is controlled by kinases and phosphatases that add and remove the phosphate group from those residues, respectively [116]. The anaphase dynamics of protein phosphorylation by aurora B kinase have been examined by using CFP/YFP FRET-based sensors in living HeLa cells. The quantitative analysis of phosphorylation dynamics revealed that midzone activation of aurora

B kinase in anaphase produces an intracellular phosphorylation gradient [117]. For histone methylation, it is controlled by methyltransferases and demethylases that can add and remove the methyl group, respectively. Ting and coworkers first developed CFP/YFP FRET-based histone methylation reporters to visualize methylation of H3K9 and H3K27 *in vitro* and in single living cells [118]. SUV39H1 methyltransferase activity in mitosis was monitored by another Cerulean/Venus FRET-based methylation biosensor in living HeLa cells [119]. However, the usage of current FRET-based epigenetic biosensors is relatively limited, possibly due to improper subcellular localization [120]. Therefore, improvement of epigenetic FRET biosensors is needed by using protein-engineering approaches such as creating large biosensor variants (e.g. mutation on binding domain) and screening for the variant with desired function [101].

15.3.2 Application in Diagnostics and Medicine

Biosensors based on FPs or protein scaffolds have proven to be powerful imaging tools to visualize localization of biomolecules and cells signaling in living cells with high spatiotemporal resolution. They can also be further used as detection tools for disease diagnostics and drug screening tools for therapeutic applications.

15.3.2.1 Detection

Metalloproteinases (MMPs), a class of enzymes that cleave matrix proteins, play important roles when tumor cells remodel the extracellular environment and prepare for metastasis. Among these MMPs, the membrane-type 1 matrix metalloproteinase (MT1-MMP) has been considered as a biomarker to detect cancer cell invasion (metastasis) [121, 122]. Therefore, researchers have developed FRET-based biosensors to monitor proteolytic activity of MT1-MMP, which can then be used as a detection method for invasive cancer cells and potentially in disease diagnosis [123, 124].

The engineered FRET-based biosensor for monitoring MT1-MMP activity contains an enhanced cyan fluorescent protein (ECFP) (a donor), a substrate peptide with flanking linker sequences, and a variant of yellow fluorescent protein (YPet) (an acceptor) (Figure 15.3). At rest, ECFP and YPet form a weak dimer, and the biosensor has a strong energy transfer, resulting in low ECFP/FRET emission ratio. Active MT1-MMP enzymes can cleave the specifically designed substrate peptide sequence and disrupt energy transfer, which leads to high ECFP/FRET emission ratio [123, 124]. To enhance sensitivity and specificity of the substrate to MT1-MMP cleavage, which largely determines the detection ability of the probe, the substrate peptide sequence AHLR was selected by substrate cleavage assay in a phage library. This peptide showed five times the sensitivity of the native substrate from MMP2 in the biosensor. This AHLR MT1-MMP biosensor was then expressed at the extracellular surface of different cancer cells (e.g. highly invasive HT1080 fibrosarcoma and MDA-MB-231 breast carcinoma cells, and several non-invasive breast cancer cells, including SK-BR-3, CRL2314, ZR-75-1, MCF-7, and CRL4010 to detect and visualize single-cell MT1-MMP activity (Figure 15.4)). HeLa cells with and without

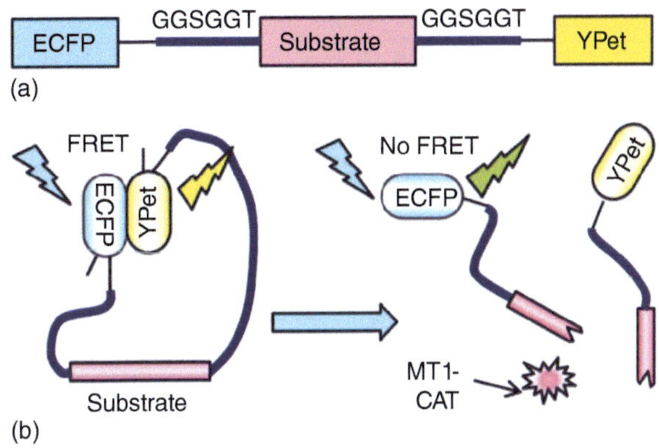

Figure 15.3 The design principle of MT1-MMP FRET biosensors. (a) The schematics of the MT1-MMP biosensor. (b) The activation mechanism of the MT1-MMP biosensor. Source: Adapted from Lu et al. [124].

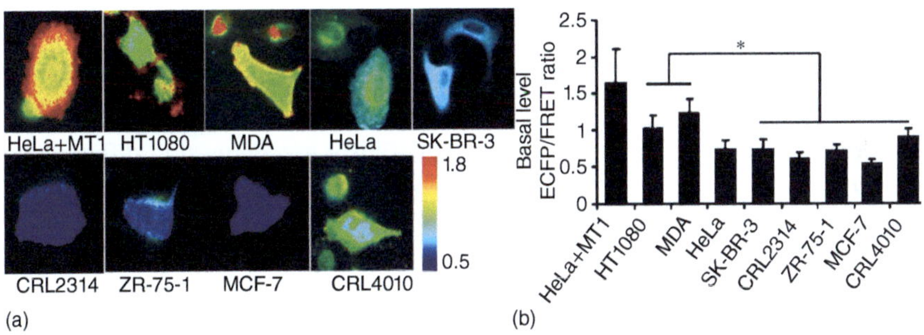

Figure 15.4 Imaging and quantification of basal level of apparent MT1-MMP activity in cancer cell lines. (a) The ECFP/FRET ratio images of representative cancer cells with the AHLR biosensor; (b) the quantified ECFP/FRET ratio values of the cancer cell lines (mean ± standard error of the mean (SEM)). (*) indicates statistically significant difference by the two-tail t-test with unequal variance. For example, between MDA-MB-231 ($n = 26$) and CRL4010 ($n = 21$) cells, p-value $= 5.0 \times 10^{-10}$. Between HT1080 ($n = 33$) and CRL4010 ($n = 21$) cells, p-value $= 9.5 \times 10^{-4}$. Source: Lu et al. [124]. CC BY 4.0.

MT1-MMP expression were used as positive and negative control, respectively. The quantified results show that the AHLR MT1-MMP biosensor can be used to detect MT1-MMP activity, which is later shown to be highly correlated with the cancer cell invasiveness [124]. Therefore, the engineered AHLR MT1-MMP proteolytic FRET biosensor can be used to successfully detect invasive cancer marker in single cells.

Apart from FPs, fluorescent nanocrystals called quantum dots (QDs) are sometimes preferred over FPs as FRET donors because of their brightness, photostability, and narrow and tunable spectra [125, 126]. The highly sensitive and specific substrate peptide, AHLR, was further utilized to engineer a quantum-dot-based nanosensor (Figure 15.5) [126]. Active MT1-MMP protease at the surface of cancer cells can cleave the substrate and then allow the penetration of the QD into the cytoplasm of cancer cells. Therefore, the QD/FRET intensity ratio signal is representative of the MT1-MMP activity at the cancer cell surface, while the intracellular QD intensity indicates the ability of cells to endocytose the cleaved QD components.

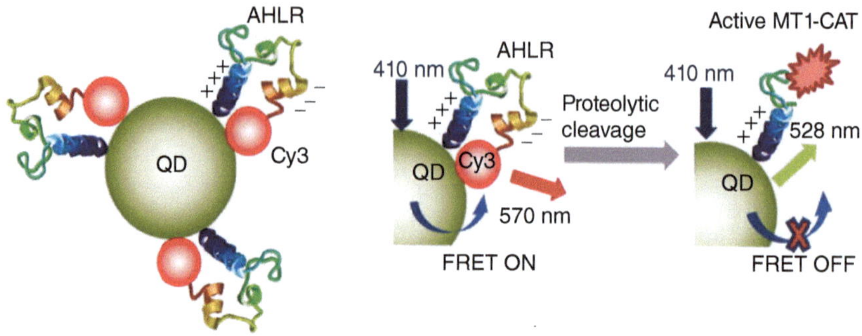

Figure 15.5 Schematic illustrations of the design and activation mechanism of the quantum dot-based FRET (QD-FRET) nanosensor. (a) Nanosensor contains a QD coupled to multiple Cy3-peptides bent to a position which allows high FRET between the QD and Cy3. (b) Activation mechanism of the FRET nanosensor upon the cleavage by MT1-CAT. Source: Adapted from Chung et al. [126].

Indeed, single-cell imaging results of this nanosensor can be used to differentiate the invasiveness and endocytic ability of different cancer cells.

15.3.2.2 Screening for Drugs

Kinase Biosensors in Drug Effectiveness Cancer invasion occurs in the 3D tissue environment through a complex network of vasculature and matrix proteins. New intravital imaging methods have been developed to monitor the function and behavior of cancer cells in 3D matrix and *in vivo*. For *in vivo* imaging, anesthetized mice with exposed subcutaneous tumors were imaged on a heated stage using multiphoton microscopes to allow for 150 μm imaging depth. Using fluorescence lifetime imaging microscopy (FLIM) with a photomultiplier tube (PMT) and a new and improved time-correlated single photon counting method, an accurate lifetime measurement of up to 12 ns was achieved [127, 128].

With this system, an FLIM-FRET Src biosensor was used to monitor spatiotemporal Src activity in response to a small-molecule Src kinase inhibitor, dasatinib, and study drug-targeting efficacy and clearance in a live mouse model of pancreatic cancer [127, 129]. Pancreatic ductal adenocarcinoma (PDA) is characterized by an *in vivo* environment of poor vasculature and extensive extracellular matrix with proteins and hyaluronic acid (HA, also known as hyaluronan, which often hampers the delivery of drugs [130]. Using the HA images as reference, the researchers showed a spatial gradient of Src kinase activity in invading tumor cells when they penetrate into 3D matrix and tissue of live tumors. Cancer cells were observed to have low Src kinase activity at shallow matrix and high activity at deep matrix. Meanwhile, the inhibition of cellular Src activity gradient by dasatinib was found to be dependent on the location of the cells relative to their distance to the blood vessels. The results show that dasatinib inhibition can be effective within 100 μm of the blood vessel, but not outside of this range [127].

The emerging standard of care for pancreatic cancer patients is the combined therapy of the cytotoxic drugs gemcitabine (Gem) and Abraxane, which is a nanoparticle albumin-bound paclitaxel. To study the effect of these drugs in combination with tissue priming, further experiments using this live FLIM-FRET imaging technology were used to visualize biochemical activity inside cancer cells [131]. Briefly, patient-derived cancer cells and xenographs were engineered to express the cyclin-dependent kinase 1 (CDK1) FRET biosensor, whose signals were used as a marker of M-phase cell cycle arrest induced by Gem/Abraxane. The Rock inhibitor, Fasudil, was used to treat the matrix and was found to disrupt matrix integrity and reduce tension, as well as increase the permeability of vascular wall in animals. Next, FLIM-FRET intravital imaging technology was used to monitor CDK1 kinase activity of pancreatic cancer cells treated with Gem/Abraxane. The results are very promising that the transient pre-manipulation of tissue tension and tumor vasculature with Fasudil resulted in a reduction of CDK1 activity in invading cancer cells and multisite improvement of metastases in response to chemotherapy [131]. A phase 1 clinical trial can be further designed to determine the safety of this tissue-priming drug in patients [123]. Altogether, these FLIM-FRET imaging results show that engineered biosensors can be used in preclinical animal studies to investigate the effect of cellular response to drug penetration *in vivo*.

Kinase Biosensors in Drug Resistance In addition to monitoring drug effectiveness, kinase biosensors have also been used to monitor chronic myeloid leukemia (CML) cells' resistance to drug treatment. In CML, mutational fusion led to a novel gene Brc-abl and the subsequent expression of the chimeric BCR-ABL tyrosine kinase, which is constitutively active and pathogenic [132]. The tyrosine kinase inhibitor, imatinib, has been introduced to treat CML with success in patients with chronic phase CML, but less so in patients with the advanced disease. To evaluate the efficacy of new kinase inhibitors, such as nilotinib and dasatinib, an FRET biosensor named "pickles" has been engineered to report the BCR-ABL kinase activity in single cells. The phosphorylation on the BCR-ABL tyrosine substrate of this biosensor causes the intramolecular binding of the phosphotyrosine to SH2 domain, which brings Venus (a variant of YFP) to close proximity of ECFP and results in high FRET efficiency. FRET imaging results indicate that the biosensor can be used to detect patients' cells with high ABL kinase activity and differentiate the cells' response to inhibitor treatment. The responsive cells belonged to patients who showed consistent clinical response to treatment, while the non-responsive cells belonged to drug-resistant patients. The results indicate that the engineered FRET biosensors can be used to screen the efficacy of kinase inhibitors and other drugs in live cells, as well as for diagnostics and monitoring of the effectiveness of therapy [132, 133].

15.4 Conclusion and Perspectives

Protein engineering is an important approach to develop cellular imaging tools, including FPs and protein scaffolds, that have revolutionized our ability to monitor biological processes, particularly at the cellular level, using fluorescence microscopy and related imaging techniques such as FRET. The main advantage of FPs as imaging tools is that they can be genetically encoded and functional without additional cofactors. Similar to FPs, protein scaffolds are also genetically encoded, but they require additional cofactors such as fluorescent dyes or other reagents. With new photophysical and biochemical properties of FPs and diverse functions of protein scaffolds, these will increase flexibility of imaging paradigm to further investigate complex biological processes.

Fluorescent biosensors are increasingly used for studying cellular processes and dynamics at the single-cell level. In the past decades, fluorescent biosensors have been extensively developed on several platforms, such as FPs or combinations of protein scaffolds and fluorescent dyes, using fluorescence microscopy to examine cellular localization and cell signaling of important biomolecules. With the advent of photocontrollable FPs, it enables more sophisticated visualization of proteins localization and activity at the subcellular level (less than 200 nm), using super-resolution microscopy. Moreover, these fluorescent biosensors can also be used in medical research for both diagnostic and therapeutic applications. Therefore, the huge diversity of FPs and protein scaffolds that are constantly developed through protein engineering allows for large-scale construction of cellular sensors with many applications.

References

1 Tsien, R.Y. (1998). The green fluorescent protein. *Annu. Rev. Biochem.* 67: 509–544.

2 Shaner, N.C., Campbell, R.E., Steinbach, P.A. et al. (2004). Improved monomeric red, orange and yellow fluorescent proteins derived from *Discosoma* sp. red fluorescent protein. *Nat. Biotechnol.* 22 (12): 1567.

3 Keppler, A., Gendreizig, T., Gronemeyer, T. et al. (2003). A general method for the covalent labeling of fusion proteins with small molecules in vivo. *Nat. Biotechnol.* 21 (1): 86–89.

4 Holliger, P. and Hudson, P.J. (2005). Engineered antibody fragments and the rise of single domains. *Nat. Biotechnol.* 23 (9): 1126.

5 Shcherbakova, D.M., Sengupta, P., Lippincott-Schwartz, J. et al. (2014). Photocontrollable fluorescent proteins for superresolution imaging. *Annu. Rev. Biophys.* 43: 303–329.

6 Gunewardene, M.S., Subach, F.V., Gould, T.J. et al. (2011). Superresolution imaging of multiple fluorescent proteins with highly overlapping emission spectra in living cells. *Biophys. J.* 101 (6): 1522–1528.

7 Lippincott-Schwartz, J. and Patterson, G.H. (2009). Photoactivatable fluorescent proteins for diffraction-limited and super-resolution imaging. *Trends Cell Biol.* 19 (11): 555–565.

8 Rodriguez, E.A., Campbell, R.E., Lin, J.Y. et al. (2017). The growing and glowing toolbox of fluorescent and photoactive proteins. *Trends Biochem. Sci* 42 (2): 111–129.

9 Lauterbur, P.C. (1973). Image formation by induced local interactions – examples employing nuclear magnetic-resonance. *Nature* 242 (5394): 190–191.

10 Rao, J.H., Dragulescu-Andrasi, A., Yao, H. et al. (2007). Fluorescence imaging in vivo: recent advances. *Curr. Opin. Biotechnol.* 18 (1): 17–25.

11 Contag, C.H. and Bachmann, M.H. (2002). Advances in vivo bioluminescence imaging of gene expression. *Annu. Rev. Biomed. Eng.* 4: 235–260.

12 Ferrara, K., Pollard, R., and Borden, M. (2007). Ultrasound microbubble contrast agents: fundamentals and application to gene and drug delivery. *Annu. Rev. Biomed. Eng.* 9: 415–447.

13 Goel, S., England, C.G., Chen, F. et al. (2017). Positron emission tomography and nanotechnology: a dynamic duo for cancer theranostics. *Adv. Drug Delivery Rev.* 113: 157–176.

14 Wadas, T.J., Wond, E.H., Weisman, G.R. et al. (2010). Coordinating radiometals of copper, gallium, indium, yttrium, and zirconium for PET and SPECT imaging of disease. *Chem. Rev.* 110 (5): 2858–2902.

15 Wang, Y., Shyy, J.Y.-J., and Chien, S. (2008). Fluorescence proteins, live-cell imaging, and mechanobiology: seeing is believing. *Annu. Rev. Biomed. Eng.* 10: 1–38.

16 Subach, O.M., Cranfill, P.J., Davidson, M.W. et al. (2011). An enhanced monomeric blue fluorescent protein with the high chemical stability of the chromophore. *PLoS One* 6 (12): e28674.

17 Subach, O.M., Gundorov, I.S., Yoshimura, M. et al. (2008). Conversion of red fluorescent protein into a bright blue probe. *Chem. Biol.* 15 (10): 1116–1124.

18 Goedhart, J., Stetten, D.V., Noirclerc-Savoye, M. et al. (2012). Structure-guided evolution of cyan fluorescent proteins towards a quantum yield of 93%. *Nat. Commun.* 3: 751.

19 Kremers, G.-J., Goedhart, J., Munster, E.B.V. et al. (2006). Cyan and yellow super fluorescent proteins with improved brightness, protein folding, and FRET Förster radius. *Biochemistry* 45 (21): 6570–6580.

20 Goedhart, J., Weeren, L.V., Hink, M.A. et al. (2010). Bright cyan fluorescent protein variants identified by fluorescence lifetime screening. *Nat. Methods* 7 (2): 137.

21 Lam, A.J., St-Pierre, F., Gong, Y. et al. (2012). Improving FRET dynamic range with bright green and red fluorescent proteins. *Nat. Methods* 9 (10): 1005–1012.

22 Bajar, B.T., Wang, E.S., Lam, A.J. et al. (2016). Improving brightness and photostability of green and red fluorescent proteins for live cell imaging and FRET reporting. *Sci. Rep.* 6: 20889.

23 Shaner, N.C., Lambert, G.G., Chammas, A. et al. (2013). A bright monomeric green fluorescent protein derived from *Branchiostoma lanceolatum*. *Nat. Methods* 10 (5): 407.

24 Kredel, S., Oswald, F., Nienhaus, K. et al. (2009). mRuby, a bright monomeric red fluorescent protein for labeling of subcellular structures. *PLoS One* 4 (2): e4391.

25 Shcherbo, D., Murphy, C.S., Ermakova, G.V. et al. (2009). Far-red fluorescent tags for protein imaging in living tissues. *Biochem. J* 418 (3): 567–574.

26 Shemiakina, I.I., Ermakova, G.V., Cranfill, P.J. et al. (2012). A monomeric red fluorescent protein with low cytotoxicity. *Nat. Commun.* 3: 1204.

27 Bindels, D.S., Haarbosch, L., Weeren, L.V. et al. (2017). mScarlet: a bright monomeric red fluorescent protein for cellular imaging. *Nat. Methods* 14 (1): 53.

28 Dickson, R.M., Cubitt, A.B., Tsien, R.Y. et al. (1997). On/off blinking and switching behaviour of single molecules of green fluorescent protein. *Nature* 388 (6640): 355.

29 Lukyanov, K.A., Chudakov, D.M., Lukyanov, S. et al. (2005). Photoactivatable fluorescent proteins. *Nat. Rev. Mol. Cell Biol.* 6 (11): 885.

30 Subach, F.V., Patterson, G.H., Manley, S. et al. (2009). Photoactivatable mCherry for high-resolution two-color fluorescence microscopy. *Nat. Methods* 6 (2): 153.

31 Subach, F.V., Patterson, G.H., Renz, M. et al. (2010). Bright monomeric photoactivatable red fluorescent protein for two-color super-resolution sptPALM of live cells. *J. Am. Chem. Soc.* 132 (18): 6481–6491.

32 McKinney, S.A., Murphy, C.S., Hazelwood, K.L. et al. (2009). A bright and photostable photoconvertible fluorescent protein. *Nat. Methods* 6 (2): 131.

33 Zhang, M., Chang, H., Zhang, Y. et al. (2012). Rational design of true monomeric and bright photoactivatable fluorescent proteins. *Nat. Methods* 9 (7): 727.

34 Hoi, H., Shaner, N.C., Davidson, M.W. et al. (2010). A monomeric photoconvertible fluorescent protein for imaging of dynamic protein localization. *J. Mol. Biol.* 401 (5): 776–791.

35 McEvoy, A.L., Hoi, H., Bates, M. et al. (2012). mMaple: a photoconvertible fluorescent protein for use in multiple imaging modalities. *PLoS One* 7 (12): e51314.

36 Grotjohann, T., Testa, I., Leutenegger, M. et al. (2011). Diffraction-unlimited all-optical imaging and writing with a photochromic GFP. *Nature* 478 (7368): 204.

37 Grotjohann, T., Testa, I., Reuss, M. et al. (2012). rsEGFP2 enables fast RESOLFT nanoscopy of living cells. *Elife* 1: e00248.

38 Brakemann, T., Stiel, A.C., Weber, G. et al. (2011). A reversibly photoswitchable GFP-like protein with fluorescence excitation decoupled from switching. *Nat. Biotechnol.* 29 (10): 942.

39 Stiel, A.C., Andresen, M., Bock, H. et al. (2008). Generation of monomeric reversibly switchable red fluorescent proteins for far-field fluorescence nanoscopy. *Biophys. J.* 95 (6): 2989–2997.

40 Subach, F.V., Zhang, L., Gadella, T.W.J. et al. (2010). Red fluorescent protein with reversibly photoswitchable absorbance for photochromic FRET. *Chem. Biol.* 17 (7): 745–755.

41 Baird, G.S., Zacharias, D.A., and Tsien, R.Y. (1999). Circular permutation and receptor insertion within green fluorescent proteins. *Proc. Natl. Acad. Sci. U.S.A.* 96 (20): 11241–11246.

42 Nagai, T., Sawano, A., Park, E.S. et al. (2001). Circularly permuted green fluorescent proteins engineered to sense Ca^{2+}. *Proc. Natl. Acad. Sci. U.S.A.* 98 (6): 3197–3202.

43 Shui, B., Wang, Q., Lee, F. et al. (2011). Circular permutation of red fluorescent proteins. *PLoS One* 6 (5): e20505.

44 Carlson, H.J., Cotton, D.W., and Campbell, R.E. (2010). Circularly permuted monomeric red fluorescent proteins with new termini in the β-sheet. *Protein Sci.* 19 (8): 1490–1499.

45 Cabantous, S., Terwilliger, T.C., and Waldo, G.S. (2005). Protein tagging and detection with engineered self-assembling fragments of green fluorescent protein. *Nat. Biotechnol.* 23 (1): 102.

46 Kamiyama, D., Sekine, S., Barsi-Rhyne, B. et al. (2016). Versatile protein tagging in cells with split fluorescent protein. *Nat. Commun.* 7: 11046.

47 Correia, I.R. (2010). Stability of IgG isotypes in serum. *mAbs* 2 (3): 221–232.

48 Park, H.I., Yoon, H.W., and Jung, S.T. (2016). The highly evolvable antibody Fc domain. *Trends Biotechnol.* 34 (11): 895–908.

49 Levin, D., Golding, B., Strome, S.E. et al. (2015). Fc fusion as a platform technology: potential for modulating immunogenicity. *Trends Biotechnol.* 33 (1): 27–34.

50 Dennis, M.S., Jin, H., Dugger, D. et al. (2007). Imaging tumors with an albumin-binding Fab, a novel tumor-targeting agent. *Cancer Res.* 67 (1): 254–261.

51 Mazzocco, C., Fracasso, G., Germain-Genevois, C. et al. (2016). In vivo imaging of prostate cancer using an anti-PSMA scFv fragment as a probe. *Sci. Rep.* 6: 1–10.

52 Leyton, J.V., Olafsen, T., Sherman, M.A. et al. (2009). Engineered humanized diabodies for microPET imaging of prostate stem cell antigen-expressing tumors. *Protein Eng. Des. Sel.* 22 (3): 209–216.

53 Leyton, J.V., Olafsen, T., Lepin, E.J. et al. (2008). Humanized radioiodinated minibody for imaging of prostate stem cell antigen-expressing tumors. *Clin. Cancer Res.* 14 (22): 7488–7496.

54 Chakravarty, R., Goel, S., and Cai, W.B. (2014). Nanobody: the "magic bullet" for molecular imaging? *Theranostics* 4 (4): 386–398.

55 Limsakul, P., Peng, W., Wu, Y.Q. et al. (2018). Directed evolution to engineer monobody for FRET biosensor assembly and imaging at live-cell surface. *Cell Chem. Biol.* 25 (3): 1–10.

56 Rudnick, S.I. and Adams, G.P. (2009). Affinity and avidity in antibody-based tumor targeting. *Cancer Biother. Radiopharm.* 24 (2): 155–161.

57 Koide, A., Wojcik, J., Gilbreth, R.N. et al. (2012). Teaching an old scaffold new tricks: monobodies constructed using alternative surfaces of the FN3 scaffold. *J. Mol. Biol.* 415 (2): 393–405.

58 McCullum, E.O., Williams, B.A.R., Zhang, J. et al. (2010). Random mutagenesis by error-prone PCR. In: *In Vitro Mutagenesis Protocols*, 3e, vol. 634, 103–109. Totowa, NJ: Humana Press.

59 Rosenfeld, L., Heyne, M., Shifman, J.M. et al. (2016). Protein engineering by combined computational and in vitro evolution approaches. *Trends Biochem. Sci.* 41 (5): 421–433.

60 Hanes, J. and Pluckthun, A. (1997). In vitro selection and evolution of functional proteins by using ribosome display. *Proc. Natl. Acad. Sci. U.S.A.* 94 (10): 4937–4942.

61 Mccafferty, J., Griffiths, A.D., Winter, G. et al. (1990). Phage antibodies – filamentous phage displaying antibody variable domains. *Nature* 348 (6301): 552–554.

62 Boder, E.T. and Wittrup, K.D. (1997). Yeast surface display for screening combinatorial polypeptide libraries. *Nat. Biotechnol.* 15 (6): 553–557.

63 Boder, E.T., Midelfort, K.S., and Wittrup, K.D. (2000). Directed evolution of antibody fragments with monovalent femtomolar antigen-binding affinity. *Proc. Natl. Acad. Sci. U.S.A.* 97 (20): 10701–10705.

64 Herwig, L., Rice, A.J., Bedbrook, C.N. et al. (2017). Directed evolution of a bright near-infrared fluorescent rhodopsin using a synthetic chromophore. *Cell Chem. Biol.* 24 (3): 415–425.

65 Plamont, M.A., Billon-Denis, E., Maurin, S. et al. (2016). Small fluorescence-activating and absorption-shifting tag for tunable protein imaging in vivo (vol 113, pg 497, 2015). *Proc. Natl. Acad. Sci. U.S.A.* 113 (10): E1412–E1412.

66 Li, C., Tebo, A.G., and Gautier, A. (2017). Fluorogenic labeling strategies for biological imaging. *Int. J. Mol. Sci.* 18 (7): 1–11.

67 Brustad, E.M., Lelyveld, V.S., Snow, C.D. et al. (2012). Structure-guided directed evolution of highly selective P450-based magnetic resonance imaging sensors for dopamine and serotonin. *J. Mol. Biol.* 422 (2): 245–262.

68 Sun, X.L., Zhang, A., Baker, B. et al. (2011). Development of SNAP-tag fluorogenic probes for wash-free fluorescence imaging. *ChemBioChem* 12 (14): 2217–2226.

69 Jing, C.R. and Cornish, V.W. (2013). A fluorogenic TMP-tag for high signal-to-background intracellular live cell imaging. *ACS Chem. Biol.* 8 (8): 1704–1712.

70 Howarth, M., Takao, K., Hayashi, Y. et al. (2005). Targeting quantum dots to surface proteins in living cells with biotin ligase. *Proc. Natl. Acad. Sci. U.S.A.* 102 (21): 7583–7588.

71 Wu, P., Shui, W., Carlson, B.L. et al. (2009). Site-specific chemical modification of recombinant proteins produced in mammalian cells by using the genetically encoded aldehyde tag. *Proc. Natl. Acad. Sci. U.S.A.* 106 (9): 3000–3005.

72 Clark, P.M., Dweck, J.F., Mason, D.E. et al. (2008). Direct in-gel fluorescence detection and cellular imaging of *O*-GlcNAc-modified proteins. *J. Am. Chem. Soc.* 130 (35): 11576.

73 Frommer, W.B., Davidson, M.W., and Campbell, R.E. (2009). Genetically encoded biosensors based on engineered fluorescent proteins. *Chem. Soc. Rev.* 38 (10): 2833–2841.

74 Okumoto, S., Jones, A., and Frommer, W.B. (2012). Quantitative imaging with fluorescent biosensors. *Annu. Rev. Plant Biol.* 63: 663–706.

75 Tsien, R.Y. (2010). Nobel lecture: constructing and exploiting the fluorescent protein paintbox. *Integr. Biol.* 2 (2–3): 77–93.

76 Wang, H.X., Nakata, E., and Hamachi, I. (2009). Recent progress in strategies for the creation of protein-based fluorescent biosensors. *ChemBioChem* 10 (16): 2560–2577.

77 Zhang, J., Campbell, R.E., Ting, A.Y. et al. (2002). Creating new fluorescent probes for cell biology. *Nat. Rev. Mol. Cell Biol.* 3 (12): 906–918.

78 Tamura, T. and Hamachi, I. (2014). Recent progress in design of protein-based fluorescent biosensors and their cellular applications. *ACS Chem. Biol.* 9 (12): 2708–2717.

79 Ai, H.W., Baird, M.A., Shen, Y. et al. (2014). Engineering and characterizing monomeric fluorescent proteins for live-cell imaging applications. *Nat. Protoc.* 9 (4): 910–928.

80 Hattori, T., Lai, D., Dementieva, I.S. et al. (2016). Antigen clasping by two antigen-binding sites of an exceptionally specific antibody for histone methylation. *Proc. Natl. Acad. Sci. U.S.A.* 113 (8): 2092–2097.

81 Hattori, T., Taft, J.M., Swist, K.M. et al. (2013). Recombinant antibodies to histone post-translational modifications. *Nat. Methods* 10 (10): 992.

82 Chudakov, D.M., Matz, M.V., Lukyanov, S. et al. (2010). Fluorescent proteins and their applications in imaging living cells and tissues. *Physiol. Rev.* 90 (3): 1103–1163.

83 Betzig, E., Patterson, G.H., Sougrat, R. et al. (2006). Imaging intracellular fluorescent proteins at nanometer resolution. *Science* 313 (5793): 1642–1645.

84 Hess, S.T., Girirajan, T.P.K., and Mason, M.D. (2006). Ultra-high resolution imaging by fluorescence photoactivation localization microscopy. *Biophys. J.* 91 (11): 4258–4272.

85 Rust, M.J., Bates, M., and Zhuang, X.W. (2006). Sub-diffraction-limit imaging by stochastic optical reconstruction microscopy (STORM). *Nat. Methods* 3 (10): 793–795.

86 Sengupta, P., Van Engelenburg, S., and Lippincott-Schwartz, J. (2012). Visualizing cell structure and function with point-localization superresolution imaging. *Dev. Cell* 23 (6): 1092–1102.

87 Nienhaus, K. and Nienhaus, G.U. (2014). Fluorescent proteins for live-cell imaging with super-resolution. *Chem. Soc. Rev.* 43 (4): 1088–1106.

88 Subach, F.V., Patterson, G.H., Manley, S. et al. (2009). Photoactivatable mCherry for high-resolution two-color fluorescence microscopy (vol 6, pg 153, 2009). *Nat. Methods* 6 (4): 311–311.

89 Feng, S.Y., Sekine, S., Pessino, V. et al. (2017). Improved split fluorescent proteins for endogenous protein labeling. *Nat. Commun.* 8: 1–11.

90 Barr, V.A., Yi, J., and Samelson, L.E. (2017). Super-resolution analysis of TCR-dependent signaling: single-molecule localization microscopy. *Methods Mol. Biol.* 1584: 183–206.

91 Clapham, D.E. (2007). Calcium signaling. *Cell* 131 (6): 1047–1058.

92 Suzuki, J., Kanemaru, K., and Iino, M. (2016). Genetically encoded fluorescent indicators for organellar calcium imaging. *Biophys. J.* 111 (6): 1119–1131.

93 Miyawaki, A., Llopis, J., Heim, R. et al. (1997). Fluorescent indicators for Ca^{2+} based on green fluorescent proteins and calmodulin. *Nature* 388 (6645): 882–887.

94 Heim, N. and Griesbeck, O. (2004). Genetically encoded indicators of cellular calcium dynamics based on troponin C and green fluorescent protein. *J. Biol. Chem.* 279 (14): 14280–14286.

95 Nakai, J., Ohkura, M., and Imoto, K. (2001). A high signal-to-noise Ca(2+) probe composed of a single green fluorescent protein. *Nat. Biotechnol.* 19 (2): 137–141.

96 Ni, Q., Mehta, S., and Zhang, J. (2018). Live-cell imaging of cell signaling using genetically encoded fluorescent reporters. *FEBS J.* 285 (2): 203–219.

97 Akerboom, J., Chen, T.W., Wardill, T.J. et al. (2012). Optimization of a GCaMP calcium indicator for neural activity imaging. *J. Neurosci.* 32 (40): 13819–13840.

98 Dutta, S., Chen, T.S., and Keating, A.E. (2013). Peptide ligands for pro-survival protein Bfl-1 from computationally guided library screening. *ACS Chem. Biol.* 8 (4): 778–788.

99 Shen, Y., Dana, H., Abdelfattah, A.S. et al. (2018). A genetically encoded Ca(2+) indicator based on circularly permutated sea anemone red fluorescent protein eqFP578. *BMC Biol.* 16 (1): 9.

100 Horikawa, K., Yamada, Y., Matsuda, T. et al. (2010). Spontaneous network activity visualized by ultrasensitive Ca(2+) indicators, yellow Cameleon-Nano. *Nat. Methods* 7 (9): 729–732.

101 Thestrup, T., Litzlbauer, J., Bartholomäus, I. et al. (2014). Optimized ratiometric calcium sensors for functional in vivo imaging of neurons and T lymphocytes. *Nat. Methods* 11 (2): 175–182.

102 Park, J.G. and Palmer, A.E. (2015). Properties and use of genetically encoded FRET sensors for cytosolic and organellar Ca^{2+} measurements. *Cold Spring Harbor Protoc.* 2015 (1): pdb.top066043.

103 Bolbat, A. and Schultz, C. (2017). Recent developments of genetically encoded optical sensors for cell biology. *Biol. Cell* 109 (1): 1–23.

104 Gulyani, A., Vitriol, E., Allen, R. et al. (2012). A biosensor generated via high-throughput screening quantifies cell edge Src dynamics (vol 7, pg 437, 2011). *Nat. Chem. Biol.* 8 (8): 737–737.

105 Seong, J.Y., Ouyang, M., Kim, T. et al. (2011). Detection of focal adhesion kinase activation at membrane microdomains by fluorescence resonance energy transfer. *Nat. Commun.* 2: 1–9.

106 Hayashi, I., Vuori, K., and Liddington, R.C. (2002). The focal adhesion targeting (FAT) region of focal adhesion kinase is a four-helix bundle that binds paxillin. *Nat. Struct. Biol.* 9 (2): 101–106.

107 Schlaepfer, D.D. and Mitra, S.K. (2004). Multiple connections link FAK to cell motility and invasion. *Curr. Opin. Genet. Dev.* 14 (1): 92–101.

108 Wu, Y.Q., Zhang, K., Seong, J. et al. (2016). In-situ coupling between kinase activities and protein dynamics within single focal adhesions. *Sci. Rep.* 6: 1–13.

109 Strahl, B.D. and Allis, C.D. (2000). The language of covalent histone modifications. *Nature* 403 (6765): 41–45.

110 Kouzarides, T. (2007). Chromatin modifications and their function. *Cell* 128 (4): 693–705.

111 Bannister, A.J. and Kouzarides, T. (2011). Regulation of chromatin by histone modifications. *Cell Res.* 21 (3): 381–395.

112 Nakaoka, S., Sasaki, K., Ito, A. et al. (2016). A genetically encoded FRET probe to detect intranucleosomal histone H3K9 or H3K14 acetylation using BRD4, a BET family member. *ACS Chem. Biol.* 11 (3): 729–733.

113 Sasaki, K., Ito, T., Nishino, N. et al. (2009). Real-time imaging of histone H4 hyperacetylation in living cells. *Proc. Natl. Acad. Sci. U.S.A.* 106 (38): 16257–16262.

114 Ito, T., Umehara, T., Sasaki, K. et al. (2011). Real-time imaging of histone H4K12-specific acetylation determines the modes of action of histone deacetylase and bromodomain inhibitors. *Chem. Biol.* 18 (4): 495–507.

115 Dancy, B.M., Crump, N.T., Peterson, D.J. et al. (2012). Live-cell studies of p300/CBP histone acetyltransferase activity and inhibition. *ChemBioChem* 13 (14): 2113–2121.

116 Oki, M., Aihara, H., and Ito, T. (2007). Role of histone phosphorylation in chromatin dynamics and its implications in diseases. *Subcell. Biochem.* 41: 319–336.

117 Fuller, B.G., Lampson, M.A., Foley, E.A. et al. (2008). Midzone activation of aurora B in anaphase produces an intracellular phosphorylation gradient. *Nature* 453 (7198): 1132–1136.

118 Lin, C.W., Jao, C.Y., and Ting, A.Y. (2004). Genetically encoded fluorescent reporters of histone methylation in living cells. *J. Am. Chem. Soc.* 126 (19): 5982–5983.

119 Chu, L., Zhu, T., Liu, X. et al. (2012). SUV39H1 orchestrates temporal dynamics of centromeric methylation essential for faithful chromosome segregation in mitosis. *J. Mol. Cell. Biol.* 4 (5): 331–340.

120 Peng, Q., Lu, S., Shi, Y. et al. (2018). Coordinated histone modifications and chromatin reorganization in a single cell revealed by FRET biosensors. *Proc. Natl. Acad. Sci. U.S.A.* 115 (50): E11681–E11690.

121 Kessenbrock, K., Plaks, V., and Werb, Z. (2010). Matrix metalloproteinases: regulators of the tumor microenvironment. *Cell* 141 (1): 52–67.

122 Sabeh, F., Ota, I., Holmbeck, K. et al. (2004). Tumor cell traffic through the extracellular matrix is controlled by the membrane-anchored collagenase MT1-MMP. *J. Cell Biol.* 167 (4): 769–781.

123 Ouyang, M., Lu, S., Li, X.Y. et al. (2008). Visualization of polarized membrane type 1 matrix metalloproteinase activity in live cells by fluorescence resonance energy transfer imaging. *J. Biol. Chem.* 283 (25): 17740–17748.

124 Lu, S., Wang, Y., Huang, H. et al. (2013). Quantitative FRET imaging to visualize the invasiveness of live breast cancer cells. *PLoS One* 8 (3): e58569.

125 Michalet, X., Pinaud, F.F., Bentolila, L.A. et al. (2005). Quantum dots for live cells, in vivo imaging, and diagnostics. *Science* 307 (5709): 538–544.

126 Chung, E.Y., Ochs, C.J., Wang, Y. et al. (2015). Activatable and cell-penetrable multiplex FRET nanosensor for profiling MT1-MMP activity in single cancer cells. *Nano Lett.* 15 (8): 5025–5032.

127 Nobis, M., McGhee, E.J., Morton, J.P. et al. (2013). Intravital FLIM-FRET imaging reveals dasatinib-induced spatial control of src in pancreatic cancer. *Cancer Res.* 73 (15): 4674–4686.

128 Timpson, P., McGhee, E.J., Morton, J.P. et al. (2011). Spatial regulation of RhoA activity during pancreatic cancer cell invasion driven by mutant p53. *Cancer Res.* 71 (3): 747–757.

129 Wang, Y., Botvinick, E.L., Zhao, Y. et al. (2005). Visualizing the mechanical activation of Src. *Nature* 434 (7036): 1040–1045.

130 Jacobetz, M.A., Chan, D.S., Neesse, A. et al. (2013). Hyaluronan impairs vascular function and drug delivery in a mouse model of pancreatic cancer. *Gut* 62 (1): 112–120.

131 Vennin, C., Chin, V.T., Warren, S.C. et al. (2017). Transient tissue priming via ROCK inhibition uncouples pancreatic cancer progression, sensitivity to chemotherapy, and metastasis. *Sci. Transl. Med.* 9 (384): 1–17.

132 Mizutani, T., Kondo, T., Darmanin, S. et al. (2010). A novel FRET-based biosensor for the measurement of BCR-ABL activity and its response to drugs in living cells. *Clin. Cancer Res.* 16 (15): 3964–3975.

133 Lu, S. and Wang, Y. (2010). Fluorescence resonance energy transfer biosensors for cancer detection and evaluation of drug efficacy. *Clin. Cancer Res.* 16 (15): 3822–3824.

Index

Protein Engineering: Tools and Applications, First Edition. Edited by Huimin Zhao.
© 2021 WILEY-VCH GmbH. Published 2021 by WILEY-VCH GmbH.

d

e

f

t

u